HIGH POLYMERS

HIGH POLYMERS

A SERIES OF MONOGRAPHS ON THE CHEMISTRY, PHYSICS, AND

TECHNOLOGY OF HIGH POLYMERIC SUBSTANCES

VOLUME XXIV

VINYL AND DIENE MONOMERS

Part 3

Part 3

EDITED BY

EDWARD C. LEONARD

Kraftco Corporation
Glenview, Illinois

WILEY-INTERSCIENCE

A Division of John Wiley & Sons,
New York · London · Sydney · Toronto

Library of Congress Catalogue Card Number: 77–94013

SBN 471   39330   4

Printed in the United States of America

10 9 8 7 6 5 4 3 2 1

CONTRIBUTORS TO VOLUME XXIV

WILLIAM J. BAILEY, *University of Maryland, College Park, Maryland*

P. S. BAUCHWITZ, *E. I. du Pont de Nemours & Co., Wilmington, Delaware*

NORBERT M. BIKALES, *Chemical Consultant, Livingston, New Jersey*

KENNETH E. COULTER, *The Dow Chemical Company, Midland, Michigan*

NATHAN D. FIELD, *Central Research Laboratories, GAF Corporation, Easton, Pennsylvania*

J. B. FINLAY, *E. I. du Pont de Nemours & Co., Wilmington, Delaware*

R. H. FISACKERLY, *The Dow Chemical Company, Freeport, Texas*

D. E. HAMILTON, *The Dow Chemical Company, Freeport, Texas*

ROY E. HANEY, *Esso Research and Engineering Company, Florham Park, New Jersey*

BRUCE F. HISCOCK, *The Dow Chemical Company, Midland, Michigan*

HOWARD KEHDE, *The Dow Chemical Company, Midland, Michigan*

JOSEPH P. KENNEDY, *University of Akron, Akron, Ohio*

ISIDOR KIRSHENBAUM, *Esso Research and Engineering Company, Linden, New Jersey*

EDWARD C. LEONARD, *Kraftco Corporation, Glenview, Illinois*

MARTIN K. LINDEMANN, *Chas. S. Tanner Company, Greenville, South Carolina*

DONALD H. LORENZ, *Central Research Laboratories, GAF Corporation, Easton, Pennsylvania*

LEO S. LUSKIN, *Industrial Chemicals Department, Rohm and Haas Company, West Philadelphia, Pennsylvania*

DOROTHY C. SCHROEDER, *University of North Carolina, Greensboro, North Carolina*

J. P. SCHROEDER, *University of North Carolina, Greensboro, North Carolina*

L. G. SHELTON, *The Dow Chemical Company, Freeport, Texas*

C. A. STEWART, Jr., *E. I. du Pont de Nemours and Co., Wilmington, Delaware*

v

KENNETH STUEBEN, *Union Carbide Corporation, Bound Brook, New Jersey*

B. E. TATE, *Chemical Research and Development Department, Chas. Pfizer & Company, Inc., Groton, Connecticut*

L. E. WOLINSKI, *Former Lecturer in Polymer Chemistry, Canisius College, Buffalo, New York*

PREFACE

The three volumes of *Vinyl and Diene Monomers* were initiated in the fall of 1966 with a suggestion by Professor Herman Mark that it would be an appropriate time to record the chemistry of the commercially important vinyl monomers—the organic compounds $CH_2{=}CHX$ or $CH_2{=}CX_2$. The collection *Monomers* edited by Blout, Hohenstein, and Mark had appeared more than fifteen years earlier and *Vinyl and Related Polymers* by Schildknecht nearly as long ago. Since that time, there have been monographs on a few special monomers and substantial articles in encyclopedias on most, but no comprehensive, systematic, and uniform treatment in a single place. The aim of this work is to provide a reasonably up-to-date and complete treatise.

This book emphasizes the commercial manufacture, laboratory synthesis, purification, and physical and chemical properties of vinyl monomers. It is not meant to cover exhaustively the physics and chemistry of polymerization, or polymer manufacture and applications. A small portion of each chapter, however, is devoted to polymerization and polymer properties, since the commercial importance of the polymers is so overwhelming. Even though the intent of the book is to cover only commercially important monomers, there is some description of monomers of merely marginal market interest.

A number of people have contributed generously to the preparation of these volumes in addition to the authors of the individual chapters. J. P. Schroeder of the University of North Carolina at Greensboro has read almost all of the manuscripts and has made many suggestions and provided meticulous criticism.

Kathryn Parsons has been responsible for checking the accuracy of most of the thousands of references and for a substantial portion of the typing. Patricia Estes, Emma Moesta, and Evelyn Leonard have also made stenographic contributions. Corporate and individual acknowledgments are made in the preface associated with each volume.

There are now, with this work, two dozen titles in the Interscience Series "High Polymers" that began with *The Collected Papers of Wallace Hume Carothers on High Polymeric Substances* in 1940. Almost the entire growth of the multibillion pound addition polymer industry has been accomplished in the thirty years since then. This growth is a tribute to the skill of the scientists and technicians who engineered it and to the patience and courage of the businessmen who financed it.

E. C. Leonard

Park Ridge, Illinois
July 1969

PREFACE TO PART 3

Part 3 describes the manufacture, physical and chemical properties, and polymerization and polymer characteristics of vinyl and vinylidene chloride, the fluorocarbon monomers, and certain miscellaneous monomers. This last includes N-vinyl compounds, vinylpyridines, vinylsilanes, acrolein, methacrolein and vinyl ketones, vinyl sulfur compounds, vinylphosphorus compounds and vinylfuran, vinylthiophene, and certain substituted styrenes.

The permission of The Dow Chemical Company to publish the chapter on vinyl and vinylidene chloride is acknowledged along with the assistance of W. J. Frissell, R. F. Huston, M. R. Meeks, and W. D. Shellburg in preparation of the polymerization section of that chapter.

E. C. Leonard

Park Ridge, Illinois
December 1969

CONTENTS OF PART 3

1. VINYL AND VINYLIDENE CHLORIDE

L. G. SHELTON, D. E. HAMILTON, and R. H. FISACKERLY, *The Dow Chemical Company, Freeport, Texas*

Contents

1205

PART A. VINYL CHLORIDE

I. INTRODUCTION

Vinyl chloride monomer and its polymers occupy a unique place in the history of plastics. Arriving early in the technology of synthetic resins, they have not been displaced by newer polymers. To the contrary, they have become increasingly important. Approximately 2,500 million lb of monomer were produced in 1966 compared with 528.6 million lb in 1955 (1). Practically all this monomer was used to produce PVC and various copolymers.

Vinyl chloride had its beginning in the laboratory of the French chemist Regnault (2). In 1835 he produced vinyl chloride by mixing ethylene dichloride with an alcoholic solution of potassium hydroxide. The material remained a laboratory curiosity, however, until Ostromislensky (3) investigated it in 1912.

The First World War resulted in heavy demands on Germany's chemical industry. Because of a rubber shortage, German chemists again investigated Regnault's work, and in 1917 Klatte and Rollett (4) developed the first practical method of polymerizing vinyl chloride. The polymer was called polyvinyl chloride, abbreviated PVC, and it found wide acceptance as a substitute for rubber as well as building materials.

Use of PVC in the United States remained rather small until about 1927, when the promising properties of plasticized formulations were realized. At this time, the major chemical companies became aware of the possibilities for market development for polyvinyl chloride within the United States. By 1933 vinyl chloride-based resins were available in commercial quantities, although the growth rate until the Second World War was only moderate. The growth since 1946 has been 15 percent per year for both monomer and polymer. Many new applications have been developed yearly.

About 95 percent of the vinyl chloride monomer produced in the United States is used for the production of PVC and the various copolymers. The remaining 5 percent is used as an intermediate to produce chemicals such as 1,1,2-trichloroethane, vinylidene chloride, 1,1,1-trichloroethane, and others. Some vinyl chloride is used as a propellant for aerosol sprays.

The success of polyvinyl chloride as a plastic is due to its unique combination of properties. Formulated with plasticizers and fillers, it ranges from soft and elastic to hard and rigid. It can be produced in a wide range of colors and patterns, may exhibit good elongation and impact resistance, and is adaptable in its applications. The earliest products were made by calendering, extrusion, and molding. More recently lacquer, latex coating, and organosol and plastisol casting have become important.

Copolymers of vinyl chloride with vinylidene chloride, acrylonitrile, and vinyl acetate are important in the textile and paper-coating fields and in base resins for plastic flooring. In addition, they are used in a wide variety of protective coatings.

II. COMMERCIAL DEVELOPMENTS

Polyvinyl chloride has maintained a steady growth and is still one of the three largest-volume plastics in spite of competition from newer products. For a number of years polyvinyl chloride and polystyrene shared the lead as the largest-volume plastic materials. Both of these have now been passed by polyethylene, but PVC maintained a 15 percent annual growth from 1955 to 1964. In 1964, production in the United States was over 1.5 billion, in Japan 1 billion, and in western Europe about 2.5 billion lb. Monomer capacity approximately doubled between 1960 and 1964 in the United States, and monomer production increased 25 percent annually from 1964 to 1966, reaching 2.5 billion lb in the latter year. Figure 1 shows United States production since 1952. New plants and expansions

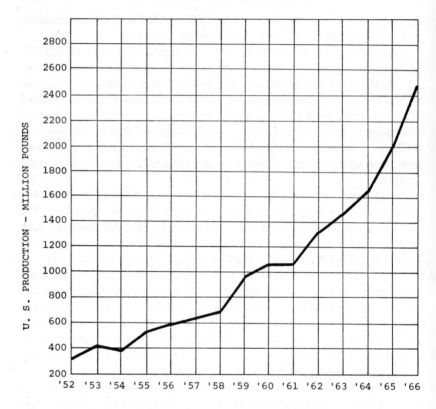

Fig. 1. Vinyl chloride production. From U.S. Tariff Commission.

that have been announced will add about 60 percent to the monomer capacity in the United States and 100 percent in Japan. These are remarkable increases for an already large-volume product. It does not seem probable that this growth rate could be continued for many years, but vinyl chloride has entered new markets on the basis of price and properties. Figure 2 shows the decline in the price of vinyl chloride in the United States since 1952. Contract or long-term sales prices have always been significantly lower than the prices quoted by the United States Tariff Commission. Price declines in vinyl chloride have been possible because of availability of cheaper ethylene, development of plants balanced with oxychlorination and acetylene-ethylene mixed-gas plants, and construction of larger plants that give a lower unit volume price.

The first vinyl chloride production in most countries was based on

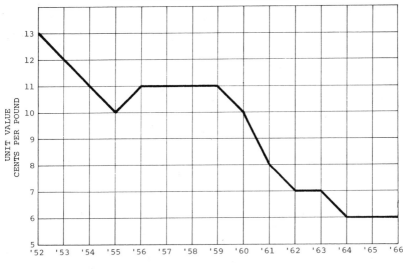

*Estimated - U. S. Tariff
Commission

Fig. 2. Vinyl chloride unit value. Data estimated by the U.S. Tariff Commission.

acetylene derived from calcium carbide, but in recent years processes using petrochemical acetylene have appeared. As ethylene has become available, the ethylene dichloride (EDC) cracking process has been developed and either balanced with the acetylene process or built as a separate plant, depending on the economics and outlets for the HCl. The caustic cracking of EDC was used to a small extent in the United States for a 10-year period but was abandoned because of the poor economics involved in salt formation. During the time that vinyl chloride and ethylene prices were dropping, acetylene prices continued to rise. Greater effort was then put into developing a process totally based on ethylene. The balanced oxychlorination route was the result and is being used in many of the major vinyl chloride expansions throughout the world. In 1965 only about 40 percent of the vinyl chloride produced was from acetylene. In other countries where naphtha rather than LPG was available, the acetylene-ethylene mixed-gas route was developed and is now competing with the balanced oxychlorination route. Processes are under development for the oxychlorination of ethylene and chlorination of ethylene directly to vinyl chloride, but are not in commercial production at this time.

Vinyl chloride monomer plants throughout the world range in capacity from 12 million to 400 million lb annually. Plants are under construction that will produce 700 million lb per year. Plants in the United States are generally larger than 100 million lb per year, and those announced since 1965 have been significantly larger than 100 million lb. Plants in other parts of the world, except Japan, tend to be smaller. Tables 1 and 2 illustrate vinyl chloride and PVC producers and capacities on a worldwide basis. The capacities shown are for those plants which are believed to be on stream as of September, 1967. Many new and larger facilities have been announced and may be under construction at this time.

TABLE 1

United States Vinyl Chloride Producers [5]

(Million lb per year)

Producer	Location	Capacity	Process[a]
Air Reduction	Calvert City, Ky.	115	Acetylene
Allied	Moundsville, W.Va.	150	Acetylene
American	Watson, Calif.	170	Ethylene (oxy)
Diamond Alkali	Deer Park, Tex.	100	Acetylene
Ethyl	Baton Rouge, La.	180	Ethylene
	Houston, Tex.	150	Ethylene
General Tire	Ashtabula, Ohio	75	Acetylene
Goodrich	Calvert City, Ky.	400	Ethylene (oxy)
	Niagara Falls, N.Y.	40	Acetylene
Goodyear	Niagara Falls, N.Y.	70	Acetylene
Monochem	Geismar, La.	250	Acetylene
Monsanto	Texas City, Tex.	120	Acetylene-ethylene and ethylene (oxy)
Tenneco	Houston, Tex.	200	Acetylene
Union Carbide	South Charleston, W.Va.	150	Acetylene-ethylene
	Texas City, Tex.	200	Acetylene-ethylene
Dow	Freeport, Tex.	200	Ethylene (oxy)
	Plaquemine, La.	250	Ethylene (oxy)

Along with expansion of capacity, improvements in monomer quality have been significant. The typical monomer in 1955 had an overall purity of 99.8 to 99.9 percent; the 1967 monomer had a purity of 99.99+ percent. A comparison of monomer specifications in 1955 and 1967 is given in Table 3. Gas-liquid chromatography has made it possible to identify the offending impurities in the vinyl chloride monomer.

TABLE 2
Foreign Vinyl Chloride Producers [5]
(Metric tons per year)

Producer	Location	Capacity	Process[a]
Argentina:			
Electrochlor	Capitan	11,000	Ethylene
Industrias Patagonicas	Cinco Saltos	2,500	Acetylene
Monsanto Andes	Mendoza	8,500	Acetylene
Australia:			
CSRC-Goodrich	Altona	14,000	Ethylene
ICIANZ	Botany	34,000	Acetylene-ethylene
Austria:			
Halvic Kunstoffwerke	Hallein	20,000	Acetylene
Belgium:			
Basant	Antwerp	110,000	Ethylene (oxy)
Solvay & Cie	Jemeppe-sur-Sambre	100,000	Ethylene (oxy)
Brazil:			
Electrocloro		28,000	Acetylene
Geon de Brazil	Sao Paulo	12,000	Acetylene
Britain:			
British Geon	Barry, S. Wales	130,000	Acetylene-ethylene
ICI	Runcorn	280,000 (est.)	Acetylene
Bulgaria:			
Devnya Reka		24,000	Ethylene (oxy)
Canada:			
Shawinigan	Varennes	27,000	Ethylene (oxy)
Dow	Sarnia	70,000	Ethylene
China:			
State Authority	Peking	25,000	Acetylene
Colombia:			
Columbia de Carburo y Derivados	Bogota	8,000	Acetylene
Formosa:			
Cathay Plastics	Taipei	9,000	Acetylene
China Plastics	Kaohsiung	12,000	Acetylene
Taiwan Plastics		30,000	Acetylene
France:			
Daufac	Jarrie	90,000	Ethylene (oxy)
Kuhlmann	Brignoud	50,000	Acetylene
Pechiney–St. Gobain	Montlucon	80,000	Acetylene and ethylene
Solvay	Tavaux	200,000	Acetylene
Vinylacq	Lacq	50,000	Acetylene

TABLE 2 (*continued*)

Germany (East):			
Buna Werke	Schkopau	110,000	Acetylene
Electrochem Bitterfield	Bitterfield		Acetylene
Germany:			
BASF	Ludwigshafen	122,000	Acetylene-ethylene
Dynamit Nobel	Luelsdorf	80,000	Ethylene (oxy)
Hoechst	Frankfurt	12,000	Acetylene
	Gendorf	24,000	Acetylene
Huls	Marl	170,000	Acetylene-ethylene
Knapsack-Griesheim	Cologne	100,000	Acetylene-ethylene
Lonza Elektrochemie	Waldshut	15,000	Acetylene
Greece:			
Ethyl Hellas	Thessaloniki	17,000	Ethylene
Hungary:			
Borsed Chemical	Berente	30,000	Acetylene
India:			
Calico Mills	Bombay	20,000	Acetylene
DCM Chemical	Bombay	6,000	Acetylene
Dhrangada Chemical	Sahupeuam	12,000	Mixed gas
National Organic Chemical	Bombay	30,000	Acetylene
Plastic Resin & Chemical	Madras	6,000	Ethylene
SREE Ram Vinyls	Kotah	20,000	Acetylene
Israel:			
Electrochemical Industries	Acre	9,000	Ethylene (oxy)
Italy:			
Chemical Ravenna (ANIC)	Ravenna	50,000	Acetylene
Ceramica Pozzi	Ferrandina	48,000	Acetylene
Montecatini-Edison	Brindisi	200,000	Acetylene-ethylene and ethylene (oxy)
	Porto Marghera	150,000	Acetylene-ethylene
Solvay	Ferrara	70,000	Acetylene-ethylene
U.S. Rubber Rumianca	Pieve Vergonte	22,000	Acetylene
Japan:			
Denki Chemical	Ohme	6,000	Acetylene
Gunma Kagaku	Shibukau	6,000	Acetylene
Japanese Geon	Takaoka	85,200	Acetylene
Kanegafuchi	Takasago	44,500	Acetylene-ethylene
Kureha	Nisiki	60,000	Mixed gas
Mitsubishi-Monsanto	Yokkaichi	50,000	Acetylene
Mitsui	Omuta	45,000	Ethylene (oxy)
Nippon Carbide	Uotsu and Hayatsuhi	36,960	Acetylene-ethylene
Nissan	Goi	30,000	Acetylene
Nissin	Takefu	50,000	Acetylene

TABLE 2 (*continued*)

San Arrow	Tokuyama	60,000	Ethylene (oxy)
Sekisui	Tokuyama	36,000	Acetylene
Shin-Etsu	Naoetsu	48,000	Acetylene
Sumitomo	Kikumoto	35,000	Acetylene-ethylene
	Shizura	18,000	Acetylene-ethylene
Tekkosha	Sakata	30,000	Acetylene-ethylene
Toa Gosei	Tokushima	6,000	Acetylene
	Nagoya	20,000	Acetylene-ethylene
Toyo Soda	Tokuyama	50,000	Ethylene (oxy)
Korea:			
Daehan Plastics	Fukouri	7,200	Acetylene
Netherlands:			
Royal Dutch Shell	Pernis	75,000	Ethylene
Norway:			
Norsk Hydro	Heroya	50,000	Acetylene
Pakistan:			
Arokey	Karachi	6,000	Acetylene
Peru:			
W. R. Grace	Paramonga	5,000	Ethylene (oxy)
Poland:			
State Authority	Tarnow	20,000	Acetylene
	Oswiecim	25,000	Acetylene
Portugal:			
Ceres	Oporto	12,000	Acetylene
Rumania:			
State Authority	Borzesti	36,000	Acetylene
Singapore:			
Allied Chemical		not known	Acetylene
Spain:			
Etino Quimica	Monzon	22,500	Acetylene
Hispavic	Torrelavega	30,000	Acetylene
Resinas Poliesteres	Hernani	25,000	Acetylene
Sweden:			
Fosfatbolaget	Stockwikswerken	30,000	Acetylene
Switzerland:			
Lonza	Sins	20,000	Acetylene
Yugoslavia:			
Yugovinil	Kastel Sucurac	8,000	Acetylene

[a] Acetylene indicates hydrochlorination of acetylene process.

Ethylene indicates direct chlorination route to ethylene dichloride and cracking.

Ethylene (oxy) indicates direct chlorination and oxychlorination of ethylene to ethylene dichloride and cracking.

Acetylene-ethylene indicates direct chlorination of ethylene to ethylene dichloride and cracking and using by-product HCl for hydrochlorination of acetylene to vinyl chloride.

Mixed gas indicates the process starting with a dilute mixed stream of acetylene and ethylene.

TABLE 3
Typical Monomer Specifications for Vinyl Chloride

	1955	1967
Water, percent, max	0.1	0.005–0.03
Polymerization test	80–90%	...
Dorell, T°C at:		
3.0 ml	0.20 max	...
2.0 ml	0.40	...
1.0 ml	0.80	...
0.5 ml	1.50	...
Acetylene, ppm, max	10	2
Acetaldehyde, ppm, max	50	5
Inhibitor (phenol), ppm	50–500	0–100
Nonvolatiles, percent, max	0.1	0.01
Sulfur, ppm, max	25	5
Iron, ppm, max	5	1
Acidity, ppm, max	50	5
Peroxides (as H_2O_2), ppm, max	0.1	0.1
Organics by VPC, ppm:		
Acetylenic compounds	...	10
1,2-Dichloroethylene	...	10
Butadiene	...	5
Methyl chloride	...	25
Total organic impurities	...	75–100

III. COMMERCIAL PRODUCTION PROCESSES

There follows a general discussion of the various commercial production methods for vinyl chloride in the chronological order of appearance as commercial processes.

A. Hydrochlorination of Acetylene

Electrochemically derived calcium carbide and caustic established the early importance of the acetylene route to vinyl chloride. Calcium carbide could be readily transported to the location where the by-product chlorine from caustic production was available:

$$Cl_2 + H_2 \longrightarrow 2HCl \tag{1}$$

$$CaC_2 + 2H_2O \longrightarrow C_2H_2 + Ca(OH)_2 \tag{2}$$

$$C_2H_2 + HCl \longrightarrow CH_2{:}CHCl + 24{,}500 \text{ cal/mole (6)} \tag{3}$$

The first recorded application of Eq. (3) for the production of vinyl chloride was by Griesheim (7) in 1913. Numerous attempts have been made to conduct this reaction using techniques other than that described below, which is merely an extension of Griesheim's process. Extensive investigations have been made into the promotion of mercury-based catalyst or the improvement of the basic catalyst system and process. Work continues in this direction, with one of the most recent contributions, by Czarny (8), being particularly valuable.

Vertical fixed-tube and shell heat exchangers, having about 200 to 1,000 tubes $1\frac{1}{2}$ to 3 in. in diameter and 8 to 20 feet long, are packed with 4 to 10 mesh activated carbon impregnated with about 10 percent mercuric chloride. In numerous cases this catalyst may be promoted with other components. The reactor pressure is maintained at about 15 to 25 psig. Higher pressures are not practical, because of the dangers associated with the compression of concentrated acetylene streams and difficult temperature control due to hot spots.

Flow charts of the reactor portions of two types of commercial acetylene hydrochlorination plants are shown in Figs. 3 and 4. Figure 3 is based on a German process (9) and Fig. 4 on information from General Tire and Rubber Company (10). A more detailed description of the operating variables for the process represented by Fig. 4 is available (11).

The activated carbon used as the carrier can be derived from coal, nut hulls, or petroleum coke. The content of mercuric chloride or mercurous chloride can be increased to above 10 percent to extend the life of the catalyst, which depends very significantly on operating conditions. Dryness and the purity of the feed gases are very important, with the presence of higher acetylenes, sulfides, and phosphorous and arsenic hydrides being particularly deleterious. Purification of the acetylene for this process is of major concern and an important factor in capital and operating expenses. Initially the catalyst is quite active, requiring a temperature of about 80 to 100°C for satisfactory yields and conversions. As the catalyst ages, the temperature must be gradually increased to maintain acceptable yields. A final maximum operating temperature of about 150 to 170°C is dictated because of the volatility of mercuric chloride above about 180°C and the formation of tars that foul and deactivate the catalyst. The normal range of life of about 6 months to somewhat over a year can be significantly influenced by the quality of temperature control. Control is maintained by boiling or circulating a heat-transfer medium, such as water, oil, or Dowtherm, through the shell. Materials of construction for this process are all carbon steel with the exception of distillation reboiler tubes, which should be nonferrous. The process is anhydrous, and all feed streams should contain no more than 50 ppm of water.

Feed gases are mixed at a 1:1 ratio. Normally, the acetylene conversion averages between about 90 and 98 percent per pass. In the case of lower

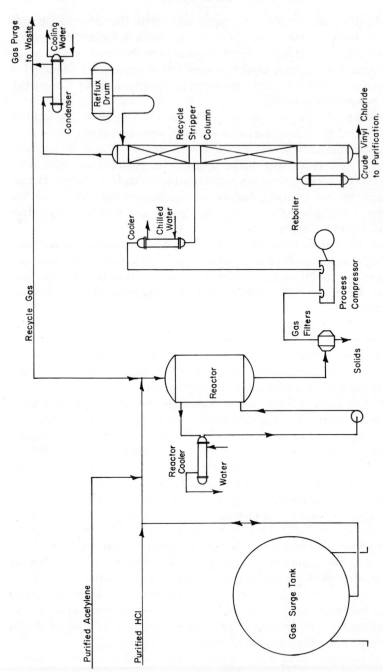

Fig. 3. Vinyl chloride by acetylene hydrochlorination.

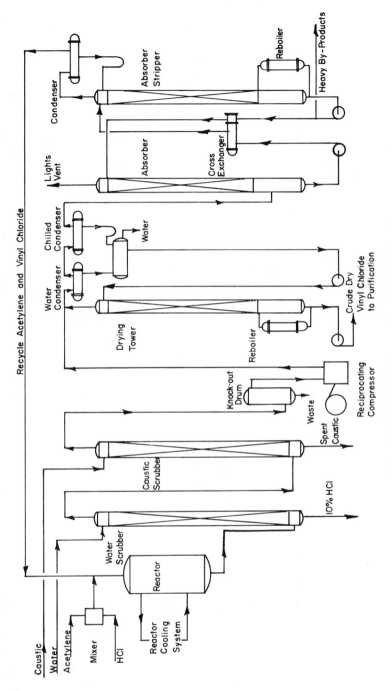

Fig. 4. Vinyl chloride by acetylene hydrochlorination.

1217

conversion, some recirculation of the unreacted gases is required. HCl conversions of about 95 to 99 percent are attained. Overall yields to vinyl chloride for both raw materials can be maintained at above 96 percent, with yields based on HCl of 99 percent being possible (19). Principal byproducts are dichloroethane and acetaldehyde.

Attempts have been made to improve the temperature control by dissolving or suspending the active catalyst in an inert solvent, such as perchloroethylene. A major drawback to this approach, however, is the much lower productivity of the catalyst system, and it has received no commercial use of consequence.

Reactant conversion per pass and reactor productivity affect catalyst life. Higher conversions are attainable at any flow rate by elevation of the temperature with a consequent reduction in the life of the catalyst. Reaction rates from 17 to 273 g vinyl chloride per liter of catalyst per hr are cited by Lynn (12) in pilot studies on dilute acetylene streams. Production facilities operate under widely differing conditions with a range of about 30 to 300 g of vinyl chloride produced per liter of catalyst per hr (13). A very general relationship between reactor productivity and catalyst life has been developed from a survey of available data and is shown in Fig. 5 (9,13–17).

Barton and Mugdan (18) have made a study of this process for the

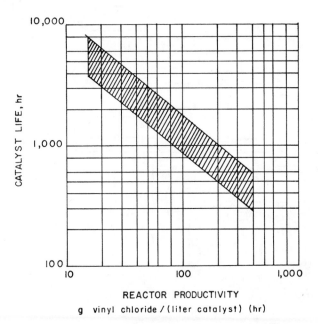

Fig. 5. Effect of reactor productivity on catalyst life.

production of vinyl chloride and present convincing arguments for the mechanism shown in Eqs. 4 and 5:

$$C_2H_2 + HgCl_2 \longrightarrow CHCl:CH \cdot HgCl \qquad (4)$$

$$CHCl:CH \cdot HgCl + HCl \longrightarrow CHCl:CH_2 + HgCl_2 \qquad (5)$$

The kinetic data for this reaction, which was developed by Wesselhoft (6), does not seem to relate to the data obtained from commercial plants.

Henri Gibello, in the French publication "Vinyl Chloride and Its Polymers" (19), discusses in considerable depth the patent and industrial situation in Europe in 1955. It is evident from his work that virtually all the vinyl chloride manufactured in Europe at this time was by hydrochlorination of acetylene. Other sources of information on this process are available (20). The last large facility based on acetylene and HCl that was built in the United States was a Tenneco plant, which began operating in 1962.

As the cost of petroleum-based acetylene has decreased and the availability of by-product HCl from chlorinated hydrocarbon production has increased, the source of the two raw materials has correspondingly shifted. Today virtually no HCl is produced directly from chlorine in the United States, and calcium carbide has been replaced, or plans are being made to replace it, with petroleum as the source of acetylene. The prospect of lower-cost concentrated acetylene from petroleum-based plants has extended the usefulness of the acetylene–vinyl chloride process. As the EDC cracking process became significant, so did the quantity of by-product HCl. Outlets for aqueous solutions of HCl were insufficient. A natural combination resulted in which the HCl by-product from the cracked ethylene dichloride became the raw material for the acetylene production of vinyl chloride. When more vinyl chloride capacity was needed by a producer using the acetylene process, it became economically sound to obtain it by installing an ethylene dichloride cracking facility to operate in balance with the original acetylene-HCl unit. Recent economic developments, however, have changed the picture. Instances are known in which acetylene-based producers are abandoning their facilities and either purchasing vinyl chloride or converting entirely to the less costly ethylene-based process.

A comparison of the various ethylene- and acetylene-based processes, both singly and balanced, has been made (21). Using very generous allowances, the conclusion is drawn that ethylene is certain to displace acetylene as a raw material for the production of vinyl chloride.

Although there is a possibility (22) of new technology that will reduce acetylene costs, the trend away from acetylene is now so strong as to be difficult to reverse.

B. Thermal Cracking of Ethylene Dichloride

The large-scale use of ethylene for manufacture of ethyl chloride, ethylene glycol, styrene, and polyethylene in the period 1940 to 1955 provided a cheap raw material for a new vinyl chloride process—Eqs. (6) and (7):

$$Cl_2 + CH_2:CH_2 \longrightarrow CH_2ClCH_2Cl + 79{,}020 \text{ Btu/lb-mole (25°C)} \quad (6)$$

$$CH_2ClCH_2Cl \longrightarrow CH_2:CHCl + HCl \quad (7)$$

The first step was already well developed by 1945. Yields based on both chlorine and ethylene are above 98 percent, and low-purity ethylene and chlorine can be used, with the ethylene dichloride being subsequently purified by simple distillation techniques. The availability of ethylene dichloride was already established, since it was a by-product from the production of ethylene glycol by the chlorohydrin route. Initially, this ethylene dichloride was converted to vinyl chloride by caustic dehydrochlorination. The number of companies producing vinyl chloride in this manner was small, and it had ceased to be a process of commercial consequence by 1955.

1. ETHYLENE DICHLORIDE BY CHLORINATION OF ETHYLENE

A discussion of this commercial process is included here in view of its close relationship to the production of vinyl chloride.

Ethylene dichloride is produced in a liquid-phase reaction with a Friedel-Crafts catalyst, such as ferric chloride. This material at a concentration of 0.02 to 0.2 percent under the conditions of operation is in solution. Production can be in a liquid bed that is cooled with either an internal or an external heat exchanger or by evaporation of ethylene dichloride. The gaseous product produced by this latter method is externally condensed and returned as a liquid to the reactor. An advantage of this approach is that the product is not contaminated with catalyst (23). The presence of oxygen has been shown (24) to be beneficial in preventing substitutive chlorination, which leads to the formation of 1,1,2-trichloroethane. This effect is more pronounced at the higher temperature experienced in the boiling liquid bed. The mechanism has not been defined, although it is probably due to the chain-stopping ability of oxygen. The presence of metal walls or packing promotes substitution side reactions (25). The reaction appears to be controlled by the absorption of ethylene (26). In industrial applications the feed gases are normally introduced as a gas through spargers into a liquid bed several feet deep. The major factors affecting yields are the presence of other unsaturates in the ethylene and the undesired substitution chlorination mentioned previously.

The productivity of a liquid-phase boiling kind of bed has been shown in a patent (27) to be between 0.28 and 0.56 mole of ethylene dichloride per hr per cu ft of liquid bed.

Where streams with unusually low concentrations of either chlorine or ethylene are available, a gas-phase reaction has been tried (28–30). Precise temperature-control requirements, low yields due to overchlorination (substitution), difficult product recovery, and high cost make this variation less desirable. The liquid-phase process must be anhydrous, and all feed streams should contain less than 30 ppm of water. A steel reactor can be used if a 2 to 5 year life is acceptable, but a nickel-based nonferrous alloy is better. Steel construction is satisfactory for the remaining equipment. A flow chart for one variation of this process is shown in Fig. 6. Normal operating pressure can vary from atmospheric to about 35 psig. The upper limit of pressure is set by the initiation of cracking of the higher-chlorinated derivatives in the bottom of the reactor. Normal operating temperatures are 50 to 130°C.

The operation may be conducted with a slight excess of either chlorine or ethylene. Complete reaction of either chlorine or ethylene is not realized in the commercial units, and scrubbing of the small amounts of chlorine and HCl must be conducted. The limiting quantity of air is determined more by the need to keep the vent gases outside the explosive range than by the catalytic requirements. Other inert components also affect the quantity of air used. Further complicating the picture of vent control is the presence of substantial quantities of hydrogen in chlorine produced from mercury cells.

The mechanism that best explains this process is a polar reaction (31–35) shown in Eqs. (8) to (10):

$$FeCl_3 + Cl_2 \longrightarrow [FeCl_4{}^- Cl^+] \tag{8}$$
$$[FeCl_4{}^- Cl^+] + C_2H_4 \longrightarrow [CH_2ClCH_2{}^+ FeCl_4{}^-] \tag{9}$$
$$[CH_2ClCH_2{}^+ FeCl_4{}^-] \longrightarrow CH_2ClCH_2Cl + FeCl_3 \tag{10}$$

Purification of the ethylene dichloride can be handled by installing a rectification section above the liquid bed for removal of the heavies (27). The light components are then stripped out in a lights-removal column. A small flow must be taken out of the bottom of the reactor to remove accumulated heavies. The ethylene dichloride present in this stream is recovered in a heavies-removal column run at atmospheric or reduced pressure. Alternatively, the reaction can be run at a lower temperature with a liquid flow being removed from the reactor, cooled, and returned to control the temperature. The liquid-product stream withdrawn from this reactor is treated for removal of catalyst and distilled through a lights- and

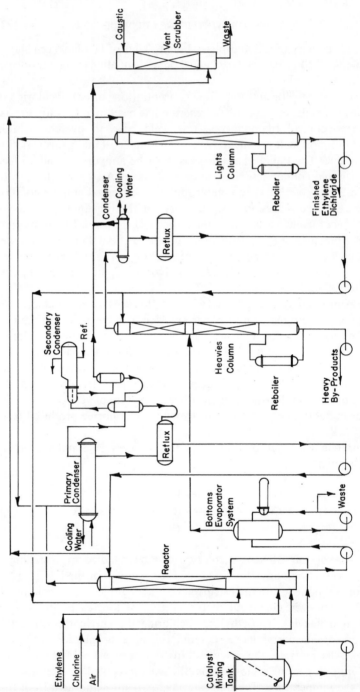

Fig. 6. Ethylene dichloride by ethylene chlorination.

heavies-removal column. Utility consumption and equipment cost of the latter variation are somewhat higher.

Unconverted ethylene dichloride from the vinyl chloride–producing unit must also be purified. After stripping of dissolved monomer and HCl, it is recovered from heavy impurities by fractional distillation. Certain components in the ethylene dichloride recycle stream can build up and present operating problems.

Literature and other sources indicate that the purer the ethylene dichloride, the less involved and more reliable the production of vinyl chloride. The purity of furnace feed-grade ethylene dichloride has increased over the years and currently is between 99.7 and 99.95 percent with the type of furnace, the operating condition, and the process scheme being the factors that dictate the purity.

2. THERMAL DEHYDROCHLORINATION OF ETHYLENE DICHLORIDE

The overwhelmingly important development in the commercial production of vinyl chloride was the thermal dehydrochlorination of ethylene dichloride. The process involves relatively small capital expenditure and has been developed into a very efficient process, with yields of about 99 percent. The importance of this process was magnified by the simplicity and low cost of the production of ethylene dichloride by direct chlorination, and the evolution of very large plants producing inexpensive ethylene.

The acetylene route for producing vinyl chloride has, since shortly before 1960, represented an insignificant portion of the growth of the vinyl chloride industry in the United States. The other more advanced industrial countries, such as those in western Europe, Japan, and Russia, are experiencing the same swing to ethylene as large natural gas- and naphtha-based ethylene plants are introduced.

Biltz and Küppers (34) reported their work on the effect of volatile substances on the thermal dehydrochlorination of ethylene dichloride in 1904. Imperial Chemical Industries (36) patented the use of steam or HCl as a diluent in the reaction. In 1940, Barton (37), under the auspices of the Distillers Company, investigated the induction of the dehydrochlorination of chlorinated ethanes. The information obtained in this work is of importance even today. The Distillers Company patented aspects of this work in 1945.

The thermal dehydrochlorination of ethylene dichloride is a free-radical reaction, and a probable mechanism is shown by the following equations:

Initiation

$$CH_2ClCH_2Cl \longrightarrow CH_2ClCH_2\cdot + Cl\cdot \qquad (11)$$

$$CH_2ClCH_2Cl \longrightarrow C_2H_4 + Cl_2 \qquad (12)$$

$$Cl_2 \longrightarrow 2Cl\cdot \qquad (13)$$

Propagation

$$CH_2ClCH_2Cl + Cl\cdot \longrightarrow CH_2ClCHCl\cdot + HCl \tag{14}$$

$$CH_2ClCHCl\cdot \longrightarrow CH_2:CHCl + Cl\cdot \tag{15}$$

$$C_2H_4 + Cl\cdot \longrightarrow C_2H_3\cdot + HCl \tag{16}$$

$$C_2H_3\cdot \longrightarrow CH\vdots CH + H\cdot \tag{17}$$

$$CH_2ClCH_2\cdot \longrightarrow CH_2:CHCl + H\cdot \tag{18}$$

Termination

$$CH_2ClCHCl\cdot + Cl\cdot \longrightarrow CH_2ClCHCl_2 \tag{19}$$

$$H\cdot + Cl\cdot \longrightarrow HCl \tag{20}$$

$$\text{Wall effects, dimerization} \tag{21}$$

Equations (11), (13), (14), (15), (19), (20), and (21) appear to be favored at temperatures below 480°C, based on the products formed under these conditions. At higher temperatures, Eqs. (12), (16), (17), and (18) become increasingly important. Equations (12), (16), and (17) explain the production of acetylene, although Barton (37) states that it may be formed by the heterogeneous decomposition of vinyl chloride, as shown in Eqs. (22) and (23):

$$CHCl:CH_2 + Cl\cdot \longrightarrow CHCl:CH\cdot + HCl \tag{22}$$

$$CHCl:CH\cdot \longrightarrow CH\vdots CH + Cl\cdot \tag{23}$$

Equations (24) and (25) account for the formation of chloroprene and butadiene:

$$CH_2:CHCl + CH_2:CH\cdot \longrightarrow CH_2:CClCH:CH_2 + H\cdot \tag{24}$$

$$2CH_2:CH\cdot \longrightarrow CH_2:CHCH:CH_2 \tag{25}$$

Chloroprene could also be produced during cooling by the hydrochlorination of vinylacetylene, an impurity that is probably formed by the dimerization of acetylene. Vinylacetylene is found in products from the cracking furnace in varying quantities. Alternatively, a dimerization reaction of the dichloroethyl radical with subsequent bimolecular expulsion of HCl may be of significance, although an explanation of the dependence of this mechanism on temperature is more difficult. Since the amounts of acetylene and butadiene that are produced increase significantly with vinyl chloride concentration (conversions of ethylene dichloride above 75 percent), the indicated routes are favored. Dimerization of chlorine free radicals is of little consequence.

The reaction-rate constant was found to be $1.59 \times 10^6 e$ exp $(-27,000/RT)$ by Barton (37).

The activation energy has been determined to be 78 ± 10 kcal per mole.

As a free-radical reaction, this process is especially sensitive to promoters and inhibitors. Barton (37) demonstrated the beneficial effects of chlorine and oxygen and the detrimental effects of packing and walls. Equation (13) is a very rapid reaction at the temperatures used in this process.

Doraiswamy (38) studied the kinetics and thermodynamics of this process. Unfortunately, he used an impure and unanalyzed grade of ethylene dichloride despite his knowledge of the importance of inhibitors and purity on conversion and reproducibility. His values for the reaction rate constant are shown below:

$$k = 1.45 \times 10^{20}e \exp(-59,000/RT) \qquad 0\text{--}80\% \text{ conversion}$$
$$k = 9.8 \times 10^{7}e \exp(-20,000/RT) \qquad 80\text{--}100\% \text{ conversion}$$

The heat of reaction for the formation of vinyl chloride from ethylene dichloride—Eq. (7)—has been calculated to be 28,300 Btu per lb-mole at 25°C, based on the data of McGovern (39), and 31,680 Btu per lb-mole at 25°C and 32,170 Btu per lb-mole at 475°C, based on the data of Sinke and Stull (40).

There are two types of processes for the thermal cracking of ethylene dichloride. By far the largest quantity of vinyl chloride is produced by the noncatalytic or homogeneous catalytic process, with only a few reported processes that utilize a heterogeneous catalyst.

a. Heterogeneous Catalyst

Braconier (28) reports, that in the Société Belge de l'Azote (SBA) process, the thermal dehydrochlorination of ethylene dichloride is conducted catalytically. SBA patents (29,41) show that the catalyst is of a carbon type and describe it as "deashed activated." The operating conditions of this process are 450°C and 120 psig. Conversions of 65 percent and yields of 99 percent are claimed.

Wacker-Chemie (42,43) claims the development of a catalyst with sufficient activity to allow operation at 200 to 350°C. Operation of this packed reaction tube for up to 1 year is claimed to be possible because of reduced carbon formation. The favored catalyst is activated carbon impregnated with barium chloride. Elevation of the temperature over the range stated is necessary during the life of the catalyst to maintain acceptable conversion. Also, a diluent gas, such as HCl, is used to lower the vinyl chloride in the product to a level below 40 percent to reduce carbon formation.

Cracking furnaces for these processes are normally direct-fired.

Although there is a significantly lower operating temperature for this kind of process, it has not received any general acceptance primarily because of three factors:

1. There are additional costs of charging and downtime required for changing the catalyst. This, coupled with the sensitivity of the catalyst to fouling due to tars and carbon deposition, requires that spare capacity be provided through either more or larger units.

2. The actual selectivity of the process to vinyl chloride is not appreciably better than the more conventional process, although a lower temperature is utilized.

3. Improvements in the conventional thermal cracking process with regard to coking rate have made a lower operating temperature less important.

b. Noncatalytic or Homogeneous Catalyst

These processes all operate in a similar fashion. The more important distinctions will be discussed below. A flow chart of the general process is shown in Fig. 7. Ethylene dichloride is pumped under pressure into an evaporator and there vaporized by steam. The gaseous ethylene dichloride is preheated from 150 to 400°C in the convection section of a direct-fired, radiant-wall kind of furnace. The preheated gases are then heated in the radiant section up to the cracking temperature of 500°C. The geometry of the radiant section, fluid-flow dynamics, and the temperature profile (firing pattern of the furnace) are considered highly confidential information in the industry. The most detailed description of an industrial kind of ethylene dichloride cracking furnace is that described by B. F. Goodrich (44).

Residence time in the radiant section of a cracking furnace may vary significantly, but is normally in the range of 3 to 10 sec. The reactor gaseous product contains, on a molar basis, 39 percent vinyl chloride, 39 percent hydrogen chloride, 22 percent unconverted ethylene dichloride, and 0.5 percent by-products, consisting principally of acetylene, vinylacetylene, 1,1-dichloroethane, 2-chloro-1,3-butadiene, and 1,3-butadiene. These hot gases are quickly cooled, usually by direct contact, to a point where most of the unconverted ethylene dichloride is condensed. This condensate is externally cooled and used as the liquid for the quenching operation. The unconverted ethylene dichloride is stripped of its dissolved vinyl chloride and hydrogen chloride and returned to the ethylene dichloride purification facilities for treatment.

There are many variations to the process described above. Some of the more important are as follows.

c. Pressure

Furnaces operate with an outlet pressure of 20 to 400 psig. The lower pressures are usually used in the older plants and in plants that remove

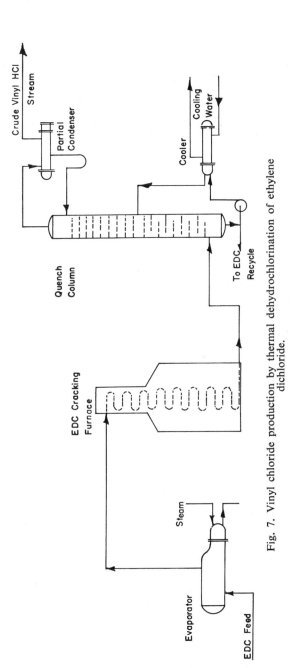

Fig. 7. Vinyl chloride production by thermal dehydrochlorination of ethylene dichloride.

by-product HCl by absorption. The gases must then be compressed and/ or dried for purification. These types of systems are no longer being built in the more industrially developed countries because of the expense of operation. To achieve a satisfactory conversion (60 to 65 percent) of a high-quality ethylene dichloride stream, cracking temperatures of 450 to 480°C are required. Increased pressure of about 100 to over 200 psig is being utilized in the present generation of furnaces. The major reason for this change is the elimination of the high-maintenance compression step required in the process in which hydrogen chloride is recovered in the more useful anhydrous form. To achieve satisfactory conversions under these conditions, a slightly higher temperature of 470 to 500°C is required.

Very high-pressure furnaces (375 to 450 psig) are operated by Goodrich, Hoechst and their licensees (44–46). Conversions are about 60 percent, and temperatures from 500°C to 515°C are cited. Increased pressure requires small increases in temperature to secure acceptable conversions. This increased severity, as well as the indicated method of vaporization (44), would require a very high-purity feedstock.

d. Feed Evaporization

The ethylene dichloride may be fed to the furnace as a vapor, or it may be vaporized within the furnace in a portion of the coil. The former method is more capable of handling feeds of varying quality; the latter is more economical, although shorter run times between decokings may result.

e. Quenching

There are several methods of quenching, the majority of which involve the direct contacting of the hot furnace off-gas with a portion of the condensed uncracked ethylene dichloride. Probably the most common technique is use of a quench tower. Because of the possible presence of carbon in the furnace outlet, the tower is seldom of the packed or trayed type but rather the spray or splash-tray type. Another quenching method that involves direct contact is the quench nozzle.

Recently, some mention has been made of an indirect quenching method in which the hot gases are contacted with cooled metal tubes, such as in an exchanger.

f. Catalysis

Barton (37) investigated the catalytic effects of various materials and found that chlorine and oxygen were both beneficial in increasing conversion at a given temperature or in reducing, by 50 to 100°C, the required temperature to achieve a given conversion. The level of catalyst

TABLE 4

Inhibitors of the Thermal Dehydrochlorination of
Ethylene Dichloride [37,178]

Compound	Qualitative inhibitive effect[a]
Alcohols	Strong
Aliphatics	Strong
Ethylene chlorohydrin	Strong
Propylene	Strong
Propylene dichloride	Strong
Toluene	Strong
Aldehydes	Mild
Chlorobutanes	Mild
Ethers	Mild
Ethylene	Mild
Ketones	Mild
2-Methyl-1,3-dioxolane	Mild
Benzene	Weak
1,1-Dichloroethane	Weak

[a] This designation refers to the concentration of compound required to produce a significant reduction in cracking rate, as follows:

Strong	Less than 0.02%
Mild	0.02–0.05%
Weak	More than 0.05%

required was about 0.2 to 0.5 percent. There is little evidence that the benefits of this effect have been utilized to a significant degree, but the Distillers Company (47) patented Barton's findings in 1945. Recently, Hüls (48) has obtained a patent in Germany for the use of chlorine. The extent to which chlorine is used as a catalyst in the industry is not known.

It has been shown (37) that the vessel walls do not inhibit the reaction initially. Some observations, in fact, indicate that they may promote the reaction in a freshly decoked furnace. After a short time the carbon deposited on the walls shows significant inhibitive effects that can be reduced somewhat by increasing the diameter of the reactor coil. Development of some method to prevent carbon deposition might permit higher conversions or use of lower temperatures. Other inhibitors of the reaction are listed in Table 4.

g. Ethylene Dichloride Purity

It was recognized early in the development of the process that ethylene dichloride quality influenced the cracking rate and the purity of the vinyl chloride produced. Today, the use of very pure ethylene dichloride makes operation for several months between decokings feasible even under severe conditions, and conversions have been raised to the 70 percent level. By-product formation has decreased to a point that furnace geometry and flow dynamics are the only remaining areas for significant improvement. Pressures higher than about 200 psig are felt to be of little value, since the benefits, which are principally the lowered refrigeration requirements necessary to recover the anhydrous hydrogen chloride, become increasingly small.

3. Oxychlorination Processes

The advantages of high efficiencies, simplicity of operation, and low capital requirements in the production of vinyl chloride by dehydrochlorination of ethylene dichloride are somewhat offset by the loss as HCl of half the chlorine in the EDC. This loss is negated if the dehydrochlorination is operated in conjunction with an acetylene-based vinyl chloride process that consumes the by-product HCl. As pointed out earlier, however, the latter production method has become unattractive because of the relatively high price of acetylene. A way out of this apparent impasse is to oxidize the HCl by the well-known Deacon process—Eq. (26) —and use the resulting chlorine to chlorinate more ethylene. In effect, then, the raw materials become ethylene, chlorine, and oxygen, and the only by-product is water—Eq. (27):

$$2HCl + \tfrac{1}{2}O_2 \xrightarrow{\text{catalyst}} Cl_2 + H_2O \tag{26}$$

$$CH_2{=}CH_2 + HCl + \tfrac{1}{2}O_2 \longrightarrow CH_2{=}CHCl + H_2O \tag{27}$$

Producers of vinyl chloride have recognized the advantage of this approach and have developed it commercially.

The Deacon process is based on a cupric chloride catalyst, and the mechanism appears to be as shown in Eqs. (28) to (30) (49):

$$2CuCl_2 \longrightarrow Cu_2Cl_2 + Cl_2 \tag{28}$$

$$Cu_2Cl_2 + \tfrac{1}{2}O_2 \longrightarrow CuO \cdot CuCl_2 \tag{29}$$

$$CuO \cdot CuCl_2 + 2HCl \longrightarrow 2CuCl_2 + H_2O \tag{30}$$

Net Reaction

$$2HCl + \tfrac{1}{2}O_2 \longrightarrow Cl_2 + H_2O$$

Equation (29) has been suggested as the rate-controlling step (50).

a. Chlorine by the Deacon Process

Shell (51,52) originally took the approach of converting the hydrogen chloride directly to chlorine. The chlorine is produced at about a 40 percent

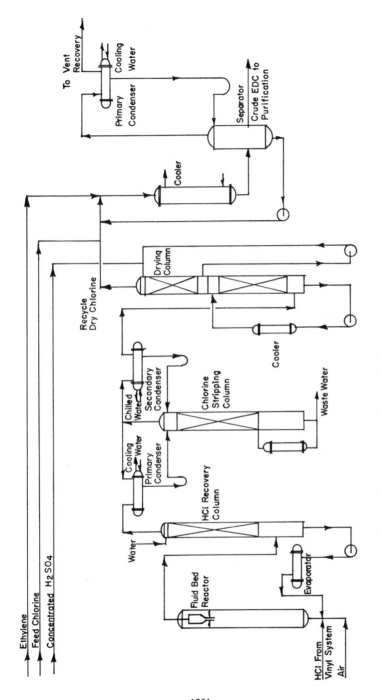

Fig. 8. Flow diagram of the Shell chlorine and ethylene dichloride processes.

1231

gaseous concentration, dried, and then used to chlorinate ethylene to ethylene dichloride. A flow chart of a method of conducting this process is shown in Fig. 8. The Shell chlorine process reacts a stream of HCl and air in a fluidized bed. The conversion of the HCl to chlorine is about 75 percent. The reactor gases are cooled sufficiently to condense the 20 percent hydrochloric acid solution, which is recycled to the process for temperature control and conversion to chlorine. The vapor, which consists of chlorine, water, oxygen, and inerts, is cooled to 5°C to condense most of the water. The chlorine-rich gas is dried with sulfuric acid and mixed with the chlorine feed to the ethylene chlorination unit. The condensed water is stripped of chlorine, which is returned to the wet chlorine product stream.

Shell states that an advantage of their process is that it can utilize the dilute stream of chlorine as well as a dilute stream of ethylene. The cost of recovery of the ethylene dichloride from this type of system is probably quite high for both capital and refrigeration. The cost of drying the large volume of wet dilute chlorine stream is probably also significant. There are five reported facilities (52) in which the Shell chlorine process is integrated into the production of vinyl chloride. The catalyst support is believed to be silica gel having a surface area of 300 sq m per g. The material of construction for the reactor is probably nickel, monel, or a nonferrous alloy. The remainder of the equipment is probably of acid-resistant construction to the point where the chlorine has been dried.

A Shell patent (53) states that Eqs. (28) and (29) may be carried out simultaneously while Eq. (30) is run in another reactor. No example of commercial application of this approach has been found. A patent (54) assigned to Solvay et Cie describes a similar process.

Recent Shell patents have indicated an interest in the utilization of a fixed bed of copper chloride on silica gel to effect the chlorination of unsaturates such as ethylene (53,55).

b. Ethylene Dichloride from Ethylene

The process for utilizing the chlorine produced from HCl and oxygen *in situ* to chlorinate hydrocarbons has involved a great amount of research. Initial attempts were made about 1922 (56), but it was not until 1958 that the Dow Chemical Company came on-stream with the first large-scale commercial oxychlorination process. This plant produces ethylene dichloride from ethylene and utilizes a fixed-tube reactor containing a catalyst consisting of cupric chloride on an active carrier. Moderation of the reaction is accomplished with a diluent having suitable thermal conductivity characteristics (57).

Goodrich has developed a fluidized-bed oxychlorination reactor (58) that is claimed to provide very close temperature control and does not require recycle for high conversions of ethylene and hydrogen chloride. Some type of vent recovery system is apparently used, possibly a liquid absorption system. There are at least three facilities that are operating or close to operation that are licensing the Goodrich oxychlorination process (59).

Other companies, such as Asahi Chemical, Monsanto, Shell, Stauffer, Toyo Soda, and Vulcan Materials, have developed oxychlorination processes that are similar to one or the other of the above, the major difference being catalyst formulation. Normally the catalyst consists of 5 to 20 percent copper chloride on a support. Supports vary from high-surface-area materials, such as activated alumina and silica gel, to low-surface-area materials such as Celite. Usually the latter require high concentrations of copper chloride to achieve operation at satisfactorily low temperatures. High-surface-area catalyst could produce significant oxidation of ethylene if proper temperature control were not maintained. Normal losses to oxidation amount to about 2 to 8 percent.

Many promoters have been tried, but only potassium chloride seems to be effective. The mechanism of this contribution is explained by Fontana (60) as the production of a low-melting (150°C) eutectic in contrast to the 375°C melting point of the Cu_2Cl_2–$CuCl_2$ system. The requirement of a liquid phase has been demonstrated.

Kellogg has developed a high-pressure (250 psi) process in glass-lined equipment for which yields of 96 and 98 percent, based on ethylene and HCl, respectively, are claimed (49). It would appear that ethylene is recycled at a rapid rate in order to achieve such a high conversion of HCl and to remove water from the reactor.

A flow chart for an oxychlorination process with a fixed-bed reactor is shown in Fig. 9. A fluid-bed may also be used if internal cooling or cooling by recycle gases or liquids is provided. The heat of reaction for the process—Eq. (31)—at 25°C and with all components in the gaseous state is 49,835 Btu per lb-mole:

$$CH_2{=}CH_2 + 2HCl + \tfrac{1}{2}O_2 \longrightarrow CH_2ClCH_2Cl + H_2O \qquad (31)$$

Since side reactions (oxidation, chlorination, and reaction of impurities) take place, a more realistic heat of reaction would be about 53,000 Btu per lb-mole. Further, since the gases enter the reactor in a preheated condition and leave at a temperature of about 250°C, the approximate usable heat release is 47,500 Btu per lb-mole.

In virtually every commercial process, this heat is recovered in the form of steam. In the fixed-bed reactor, heat may be removed by circulation of a heat-transfer fluid through a steam-generation unit as shown or by boiling

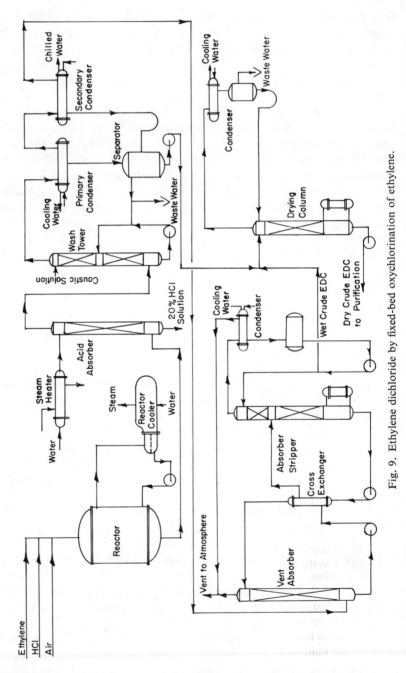

Fig. 9. Ethylene dichloride by fixed-bed oxychlorination of ethylene.

a heat-transfer fluid in the shell side of the reactor and condensing in a steam-generation unit.

The feed gases are mixed in a nearly stoichiometric ratio with a slight excess of air, which is required since some oxidation of ethylene occurs. The temperature of the feed gases must be carefully controlled. If it is too low, aqueous hydrochloric acid condenses, and if too high, reaction takes place too early in the bed. The reaction is initiated satisfactorily at 180 to 200°C, and a fixed-catalyst bed operates normally at 230 to 290°C. Hot spots may occur because of improper catalyst strength, insufficient cooling, or improperly adjusted flows. Because of better temperature control, a fluid bed usually operates at 200 to 250°C. Reactor materials must be resistant to corrosion and should not cause decomposition of the product. Nickel, copper-nickel alloys, and high-nickel or -chromium steels have been recommended.

The lowest possible operating temperature is best for the following reasons:

1. Cracking of ethylene dichloride is minimized, resulting in higher yields and fewer impurities.
2. Loss of copper chloride by vaporization is minimized.
3. The materials of construction have a longer life.

Normal operating pressures are 40 to 200 psi. It would be expected that higher pressure in the fluidized-bed process would increase the productivity of the catalyst; however, various Goodrich patents (61) have cited values of 10 to 50 psig.

Overall yields and conversions for this process are in the range of 90 to 99 percent, based on both ethylene and HCl. With catalysts normally used in the industry it has not been possible to achieve the higher yields for both raw materials. Usually, the higher HCl conversions are accompanied by the lower ethylene conversions, and vice versa. The crude ethylene dichloride product is 95 to 99 percent pure, depending on the purity of the raw materials and the reaction temperature. If the latter is relatively low and well controlled, the product is purer.

With the establishment of large tonnages of vinyl chloride, based on the thermal cracking of ethylene dichloride, along with the lack of improvement in acetylene costs (62), considerable emphasis is being placed on the oxychlorination process throughout the world. Major effort at this time is being directed toward improvement of the catalyst and the capacity of the reactor system. Continued improvement of catalyst is predicted (63–66,79).

The heat of reaction for the oxychlorination of ethylene is very high, and its removal has limited the productivity of the currently utilized reactor systems to a range of about 250 lb of EDC per day per cu ft of reactor volume. The application of known engineering techniques may permit extending this productivity to about twice this rate in the near future. Further increases will probably require the development of new reactor technology.

Utilization of oxygen in the process rather than air has received some attention and offers the significant advantage of not having to separate the ethylene dichloride produced from the very large quantity of inert gases that are present in air. There are literature references to the use of oxygen, but no commercial process is known to practice this feature. Problems of explosive mixtures (67) and the difficulty of removal of the heat of reaction have been the major deterrents. Costs of oxygen from very large air-separation plants are expected to improve and may reach the point of being competitive with air. With the development of suitable equipment, this could have a significant economic impact on the oxychlorination process.

By operation of this process at a temperature high enough either to crack the ethylene dichloride produced or to allow the reaction to proceed by substitution rather than addition, vinyl chloride may be obtained directly. A temperature in the range of 350 to 450°C would be expected to be necessary, although lower temperatures have been cited. Several processes have been proposed (68–71).

4. Balanced Process—Oxychlorination

In the period since 1964, there have been a number of announcements of plants that combine thermal dehydrochlorination of ethylene dichloride with the direct chlorination and oxychlorination of ethylene to produce the ethylene dichloride.

This combination may be represented by Eqs. (32) to (34):

$$Cl_2 + C_2H_4 \longrightarrow CH_2ClCH_2Cl \tag{32}$$

$$C_2H_4 + 2HCl + \tfrac{1}{2}O_2 \longrightarrow CH_2ClCH_2Cl + H_2O \tag{33}$$

$$2CH_2ClCH_2Cl \longrightarrow 2CH_2{:}CHCl + 2HCl \tag{34}$$

The principal strong points of this combination are that it is based on ethylene rather than the more expensive acetylene, each step proceeds in high yield, and almost all the chlorine ends up in the product. In practice, the ethylene dichloride from both chlorination reactions may be purified in the same distillation train. The raw-material requirements can be determined by the application of the yield figures given in the discussion of the individual processes.

5. BALANCED PROCESS—MIXED GAS

One of the major deterrents to the use of an acetylene balanced process is the expense involved in the recovery, concentration, and purification of the acetylene. A major fraction of this expense has been eliminated by the development of processes by Kureha (72) and Société Belge de l'Azote (73). Both of these processes also reduce the cost of purification and concentration of ethylene by processing a dilute gas stream of mixed acetylene and ethylene to the desired vinyl chloride. The two processes are similar, and since considerably more information has been released concerning the Kureha process, which has been licensed in Russia, only this route will be considered here. A flow chart is shown in Figs. 10 and 11. A considerably more detailed discussion is available in the cited reference.

The principal advantage of the process is that the costs of acetylene and ethylene are both considerably reduced. Costs of $0.0758 per lb of acetylene and $0.0378 per lb of ethylene are given. The procedure is based on naphtha as the raw material, although other hydrocarbons could possibly be used. The principal disadvantage of this route is that a very dilute gas—9 percent acetylene and 11 percent ethylene—must be handled. Therefore, recovery and purification of the vinyl chloride and ethylene dichloride would probably involve high refrigeration costs. Absorption of the vinyl chloride produced in the acetylene hydrochlorination portion by chilled ethylene dichloride is probably practiced. Any vinyl chloride passing this step into the reactor for direct chlorination of ethylene would be lost. Somewhat higher pressure than that which is tolerable for concentrated acetylene streams may be used in the portion of the facilities that processes the dilute acetylene, since temperature control and danger of explosion are significantly less.

The SBA literature states that yields of 92, 83, and 88 percent are realized, based on acetylene, ethylene, and chlorine respectively. Not surprisingly, these values are significantly lower than those for the concentrated ethylene and acetylene balanced case or for the straight ethylene oxychlorination balanced process.

C. Potential Processes

Several other processes for the production of vinyl chloride have been cited in the literature. Although none of these has attained more than the pilot-plant stage, they are briefly discussed here.

1. THERMAL CHLORINATION OF ETHYLENE

At elevated temperatures, vinyl chloride may be produced directly by reaction of ethylene with chlorine, as shown in Eq. (35):

$$CH_2:CH_2 + Cl_2 \longrightarrow CH_2:CHCl + HCl \tag{35}$$

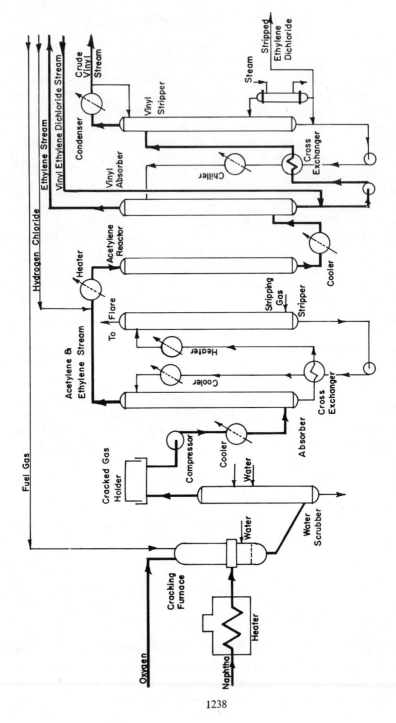

Fig. 10. Kureha process; naphtha and acetylene reacting system.

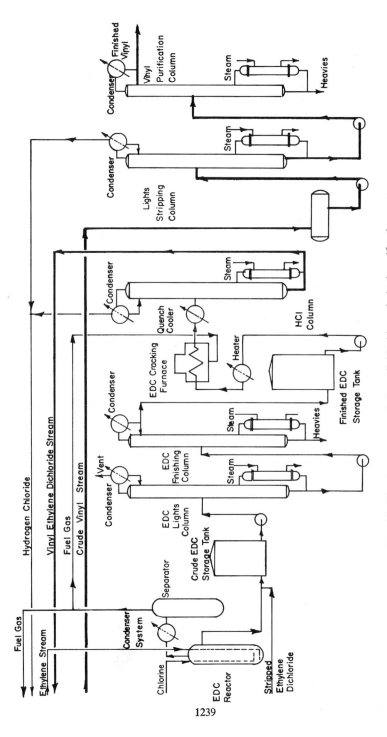

Fig. 11. Kureha process; ethylene dichloride and vinyl purification systems.

1239

This process is attractive relative to others based on ethylene in that the capital requirements and manufacturing costs for production, purification, and cracking of ethylene dichloride are eliminated. There are, however, significant disadvantages:

1. Conversions of ethylene per pass would most probably need to be less than 20 percent to provide selectivity to vinyl chloride of 95 percent. The costs of vinyl chloride recovery and HCl removal from this dilute stream as well as the cost of recycling the ethylene would be significant.

2. Selectivity to vinyl chloride would not be very high. Recovery and use of the major expected by-products, 1,1-dichloroethylene and *cis-* and *trans-*1,2-dichloroethylene, however, would mitigate this disadvantage.

3. Less than half of the chlorine feed ends up in the product. The remainder is contained in the by-products, including HCl.

4. For thermal chlorination, a liquefied chlorine source is needed. Dried cell gas, which normally contains significant quantities of oxygen, would not be suitable, since, at the temperature involved, water would be produced creating a very corrosive mixture.

Initial work on the process was done in about 1938 by Shell Development (74). Union Carbide (75) and Toyo Koatsu (76,77) have investigated it more recently. The approach taken is simultaneously to chlorinate the ethylene and dehydrochlorinate ethylene dichloride with the heat released by the chlorination. The by-product hydrogen chloride and unconverted ethylene are then converted to ethylene dichloride by the oxychlorination route. In this approach the major problem should still be the lack of selectivity to the desired vinyl chloride. It is doubtful that this process or related ones proposed by Solvay (78) and Subbotin (79) could be developed to the degree to permit general commercialization.

2. THERMAL CHLORINATION OF HYDROCARBONS

Reduction in the major raw-material cost could be realized by either Eq. (36) or (37):

$$2CH_4 + 3Cl_2 \xrightarrow{600°C} CH_2{=}CHCl + 5HCl \tag{36}$$

$$CH_3CH_3 + 2Cl_2 \longrightarrow CH_2{=}CHCl + 3HCl \tag{37}$$

Equation (36) can be called chloropyrolysis. The problems in selectivity and purification in such a reaction would be appalling. References are cited (80–84). Equation (37) may proceed at less than 600°C, since vinyl chloride may be produced by a succession of chlorination and dehydrochlorination steps. This equation would appear to give a better chance of higher selectivity than Eq. (36).

3. ETHANE OXYCHLORINATION

Interest in oxychlorination of ethane—Eq. (38)—as a means of producing vinyl chloride has been aroused by the significant price advantage of ethane over ethylene (85,86):

$$CH_3CH_3 + 2Cl_2 \longrightarrow CH_2{=}CHCl + 3HCl$$

$$4HCl + O_2 \longrightarrow 2Cl_2 + 2H_2O \qquad (38)$$

Net Reaction

$$CH_3CH_3 + HCl + O_2 \longrightarrow CH_2{=}CHCl + 2H_2O$$

Although it is expected that selectivities would be less than desirable, the inexpensive hydrocarbon source coupled with the utilization of by-product hydrogen chloride makes this type of process interesting. Principal liabilities, other than poor selectivity, are the possibility of oxidation losses, and problems in materials of construction and in catalyst life and activity. Nearly all these relate to the temperature that would be required, probably 400 to 450°C.

IV. COMMERCIAL PURIFICATION PROCESS

The purity of commercial vinyl chloride has increased significantly over the years. The most common method of achieving this purity is by distillation, whether the vinyl chloride is produced from acetylene or from ethylene dichloride.

Flow charts of two methods of purifying vinyl chloride are shown in Figs. 12 and 13. Both are for ethylene dichloride–based processes, but the by-product HCl is obtained as an aqueous solution in one (Fig. 12) and as the anhydrous gas in the other (Fig. 13).

In Fig. 12, one stage of compression is shown, since conventional acid-absorption equipment is not suitable for pressures much above atmospheric. The operating pressure of the azeotrope drying column is 60 psig in order to avoid the need for refrigeration. This column is operated so that it also removes light components. Polymerization in the condenser reflux separator and the top of the column occurs in the presence of oxygen. Oxygen comes from the air stripped from the water used in the acid absorbers, and its concentration can be reduced by a sufficiently high vent. The column reboiler must provide good circulation, or polymerization and degradation will occur.

Impurities, such as vinylacetylene, butadiene, and chloroprene, which are normally present in crude vinyl chloride, contribute significantly to polymer buildup and decomposition products.

The need for the butadiene removal system depends on several factors: the quality and source of the crude vinyl chloride, the specifications of the

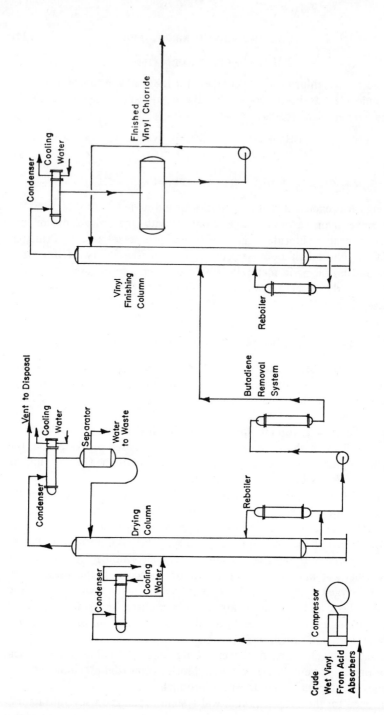

Fig. 12. Purification of wet vinyl chloride.

1242

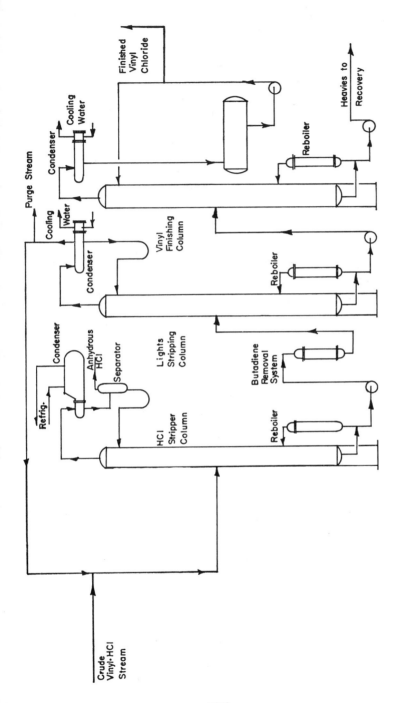

Fig. 13. Purification of vinyl chloride containing anhydrous HCl.

1243

user, and the capacity of the finishing column for removal of butadiene. The most practical chemical method seems to be liquid-phase hydro-chlorination (87). Vapor-phase methods have also been proposed (88,89).

The finishing column may be operated at pressures of up to 60 psi with varying levels of refrigeration required. The composition of the bottoms is adjusted so that the temperature does not rise over 90°C. This is to prevent cracking and the resulting impurities in the overhead product. The purified vinyl chloride is placed in storage tanks. Inhibitors, such as phenol, may be added to the finished product.

All the equipment used in the purification of vinyl chloride may be steel except the reboiler tubes, which should be a nonferrous alloy to minimize induced decomposition.

The proper and efficient operation of the acid-absorption equipment preceding this purification system is important. Any hydrogen chloride passing through this unit causes serious corrosion in the drying column by the formation of aqueous hydrogen chloride.

In the case shown in Fig. 13, it is presumed that the furnace and quench system are operated at a sufficiently high pressure to permit elimination of a compression step.

The gaseous stream of vinyl chloride, hydrogen chloride, and some ethylene dichloride leaving the quench system may be partially condensed or fed directly into a hydrogen chloride stripping column. Since the hydrogen chloride is frequently used for other processes, such as oxy-chlorination and hydrochlorination of acetylene, the column is operated under reflux to yield maximum purity of overhead product. The liquid from the bottom of the hydrogen chloride stripping column is crude (90 to 98 percent) vinyl chloride. The requirement for the butadiene removal system again depends on the quality of the crude vinyl chloride and the ethylene dichloride feed to the cracking furnace.

As vinyl chloride purity requirements have become more stringent, im-proved distillation efficiency has become necessary. Currently, a typical finishing column has 30 to 50 trays and operates at a reflux ratio of between 2/1 and 3/1. Another improvement has been the addition of a lights-stripping column to separate small amounts of impurities boiling between HCl and vinyl chloride that formerly were taken off overhead with the HCl. By this modification, 99.8 percent HCl, for direct use in other processes, notably oxychlorination, is obtained.

Barr (90) states that vinyl chloride is a monomer that is relatively insensitive to polymerization if oxygen is carefully excluded. Water and acidic components accelerate the formation of peroxides and reduce the stability of the vinyl chloride. Any purification system must be capable of eliminating these components. The magnitude of the purification problem is shown by a list of components that may be found under varying operating conditions in the crude vinyl chloride stream from an ethylene dichloride cracking furnace:

Hydrogen	Vinyl Bromide	Chlorobutadienes
Nitrogen	Methyl Chloride	Butadiene
Oxygen	Methylene Chloride	Propadiene
Carbon	Chloroform	Ethylene Dichloride
Carbon Monoxide	Carbon Tetrachloride	Trichloroethylene
Carbon Dioxide	Vinylacetylene	Benzene
Methane	Allyl Chloride	1,1,2-Trichloroethane
Ethane	Vinylidene Chloride	Dichlorobutadiene
Ethylene	1-Chloropropene	Ethyl Chloride
Acetylene	1,2-Dichloroethylene	Butene-2
Phosgene	1,1-Dichloroethane	Chlorobenzene
Propylene		

Analysis for the majority of these components has been made possible by vapor-phase chromatography.

Of the list above, certain compounds are important because they are polymerization inhibitors. These are principally butadiene, chloroprene, acetylene, and vinylacetylene. The properties of the polymer are influenced by halogen-containing compounds, such as ethylene dichloride, vinylidene chloride, vinyl bromide, ethyl chloride, and the various chloromethanes.

V. LABORATORY METHODS OF PREPARATION

Vinyl chloride can best be prepared in the laboratory by the thermal dehydrochlorination of ethylene dichloride. Vaporized ethylene dichloride, at least 99 percent pure, is passed through an externally heated stainless steel tube $3/4$ to 1 in. in diameter and 3 to 7 ft long. The tube may be wrapped with electrical-resistance wire or heated in a furnace .The outlet temperature is measured and controlled in the range of 450 to 500°C. The flow of ethylene dichloride is adjusted for a contact time of 4 to 8 sec. Under these conditions about 50 percent of the ethylene dichloride is cracked. The hot gases are discharged into a scrubbing chamber of dry ethylene dichloride and the cooled gases from the scrubber passed through an ice-water trap and then a dry ice–acetone trap. The anhydrous HCl gas should be vented in a safe location. The two traps may be combined, after allowing the colder one to warm to about −40°C, and stored in a freezing chest until low-temperature distillation may be carried out. The 95 percent heart cut of the fraction boiling at −13.8°C is suitable for polymerization studies.

All equipment should be purged free of oxygen with nitrogen prior to starting the ethylene dichloride flow or distillation of the product. Caution should be exercised in handling vinyl chloride because of its instability, flammability, and low boiling point. Water, acids, and air should be excluded from contact with the uninhibited product. Purified vinyl chloride may be stored for long periods of time at less than 0°C in clean glass or steel containers.

VI. MONOMER PROPERTIES

A. Physical Properties

Table 5 shows the properties of vinyl chloride.

TABLE 5
Physical Properties of Vinyl Chloride

			Reference
1. Boiling point at 760 mm, °C	−13.80		92
2. Melting point, °C	−153.69		93
3. Odor:			
Uninhibited	Faintly sweet		93
Inhibited	Faintly phenolic because of inhibitor		93
4. Color	Colorless		93
5. Flash point, Cleveland open cup	−78°C (−108°F)		91
6. Flammability limits, volume percent in air	3.6–26.4		94
7. Liquid density, g/ml	100°C	0.746	93,95
	25°C	0.9013	
	−20°C	0.9834	
	−25°C	0.9918	
	−30°C	0.999	
8. Density change with temperature, g/ml/°C, between −30 and −20°C	0.00164		93
9. Liquid viscosity, cps	100°C	0.100	93
	25°C	0.185	
	−10°C	0.248	
	−20°C	0.274	
	−30°C	0.303	
	−40°C	0.340	
10. Vapor viscosity, cps	100°C	0.0138	96
	25°C	0.0108	
11. Surface tension, dynes/cm	100°C	5.4	93
	25°C	16.0	
12. Refractive index, D line	25°C	1.3642	97

TABLE 5 (*continued*)

13. Vapor pressure, mm Hg	100°C	16,617	92
	25°C	2,943	
	0°C	1,293	
	−25°C	470	
	−50°C	130	
	−75°C	24.4	
	−100°C	2.5	
14. Change in boiling point with	25°C	0.01355	93
pressure, dt/dP, °C/mm Hg	−13.37°C	0.03423	
15. Heat capacity (liquid) C_p,	100°C	26.25	93
cal/mole/°C	25°C	20.56	
16. Heat capacity (vapor) C_p,	100°C	14.88	98
cal/mole/°C	25°C	12.82	
17. Latent heat of vaporization ΔH_v,	100°C	3,310	93
cal/mole	25°C	4,710	
	−13.8°C	5,250	
18. Latent heat of fusion ΔH_m, cal/mole	1,172		93
19. Heat of formation ΔH_f, 298°K,	8,480		93,99
kcal/mole			
20. Free energy of formation ΔF_f,	12,386		93
298°K, kcal/mole			
21. Heat of polymerization ΔH_p,	−25.3 ± 0.5		93
kcal/mole			
22. Critical temperature t_c, °C	158.4		100
23. Critical pressure P_c, atm	56		100
24. Critical volume V_c, cc/mole	179		100
25. Critical compressibility $(PV/RT)K$	0.29		100
26. Volumetric shrinkage on poly-	35		93
merization, percent			
27. Dielectric constant at frequency of	17.2°C	6.26	93
10^5 Hz	−21°C	7.05	
28. Dissipation factor at frequency of	21°C	0.0011	93
10^5 Hz			
29. Solubility—soluble in	Carbon tetrachloride		
	Ether		
	Ethanol		
30. Solubility of H_2O in vinyl chloride,	0.11		
percent, 25°C			

B. Hazards and Their Control (93)

1. TOXICITY

When present in sufficiently high vapor concentrations, vinyl chloride has an anesthetic action. The threshold limit value (TLV) for repeated 7- to 8-hr daily exposures is presently set at 500 ppm, but there are indications that, on the basis of more recent data, this level may be lowered. Concentrations of greater than 500 ppm may cause warning symptoms of dizziness, disorientation, and disturbances of equilibrium and coordination similar to drunkenness. Vapor concentrations approaching 3.6 percent may cause helplessness and unconsciousness on very short exposure.

Vinyl chloride is not absorbed by the skin but may cause local irritation, frostbite, or freezing because of rapid vaporization. Inhibited vinyl chloride on the skin may also cause burns because of the phenol inhibitor. Any skin area that has been in contact with the liquid monomer should be washed with generous amounts of water. Eyes into which liquid monomer has splashed should be washed with water for at least 15 min and medical attention obtained as soon as possible. If, after 15 min of washing, signs of tissue damage to the eyes are still apparent, the washing should be continued.

Clothes or bandages wet with monomer should be removed immediately to prevent phenol burns or the refrigeration effect caused by the absorbed vinyl chloride. Contact with inhibited monomer should be treated in the same manner as phenol contamination.

Anyone displaying evidence of vinyl chloride intoxication should be put at rest in an uncontaminated atmosphere. If confined in an atmosphere of high concentration where escape is impossible, deep anesthesia may result. In such an instance, the patient should be placed at bed rest, preferably with the head slightly lower than the feet. If breathing has ceased, artificial respiration should be given. Medical attention should be obtained immediately.

2. FIRE

Vinyl chloride should always be handled with full recognition of its volatility and flammability. Vinyl chloride is flammable in the range of 3.6 to 26.4 percent by volume in air and can form flammable mixtures with air at all temperatures above $-78°C$. In general, precautions should be taken to keep the material enclosed and to eliminate all sources of ignition. In the laboratory, where vinyl chloride vapors may escape, adequate ventilation must be maintained and sources of ignition eliminated.

Fires involving large quantities of liquid are difficult to extinguish, since vinyl chloride is immiscible with and less dense than water. Most small fires can be extinguished with carbon dioxide or dry chemical agents. Fixed or portable CO_2 or dry chemical extinguishers should be provided. Water spray can be used for the control of fires and the protection of equipment and vessels involved. Diking and drainage should be provided for confining and disposing of the liquid in case of tank rupture or spills.

Precautions should be taken to guard against vinyl chloride entering the general sewer system.

In event of a fire, no unauthorized person should be permitted to enter an unventilated area until the space has been thoroughly sprayed with water to remove gases, such as hydrogen chloride, phosgene, and carbon monoxide, generated from the fire.

3. EXPLOSION

Vinyl chloride is a gas at normal atmospheric temperature and pressure and can form explosive mixtures with air or oxygen. An explosion hazard can exist when drawing samples or venting to the atmosphere. Open flames, local hot spots, friction, any spark-producing equipment, and static electricity are to be avoided when handling this material.

4. POLYMERIZATION

Pure vinyl chloride monomer is quite stable and offers essentially no hazard from autocatalytic polymerization when stored properly in the absence of air (oxygen), moisture, and free-radical or other polymerization initiators.

5. SPILLS AND LEAKS

If spills or leaks occur, all sources of ignition must be removed from the area immediately and only properly protected personnel allowed to remain. Spills, unless very large, usually evaporate rapidly, but ample ventilation must be provided to prevent the accumulation of toxic or explosive mixtures. Spills and leaks must not be allowed to enter the sewer system because of the explosion hazard involved. An approved flammable gas indicator can be used in testing for vinyl chloride leaks. Clothing and shoes contaminated with vinyl chloride should be removed immediately and the body washed thoroughly to remove any material that may have penetrated to the skin. Clothing and shoes should not be worn again until free of vinyl chloride and any inhibitor. Disposal of contaminated shoes is recommended if the spilled monomer was phenol-inhibited.

Equipment leaks involving the inhibited monomer can create a special hazard, because phenol could accumulate by evaporation at the site of the leak in sufficient quantities to cause skin burns on contact. In case of such accumulation of inhibitor, the area should be thoroughly hosed down with water after all traces of monomer have evaporated. In event of contact with phenol, the person should immediately be washed with copious amounts of water and medical aid summoned.

6. PROTECTIVE EQUIPMENT

Self-contained breathing apparatus or air- or oxygen-supplied masks equipped with full face pieces and approved by the U.S. Bureau of Mines for this purpose should be used under the following conditions:

1. In emergencies when the vapor concentration is not definitely known.
2. When the vinyl chloride concentration is greater than 2 percent by volume.
3. When the oxygen content of the air may be less than 16 percent by volume.
4. When the exposure period will last more than 30 min.
5. In tank and equipment cleaning and repair work under conditions outlined in items 1 to 4.

Industrial canister masks equipped with full face piece and approved by the U.S. Bureau of Mines and fitted with the proper canister for absorbing vinyl chloride vapor are available. These afford protection against concentrations known not to exceed 2 percent by volume when the oxygen content of the air is greater than 16 percent by volume. In all other cases, a self-contained breathing apparatus or air- or oxygen-supplied mask should be used.

Chemical cartridge respirators approved by the U.S. Bureau of Mines may be used to avoid inhaling disagreeable but harmless concentrations of vinyl chloride vapor for short periods of time.

VII. ANALYTICAL METHODS

The quality of vinyl chloride monomer is determined by both wet and dry chemical analysis. Each company that produces vinyl chloride monomer for sale makes the analytical methods for each specified impurity available to its customers. The analytical methods discussed are not specific but give the general outline for the analysis of the impurities found in vinyl chloride.

A. Monomer Purity by Chromatographic Analysis

The development of gas-liquid chromatography has made it possible to detect impurities at the part-per-million level and determine monomer purity in less than 1 hr. There are many differences in the details of the methods. Column selection and detector selection are the key to effective analyses. If the whole spectrum of impurities is of interest, two columns, one for low-boiling impurities and one for high-boiling impurities, are usually needed. Thermal-conductivity detectors with high-sensitivity filaments are generally adequate for detection of low-boiling impurities down to the 5 ppm level. Techniques for the determination of high-boiling impurities at the 5 ppm level usually involve some type of concentration step. This might be a precut column incorporated into the chromatograph or a concentration step in a high-boiling solvent prior to gas-chromatographic analysis. When thermal-conductivity detectors are not adequate, hydrogen-flame-ionization detectors are useful. In many cases these detectors can be used in conjunction with columns that are used with thermal-conductivity detectors. Instances are noted in which liquid-phase bleed causes noise and drift in hydrogen-flame-ionization detectors. The detection limit for many vinyl chloride impurities is in the fractional ppm range when flame-ionization detectors are used.

B. Water

Water is most often determined by the Karl Fischer method. The titration assembly should be moisture-proof or should be provided with a dry-air purge. Samples should be withdrawn from the liquid phase of the sample cylinder. Since water and vinyl chloride form a low-boiling azeotrope, samples withdrawn from the vapor phase of the cylinder would be enriched in water content and would give high results.

C. Acidity

Acidity is determined by titration with weak alcoholic caustic solutions. Aqueous titration may be performed after scrubbing the vinyl chloride through distilled water. Nonaqueous direct titrations are performed by adding alcoholic caustic directly to the liquid vinyl chloride. Alternatively, the vinyl chloride is added to a neutralized high-boiling chlorinated solvent, the vinyl chloride is evaporated, and the acidic components determined by titration of the solvent with alcoholic caustic. In each method an indicator is selected that has an equivalence point below pH 7 to avoid titration of acid anhydrides, such as CO_2.

D. Iron

The iron content is generally specified as that which is nonfilterable. This is necessary because the limits are so low at present that external contamination from sample or shipping containers is significant. Samples may be filtered through an iron-free paper filter or through a membrane filter. The sample is evaporated, and the residue taken up in iron-free hydrochloric acid. The iron content is determined spectrophotometrically after formation of a coordination compound, such as the ferrous o-phenanthroline or ferrous thioglycolate complex.

E. Peroxides

Vinyl chloride is added to a methanol solution containing ferrous sulfate. The ferrous ion is oxidized to the ferric state by any peroxides present. The ferric thiocyanate complex is formed, and the solution compared colorimetrically with standards of known peroxide content.

F. Oxygen

The oxygen content, usually determined on the vapor space above the monomer, may be determined by gas chromatography. Alternatively, this analysis can be performed using one of the commercial trace oxygen analyzers that are based on the lead-oxygen cell or paramagnetic phenomenon.

G. Acetylenes

Specific acetylenic compounds may be determined by gas chromatography, or acetylenic compounds may be determined as a group by classical wet methods. Acetylenic hydrogen reacts quantitatively with cuprous ion in an alkaline system, forming insoluble copper acetylide. This is separated and treated with acid to regenerate cuprous ion. The latter is oxidized to cupric ion, which is determined iodometrically.

H. Aldehydes

Aldehydes are extracted from vinyl chloride with aqueous ammonia under pressure. The extract is treated with α-methylindole to form the water-insoluble aldehyde complex, which is determined turbidimetrically.

I. Sulfur

Sulfur present as SO_2 may be absorbed in water, oxidized to sulfate, and determined turbidimetrically as barium sulfate. Alternatively, the SO_2 may be absorbed in sodium tetrachloromercuriate, and the sulfur content determined colorimetrically. A colored complex is formed when the solution is treated with p-rosaniline and formaldehyde. NO_2 interferes with the latter method.

J. Nonvolatile Matter

Nonvolatile matter is determined by weighing the residue remaining after evaporation of a predetermined amount of the monomer. Actual conditions of the test vary from method to method. In some, the residue is divided into two classes, soluble and insoluble. The soluble residue is determined by filtration of a known volume of monomer and weighing the filter to determine the amount of insoluble matter removed. The filtrate is then evaporated for determination of the soluble portion. A division of this kind is a good way to distinguish between polymer and nonvolatile residue which is soluble in the monomer.

VIII. CHEMICAL REACTIONS OF VINYL CHLORIDE

The primary chemical reaction of commercial interest that vinyl chloride undergoes is polymerization. This is discussed in detail subsequently.

A. Reaction with Chlorine

1. Good yields of 1,1,2-trichloroethane are obtained at 0°C by chlorination of vinyl chloride in chloroform. Substitution is avoided by using a slight excess of vinyl chloride and adding chlorine until yellow coloration indicates an excess of the gas. The excess chlorine can be quickly removed by washing with caustic soda, and the product is then fractionated (101).

2. Thermal chlorination, in the absence of any solvent or light, is effected at 100 to 250°C with yields of 90 to 95 percent of 1,1,2-trichloroethane (102).

3. Below 80°C, in the absence of solvent and the presence of light, chlorination takes place with 90 to 95 percent yields of 1,1,2-trichloroethane (103).

4. The chlorination of vinyl chloride produces 1,1,2-trichloroethane when a chlorinated hydrocarbon (CCl_4 or trichloroethane) is used as a solvent in the presence of a chlorine carrier, such as antimony pentachloride (104).

5. Essentially pure trichloroethane can be produced in 99.2 percent yield by passing vinyl chloride and dry chlorine gas through a tube packed with iron shavings (20 to 25°C) (105).

6. Vinyl chloride and chlorine react in the dark, using benzoyl peroxide as catalyst and chlorinated hydrocarbon as solvent, to produce high yields of 1,1,2-trichloroethane (106).

7. The kinetics of the photochemical chlorination and bromination of vinyl chloride have been studied (107).

B. Reactions with Other Inorganic Chemicals

1. HYDROGEN SULFIDE

Reaction of vinyl chloride and H_2S at low temperature (light-catalyzed) results in the production of 2-chloroethylmercaptan and β,β'-dichlorodiethyl sulfide (108).

2. Selenium Derivatives

Vinyl chloride reacts with selenium tetrachloride to produce $\beta,\beta,\beta',\beta'$-tetrachlorodiethylselenium dichloride and with selenium oxychloride to produce $\beta,\beta,\beta',\beta'$-tetrachlorodiethylselenium oxide. Selenium dichloride and vinyl chloride react in the presence of $AlCl_3$ to give a mixture of β,β-dichloroethylselenium chloride and $\beta,\beta,\beta',\beta'$-tetrachlorodiethylselenium (109).

3. Nitrosyl Chloride

The reaction of vinyl chloride and nitrosyl chloride produces 1,1-dichloro-2-nitrosoethane in 20 percent yield (110).

4. Sulfuryl Chloride

A stream of vinyl chloride passed into a carbon tetrachloride solution of sulfuryl chloride, with $AlCl_3$ gradually added as catalyst, resulted in β,β-dichloroethylsulfonyl chloride in 31.5 percent yield (111).

5. Silicon

A method for the production of vinylsilicon chloride consists of heating silicon with cuprous chloride or stannous chloride catalyst and treating with hydrogen. Vinyl chloride is passed upward through this catalyst mixture at 260°C or greater. Vinyltrichlorosilane and divinylchlorosilane result (112).

6. Methyldichlorophosphine

A mixture of isomers has been obtained by reaction of vinyl chloride with methyldichlorophosphine (113).

7. Ferric Carbonate

The photochemical reactions of ferric carbonate with vinyl chloride are used to produce iron tetracarbonyls (114).

C. Reactions with Organic Chemicals

1. Grignard Reagents

Condensations of vinyl chloride and substituted vinyl halides with aromatic or alkaryl Grignard reagents proceed readily in the presence of cobalt or chromium halides to form the corresponding vinylated hydrocarbons (115).

2. TRIFLUOROIODOMETHANE

Vapor-phase reaction of vinyl chloride and trifluoroiodomethane (light-catalyzed) gives high yields of 1,1,1-trifluoro-3-iodopropane (116).

3. PHENOLS

Phenols may be alkylated with vinyl chloride to give vinyl phenols, using sulfuric acid (117) or $FeCl_3$ or $AlCl_3$ (118) as catalysts. Aromatic vinyl ethers are obtained from vinyl chloride and phenols at 170 to 200°C in the presence of agents such as alkali or alkaline earth hydroxides or phenolic salts (119).

4. ISOBUTANE

Condensation of vinyl chloride with isobutane is effected with $AlCl_3$ to give, at $-10°C$, 1,1-dichloro-3,3-dimethylbutane in 40 percent yield (120).

5. CHLOROMETHYLCYCLOHEXANES

Condensation of vinyl chloride with a mixture of isomeric chloromethylcyclohexanes in the presence of purified $AlCl_3$ produces 95 percent yields of isomeric (β,β-dichloroethyl)methylcyclohexanes (121).

6. METHYL MERCAPTAN

Methyl mercaptan and vinyl chloride were placed in a sealed tube and irradiated by a mercury-vapor lamp filtered with a 5 percent copper sulfate solution. More than 95 percent of 1-chloro-2-thioethane was obtained (122).

7. ACETIC ACID

Vinyl chloride may be converted to vinyl acetate in 25 percent yield by heating the vinyl chloride in the presence of a catalyst and diluents in a mixture of an acetate and acetic acid or a glycolate (123).

D. Pyrolysis

Pyrolysis of vinyl chloride at 550 to 650°C leads to the production of chloroprene in good yield (124).

PART B.   VINYLIDENE CHLORIDE

I. HISTORICAL DEVELOPMENT AND INDUSTRIAL USES

Vinylidene chloride, a "strange new fluid," was first reported by Regnault (125) in 1838. In his paper he stated that the compound had a slight odor of garlic and that it became cloudy on standing. It was regarded as a laboratory curiosity until the late 1920s. B. T. Brooks' (125a) announcement in 1922 that halogenated ethylenes other than vinyl chloride and vinyl bromide showed a tendency toward polymerization stimulated interest in such materials and resulted in development work with 1,1-dichloroethylene.

The Dow Chemical Company became actively engaged in research and development work with vinylidene chloride during the early 1930s. This field was entirely undeveloped, practically free of patents, and rich in possibilities for commercial applications. Attempts were made to fabricate the homopolymer in a number of ways, but none was satisfactory. Dow continued to search for a marketable product and in 1935 began to investigate copolymerization. The copolymers of vinyl and vinylidene chlorides showed particular promise, and further testing showed that this was an extremely versatile polymer system. Vinylidene chloride was also copolymerized with other vinyl monomers, such as acrylonitrile, styrene, and vinyl acetate. These thermoplastic copolymers were named Saran to avoid confusion with other vinyl and vinylidene chloride polymers. The name refers only to polymers that contain 50 percent or more of vinylidene chloride.

The first Saran copolymers were marketed in early 1940. Their constant and increasing use since that time is a result of the unique properties contributed by vinylidene chloride. The Sarans are noted for their outstanding resistance to solvents, their extremely low water absorption and water-vapor transmission, and their excellent dimensional stability. They are odorless, tasteless, and nontoxic. Their toughness and abrasion resistance are outstanding.

II. COMMERCIAL DEVELOPMENT

Commercial development of vinylidene chloride monomer in the United States has been carried out mostly by the Dow Chemical Company for use in the Saran resins. Ethyl Corporation produced vinylidene chloride for sale on the merchant market for several years, but early in 1967 stopped producing the material for outside sale. In Japan vinylidene chloride is produced by Kureha Chemical Industries and Asahi-Dow Chemical Company for the production of Saran types of resins. In Germany Badische Anilin- & Soda-Fabrik A.-G. and Hüls produce vinylidene chloride. Most vinylidene chloride is manufactured commercially by the reaction of 1,1,2-trichloroethane with a base such as calcium hydroxide or sodium hydroxide and the remainder by the cracking of 1,1,1-trichloroethane, using iron as a catalyst.

The growth of vinylidene chloride manufacture in the United States has been steady but small when compared with vinyl chloride. The total world

TABLE 6

Typical Vinylidene Chloride Specifications

1. Acetylene, max	25 ppm
2. Acidity, max	25 ppm
3. Water, max	0.01 wt %
4. Iron	1.0 ppm
5. Color, APHA, max	30
6. Peroxides, max	25
7. Purity (on inhibitor-free basis)	99.5%
8. Inhibitors	Phenol (0.4–0.6%)
	MEHQ (180–220 ppm)

production in 1967 was about 100 to 150 million lb. This annual production figure has approximately doubled in the last 10 years, and there is a predicted growth rate of 5 to 10 percent per year for the next 5 years.

Typical monomer specifications for vinylidene chloride are given in Table 6.

III. COMMERCIAL PRODUCTION PROCESSES

The earliest and still the most widely used commercial process for the production of vinylidene chloride is the treatment of 1,1,2-trichloroethane with sodium hydroxide—Eq. (38)—(126–134):

$$CH_2ClCHCl_2 + NaOH \longrightarrow CH_2:CCl_2 + NaCl + H_2O \qquad (38)$$

The 1,1,2-trichloroethane starting material may be prepared by the chlorination of ethylene dichloride or the addition of chlorine to vinyl chloride. In the industry, however, the direct production of 1,1,2-trichloroethane has seldom been required, since sufficient quantities have normally been available as a by-product in the preparation of other chlorinated hydrocarbons, such as ethylene dichloride. The base required in the synthesis may be sodium hydroxide or calcium hydroxide. One source of calcium hydroxide is a calcium carbide–based acetylene plant.

The reaction system is shown in Fig. 14. An aqueous solution of NaOH or slurry of $Ca(OH)_2$ (10 to 20 percent) and liquid 1,1,2-trichloroethane are simultaneously fed into an agitated heated reactor. The reaction is conducted under a pressure of 15 to 30 psi and a temperature of 60 to 110°C. The vinylidene chloride is distilled out of the reactor as it is produced to prevent further cracking to acetylenic compounds and mixed with an inhibitor, such as phenol. A purge is taken from the condenser to prevent build-up of oxygen, which may cause polymerization and the formation of explosive peroxides. The water is disposed of after separation.

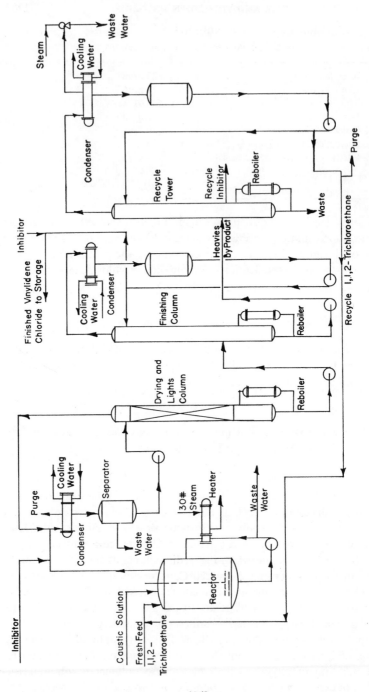

Fig. 14. Production and purification of vinylidene chloride.

The wet, inhibited vinylidene chloride is dried in an azeotropic distillation column and fed to the finishing column, where, under 10 to 20 psi of pressure, the finished vinylidene chloride is taken overhead. An inhibitor, such as phenol or the methyl ether of hydroquinone, is added as the monomer is transferred to storage.

The inhibitor, heavies, and unconverted 1,1,2-trichloroethane are removed from the bottom of the finishing column and processed through the recycle tower. Waste products are removed from the column bottom, and recycle inhibitor from the lower portion of the column. The recycle 1,1,2-trichloroethane is taken overhead. A small purge from this stream is required because of certain undesirable impurities. If appreciable quantities of these are allowed to recycle, chlorinated acetylenes are produced in the reactor, which creates explosion hazards. Yields of vinylidene chloride are 96 to 98 percent and single-pass conversions of 1,1,2-trichloroethane 80 to 85 percent. All the equipment may be steel except the reboiler tubes, which should be nonferrous.

Closely related methods are the dehydrochlorination of 1,1,2-trichloroethane with amines (135) and with alkali in the vapor phase (136). One drawback to these processes is the loss of HCl as salt. Thermal dehydrochlorinations of 1,1,2-trichloroethane (137,138) and 1,1,1-trichloroethane, which would permit the recovery of by-product HCl as the anhydrous compound, have been investigated.

IV. LABORATORY PROCESS

A distillation apparatus consisting of a flask and receiver is purged with nitrogen to remove air. The receiver is wrapped to exclude light and cooled in ice water. 1,1,1-Trichloroethane containing less than 100 ppm water and ferric chloride powder (0.1 to 0.2 percent of the trichloroethane) are charged to the flask and heated slowly until dehydrochlorination begins (about 60°C). The temperature in the flask is not allowed to rise above 100°C. The vinylidene chloride distills as it is formed and is collected in the receiver. The monomer is stored at −10°C or colder, and if the storage period is to be longer than 2 days, 0.6 to 0.8 percent of phenol is added as an inhibitor. The product may be redistilled in the same apparatus, keeping the pot temperature below 70°C.

Uninhibited monomer should be blanketed with inert gas, stored at −10°C or colder, kept out of contact with iron, and protected from light and air. If it is to be stored for 5 days or longer, 200 ppm of the methyl ether of hydroquinone (MEHQ) should be added. The MEHQ can be removed by distillation or alkaline extraction, but this is usually unnecessary, since its inhibitory action is easily overcome by conventional free-radical initiators.

V. MONOMER PROPERTIES

A. Physical Properties

Table 7 shows the properties of vinylidene chloride.

TABLE 7

Physical Properties of Vinylidene Chloride [139]

1. Odor	Pleasant, sweet
2. Appearance	Clear liquid
3. Color	10–15 APHA
4. Solubility of monomer in H_2O at 25°C, wt percent	0.021
5. Solubility of H_2O in monomer at 25°C, wt percent	0.035
6. Boiling point (760 mm Hg), °C	31.56
7. Freezing point, °C	−122.5
8. Vapor pressure, mm Hg	

$$\log P = 6.98200 - 1{,}104.29/(t + 237.697), \qquad t \text{ in } °C$$

Temperatures calculated at selected pressures are tabulated:

Pressure, mm Hg	Boiling point, °C
760	31.56
400	14.43
200	−1.78
100	−16.04
60	−25.49
40	−32.44
20	−43.31
10	−53.10
5	−61.94
1	−79.54

9. Liquid density

Temperature, °C	g/cc	lb/gal
−20	1.2902	10.768
0	1.2517	10.446
+20	1.2132	10.124

10. Index of refraction

Temperature, °C	
10	1.43062
15	1.42777
20	1.42468

11. Absolute viscosity

Temperature, °C	cps
−20	0.4478
0	0.3939
+20	0.3302

TABLE 7 (*continued*)

12. Flash point:	
Cleveland open cup, °F	5
Tag. closed cup, °F	55
13. Explosive limits in air (28°C), percent	7.3–16.0
14. Autoignition temperature, °F	955–1,031
15. Q Value	0.22
16. e Value	0.36
17. Latent heat of vaporization ΔH_v, cal/mole:	
At 25°C	$6,328 \pm 0.3\%$
At boiling point	$6,257 \pm 0.3\%$
18. Latent heat of fusion ΔH_m, cal/mole	1,557
19. Heat of polymerization ΔH_p:	
kcal/mole at 25°C	-18.0 ± 0.9
Btu/lb at 77°F	334 (exothermic)
20. Heat of combustion, liquid monomer, ΔH_c, kcal/mole	261.93 ± 0.3
21. Heat of formation:	
Liquid monomer, ΔH_f, kcal/mole	-6.0 ± 0.3
Gaseous monomer, ΔH_f, kcal/mole	$+0.3 \pm 0.3$
22. Heat capacity, liquid monomer, C_p, cal/mole/°C at 25.15°C	26.745
23. Heat capacity, ideal gas state, C_p, cal/mole/°C at 25.15°C	16.04
24. Critical temperature T_c, °C	222
25. Critical pressure P_c, atm	51.3
26. Critical volume V_c, cm³/mole	219

B. Hazards and Their Control (116)

1. Toxicity

If the monomer is accidentally swallowed, vomiting should be induced by tickling the back of the tongue or by giving an emetic, such as 2 tablespoons of table salt in a glass of warm water. Medical attention should be obtained immediately.

Inhibited vinylidene chloride is moderately irritating to the eyes. Contact can be expected to cause pain and conjunctival irritation and possibly some transient corneal injury and iritis; permanent damage is not likely. Concentrated phenolic inhibitor, however, may cause serious and permanent injury. If the eyes become contaminated, they should be flushed immediately with copious amounts of flowing water for 15 min or more. Medical attention should be obtained immediately.

Liquid vinylidene chloride monomer is irritating to the skin after direct contact of only a few minutes. The inhibitor content of the vinylidene chloride may be partly responsible for this. If leaks in pipes or containers occur, the vinylidene chloride monomer evaporates, and the inhibitor may accumulate until it has reached a concentration capable of causing local irritation (MEHQ) or burns (phenol) of the skin. There appears to be little likelihood of systemic injury from skin absorption of the amounts of phenol inhibitor encountered ordinarily in the handling of vinylidene chloride.

Precautions should be taken to prevent skin contact with vinylidene chloride. Any time such contact is possible, protective clothing (impervious gloves, aprons, shoes) should be worn. If, in spite of these precautions, contact does occur, all contaminated clothing, including shoes, should be removed immediately. The affected skin area should be washed with soap and water and any injuries or irritations should receive prompt medical attention. The contaminated clothing should be thoroughly cleaned and aerated before reuse.

Experimental studies on laboratory animals have shown that the quantitative vapor toxicity of vinylidene chloride is slightly greater than that of ethylene dichloride and slightly less than that of carbon tetrachloride. Vinylidene chloride is highly volatile, and its odor not sufficiently strong to serve as a warning, Therefore, this material presents a definite hazard from inhalation of vapor. A single exposure for a few minutes to a high concentration of vinylidene chloride vapor (such as 4,000 ppm) rapidly produces a "drunkenness" that may progress to unconsciousness if the exposure is continued. Animal experiments and industrial human experience have shown that prompt, complete recovery from the anesthetic effect results when the exposure is of short duration.

The principal cause of concern with the inhalation of vinylidene chloride vapor is that animal experimentation has shown that both a single prolonged exposure and repeated short-term exposures can be dangerous. Even when the concentrations of vapor are too low to cause an anesthetic effect, they may produce organic injury to the liver and kidneys. The quantitative data indicate that the maximum single exposure permitting a reasonably high probability of no injury is about 1,000 ppm for up to 1 hr and 200 ppm for up to 8 hr. For repeated exposures (8 hr per day, 5 days a week), the vapor concentration of vinylidene chloride should be maintained below 25 ppm. Persons who are known to have kidney or liver disease, or who are known to be excessive users of alcohol, should be forbidden to work in areas where they are exposed even to mild concentrations of vinylidene chloride vapor.

The vapor of freshly prepared vinylidene chloride free of decomposition products has a characteristic "sweet" smell that resembles that of carbon tetrachloride and chloroform. Most persons find that 1,000 ppm in air has a mild but definite odor, and many persons can detect 500 ppm. Vapors containing decomposition products have a disagreeable odor and can be detected at concentrations considerably less than 500 ppm. It must be recognized that the odor and irritating properties of vinylidene chloride vapor are ordinarily inadequate to prevent excessive exposure. Analysis must be relied on for knowledge of the intensity of exposures.

To prevent inhalation, large quantities of monomer should be handled in a closed system or where there are special provisions to control vapor. Although small amounts can be used safely with adequate ventilation in the manner customarily used with many volatile solvents, it is preferable to work with even small quantities of monomer only in very well-ventilated areas. Because vinylidene chloride vapors are heavy and may concentrate in low areas, special attention should be paid to low-level ventilation. If a person is affected or overcome from breathing vinylidene chloride vapors, he should be removed to fresh air at once, made to rest, kept warm, and medical attention should be obtained immediately. Artificial respiration should be administered immediately if breathing stops.

2. Spills and Leaks

If spills or leaks of vinylidene chloride monomer occur, all possible sources of ignition should be eliminated immediately, and only properly protected and trained personnel should remain in the vicinity. In cleaning up spills or leaks, personnel should wear a full-face gas mask with a canister for organic vapors. The volatility of vinylidene chloride is such that vapor concentrations exceeding 2 percent may readily occur in instances of large spills and in confined and unventilated spaces. In such situations and locations, and where an oxygen deficiency might exist, an air-supplied mask or self-contained breathing apparatus should be used.

Wearing apparel contaminated by vinylidene chloride monomer should be washed thoroughly before reuse. Shoes in particular should be washed and aerated before reuse. Personnel should bathe thoroughly to remove any monomer that may have penetrated to the skin.

3. Fire and Explosion

Vinylidene chloride vapor is flammable at concentrations between 7 and 16 percent by volume in air. Vapors of the liquid monomer, once ignited, burn strongly but not violently. Vinylidene chloride fires are relatively more difficult to start than hydrocarbon fires, but the rate of burning is similar. The products of combustion are to be avoided, and possible sources of ignition in the area of use should be eliminated or controlled. It is important to minimize the liberation and accumulation of vapors. Dry powder, foam, or carbon dioxide extinguishers may be used to extinguish vinylidene chloride fires. Water spray may be effective where cooling is required. Fire fighters should be cautioned about exposure to vinylidene chloride vapors and other poisonous fumes, smoke, or gases that may result from a vinylidene chloride fire.

4. PEROXIDE COMPOUNDS

In the presence of air or oxygen and at temperatures as low as $-40°C$, uninhibited vinylidene chloride may form a complex peroxide compound that is violently explosive. Reaction products formed with ozone are particularly dangerous. The decomposition products of vinylidene chloride peroxides are formaldehyde, phosgene, and hydrochloric acid. The presence of a sharp acid odor thus indicates oxygen exposure and the possible presence of peroxides. Since the peroxide is a polymerization initiator, formation of insoluble polymer in stored vinylidene chloride monomer may also indicate peroxide formation and a potentially hazardous condition. The peroxides are absorbed on the precipitated polymer, and its separation from monomer by filtration, evaporation, or drying may result in an explosive composition. If the peroxide content is more than 15 percent, this solid detonates from a slight mechanical shock or from heat. The presence of peroxides may be confirmed by the liberation of iodine from a slightly acidified dilute potassium iodide solution. It should be noted that peroxide-free polyvinylidene chloride is not explosive. Vinylidene chloride that contains peroxides may be purified by washing several times with 10 percent aqueous sodium hydroxide or freshly prepared 5 percent aqueous sodium bisulphite solution. In research work and in plants using vinylidene chloride, residues may be found in bottles, distillation flasks, condensers, filters, pipe lines, drums, and tanks. These should be handled with great care, and the peroxides destroyed by contacting with water at room temperature. Large-scale equipment that is used intermittently should be filled with water during shutdown periods.

The storage of vinylidene chloride monomer in the absence of air or oxygen is the best method to prevent peroxide formation. The use of a blanket of dry nitrogen under positive pressure is recommended. Phenolic polymerization inhibitors that delay or prevent the formation of peroxide compounds are also used.

VI. ANALYTICAL METHODS

Vinylidene chloride analysis is designed to determine monomer purity and specific contaminants that affect monomer quality.

A. Monomer Purity by Infrared Analysis

The major organic impurities can be determined from the infrared spectrum of the undiluted sample. The lower limit of detection using ordinary techniques is approximately 0.05 percent. As monomer purity has increased, the utility of this method has decreased.

B. Monomer Purity by Chromatographic Analysis

The development of gas-liquid chromatography has made it possible to detect organic impurities at the part-per-million level and determine monomer purity in less than 30 min. Column selection is the key to high-quality analyses. Thermal-conductivity detectors with high-sensitivity filaments are generally adequate for detection of impurities down to the 20-ppm level.

C. Specific Gravity

Specific gravity is determined with a Westphal balance at 25°C. Precise standardization methods and accurate control of sample temperature are necessary for maximum accuracy when using this method.

D. Appearance and Color

Appearance is determined visually, and color is determined by comparison with APHA color standards.

E. Acidity

Acidity is determined by titration of the acidic components with weak alcoholic caustic solutions. Aqueous titrations are performed by extracting the acid constituents with distilled water. An indicator is selected that has an equivalence point below pH 7 to avoid titration of acid anhydrides, such as CO_2.

F. Water

Water is most often determined by the Karl Fischer method, as in vinyl chloride analysis.

G. Acetylenes

Specific acetylenic compounds may be determined by gas chromatography, or acetylenic compounds may be determined by classical wet methods, as in vinyl chloride analysis.

H. Peroxides

Vinylidene chloride is added to acidified isopropanol saturated with sodium iodide. Iodide ion is oxidized by any peroxide present to free iodine, which is titrated with standard sodium thiosulfate solution.

VII. CHEMICAL REACTIONS OF VINYLIDENE CHLORIDE

The primary chemical reaction of commercial interest that vinylidene chloride undergoes is polymerization. A number of other reactions, however, have been reported.

A. Hydrogen Bromide

The light-catalyzed hydrobromination of vinylidene chloride yields 61.5 percent 1-bromo-2-dichloroethane and 31.6 percent 1-bromo-2,2,4,4-tetrachlorobutane (140).

B. Oxygen

Vinylidene chloride and oxygen react in the range of -40 to 25°C to form an explosive peroxide and polyvinyl chloride (141).

C. Disodium Cysteinate

Both vinylidene chloride and cis-dichloroethylene react with disodium cysteinate in liquid ammonia to give a 63 percent yield of the same dithioether, cis-1,2-bis-(1-amino-1-carboxy-2-ethylthio)ethylene (142).

D. Sodium n-Propylthiolate

Sodium n-propylthiolate reacts with vinylidene chloride under alkaline conditions in the presence of n-amyl alcohol to form cis-1,2-bis-(n-propylthio)ethylene (143).

E. Hydrogen Chloride

Hydrogen chloride adds to vinylidene chloride under a variety of conditions to form methyl chloroform (144–148).

PART C. INDUSTRIAL STORAGE AND HANDLING

The hazards and safety precautions associated with the handling of vinyl or vinylidene chloride should be realized. Reference is made to the section "Hazards and Their Control" for each of these monomers.

Vinyl chloride and vinylidene chloride are transported by tank truck, tank car, and barge, the most common container being the "jumbo" tank car, which contains 20,000 to 30,000 gal. Diagrams of typical unloading stations are shown in Figs. 15 and 16. Vinyl chloride and vinylidene chloride are usually stored in horizontal "bullets" or in spheres. The storage tanks for vinyl chloride should be designed for a working pressure from about 25 to 100 psig, depending on anticipated maximum storage temperature. Tanks for vinylidene chloride may be designed for only 25 psig. The tanks should be equipped with a pressure-relief valve, level gauge, pressure gauge, and remote shutoff valves. The monomers must be protected from contact with oxygen, sunlight, water, and other polymerization initiators. There should also be a water-spray system to keep the tanks cool in the event of a fire. Adequate dikes and drainage should be provided to confine and dispose of the liquid in case of a tank rupture. A typical tank farm is shown in Fig. 17.

Dry vinyl chloride, inhibited or uninhibited, and dry inhibited vinylidene chloride at normal ambient temperatures may be handled in steel without corrosion or deterioration of quality. Stainless steel, nickel, glass-lined

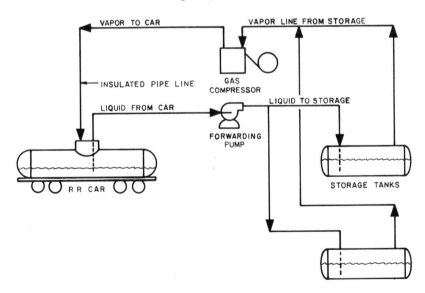

Fig. 15. Vinyl chloride unloading station.

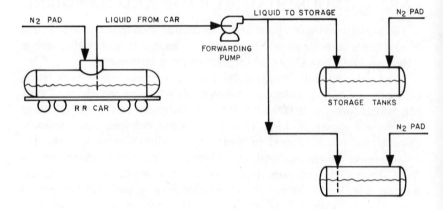

Fig. 16. Vinylidene chloride unloading station.

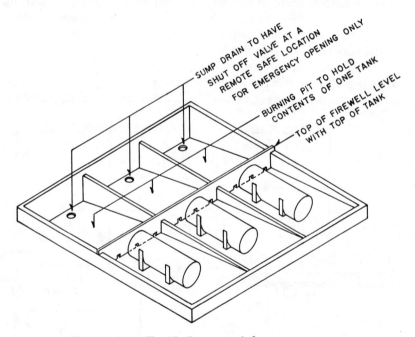

Fig. 17. Storage tank farm.

steel, and baked-phenolic–lined steel are also satisfactory. Wet vinyl or vinylidene chloride and dry uninhibited vinylidene chloride may be corrosive or unstable in the presence of steel. Nickel, baked-phenolic, or glass linings should then be used. Aluminum, magnesium, or their alloys should never be used with vinyl or vinylidene chloride.

Vinyl chloride is usually shipped uninhibited but may be shipped phenol-inhibited if this is required by the user. Vinylidene chloride is usually shipped with either 200 ppm of MEHQ or 5,000 ppm of phenol as an inhibitor. Vinyl chloride with less than 5 ppm of inhibitor and MEHQ-inhibited vinylidene chloride are usually polymerized without removing the inhibitor. If inhibitor removal is required, it may be done by either distillation or aqueous-caustic extraction. The wastes from the inhibitor removal system contain phenolic compounds and must be disposed of safely.

Vinyl and vinylidene chloride monomers may form acidic compounds or pick up iron contamination when stored for extended periods. This is particularly true when water or oxygen is present. Therefore, it is usually advantageous to treat the monomers for removal of these components immediately prior to use.

PART D. POLYMERIZATION

I. SUSPENSION POLYMERIZATION

More polyvinyl chloride (PVC) is made by this method than by any other, both in this country and abroad. A recent patent (117) describes a continuous-suspension process, but batch operation is the usual manu-facturing procedure.

A simplified flow chart for such a process is shown in Fig. 18. Because of the high purity of today's vinyl chloride, it can be stored without in-hibitors or refrigeration, and the time-consuming step of washing or distilling to remove inhibitor is no longer necessary. Storage tanks are steel, but the lines and vessels in the rest of the system are commonly glass-lined, resin-lined, or stainless steel. The reactors are generally glass-lined and have capacities of up to 8,000 gal.

Suspension-polymerization recipes differ widely, but a typical basic one is given below:

	Parts by weight
Vinyl chloride	100
Deionized water	200
Lauroyl peroxide	0.04
Polyvinyl alcohol	0.01

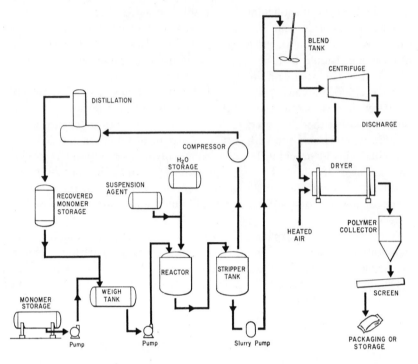

Fig. 18. Suspension polymerization of vinyl chloride.

An emulsifier is added in an amount approximately one-tenth that of the polyvinyl alcohol to enhance the porosity of the polymer particles and hence their ability to absorb plasticizer. Being porous, the particles take up a large amount of water. In order to maintain a free-flowing suspension and adequate temperature control (good heat transfer), the water-to-monomer ratio can be increased to a maximum consistent with economical operation, sometimes as high as 4:1.

The charge for suspension polymerization is often added to the reactor at ambient temperature. One hour may be needed to raise the mixture to polymerization temperature by discharge of steam into the reactor jacket. Heated water added initially to the vessel reduces the time necessary to reach operating temperature but may also change the properties of the polymer obtained. Typical temperatures for polymerization range from 45 to 60°C. The molecular weight, and therefore the physical properties, of PVC is highly sensitive to temperature changes and only slightly dependent on initiator concentration. Proper temperature control throughout the polymerization is, therefore, essential and must be obtained by sensitive instrumentation and, in some cases, with the use of refrigerated cooling water. Producers of low-molecular-weight polymer benefit from the relatively high temperatures necessary in this polymerization because of increased polymerization rates. This is limited, of course, by the pressure rating of the reactors. Alternatively, use is often made of chain-transfer agents, typically halogenated hydrocarbons in small concentrations, which reduce the molecular weight of the polymer. The production of high-molecular-weight PVC is carried out at lower temperatures, which results in reduced polymerization rates. Lauroyl peroxide produces free radicals much too slowly to be of economic importance in low-temperature polymerizations, but there are other commercially available initiators that are effective under these conditions.

Of prime importance in the production of high-quality PVC is the cleanliness of the polymerization vessel. Polymer buildup on reactor walls may occur, especially at the vapor-liquid interface. Removal of this material is accomplished by the laborious scraping of surfaces, the use of solvents, or the use of extremely high-pressure water streams.

Buildup allowed to remain in the reactor soon becomes hard, dense, and nonabsorbent and may give rise to imperfections and "fish eyes" in finished product, particularly in calendered sheets.

Toward the end of the polymerization cycle the pressure in the system begins to drop, followed very shortly by a peak in the polymerization rate. Beyond the peak the rate begins to drop sharply, and the polymer beads become less porous. The experience of the manufacturer determines the point at which polymerization is terminated to give high-quality polymer and acceptable economy of production. At this point the polymer slurry is discharged into an evacuated stripper tank, quickly reducing the pressure and temperature and effectively stopping polymerization. Unreacted monomer is pumped from the slurry, condensed, and, after processing, fed back into the system.

The blend system may have sufficient volume to contain product from several polymerization vessels and provide uniformity of product by a continual circulation of the slurry. Beyond this stage the operation may be continuous. Removal of water from the slurry is accomplished by the use of a continuous centrifuge in which washing of the cake is often done to remove soluble electrolytes. Final drying of the wet cake, which in some cases may contain as much as 30 percent water, may be by rotary dryer, spray dryer, one- and two-stage flash dryer, or combinations of these. Moving streams of heated air remove moisture to a level below 0.25 percent. Careful temperature control must be maintained in the drying step in order to produce quality resin. Any heat above that necessary to remove moisture results in loss of heat stability and discoloration of polymer.

After screening, packaging has traditionally been done in 50-lb bags, but more recently bulk shipments are being made to fabricators' plants. With proper attention to plant cleanliness, removal of buildup in reactors, and drying temperatures, manufacturers have sought to supply the demands of fabricators for high-quality, contamination-free PVC. That they have succeeded is shown by the continued rise in PVC consumption in recent years.

II. EMULSION POLYMERIZATION

Emulsion polymerization is used commercially for both PVC polymers and vinylidene chloride copolymers. Some products are used directly as the polymer latex, and others are converted to the dried polymer form. The reasons for using emulsion polymerization for certain products or applications will be explained later, when the advantages and disadvantages of this polymerization technique are discussed.

Although other theories have been proposed and specific cases of variance from the basic mechanism have been detailed, currently the most widely accepted mechanism of emulsion polymerization was proposed by W. D. Harkins in a series of papers published between 1945 and 1950 (150–152). The elements of the Harkins theory are that (1) the principal source of polymer particles is initiation of polymerization within monomer swollen soap micelles and (2) the major site of polymer formation is within the polymer particles that absorb more monomer after their formation within the soap micelles.

The essential ingredients of an emulsion polymerization are the following:

1. A monomer capable of undergoing addition polymerization.
2. A source of free radicals (initiator or catalyst).
3. A surfactant (above its critical micelle concentration).
4. A dispersing phase (most commonly water, except for polymerizations conducted at or below 0°C).

The monomers under consideration here are vinyl chloride and vinylidene chloride with small concentrations of other monomers amenable to addition polymerization added to obtain some specific modification of the polymer, for example, adhesion, solubility, or flexibility.

In emulsion-polymerization recipes, the initiator must be soluble in the dispersing phase and capable of being decomposed into free radicals at reasonable temperatures. Examples of commonly used compounds that meet these conditions are sodium, potassium and ammonium persulfate, hydrogen peroxide, alkyl hydroperoxides, and various perborates and percarbonates. The rates of decomposition of these water-soluble initiators are temperature-dependent, as are those of the commonly used oil-soluble initiators.

The rate of decomposition of the initiator into free radicals can be accelerated through the addition of an activator or promoter. Excellent examples of this type of system are the redox initiator pairs, such as either persulfate or perchlorate with bisulfite. The use of metallic ions, such as ferric, as accelerators of hydrogen peroxide decomposition is also known. Activators have the effect of lowering the temperature at which free radicals are produced. Lower temperatures may be desired in vinyl chloride and vinylidene chloride polymerizations to keep operating pressures down and to obtain higher molecular weight.

Anionic emulsifiers, alone or in combination with nonionics, are generally used for emulsion polymerization of vinyl chloride and vinylidene chloride. Representative of the types of surfactants used are the sodium alkylarylsulfonates, such as sodium dodecylbenzenesulfonate, the alkyl esters of sodium sulfosuccinic acid, and the sodium and ammonium salts of fatty alcohol sulfates, such as sodium lauryl sulfate.

A typical recipe and reaction conditions for the emulsion polymerization of vinylidene chloride or vinyl chloride are given below and a simplified flow chart is shown in Fig. 19:

	Parts by weight	Conditions
Monomer	100	
Water	100–200	
Sodium persulfate	0.4	
Sodium dodecylbenzenesulfonate	0.5–2.0	
Sodium bicarbonate		To pH of 4–9
Temperature		50–55°C
Reaction time		10–18 hr
Latex particle diameter (peak)		800–1,600°A

In addition to these ingredients, modifiers may be present, such as colloidal protective agents (gelatin or carboxymethyl cellulose), chelates, plasticizers, stabilizers, and chain-transfer agents.

The temperature of the polymerization is selected on the basis of the decomposition kinetics of the particular initiator used and the desired polymer molecular weight. Since chain transfer to monomer is an important termination mechanism in both vinyl chloride and vinylidene chloride polymerization, the use of high temperatures results in a decrease in the average molecular weight of the polymer produced.

In copolymerization of vinylidene chloride and vinyl chloride, the reaction time given above may be exceeded significantly unless an

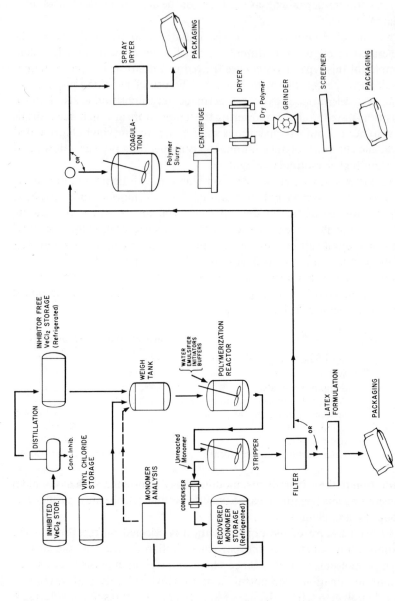

Fig. 19. Simplified flow diagram for emulsion polymerization process for vinyl chloride/vinylidene chloride copolymers.

initiator-activator combination is used. The polymerization rate for mixtures of vinyl chloride and vinylidene chloride over the concentration range of 30 to 80 percent vinyl chloride is about one-sixth that of either monomer alone (153).

Emulsion polymerization of vinyl chloride monomer to PVC is the commercial route to *paste resins*. The paste, or plastisol grade PVC resins, are used as slurries in plasticizers. Sometimes a solvent is also added, in which case the composition is called an *organosol*. Emulsion PVC resins are particularly well suited to this application, having a smaller particle size than that of normal suspension PVC and being relatively impervious to plasticizer absorption. The small particle size is necessary to obtain good fusion in fabricating the plastisol, which is done with practically no shear applied to the particles. Excessive plasticizer absorption results in undesirably high plastisol viscosities, because the continuous (plasticizer) phase is reduced in volume and the polymer particles are swollen. To prepare a PVC polymer having the desired nonabsorbing characteristics and small particle size, emulsion polymerization is conducted with a recipe and under conditions that give a particle size of approximately 1 μ in diameter. The polymer is dried, usually by spray drying, to a free-flowing powder.

Since all the recipe ingredients except those which volatilize in the drying operation remain in the resin, the polymers produced by this method are not suitable for electrical applications. It should be pointed out, however, that this limitation pertains to most emulsion-produced polymers, not just those which are spray-dried.

Vinyl and vinylidene chloride emulsion polymers are used commercially in two forms. In some applications, the latex, after addition of stabilizers, is used directly, for example, as a coating composition for applying the polymer to a substrate. In other applications, the polymer is recovered from the latex in dry form before use, either by volatilizing the water, as described earlier, or by breaking the emulsion and isolating the coagulated resin by filtration or centrifugation. The latter method yields a cleaner, more stable product, because some of the water-soluble components have been removed.

The principal advantages of emulsion polymerization as a method of making vinyl chloride and vinylidene chloride polymers for nonlatex applications are as follows:

1. High-molecular-weight products, especially VCl–VCl$_2$ copolymers, can be produced in reasonable reaction times. The initiation and propagation steps can be controlled more independently than is the case for mass or suspension methods.

2. Monomer can be added during the polymerization to maintain polymer composition control.

3. The polymer particles have relatively little internal void volume as compared with suspension- and mass-polymerized products.

The disadvantages of emulsion polymers are the result of the relatively high concentrations of additives that are required in the polymerization recipe. The residual surfactants and ionic ingredients give products that have greater water sensitivity, poorer electrical properties, and generally poorer heat and light stability.

In copolymerizations, vinylidene chloride reacts much more readily than vinyl chloride. In bulk, this results in a heterogeneous product, the first copolymer formed being high in vinylidene chloride content and each successive increment of copolymer containing less and less. In emulsion, the monomer droplets are away from the initiation and propagation sites, which they reach by diffusion through the aqueous phase. Therefore, an essentially constant copolymer composition can be achieved by adding the more reactive vinylidene chloride gradually during the copolymerization. One method is to introduce this monomer at such a rate that the pressure in the reactor remains constant (122).

Almost no PVC homopolymer is used in latex applications, since fusion of the resin particles requires high temperature or plasticization. Copolymerization of small quantities (15 percent or less) of other monomers, such as alkyl acrylates and vinyl acetate, with vinyl chloride lowers the fusion temperature to a practical level. These products are used principally as fabric coatings.

Vinylidene chloride copolymer latex compositions have been receiving increasing attention in recent years as barrier coatings for packaging materials, such as paper, boxboard, and plastic films. These copolymers are usually 90 percent or more vinylidene chloride with the balance made up of alkyl acrylates or acrylonitrile to facilitate film formation and provide flexibility in the finished coating. In addition, these products, especially those intended for coatings on plastic films, also contain a component for improved adhesion of the coating to the substrate. Monomers that are used to improve coating adhesion include acrylic, methacrylic, and itaconic acids (155–157).

III. BULK POLYMERIZATION

The polymerization of vinyl chloride in bulk, or mass, has been known and practiced, at least in the laboratory, for a number of years (158–164). Commercial utilization of this technique, however, has been fairly recent. So far as is known, only one bulk process, that developed by Pechiney–St.

Gobain, is used to manufacture polyvinyl chloride on a commercial scale.

The bulk polymerization of vinyl chloride is heterogeneous; that is, the polymer is insoluble in the monomer and precipitates as it is formed. The theory of heterogeneous polymerization has been discussed in the literature (165–167). Briefly, these discussions suggest that many of the growing polymer radicals precipitate, before they are terminated, as "living polymers" in which the radical is buried in the tightly coiled chain and, therefore, is available for further reaction under the proper conditions.

The commercial method for bulk polymerization of vinyl chloride developed by Pechiney–St. Gobain is a two-stage batch process and is described in detail in a number of articles (168–171) and patents (172–176). Pechiney–St. Gobain operates a 5,000 metric ton per month plant at St. Fons, France. In addition, licenses have been awarded to Wacker Chemie and Farbwerke Hoechst in West Germany; to Reposa in Spain; to Hooker Chemical, Dow Chemical, and B. F. Goodrich Chemical in the United States; to Toa Gosei in Japan; and to the U.S.S.R. A flow chart of the process is shown in Fig. 20.

The first-stage reactor, or prepolymerizer, is a 2,000-gal stainless-steel vertical autoclave equipped with a flat-blade turbine and baffles to give turbulent agitation. Vinyl chloride and a peroxide initiator that is soluble in it are charged and heated to 40 to 70° at a pressure of 75 to 175 psi. The insoluble polymer begins to precipitate immediately, forming granules about 0.1 μ in diameter. Once formed, the number of granules does not change, but their average diameter increases to 1 μ at 1 percent conversion. This is consistent with initial formation of insoluble "living polymer" radicals, which undergo further polymerization. Subsequently, the granules agglomerate into beads of about 50 μ in diameter at 2 to 3 percent conversion. Polymerization in the first-stage reactor is continued to 7 to 15 percent conversion to allow completion of the bead-forming process. Beyond this the mixture becomes too viscous to stir, and, at about 20 percent conversion, it is a wet powder.

The material from the prepolymerizer is transferred to the second-stage reactor, a 4,000-gal stainless-steel horizontal autoclave stirred with a frame type of agitator turning at about 9 rpm, more vinyl chloride and initiator are added, and the polymerization is continued. Monomer conversion is followed by determining the heat evolved. As the reaction proceeds, the beads grow larger (final diameter of 80 to 200 μ), and the mixture takes on the appearance of a dry powder. At 70 to 80 percent conversion the polymerization is stopped by flash-distilling unreacted monomer. No further drying is required, and after screening to remove large agglomerates, the product is packaged.

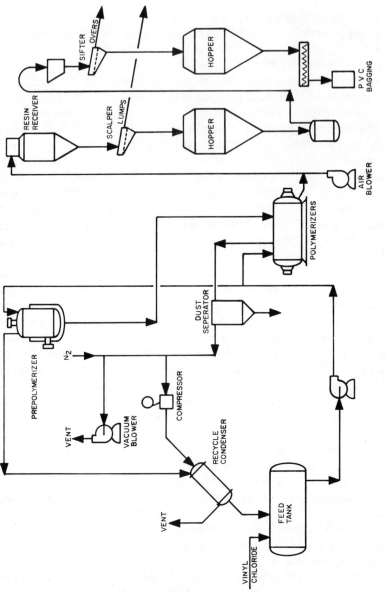

Fig. 20. Bulk polymerization of vinyl chloride.

1278

Temperature control of the highly exothermic polymerization is critical, since it determines the molecular weight of the product. The inherently poor heat-transfer properties of the thick slurries and particulate solids that are encountered in the reactors make this a challenging problem. The keys to successful operation appear to be control of polymer particle size, efficient agitation, and minimum fouling of heat-transfer surfaces.

PVC manufactured by this process is very pure, for it contains no contaminants other than initiator fragments. The particles are transparent, have a narrow size distribution, and are relatively porous. They do not, of course, have the pericellular membrane due to suspending agent, which is a characteristic of PVC particles made by the suspension process.

IV. POLYMER APPLICATIONS

Vinyl and vinylidene chloride polymers have unique properties that account for their broad range of usefulness. They are flame-resistant because of high chlorine content, can be plasticized to yield flexible materials, and are resistant to attack by chemicals. Vinylidene chloride polymers are outstanding barriers to gases and vapors, including moisture and oxygen.

A. Polymers of Vinyl Chloride

Homopolymers of vinyl chloride and copolymers containing up to 15 percent of vinyl acetate constitute a very important class of commercial products. They are second in volume of production and sale among polymers in the United States and are first on a worldwide basis. Total consumption in the United States in 1966 was approximately 2.2 billion lb. Some indication of the variety of applications can be gained from the 1966 consumption figures shown in the table below.

Suspension polymers are made in the largest volume. The homopolymers and vinyl chloride–vinyl acetate copolymers manufactured by this process are fabricated by extrusion, calendering, and molding. In general, medium- to high-molecular-weight resins are used with plasticizers to make flexible products; those with lower molecular weights go into rigid applications. Bulk polymers were available in the United States in 1968 and are expected to be used in the same applications as suspension polymers.

A fabrication process unique to vinyl chloride polymers is that involving a fluid dispersion of polymer in plasticizer called a plastisol, or organosol if a volatile diluent is present. The polymer used for this is made by emulsion polymerization. Plastisols are fabricated by slush molding, rotational molding, and dipping and are also used for coating various substrates, especially fabrics. This is the only area of use in the United States

1966 Polyvinyl Chloride and Copolymer
Consumption [177]

	Million lb
Calendered film and sheeting	385
Coatings, bonding, and adhesives, including coated fabrics	210
Floor covering	370
Wire and cable insulation	251
Flexible profile extrusions	85
Extruded film and sheeting	80
Extruded rigid products, such as pipe and house siding	180
Miscellaneous extrusions	60
Phonograph records	100
Other molded products, such as shoe soles and heels	155
All other uses	269
Exports	70
Total	2,215

for emulsion polymers of vinyl chloride. They are used in Europe, however, for hot processing, especially in extrusion.

The solution process is used only to make copolymers for coatings applications. The higher cost can be tolerated here, because it results in a product of the desired high purity and consistency. Many of the polymers used in coatings contain small amounts of a third monomer, such as maleic anhydride, to aid in adhesion to metals.

Additives are very important in the proper application of vinyl chloride polymers. Pigments, fillers, lubricants, and ultraviolet-light absorbers perform their normal functions. Two other types of additives, however, stabilizers and plasticizers, are uniquely important in polymers of vinyl chloride. Heating of PVC results in rapid loss of hydrogen chloride, with resulting development of color and loss of mechanical properties. A large number of stabilizers have been developed to reduce the rate of dehydrochlorination and to offset its effects. Suitable stabilizers are essential to the fabrication and use of vinyl chloride polymers.

About 80 percent of the applications of vinyl chloride polymers involve the addition of plasticizers to make a flexible product. These polymers are unique in their ability to accept large quantities of plasticizer, as much as 4 parts to 1 part of polymer. As a result, a continuous spectrum of stiffness is available to the product designer. It is this property more than any other which has resulted in the widespread commercial use of vinyl chloride polymers.

B. Polymers of Vinylidene Chloride

Polymers based on vinylidene chloride have several properties that have made them important commercially: very low vapor transmission, good resistance to attack by solvents and by fats and oils, high strength, and the ability to form heat-shrinkable films. The polymers are widely used in

TABLE 8

Copolymerization Reactivity Ratios of Vinyl and Vinylidene Chloride

M_2	r_1	r_2	$T, °C$	Initiator	Reference
Vinyl chloride (M_1)					
Dioctyl maleate	0.42	0	68	Bz_2O_2	180
Pentene-1	5	0.2	68	Bz_2O_2	180
Methyl methacrylate	0.1	10	68	Bz_2O_2	180
Styrene	0.02	17 ± 3	60	Bz_2O_2	181
Styrene	0.067	35.0	48	...	182
Styrene	0.077	35.0	50	...	182
Acrylonitrile	0.02 ± 0.02	3.28 ± 0.06	60	Bz_2O_2	181
Acrylonitrile	0.074	3.7	50	...	182
Diethyl fumarate	0.12 ± 0.01	0.47 ± 0.05	60	Bz_2O_2	181
Isobutylene	2.05 ± 0.3	0.08 ± 0.10	60	...	183
Diethyl maleate	0.77 ± 0.03	0.009 ± 0.003	60	...	184
Diethyl maleate	0.9	0.0	70	...	186
Diethyl maleate	0.8	0.0	...	Redox	189
t-Butylethylene	5.0	0.0	70	Bz_2O_2	186
Maleic anhydride	0.008	0.296 ± 0.07	75	Bz_2O_2	187
Isopropenyl acetate	2.2	0.25	65	Bz_2O_2	188
Dibutyl maleate	1.4	0.0	40	...	190
Vinylidene chloride	0.3	3.2	60	...	191
Vinylidene chloride	0.5	7.5	47	...	192
Vinylidene chloride	0.23	3.15	55	...	193
Vinyl acetate	1.68 ± 0.08	0.23 ± 0.02	60	...	185
Vinylidene chloride (M_1)					
Acrylonitrile	0.37 ± 0.10	0.91 ± 0.10	60	...	194
Allyl chloride	3.8	0.26	68	...	195
Allyl chloride	4.5	0.0	60	...	196
Diethyl fumarate	12.2 ± 2	0.046 ± 0.015	60	...	197
Diethyl maleate	12.5	...	60	...	196
Butadiene	1.0 ± 0.2	20.05	5	...	198
Chlorotrifluoroethylene	17.14	0.02	60	...	199

films, monofilaments, pipe linings, and coatings for other packaging materials, such as paper and cellophane. Success in these applications has been achieved in spite of fairly high cost, problems in heat stabilization, and processing difficulties due to the tendency of the polymers to crystallize. The sharp drop in viscosity when the crystallites melt during processing is the major problem. This crystallinity is, of course, also the reason for some of the good properties of the polymers, such as low vapor transmission and good performance as monofilaments and fibers.

The homopolymer of vinylidene chloride is not of commercial interest, because it is extremely difficult to process. Commercial polymers contain up to 30 percent of vinyl chloride. The polymers with higher levels of vinyl chloride are used in films, and those with lower concentrations for monofilaments. Comonomers, such as acrylonitrile, acrylate esters, and vinyl acetate, are also used to modify the processability and other properties of vinylidene chloride polymers.

Additives are important here as in the polymers of vinyl chloride. Both stabilizers and plasticizers are used, although the specific materials preferred may not be the same as for polyvinyl chloride. In addition, plasticizers are used only at rather low concentrations. The high level of crystallinity precludes the use of very much plasticizer, since plasticization takes place only in the amorphous regions in the polymer.

C. Copolymerization Reactivity Ratios

The reactivity ratios of vinyl chloride and of vinylidene chloride in copolymerization reactions with a variety of vinyl monomers are shown in Table 8.

References

1. "Synthetic Organic Chemicals: U.S. Production and Sales," U.S. Tariff Commission, 1965.
2. V. Regnault, *Ann. Chem.*, **15**:28, 34, 63 (1835); *Ann. Chim. Phys.*, (2)**59**:538 (1835).
3. I. Ostromislensky, *Chem. Zentr.*, **1**:1980, 1983 (1912); *Chem. Ztg.*, **36**:199 (1912).
4. F. Klatte and A. Rollett (to Chemische Fabrik Griesheim-Elektron), U.S. Patent 1,241,738, Oct. 2, 1917; *Chem. Abstr.*, **12**:231 (1918).
5. Compiled from various magazines and periodicals by The Dow Chemical Company.
6. R. D. Wesselhoft, J. M. Woods, and J. M. Smith, *AIChE (Am. Inst. Chem. Engrs.) J.*, **5**:361 (1959).
7. British Patent 21,134 (to Chemische Fabrik Griesheim-Elektron), Sept. 18, 1913; *Chem. Abstr.*, **9**:2567 (1915).
8. Z. Czarny, *Chem. Stosowana, Ser. A*, **10**(1):79–92 (1966).
9. Belgian Patent 620,123 (to Dynamit-Nobel A.-G.), Oct. 31, 1962; *Chem. Abstr.*, **58**:10032 (1963).

10. R. R. Mittiko and R. R. Soyre, AIChE Seminar, Akron, Ohio, Oct. 16, 1964.
11. L. F. Albright, *Chem. Eng.*, **74**(7):123–30 (1967).
12. R. E. Lynn, Jr., and K. A. Kobe, *Ind. Eng. Chem.*, **46**:633–643 (1954).
13. J. M. DeBell, W. C. Goggin, and W. E. Gloor, "German Plastics Practice," Chap. 3, DeBell and Richardson, Springfield, Mass., 1946.
14. H. S. Miller (to Air Reduction Co., Inc.), U.S. Patent 2,448,110, Aug. 31, 1948; *Chem. Abstr.*, **43**:1433 (1949).
15. British Patent 709,604 (to Chemische Werke Hüls GmbH), May 26, 1954; *Chem. Abstr.*, **49**:9025 (1955).
16. British Patent 573,594 (to Distillers Co. Ltd.), Nov. 28, 1945; *Chem. Abstr.*, **43**:3024 (1949).
17. British Patent 757,661 (to Monsanto Chemical Co.), Sept. 19, 1956; *Chem. Abstr.*, **51**:9671 (1957).
18. D. H. R. Barton and M. Mugdan, *J. Soc. Chem. Ind.* (*London*), **69**:75–79 (1950).
19. H. Gibello, "Le Chlorure de Vinyle et Ses Polymères," Chap. 1, Dunod, Paris, 1959.
20. B.I.O.S. Final Reports No. 104, 1290, 811, 30 (29–62) (1951).
21. C.I.O.S. Files No. 33–31, 27–51, and 28–29.
22. F.I.A.T. Final Report No. 867.
23. Netherlands Appl. 6,504,088 (to Kureha Chemical Industry Co. Ltd.), Oct. 4, 1965; *Chem. Abstr.*, **64**:8031 (1966).
24. Y. Onue et al. (to Tokuyama Soda Co. Ltd.), Japanese Patent 41-3168, 1966.
25. H. P. A. Groll, G. Hearne, F. F. Rust, and W. E. Vaughan, *Ind. Eng. Chem.*, **31**:1239–1244 (1939).
26. S. N. Balasubramanian, D. N. Rihani, and L. K. Doraiswamy, *ibid.*, *Fundamentals*, **5**(2):184–188 (1966).
27. D. B. Benedict (to Union Carbide Corp.), U.S. Patent 2,929,852, Mar. 22, 1960; *Chem. Abstr.*, **54**:10859 (1960).
28. F. F. A. Braconier, *Hydrocarbon Process. Petrol. Refiner*, **43**:140–142 (1964).
29. F. F. A. Braconier and J. A. R. O. L. Godart (to Société Belge de L'Azote et des Produits Chimiques du Marly, S.A.), U.S. Patent 2,779,804, Jan. 29, 1957; also British Patents 889,177 and 954,791; *Chem. Abstr.*, **51**:5814 (1957).
30. S. Gomi, *Hydrocarbon Process. Petrol. Refiner*, **43**:165–170 (1964).
31. I. Roberts and G. E. Kimball, *J. Am. Chem. Soc.*, **59**:947 (1937).
32. H. P. Rothbaum, I. Ting, and P. W. Robertson, *J. Chem. Soc.*, **1948**, 980–984.
33. D. A. Evans, T. R. Watson, and P. W. Robertson, *ibid.*, **1950**, 1624.
34. H. Biltz and E. Küppers, *Chem. Ber.*, **37**:2398 (1904).
35. A. S. Bratolyubov, *Russ. Chem. Rev.* (*English Transl.*), **30**(11):602 (1961).
36. J. P. Baxter, W. A. M. Edwards, and R. M. Winter (to Imperial Chemical Industries Ltd.), British Patent 363,009, Sept. 18, 1930; *Chem. Abstr.*, **27**:1365 (1933).
37. D. H. R. Barton, *J. Chem. Soc.*, **1949**, 148–154; D. H. R. Barton and K. E. Howlett, *ibid.*, **1949**, 155–164.
38. L. K. Doraiswamy, P. H. Brahme, M. U. Pai, and S. Chidambaram, *Brit. Chem. Eng.*, **5**:618–623 (1960).
39. E. W. McGovern, *Ind. Eng. Chem.*, **35**:1230 (1943).
40. G. C. Sinke and D. R. Stull, *J. Phys. Chem.*, **62**:397 (1958).
41. French Patent 1,290,953 (to Société Belge de L'Azote et des Produits Chimiques du Marly, S.A.), Apr. 20, 1962; *Chem. Abstr.*, **58**:5513 (1963).
42. British Patent 979,309 (to Wacker-Chemie GmbH), Jan. 1, 1965.

43. German Patent 1,135,451 (to Wacker-Chemie GmbH), Appl. Nov. 25, 1960.
44. British Patent 938,824 (to The B. F. Goodrich Co.), Oct. 9, 1963; *Chem. Abstr.*, **60**:5333 (1964).
45. H. Krekeler (to Farbwerke Hoechst A.-G.), U.S. Patent 2,724,006, Nov. 15, 1955; *Chem. Abstr.*, **51**:1242 (1957).
46. *Hydrocarbon Process. Petrol. Refiner*, **44**:289 (1965).
47. M. Mugdan and D. H. R. Barton (to Distillers Co. Ltd.), British Patent 573,532, Nov. 26, 1945; *Chem. Abstr.*, **43**:2216 (1949).
48. A. Schmidt and J. Schulze (to Chemische Werke Hüls A.-G.), German Patent 1,210,800, Feb. 17, 1966; *Chem. Abstr.*, **64**:15740 (1966).
49. L. F. Albright, *Chem. Eng.*, **74**(8):219–226 (1967).
50. C. M. Fontana, E. Gorin, and C. S. Meredith, *Ind. Eng. Chem.*, **44**:373 (1952).
51. "The Shell Vinyl Chloride Process," Shell Development Company Brochure, July, 1965.
52. *Hydrocarbon Process. Petrol. Refiner*, **44**:198 (1965).
53. British Patent 907,435 (to Shell Internationale Research Maatschappij NV), Oct. 3, 1962; *Chem. Abstr.*, **58**:4424 (1963).
54. P. Metaizeau (to Solvay & Cie), U.S. Patent 3,332,742, Jul. 25, 1967.
55. W. F. Engel (to Shell Oil Co.), U.S. Patent 3,210,431, Oct. 5, 1965.
56. R. M. Crawford, *Chem. Eng. Progr.*, **46**:483–485 (1950).
57. J. L. Dunn, Jr. and B. Posey, Jr. (to The Dow Chemical Co.), U.S. Patent 2,866,830, Dec. 30, 1958; *Chem. Abstr.*, **53**:10002 (1959).
58. *Chem. Week*, **95**(9):101–108 (1964).
59. Ref. 46.
60. C. M. Fontana, E. Gorin, G. A. Kidder, and C. S. Meredith, *Ind. Eng. Chem.*, **44**:363 (1952).
61. Indian Patents 102,820–3 (to The B. F. Goodrich Co.), May 24, 1965.
62. *Chem. Week*, **101**(7):73 (1967).
63. N. Todo, M. Kurita, and H. Hagiwara, *Kogyo Kagaku Zasshi*, **69**(8):1463–1466 (1966); *Chem. Abstr.*, **66**:45930 (1967).
64. K. Sakuragi et al. (to Japanese Geon Co. Ltd.), Japanese Patent 40-21764, Sept. 28, 1965.
65. British Patent 1,016,094 (to Toyo Soda Manufacturing Co. Ltd.), Jan. 5, 1966; *Chem. Abstr.*, **64**:12547 (1966).
66. Y. Kosaka and M. Hayata, *Kogyo Kagaku Zasshi*, **69**(2):244–248 (1966); *Chem. Abstr.*, **65**:5351 (1966).
67. French Patent 1,359,016 (to Imperial Chemical Industries Ltd.), Mar. 9, 1964.
68. Belgian Patent 662,098 (to Farbwerke Hoechst A.-G.), Apr. 5, 1965.
69. K. Goto et al. (to Japanese Geon Co. Ltd.), Japanese Patent 40-20250, Sept. 9, 1965.
70. French Patent 1,410,762 (to Asahi Chemical Industry Co. Ltd.), Aug. 2, 1965.
71. R. Ukaji, K. Maruo, S. Misaki, and S. Ogawa (to Osaka Kinzoku Kogyo Co. Ltd.), U.S. Patent 3,267,161, Aug. 16, 1966; *Chem. Abstr.*, **65**:12107 (1966).
72. H. Ito and S. Gomi, *Jap. Chem. Quart.*, **1**(2):46–51 (1965).
73. *Hydrocarbon Process. Petrol. Refiner*, **44**:290 (1965).
74. H. P. A. Groll, G. Hearne, J. Burgin, and D. S. La France (to Shell Development Co.), U.S. Patent 2,167,927, Aug. 1, 1939; *Chem. Abstr.*, **33**:9327 (1939).
75. Netherlands Patent 6,401,933 (to Union Carbide Corp.), Aug. 30, 1965; *Chem. Abstr.*, **64**:6492 (1966).

76. E. Otsuka, T. Takahashi, and T. Abe (to Toyo Koatsu Industries, Inc.), Belgian Patent 637,537, Jan. 16, 1964; *Chem. Abstr.*, **63**:8197 (1965).
77. T. Takahashi, T. Abe, Y. Miyakoshi, and K. Nakamura (to Toyo Koatsu Industries, Inc.), Belgian Patent 658,457, May 17, 1965; *Chem. Abstr.*, **64**:3348 (1966).
78. Belgian Patent 572,902 (to Solvay & Cie), May 12, 1959; *Chem. Abstr.*, **54**:4385 (1960).
79. A. I. Subbotin, V. S. Etlis, V. R. Likhterov, and V. N. Antonov, *Int. Chem. Eng.*, **6**:650–655 (1966).
80. G. R. Schultze, J. L. Crützen, and K. Kossmann, German Patent 1,059,439, Jun. 18, 1959; *Chem. Abstr.*, **55**:8290 (1961).
81. British Patent 842,539 (to Wacker-Chemie GmbH), Jul. 27, 1960; *Chem. Abstr.*, **55**:3429 (1961).
82. J. W. Sprauer (to E. I. du Pont de Nemours & Co.), U.S. Patent 3,234,295, Feb. 8, 1966; *Chem. Abstr.*, **64**:19411 (1966).
83. Netherlands Patent 6,408,161 (to Shell Internationale Research Maatschappij NV), Jan. 17, 1966; *Chem. Abstr.*, **64**:19411 (1966).
84. D. W. Setser, R. Littrell, and J. C. Hassler, *J. Am. Chem. Soc.*, **87**:2062–2063 (1965).
85. French Patent 1,341,711 (to Pittsburgh Plate Glass Co.), Sept. 23, 1963.
86. French Patent 1,355,870 (to Pittsburgh Plate Glass Co.), Feb. 10, 1964.
87. E. H. Gause and P. D. Montgomery (to Monsanto Co.), U.S. Patent 3,142,709, Jul. 28, 1964; *Chem. Abstr.*, **61**:8189 (1964).
88. British Patent 836,970 (to Solvay & Cie), Jun. 9, 1960; *Chem. Abstr.*, **55**:4361 (1961).
89. D. W. McDonald (to Monsanto Chemical Co.), U.S. Patent 3,125,608, Mar. 17, 1964; *Chem. Abstr.*, **61**:8190 (1964).
90. J. T. Barr, Jr., *Advan. Petrol. Chem. Refining*, **7**:368–370 (1963).
91. R. R. Dreisbach, "Advances in Chemistry Series No. 22," p. 219, American Chemical Society, Washington, 1959.
92. R. A. McDonald, S. A. Shrader, and D. R. Stull, *J. Chem. Eng. Data*, **4**:311 (1959).
93. "Properties and Essential Information for Safe Handling and Use of Vinyl Chloride," Manufacturing Chemists' Association, Chemical Safety Data Sheet SD-56, 1954.
94. T. Numano and T. Kitagawa, *Kogyo Kagaku Zasshi*, **65**:182–184 (1962); *Chem. Abstr.*, **57**:5753 (1962).
95. "Selected Values of Properties of Chemical Compounds," Manufacturing Chemists' Association Res. Proj., Jun. 20, 1956.
96. P. Zaloudik, *Chem. Prumysl*, **12**:81–83 (1962).
97. A. W. Francis, *J. Chem. Eng. Data*, **5**:534–535 (1960).
98. K. A. Kobe and R. H. Harrison, *Petrol. Refiner*, **30**(11):151 (1951).
99. J. R. Lacher, E. Emery, E. Bohmfalk, and J. D. Park, *J. Phys. Chem.*, **60**:492–495 (1956); J. R. Lacher, H. B. Gottlieb, and J. D. Park, *Trans. Faraday Soc.*, **58**:2348–2351 (1962).
100. A. L. Lydersen, *Univ. Wisconsin Coll. Eng. Expt. Station Report No. 3*, April, 1955.
101. H. Biltz, *Chem. Ber.*, **35**:3524–3528 (1902).
102. O. Ernst and H. Lange (to I. G. Farbenindustrie A.-G.), U.S. Patent 1,833,358, Nov. 24, 1931; *Chem. Abstr.*, **26**:1301 (1932).

103. O. Ernst and H. Lange (to I. G. Farbenindustrie A.-G.), U.S. Patent 1,833,393, Nov. 24, 1931; *Chem. Abstr.*, **26**:1301 (1932).
104. British Patent 298,084 (to I. G. Farbenindustrie A.-G.), Sept. 30, 1927; *Chem. Abstr.*, **23**:2724 (1929).
105. Japanese Patent 158,669 (to Japan Celluloid Co.), Sept. 2, 1943; *Chem. Abstr.*, **43**:7035 (1949).
106. T. Fukushima et al. (to Asahi Chemical Industries Co.), Japanese Patent 6873, Oct. 30, 1951; *Chem. Abstr.*, **47**:11218 (1953).
107. H. Schmitz and H. J. Schumacher, *Z. Physik. Chem.*, **B-52**:72–89 (1942).
108. W. E. Vaughan and F. F. Rust, *J. Org. Chem.*, **7**:472–476 (1942).
109. H. Brintzinger, K. Pfannstiel, and H. Vogel, *Z. Anorg. Chem.*, **256**:75–88 (1948).
110. A. Ya. Yakubovich and A. L. Lemke, *J. Gen. Chem. USSR (Eng. Transl.)*, **19**:649–659 (1949).
111. N. K. Kochetkov, *ibid.*, **23**:777–778 (1953).
112. C. O. Strother and G. H. Wagner (to Linde Air Products Co.), U.S. Patent 2,532,430, Dec. 5, 1950; *Chem. Abstr.*, **45**:2968 (1951).
113. Yu. M. Zinov'ev and L. Z. Soborovskiĭ, *Zh. Obshch. Khim.*, **30**:1571–1573 (1960); *Chem. Abstr.*, **55**:1415 (1961).
114. E. K. von Gustorf, M. C. Henry, and C. Di Pietro, *Z. Naturforsch.*, **21**(1):42–45 (1966).
115. M. S. Kharasch and C. F. Fuchs, *J. Am. Chem. Soc.*, **65**:504–507 (1943).
116. R. N. Haszeldine and B. R. Steele, *J. Chem. Soc.*, **1953**, 1199–1206.
117. J. E. Fearey (to Imperial Chemical Industries Ltd.), U.S. Patent 2,537,337, Jan. 9, 1951; *Chem. Abstr.*, **45**:2504 (1951).
118. British Patent 409,132 (to British Celanese Ltd.), Apr. 26, 1934; *Chem. Abstr.*, **28**:6158 (1934).
119. O. Ernst and W. Berndt (to I. G. Farbenindustrie A.-G.), German Patent 513,679, May 24, 1927; *Chem. Abstr.*, **25**:1841 (1931).
120. L. Schmerling, *J. Am. Chem. Soc.*, **67**:1438–1441 (1945).
121. Sh. G. Sadykhov, Sh. T. Akhmedov, V. A. Soldatova, N. S. Guseinov, S. K. Mustafaeva, M. L. Nikhamkina, and A. K. Sadykhov, *Azerb. Neft. Khoz.*, **45**(1):42–44 (1966); *Chem. Abstr.*, **64**:15758 (1966).
122. T. Hoshino, K. Yamagishi, and Y. Ichikawa, *J. Chem. Soc. Japan, Pure Chem. Sect.*, **74**:510–513 (1953).
123. I. P. Losev and P. I. Pavlovich, U.S.S.R. Patent 64,757, May 31, 1945; *Chem. Abstr.*, **40**:5603 (1946).
124. J. H. Dunn, C. M. Neher, and P. W. Trotter (to Ethyl Corp.), U.S. Patent 2,635,122, Apr. 14, 1953; *Chem. Abstr.*, **47**:7812 (1953).
125. V. Regnault, *Ann. Phys.*, **69**:151 (1838); *J. Pract. Chem.*, **18**:80 (1839).
125a. B. T. Brooks, "The Chemistry of Non-Benzenoid Hydrocarbons," Chemical Catalog, New York, 1922.
126. British Patent 349,872 (to I. G. Farbenindustrie A.-G.), Sept. 17, 1930; *Chem. Abstr.*, **26**:5314 (1932).
127. W. N. Howell (to Imperial Chemical Industries Ltd.), British Patent 534,733, Mar. 17, 1941; *Chem. Abstr.*, **36**:1336 (1942).
128. C. J. Strosacker and F. C. Amstutz (to The Dow Chemical Co.), U.S. Patent 2,322,258, Jun. 22, 1943; *Chem. Abstr.*, **38**:114 (1944).
129. Netherlands Patent 271,762 (to Feldmühle Papier- und Zellstoffwerke A.-G.), Nov. 23, 1961.

130. H. J. Vogt (to Pittsburgh Plate Glass Co.), U.S. Patent 3,065,280, Nov. 20, 1962; *Chem. Abstr.*, **58**:11216 (1963).
131. British Patent 916,407 (to Pittsburgh Plate Glass Co.), Jan. 23, 1963.
132. British Patent 997,357 (to Dynamit-Nobel A.-G.), Jul. 7, 1965.
133. British Patent 1,019,437 (to Produits Chimiques Pechiney-St.-Gobain), Feb. 9, 1966.
134. H. Richtzenhain and P. Riegger (to Feldmühle Papier- und Zellstoffwerke A.-G.), U.S. Patent 3,290,398, Dec. 6, 1966.
135. F. Conrad and M. L. Gould (to Ethyl Corp.), U.S. Patent 2,989,570, Jun. 20, 1961; *Chem. Abstr.*, **56**:9960 (1962).
136. T. Ploetz, H. Richtzenhain, and P. Riegger (to Feldmühle Papier- und Zellstoffwerke A.-G.), German Patent 1,092,903, Nov. 17, 1960; *Chem. Abstr.*, **55**:25754 (1961).
137. Netherlands Patent 69,354 (to Produits Chimiques Pechiney-St.-Gobain), Jan. 15, 1952.
138. Netherlands Patent 67,024 (to Produits Chimiques Pechiney-St.-Gobain), Dec. 15, 1950.
139. The Dow Chemical Company, "Vinylidene Chloride Monomer," 1966.
140. J. E. Francis and L. C. Leitch, *Can. J. Chem.*, **35**:500 (1957).
141. R. C. Reinhardt, *Chem. Eng. News*, **25**:2136 (1947).
142. L. L. McKinney, A. C. Eldridge, and J. C. Cowan, *J. Am. Chem. Soc.*, **81**:1423–1427 (1959).
143. E. R. Levy and D. N. Smith, Jr. (to Chemagro Corp.), U.S. Patent 3,117,069, Jan. 7, 1964; *Chem. Abstr.*, **60**:6751 (1964).
144. Canadian Patent 670,299 (to Pittsburgh Plate Glass Co.), Sept. 10, 1963.
145. H. S. Nutting and M. E. Huscher (to The Dow Chemical Co.), U.S. Patent 2,209,000, Jul. 23, 1940; *Chem. Abstr.*, **35**:140 (1941).
146. Ref. 130.
147. Dutch Patent 268,629 (to Pittsburgh Plate Glass Co.), Jun. 25, 1964.
148. German Patent 523,436 (to I. G. Farbenindustrie A.-G.), Dec. 1, 1929; *Chem. Abstr.*, **25**:3362 (1931).
149. R. E. Bingham (to The General Tire & Rubber Co.), U.S. Patent 3,125,553, Mar. 17, 1964; *Chem. Abstr.*, **60**:14633 (1964).
150. W. D. Harkins, *J. Chem. Phys.*, **13**:381 (1945).
151. W. D. Harkins, *J. Am. Chem. Soc.*, **69**:1428 (1947).
152. W. D. Harkins, *J. Polymer Sci.*, **5**:217 (1950).
153. R. C. Reinhardt, *Ind. Eng. Chem.*, **35**(4):422 (1943).
154. P. K. Isaacs and A. Trofimow (to W. R. Grace & Co.), U.S. Patent 3,033,812, May 8, 1962; *Chem. Abstr.*, **57**:2429–2430 (1962).
155. W. R. R. Park and J. H. Stickelmeyer (to The Dow Chemical Co.), U.S. Patent 3,128,200, Apr. 7, 1964.
156. A. Trofimow and E. C. Dearborn (to W. R. Grace & Co.), U.S. Patent 3,310,514, Mar. 21, 1967; *Chem. Abstr.*, **66**:106000 (1967).
157. P. M. Hay, G. R. Mitchell, and P. P. Salatiello (to Olin Mathieson Chemical Corp.), U.S. Patent 3,041,208, Jun. 26, 1962; *Chem. Abstr.*, **57**:8754 (1962).
158. J. Prat, *Mem. Serv. Chim. Etat (Paris)*, **32**:319–345 (1945); through *Chem. Abstr.*, **42**:4392 (1948).
159. W. I. Bengough and R. G. W. Norrish, *Nature*, **163**:325–326 (1949); through *Chem. Abstr.*, **43**:5272 (1949).

160. W. I. Bengough and R. G. W. Norrish, *Proc. Roy. Soc. (London)*, A200:301–320 (1950); through *Chem. Abstr.*, 45:4962–4963 (1951).

161. E. Jenckel, H. Eckmans, and B. Rumbach, *Makromol. Chem.*, 4:15–40 (1949); through *Chem. Abstr.*, 44:2352 (1950).

162. J. W. Breitenbach and A. Schindler, *Monatsh. Chem.*, 80:429–431 (1949); through *Chem. Abstr.*, 45:899 (1951).

163. H. S. Mickley, A. S. Michaels, and A. L. Moore, *J. Polymer Sci.*, 60:121 (1962).

164. French Patent 854,115 (to Société anon. des manufactures des glaces et produits chimiques de St.-Gobain, Chauny & Cirey), Apr. 5, 1940; also French Patents 976,543 and 985,473; *Chem. Abstr.*, 40:768 (1946).

165. C. H. Bamford, W. G. Barb, A. D. Jenkins, and P. F. Onyon, "The Kinetics of Vinyl Polymerization by Radical Mechanisms," Butterworth, London, 1958, pp. 111–119.

166. V. V. Mazurek, *Polymer Sci. (USSR) (English Transl.)*, 8(7):1292–1298 (1967).

167. J. D. Cotman, Jr., M. F. Gonzales, and G. C. Claver, *J. Polymer Sci., Part A*-1, 5:1137–1164 (1967).

168. *Chem. Eng. News*, 44(24):82 (1966).

169. A. Krause, *Chem. Eng.*, 72(26):72 (1965).

170. F. Berger, Extrusion of Rigid Polyvinyl Chloride, *Ann. Tech. Conf., SPE*, 12:16 (1966).

171. J. C. Thomas, Two Step Bulk Polymerization of Vinyl Chloride, *Ann. Tech. Conf., SPE*, 13:140–145 (1967).

172. M. Thomas (to Produits Chimiques Pechiney-St.-Gobain), French Patent 1,357,736, Apr. 10, 1964; *Chem. Abstr.*, 61:3231 (1964).

173. J. C. Thomas (to Produits Chimiques Pechiney-St.-Gobain), French Patent 1,382,072, Dec. 18, 1964: Additions 84,958; 84,965; 84,966; and 85,672; *Chem. Abstr.*, 62:9259 (1965).

174. J. C. Thomas (to Produits Chimiques Pechiney-St.-Gobain), French Patent 1,427,935, Feb. 11, 1966; Additions 87,148; 87,382; 87,611; and 87,965; *Chem. Abstr.*, 65:9051 (1966).

175. J. C. Thomas (to Produits Chimiques Pechiney-St.-Gobain), French Patent 1,436,744, Apr. 29, 1966; *Chem. Abstr.*, 66:19057 (1967).

176. French Patent 1,450,464 (to Produits Chimiques Pechiney-St.-Gobain), Sept. 19, 1966.

177. *Mod. Plastics*, 44(5):118 (1967).

178. Unpublished data of The Dow Chemical Company.

179. J. A. Allen, *J. Appl. Chem. (London)*, 12:406–412 (1962).

180. P. Agron, T. Alfrey, Jr., J. Bohrer, H. Haas, and H. Wechsler, *J. Polymer Sci.*, 3:157–166 (1948).

181. K. W. Doak, *J. Am. Chem. Soc.*, 70:1525–1527 (1948).

182. E. C. Chapin, G. E. Ham, and R. G. Fordyce, *ibid.*, 70:538–542 (1948).

183. F. M. Lewis, C. Walling, W. Cummings, E. R. Briggs, and W. J. Wenisch, *ibid.*, 70:1527–1529 (1948).

184. F. M. Lewis and F. R. Mayo, *ibid.*, 70:1533–1536 (1948).

185. F. R. Mayo, C. Walling, F. M. Lewis, and W. F. Hulse, *ibid.*, 70:1523–1525 (1948).

186. T. Alfrey, Jr., J. Bohrer, H. Haas, and C. Lewis, *J. Polymer Sci.*, 5:719–726 (1950).

187. M. C. De Wilde and G. Smets, *ibid.*, 5:253–258 (1950).

188. R. Hart and G. Smets, *ibid.*, **5**:55–67 (1950).
189. T. Kimura and K. Yoshida, *Science and Ind.* (*Japan*), **27**:288–290 (1953).
190. T. Kimura and K. Yoshida, *ibid.*, **28**:158–161 (1954).
191. F. R. Mayo, F. M. Lewis, and C. Walling, *J. Am. Chem. Soc.*, **70**:1529 (1948).
192. W. I. Bengough and R. G. W. Norrish, *Proc. Roy. Soc.* (*London*), **A218**:155 (1953).
193. S. Enomoto, *J. Polymer Sci.*, **55**:95 (1961).
194. F. M. Lewis, F. R. Mayo, and W. F. Hulse, *J. Am. Chem. Soc.*, **67**:1701 (1945).
195. P. Agron, T. Alfrey, Jr., J. Bohrer, H. Haas, and H. Wechsler, *J. Polymer Sci.*, **3**:157 (1948).
196. Ref. 191.
197. Ref. 181.
198. C. Walling and J. A. Davison, *J. Am. Chem. Soc.*, **73**:5736 (1951).
199. K. Crauwels and G. Smets, *Bull. Soc. Chim. Belges*, **59**:443 (1950).

2. FLUOROVINYL MONOMERS

L. E. WOLINSKI, *Former Lecturer in Polymer Chemistry, Canisius College, Buffalo, New York*

Contents

I. INTRODUCTION

Four fluorovinyl monomers are described in this chapter—tetrafluoroethylene, vinyl fluoride, vinylidene fluoride, and chlorotrifluoroethylene. These are commercially significant, two-carbon monomers that illustrate the chemistry of vinyl compounds containing the stable carbon-fluorine bond. Other monomers, such as hexafluoropropylene and the fluorinated acrylates, are mentioned briefly but are not discussed in detail.

II. TETRAFLUOROETHYLENE

A. Introduction

Tetrafluoroethylene, $CF_2{=}CF_2$, is the largest-volume fluorinated monomer, accounting for about 85 percent of the total free-world fluoropolymer market of 30 million lb in 1966 (1). Tetrafluoroethylene was first made by O. Ruff (2) in 1933. The basis for its preparation can be traced to the pioneering work of F. Swarts (3), who prepared the first fluorocarbons in the nineteenth century, and to T. Midgley, Jr. and A. L. Henne (4), who utilized the Swarts reaction to prepare fluorocarbons on an industrial scale.

Tetrafluoroethylene was first polymerized in 1938 at the Jackson Laboratory of E. I. du Pont de Nemours and Company. Subsequently a detailed polymerization system was developed (5,6). The intractable nature of the polymer (7) required specialized techniques for fabrication (8). Poly(tetrafluoroethylene) became the choice where applications required chemical inertness, stability through a broad temperature range (-100 to $350°C$), mechanical strength, low dielectric constant, nonflammability, and a low coefficient of friction.

B. Preparation of Monomer

A commercial process for tetrafluoroethylene is the pyrolysis of chlorodifluoromethane, which can be obtained from chloroform by a Swarts reaction (9):

$$CHCl_3 \xrightarrow{\text{HF, SbCl}_5} CHF_2Cl$$

$$2CHF_2Cl \underset{\longleftarrow}{\overset{\Delta}{\longrightarrow}} CF_2{=}CF_2 + 2HCl$$

The second step is an endothermic, homogeneous gas-phase reaction conducted at 650 to 800°C under 0.5 to 4 atm pressure with 0.1 to 0.8 sec contact time in platinum, silver, or carbon tubes. Tetrafluoroethylene is produced in yields of 56 to 97 percent. The yields decrease with increasing pressure (9). Thermodynamic calculations (10) gave a ΔH at 600°C of 30.5 kcal and standard entropy change, ΔS, of 30.1 kcal/(mole) (°C). Using these data, the maximum yield of tetrafluoroethylene was calculated (7) to be 75 percent at 1 atm and 90 percent at 0.1 atm at 800°C. Yields of 89 percent (9,11) were attained with these optimized conditions. Higher yields (95 percent) were obtained by recycling over a copper catalyst at 780°C with a 0.01 sec contact time (12). The use of silicon or ferrosilicon with a catalytic amount of copper has produced tetrafluoroethylene from chlorodifluoromethane at temperatures as low as 250°C (13). Mixtures of CHF_2Cl and $CHFCl_2$ have been pyrolyzed, as have the individual compounds (14–16). Pyrolysis of $CClF_2CHF_2$, $CHF_2CF_2CClF_2$ (15), CHF_2Br (17), and CHF_3 (18) individually produce tetrafluoroethylene.

The pyrolysis of CHF_2Cl yields more than 35 by-products, some of which have been identified (19) as follows: $CF_3CF{=}CF_2$ (bp $-29°C$), perfluorocyclobutane (bp $-5°C$), CHF_2CF_2Cl (bp $-10°C$), $CHF_2 \cdot CF_2CF_2Cl$ (bp $-21°C$), $H(CF_2)_nCl$ ($n = 10$ or more) (9). Toxic perfluoroisobutylene (20) is also obtained. HCl and HF are produced as well.

Reduced pressure favors the desired reaction, presumably because 1.5 moles of products are formed per mole of CHF_2Cl, and it is an equilibrium process. The mechanism is believed to be as shown below (10):

$$CHF_2Cl \rightleftharpoons \; :CF_2 + HCl$$
$$2:CF_2 \rightleftharpoons CF_2{=}CF_2$$

The reduced partial pressure can be attained with nitrogen (14) or steam (21–25) as diluent. Reduction of the pressure is important to obtain maximum conversion, but this must be combined with high temperatures, preferably 800°C or above (7).

The first section of an apparatus used to produce tetrafluoroethylene consisted of two concentric venturi tubes, the inner one receiving 1,000°C steam and the outer 400°C CHF_2Cl (13 kg per hr) at a steam/CHF_2Cl mole ratio of 1:1. The CHF_2Cl was aspirated by the steam and mixed with it at the 6 to 10 mm diameter orifice. Tetrafluoroethylene was formed during the adiabatic expansion at a reaction temperature of 800°C. The reactor ended in a straight section with a cooling reservoir that quenched the exit gases at 150 to 200°C. The steam diluent permits the use of nickel alloys in place of platinum in reactor construction (8). Some hydrolysis to carbon monoxide and HF occurs, but this is a minor side reaction at steam/CHF_2Cl mole ratios of less than 15:1 (8). With superheated steam at 1,000°C, $CHFCl_2$ conversions of 75 percent were obtained (23).

Other techniques for preparing tetrafluoroethylene by dehydrohalogenation of chlorodifluoromethane have been used. Alkaline metal oxides with chlorides stable at the pyrolysis temperature (those of Li, Na, K, and Ca) have produced tetrafluoroethylene at up to 92 percent conversion and 90 percent yield (26,27) from CHF_2Cl. A 100 percent selectivity of tetrafluoroethylene at a 30 percent yield was obtained from $C_2Cl_2F_4$ over water, zinc, and a detergent (28). Zinc dust in an organic solvent has also been used (29). The production of tetrafluoroethylene by pyrolysis of fluorochlorocarbons is accomplished at temperatures as low as 375°C in the presence of a copper-chromium-barium catalyst (30), at 500°C with Te (31), at 525 to 580°C with copper or copper and cobalt on magnesium fluoride (32), or 650 to 750°C using Pt (33,34).

The original method used by Ruff in preparing tetrafluoroethylene (2) by passing CF_4 through a carbon arc has been developed further. For example (36), CF_4 was fed at 0.5 cu ft per min and 30 to 40 mm into the gap between a carbon anode with a 2-in. inner diameter and a carbon cathode with a 0.5-in. outer diameter. The arc was operated at 80 amp and 90 to 100 volts for a total power input of 7.6 kw and was rotated by a field of 100 gauss. A carbon bed constructed of 0.125-in. porous carbon plate and 0.25-in. carbon rods was placed 1 in. from the arc. An 87.4 percent yield of C_2F_4 was obtained corresponding to 10.3 kwhr per lb of CF_4. The possible reaction sequences of the process were discussed (38). Rapid quenching of the products to less than 500°C is necessary for the best yields (35).

Other materials have been pyrolyzed in carbon arcs to produce tetrafluoroethylene. These include binary halogen fluorides (39), fluorine (40), SF_4 (41), P_2F_5 (41–43), anhydrous HF (44), CF_4 and $(CN)_2$ (45), $C_2Cl_2F_4$ (46), Cl_2, N_2, and NaF (47), HgF_2 (48), and COF_2 (49).

Convenient laboratory methods for preparing tetrafluoroethylene are pyrolysis of the polymer, decarboxylation of a salt of a fluorinated carboxylic acid, and debromination of 1,2-dibromotetrafluoroethylene. Heating poly(tetrafluoroethylene) at 600°C at reduced pressure and condensing the products in liquid nitrogen gave 97 percent tetrafluoroethylene (50). Toxic perfluoroisobutylene is a by-product of this reaction. Hexafluorocyclopropane (15 percent yield) and octafluorocyclobutane (5 to 33 percent yield) were obtained under other conditions (51,52). Heating sodium or silver pentafluoropropionate over 200°C (53–55) in glassware produced tetrafluoroethylene up to 97 percent pure. Sodium trifluoroacetate, with NaOH, decarboxylates and forms tetrafluoroethylene (56,57) at 270°C. The free acid, pentafluoropropanoic acid, forms tetrafluoroethylene in 86 percent yield when heated under CO_2 at 650° (58). Difluorodiazirine is converted completely to tetrafluoroethylene on photolysis (59).

Perfluorocyclobutane forms traces of tetrafluoroethylene under xenon photolysis (60). It gives 95 percent tetrafluoroethylene when heated with poly(tetrafluoroethylene) at 600°C (61) and traces when heated alone (62). Exposure of perfluorocyclobutane to radiofrequency electrical discharge produces difluorocarbene ($:CF_2$), which at 95°K forms tetrafluoroethylene (63). Electrolysis of LiF–NaF at 600 to 925°C produced tetrafluoroethylene and CF_4 in a 70:30 ratio. The carbon was supplied by the electrodes (64).

Photolysis of perfluorocyclopropane (65) and perfluorocyclobutanone (66) has produced tetrafluoroethylene, as has decomposition of tetrafluoroethylene oxide (67), cyanogen fluoride (68), or $CF_2ClCOCF_2Cl$ (69). Tetrafluoroethylene was formed by the recombination of the CF_2 biradical from shock-wave decomposition of fluoroform (70) or trifluoromethyl radicals (71). The reduction of CF_4 and COF_2 with carbon monoxide above 350°C in the presence of a metal and its fluoride salt yielded tetrafluoroethylene (72).

Tetrafluoroethylene monomer prepared by the pyrolysis of CHF_2Cl may be purified by washing the cooled reaction products with water to remove HCl and HF, followed by thorough drying, compressing, and distilling to recover unconverted CHF_2Cl (73). Sulfuric acid or calcium chloride is an acceptable drying agent. Silica gel has been reported to initiate polymerization (74) and hence should be handled cautiously when used as a drying agent for tetrafluoroethylene. Several azeotropes may form (8), but they are separable by extractive distillation using toluene.

Extractive distillation with methanol (74–78), dimethylformamide (78), aliphatic perhalocarbons (79), and hexamethylphosphoric triamide (80) has been used for purification of tetrafluoroethylene. Purification with sulfuric acid or sulfur trioxide has given over 98 percent tetrafluoroethylene (81,82). Complexing $CF_2{=}CHCl$ with mercuric acetate removed it from tetrafluoroethylene (83). Zeolites (alumina silicates) of an effective pore diameter of 4Å were used to remove impurities from tetrafluoroethylene by preferential absorption (84,85).

C. Properties

1. Physical Properties

The important physical properties of tetrafluoroethylene are listed in Table 1. It is a colorless, odorless gas that may explode under certain conditions and thus requires special care in the design and safe operation of equipment. It may be stored or distilled at low temperatures. The vapor density has been compared (8) with theoretical values (90). For example, at 230 psia the actual values were 0.092 g per ml at 0°C and 0.064 g per ml at 60°C; comparable ideal-gas values were 0.070 and 0.057.

Ideal-gas entropies, specific heat, heat-content function, and free-energy function have been calculated (91,92). Tetrafluoroethylene has a complex infrared spectrum (93). There is a shortening and weakening of the C=C bond in tetrafluoroethylene when compared with ethylene. The bond lengths and bond-force constants for C_2H_4 and C_2F_4 are 1.33 and 1.31Å, and 9.6 and 9.2 × 10^5 dynes/cm, respectively. This may be due to a reduction of electron density caused by the withdrawal of electrons by the

TABLE 1

Physical Properties of Tetrafluoroethylene

Property	Value	Reference
Boiling point at 760 mm, °C	−76.3	86
Freezing point, °C	−142.5	86
Liquid density at t°C, g/ml:		
−100 < t < −40	$1.202 - 4.14 \times 10^{-3}t$	86
−40 < t < 8	$1.1507 - 6.935 \times 10^{-3}t$	
	$- 3.76 \times 10^{-5}t^2$	86
8 < t < 30	$1.1325 - 2.904 \times 10^{-3}t$	
	$- 2.556 \times 10^{-4}t^2$	86
Vapor pressure at T°K, psia:		
196.85 < T < 273.15	$\log_{10} P = 5.6210 - 875.14/T$	86
273.15 < T < 306.45	$\log_{10} P = 5.5906 - 866.84/T$	86
Critical temperature, °C	33.3	86
Critical pressure, psia	572	86
Critical density, g/ml	0.58	86
Dielectric constant at 28°C, ϵ:		
At 15 psia	1.0017	86
At 125 psia	1.015	86
Thermal conductivity at 30°C, cal/(sec)(°C/cm)(cm²)	3.7×10^{-5}	86
Heat of formation for ideal gas at 298.15°K, ΔH, kcal/mole	−151.9	87
Heat of polymerization at 298.15°K to solid polymer, ΔH, kcal/mole of TFE	−41.12	87
Flammability limits in air, percent by volume	14–43	8
Ignition temperature, °C	620	8
Molal heat of fusion at 142 ± 0.01°K, abs joules/mole	7,714.5 ± 0.2	88
Molal heat of vaporization at 197.53 ± 0.01°K, abs joules/mole	16,821 ± 4	88
Entropy at 197.53°K, 1 atm, abs joules/deg mole	270.06 ± 0.37	88
Ionization potential, ev	10.12	89

fluorine atoms and the consequent increased repulsion between the carbon nuclei (94–96). Diamagnetic susceptibilities (97), the action of ultrasonic waves (98), and nuclear-magnetic resonance studies on polymer (99,100) and monomer (101) have also been reported.

The solubility of tetrafluoroethylene in water from 0 to 70°C and at 150 to 600 mm increases linearly with pressure. The measurements are made by holding the temperature constant and varying the pressure. The heat of solution decreased from 7.5 to 2.0 kcal per mole as the temperature increased from 0 to 70°C (102). Specific values for solubilities in various solvents have been listed (8) in grams per 100 g of solvent as follows: 0.1 in water at 115 psia, 20°C; 0.25 in a 0.5 percent aqueous solution of ammonium perfluorooctanoate, a surfactant useful in polymerization; 0.6 in triethylene glycol; 6.8 in perfluorotributylamine; and 20 in acetone.

2. TOXICOLOGY

Tetrafluoroethylene monomer is comparatively nontoxic, with a lethal concentration (LC_{50} for rats exposed for 4 hr) of 40,000 ppm by volume (19,103). Preparation of tetrafluoroethylene by the pyrolytic methods previously discussed frequently produces small quantities of hexafluoropropylene (LC_{50} of 3,000 ppm) and perfluoroisobutylene (LC_{50} of 0.5 ppm) (19). The presence of the latter contaminant may account for the reported effect of tetrafluoroethylene on industrial workers (104). Threshold values for tetrafluoroethylene have not been established (105); however, exposure to the monomer should be avoided.

3. ANALYTICAL METHODS

Precise and accurate analytical techniques for determining purity are required. The infrared (93) and ultraviolet spectra (106) are not satisfactory as analytical tools for tetrafluoroethylene. Gas-liquid and gas-solid chromatography are of greater value for quality control and various such systems have been described (107–109). Molecular weights of the insoluble and infusible polymer were determined with radioactive techniques, using ^{35}S in nonhydrolyzable end groups derived from oxidative bisulfite initiation (110). Fluorine in poly(tetrafluoroethylene) has been determined, using kinetic analysis of the decay of the γ radiation after γ activation (111).

D. Storage and Handling

The flammability and explosive hazards of tetrafluoroethylene necessitate extreme care in its storage and handling. Minute thermal inputs are sufficient to ignite and explode large volumes of tetrafluoroethylene.

Tetrafluoroethylene disproportionates to carbon and carbon tetrafluoride ($\Delta H = 65.9$ kcal per mole) (8), liberating about the same amount of energy as a black-powder explosion. This disproportionation can be initiated by a variety of thermal sources. There is no known way to express either the threshold temperature or minimum ignition energy required. The danger of this disproportionation leading to an explosion increases with an increase in the loading density (number of grams of tetrafluoroethylene per milliliter of vessel volume). This is illustrated in Table 2 for monomer that was initially at 0 to 15°C (8). The explosive hazards are so great that remote-control barricaded areas are used for its preparation, purification, and polymerization.

The explosive degradation can occur during polymerization by the following general mechanism. The highly exothermic heat of polymerization, combined with the low thermal conductivity of the polymer, can produce localized overheating in the polymer particle. This, in turn, can lead to the more violent disproportionation reaction. Controlled polymerizations are normally carried out in water or other liquid media for more efficient heat transfer.

The danger of accidental polymerization is minimized by the rigorous exclusion of oxygen, which can act as an initiator, by maintaining low temperatures to decrease the rate of polymerization, by preventing localized hot spots in the reaction vessel due to poor heat transfer, and by the use of properly designed valves and piping. As an additional safeguard against free-radical polymerization, stored monomer is inhibited with a terpene such as d-limonene. Stabilizers, such as tetrahydronaphthalene, 1-octene, methyl methacrylate, and γ-pinene, have also been used (112). They function as free-radical traps to quench any radicals that could initiate and

TABLE 2

Pressure Developed in a 240-ml Vessel as a
Result of Exploding Tetrafluoroethylene

Tetrafluoroethylene loading density, g/mil	Pressure developed on explosion, psia
0.07	1,800
0.6	24,000
0.8	44,000
1.1	70,000[a]

[a] Vessel ruptured before maximum pressure developed.

propagate the polymerization. They do not prevent explosive disproportionation.

E. Chemical Reactions

The chemical reactions of tetrafluoroethylene are influenced greatly by
the presence of the strongly electronegative fluorine atoms that cause the
electron density at the olefinic carbons to be decreased significantly. Thus
tetrafluoroethylene reacts readily with nucleophilic reagents and has little
tendency to react with electrophilic reagents.

Illustrative examples of this generality are the base-catalyzed additions
of alcohols, thiols, phenols, thiophenols, and ketoximes to tetrafluoroethylene (113). The postulated mechanism is as follows:

$$B^- + CF_2{=}CF_2 \longrightarrow BCF_2CF_2^- + BH \longrightarrow BCF_2HCF_2 + B^-$$
$$B{=}C_2H_5, \quad C_6H_5O, \quad C_6H_5S, \text{ etc.}$$

Typical reactions are given in Table 3. In every instance the catalyst was
the sodium salt of the starting compound.

Earlier workers used more severe conditions or obtained poorer yields
in the absence of the preferred solvent, dimethylformamide (114–118).
The presence of carbanions from tetrafluoroethylene and sodium alkoxides
was verified by carrying out the reactions in the presence of esters to trap
the intermediates (119).

The use of dimethylformamide facilitates the reaction of certain weakly
basic secondary amines and amides with tetrafluoroethylene in the
presence of the alkali metal salts of the nitrogenous compounds (113),

TABLE 3
Nucleophilic Addition Reactions of Tetrafluoroethylene [113]

Starting compound	Solvent	Product	Yield, percent
t-Butyl alcohol	DMF	$(CH_3)_3COCF_2CF_2H$	55
2,2-Dimethyl-1,3-propanediol	Dioxane	$(CH_3)_2C(CH_2OCF_2CF_2H)_2$	75
Pentaerythritol	DMF	$C(CH_2OCF_2CF_2H)_4$	65
Phenol	DMF	$C_6H_5OCF_2CF_2H$	50
Hydroquinone	Dioxane/ DMF 1:1	$1,4\text{-}(HCF_2CF_2O)_2C_6H_4$	58
Cyclohexanone oxime	Dioxane	$HCF_2CF_2ON{=}C_6H_{10}$	44
2-Ethylhexyl mercaptan	Dioxane	$(C_2H_5)(C_4H_9)CHCH_2SCF_2CF_2H$	83
Mercaptoacetic acid	$(Et)_3N$	$HCF_2CF_2SCH_2COOH$	52
Thiophenol	Dioxane	$C_6H_5SCF_2CF_2H$	91

TABLE 4

Addition of Weakly Basic Secondary Amines and Amides to Tetrafluoroethylene in the Presence of Their Alkali Metal Salts [113]

Starting compound	Solvent	Product	Yield, percent
Pyrrole	None	$HCF_2CF_2N(C_4H_4)$	74
Phenothiazine	DMF	$HCF_2CF_2N(C_{12}H_8S)$	69
Diphenylamine	DMF	$(C_6H_5)_2NCF_2CF_2H$	63
Caprolactam	DMF	$HCF_2CF_2N(COC_5H_{10})$	29
Acetanilide	$(CH_3OCH_2)_2$	$C_6H_5N(COCH_3)(CF_2CF_2H)$	51

although dimethylformamide is not needed for pyrrole or acetanilide. Typical reactions are listed in Table 4. Strongly basic secondary amines do not require alkali metal salt catalysis. The kinetics of this reaction has been reported (120). The reaction is very exothermic with strongly basic amines and requires a solvent, in some cases, to control the reaction. The tetrafluoroethylene secondary amine adducts undergo a displacement with HCN at 0° to give α,α-dicyanoamines (113):

$$(C_2H_5)_2NH \xrightarrow[< 60°]{\text{tetrafluoroethylene}} (C_2H_5)_2NCF_2CHF_2 \xrightarrow[0°]{HCN}$$
$$(77\%)$$

$$(C_2H_5)_2NC(CN)_2CHF_2$$

Tetrafluoroethylene has been reacted with piperidine and the product hydrolyzed to give an 87 percent yield of difluoroacetic acid (121). N-Substituted fluoroacetamides were made by heating tetrafluoroethylene with primary amines (113,122). The amine may be aliphatic, aromatic, or cycloaliphatic:

$$CF_2{=}CF_2 + RNH_2 \longrightarrow CHF_2CF_2NHR \xrightarrow{-HF} CHF_2CF{=}NR \xrightarrow{RNH_2}$$

$$CHF_2CF(NHR)_2 \xrightarrow{-HF} CHF_2C({=}NR)NHR \xrightarrow{H_2O} CHF_2CONHR$$

When tetrafluoroethylene reacts with a secondary amine (113,122) and the reaction product is hydrolyzed, N,N-disubstituted α,α-difluoroacetamides are obtained:

$$CF_2{=}CF_2 + R_2NH \xrightarrow{< 60°} CHF_2CF_2NR_2 \xrightarrow{H_2O} (HCl \text{ or } NaOH)$$

$$[2HF + CHF_2C(OH)_2NR_2] \xrightarrow{-H_2O} CHF_2CONR_2$$

The ready removal of the fluorine atoms from CF_2 next to a nitrogen atom substituted with alkyl, aryl, or acyl groups is typical. This reactivity of tetrafluoroethylene is also shown by its hydrolysis and reaction with HCN and H_2S (113).

Anhydrous ammonia (-30 to $50°C$) reacts with tetrafluoroethylene to yield 2,4,6-tris(difluoromethyl)-s-triazine (123,124):

$$CF_2{=}CF_2 + NH_3 \longrightarrow [CHF_2CF_2NH_2] \xrightarrow{-2HF} CHF_2CN \longrightarrow$$
$$\text{2,4,6-tris(difluoromethyl)-s-triazine}$$

Hydrazines reacted with tetrafluoroethylene to produce s-tetrazines (125). Monohydroperfluoroalkylbis(dimethyl)hydrazines were formed from tetrafluoroethylene and dimethylhydrazine (126).

Phenyllithium and tetrafluoroethylene refluxing in ether produced difluorostilbene ($C_6H_5CF{=}CFC_6H_5$) and $(C_6H_5)_2C{=}CFC_6H_5$ (127). The results suggest that this is also a nucleophilic addition reaction. A 50 percent yield of difluorostilbene without the trisubstituted product was reported at $-80°$ (128):

$$C_6H_5Li + CF_2{=}CF_2 \longrightarrow C_6H_5CF_2CF_2Li \xrightarrow{-LiF}$$
$$C_6H_5CF{=}CF_2 \xrightarrow{C_6H_5Li} C_6H_5CFLiCF_2C_6H_5 \xrightarrow{-LiF}$$
$$C_6H_5CF{=}CFC_6H_5 \xrightarrow{C_6H_5Li} (C_6H_5)_2CFCFLiC_6H_5 \xrightarrow{-LiF}$$
$$(C_6H_5)_2C{=}CFC_6H_5$$

Other organometallic-tetrafluoroethylene reactions are listed in Table 5.

Formaldehyde adds to tetrafluoroethylene (129) as follows:

$$CF_2{=}CF_2 + CH_2O + H_2O \longrightarrow [HOCH_2CF_2CF_2OH] \longrightarrow$$
$$HOCH_2CF_2COOH + 2HF$$

Electrophilic agents do not add to tetrafluoroethylene readily, the ease of addition to an ethylenic double bond decreasing as the number of fluorine substituents increase (130). Additions that have been reported are as follows (129,130):

$$CF_2{=}CF_2 + NO_2Cl \xrightarrow{60°} CF_2ClCF_2NO_2$$
$$CF_2{=}CF_2 + I_2 \longrightarrow ICF_2CF_2I$$
$$CF_2{=}CF_2 + N_2O_4 \longrightarrow O_2NCF_2CF_2NO_2$$

Ionic mechanisms have been proposed for several additions. Nitrosyl fluoride reacted to form an oxazetidine (131,132) as well as NO_2CF_2COF and $(CF_2NO_2)_2$ (132):

$$CF_2{=}CF_2 + NOF \longrightarrow CF_3CF_2N{=}O \xrightarrow{CF_2=CF_2} CF_3CF_2N{-}O$$
$$\qquad\qquad\qquad\qquad\qquad\qquad\qquad\qquad F_2C{-}CF_2$$

Nitrosyl chloride forms mainly the nitrochloride (132):

$$CF_2{=}CF_2 + NOCl \longrightarrow CF_2ClCF_2NO_2 + (CF_2Cl)_2 + CF_2ClCOF$$
$$(62\%) \qquad\quad (36\%) \qquad\quad (2\%)$$

TABLE 5

Organometallic-Tetrafluoroethylene Reactions

Organometallic	Solvent	Product	Yield, percent	Reference
Na diethylmalonate	Dioxane	$(COOC_2H_5)_2CHCF=CFCH(COOC_2H_5)_2$	48.5	113
Na diphenylacetonitrile	Dioxane	$(C_6H_5)_2C(CN)CF=CFC(CN)(C_6H_5)_2$	46	113
Na 1-hexyne	$(CH_3OCH_2)_2$	$C_4H_9C{\equiv}CCF=CFC{\equiv}CC_4H_9$	30	113
Na hexaphenyldisilane	Ether	$(C_6H_5)_3SiCF=CFSi(C_6H_5)_3$	7	113
Ethyllithium	Ether	$C_2H_5CF=CF_2$	20	128
2-Thienyllithium	Ether	$(C_4H_3S)CF=CF(C_4H_3S)$	30	128
2-Pyridinyllithium	Ether	$(C_5H_4N)CF=CF(C_5H_4N)$	30	128
Mesityllithium	Ether	2,4,6-trimethyl-1′,2′,2′-trifluorostyrene	45	135

Carbonyl fluoride adds to tetrafluoroethylene to give the perfluoroacyl fluoride (133):

$$CF_2{=}CF_2 \xrightarrow{COF_2} CF_3CF_2COF \quad (13\% \text{ conversion})$$

Dioxygen difluoride (O_2F_2) reacts with tetrafluoroethylene at $-196°$ to form COF_2, CF_4, C_2F_6, SiF_4, and CF_3OOCF_3 (134).

The reaction of NO with tetrafluoroethylene has the characteristics of a free-radical rather than an ionic mechanism (136–138). Two products were reported, $ONCF_2CF_2NO$ and $ONCF_2CF_2NO_2$. With a Lewis acid, $FeCl_3$, the main product from NO and tetrafluoroethylene was $ClCF_2CF_2NO$ in 71 percent yield (139). Oxygen difluoride reactions have been reviewed (140). Pentafluorosulfur hypofluorite (SF_5OF) reacted quantitatively with tetrafluoroethylene with heterolysis of the O—F bond to produce $F_3CCF_2(OSF_5)$ (141). Fluorine nitrate (NO_3F) reacts almost quantitatively to form carbonyl fluoride and CF_3NO_2, the latter probably a decomposition product of the intermediate $C_2F_5ONO_2$ (143). N_2O_4 reacts explosively without a solvent, but with solvents the reaction is controllable, producing $(CF_2NO_2)_2$ (144). An ionic mechanism was suggested for nitrofluorination, using HF and fuming nitric acid at $-10°C$ to form $CF_3CF_2NO_2$. The $NO_2{}^+$ ion adds, followed by attack of F^- ion on the intermediate complex (145).

Various complexes of tetrafluoroethylene have been formed by an ionic route. Examples are $(CH_3)_3SnCF_2CF_2M_n(CO)_5$ (146), $[(C_2H_5)_3P]_2 \cdot PtCl(CF_2{=}CF_2)$ (147), $[(NC)_5CoCF_2CF_2H]^{3-}$ (148), various complexes of rhodium, iridium, and nickel (149), and cyano and cyanohydrido complexes of cobalt (150). On reaction with tetrafluoroethylene, one ethylene is retained on 2,4-pentanedionatobis(ethylene)rhodium, but the ethylene can be displaced by phosphines, amines, nitriles, or cyanide ions, although the tetrafluoroethylene is not replaced (151). On reaction of tetrafluoroethylene with (π-$C_5H_5)_2CO$, the $(CF_2)_2$ group bridges the endo positions of two π-cyclopentadienylcobaltcyclopentadiene units (152). Iron carbonyls also complex with tetrafluoroethylene (152).

Perfluoroalkylnitroso compounds react with tetrafluoroethylene to form an oxazetidine (153) at and above room temperature:

$$CF_3NO + C_2F_4 \longrightarrow \begin{array}{c} CF_3N{-}O \\ | \quad | \\ F_2C{-}CF_2 \end{array}$$

At lower temperatures (0 to $-45°$) a 1:1 copolymer is formed (154) of the following structure:

$$\begin{array}{c} (NOCF_2CF_2) \\ | \\ R \end{array}$$

Even if R is vinyl, the same results are obtained, the polymerization taking place through the nitroso group rather than the vinyl (155).

Tetrafluoroethylene, in the presence of CsF in diethylene glycol dimethyl ether, adds to hexafluoroacetone. The reaction involves the attack of the anion $CF_2CF_2^-$ at the carbonyl carbon to form the perfluoropentyl-alcohol (156). Tetrafluoroethane sulfonate, sulfonic acid, and sulfonyl chloride were prepared (157), starting with sodium sulfite. Tetrafluorobis-(fluorosulfonate)ethane was formed at room temperature with peroxydisulfuryl difluoride (158). The perfluorothiolane was obtained by reacting tetrafluoroethane with sulfur and iodine (159).

$$
\begin{array}{c}
F_2 \overset{\displaystyle F_2}{\underset{\displaystyle}{\boxed{}}} F_2 \\
F_2 \quad S \quad F_2
\end{array}
$$

Free-radical reactions with tetrafluoroethylene proceed smoothly. The most important is polymerization, which will be considered in a later section. The many other free-radical reactions studied have been aimed at producing unique perfluoroalkyl compounds.

Comparison of the reactivities of tetrafluoroethylene toward free radicals showed that tetrafluoroethylene is 10 times more reactive than ethylene toward methyl radicals (160), 1.3 times more reactive toward trichloro-methyl radicals (161), but less reactive toward trifluoromethyl radicals (162). Methylene, produced by photolysis of ketene, reacted with tetra-fluoroethylene to yield 1,1,3,3-tetrafluoropropene and tetrafluorocyclo-propane (163). Table 6 lists several typical free-radical reactions of tetrafluoroethylene.

Another important free-radical reaction of tetrafluoroethylene is telo-merization. The products are useful as heat-transfer agents, greases, lubricants, and detergents, the specific application depending on the telo-gen. Examples of telomers are given in Table 7. The variation of n, the degree of polymerization (see Table 7), determines the physical properties of the telomers. In many instances further modification is possible through reactions of functional groups. This greatly increases versatility and pro-vides many unique structures.

Cyclization of tetrafluoroethylene with heat produces octafluorocyclo-butane (188–190). At high temperature (550 to 700°C) hexafluoropropylene and perfluoroisobutylene are formed. Above 700°C, the major products are perfluoroethane and nonvolatiles (190). A proposed mechanism in-volves the formation of a diradical intermediate that subsequently dimerizes (191):

$$
2CE_2{=}CF_2 \xrightarrow{200°}
\begin{bmatrix} CF_2{-}CF_2 \cdot \\ | \\ CF_2{-}CF_2 \cdot \end{bmatrix}
\longrightarrow
\begin{array}{c} CF_2{-}CF_2 \\ | \quad\quad | \\ CF_2{-}CF_2 \end{array}
$$

TABLE 6

Free-radical Reactions of Tetrafluoroethylene

Reactant	Conditions	Products	Reference
H_2S	hv	HCF_2CF_2SH, $(HCF_2CF_2)_2S$	164
HBr	hv	CF_2BrCF_2H, $CF_2Br(CF_2)_3H$, $CF_2Br(CF_2)_5H$	165
HCl	300°C	F_2ClCCF_2H	166
PH_3	hv	$CHF_2CF_2PH_2$	167
Cl_2	hv	CF_2ClCF_2Cl	168
I_2	150°C	CF_2ICF_2I	169
S_2Cl_2	120°C	$(CF_2ClCF_2)_2S_2$, $(CF_2ClCF_2)_2S$	170
$(CH_3)_3SnSn(CH_3)_3$[a]	hv	$(CH_3)_3SnCF_2CF_2Sn(CH_3)_3$	171
NO, Cl_2	hv	CF_2ClCF_2NO	
N_2O_4		$O_2NCF_2CF_2NO_2$ + $O_2NCF_2CF_2ONO$	

hv = actinic light.

[a] This type of reaction has been successful without free-radical initiators (172,173).

TABLE 7

Telomers of Tetrafluoroethylene

Telogen	Catalyst	Product	Reference
$HSiCl_3$[a]	hv	$H(CF_2CF_2)_nSiCl_3$, $n = 1, 2, 3$	174
$H_2Si(CH_3)_2$	hv	$(CH_3)_2SiH(C_2F_4)_nH$	175
CF_3I	hv	$CF_3(CF_2CF_2)_nI$	178
C_2F_5I	hv	$C_2F_5(CF_2CF_2)_nI$	179
$HPO(OC_2H_5)_2$	Benzoyl peroxide	$H(C_2F_4)_nPO(OC_2H_5)_2$	180
C_4H_{10}	Benzoyl peroxide	$H(C_2F_4)_4C_4H_9$	181
Dioxane	Peroxide	$H(C_2F_4)_nO[CH(CH_2)_3]O$	182
CH_3OH	γ	$H(C_2F_4)_nCH_2OH$	183
SF_5Cl	Peroxide	$Cl(C_2F_4)_nSF_5$	184
$(CF_3)_2CFI$	Peroxide	$(CF_3)_2CF(CF_2CF_2)_nI$	185
CCl_4	$Fe(CO)_5$	$CCl_3(C_2F_4)_nCl$	186
$Cl_2C{=}CCl_2$	γ	$Cl_2C{=}CCl(C_2F_4)_nCl$	187

[a] Chloroplatinic acid catalyzes this reaction even in the absence of hv (176–177).

Tetrafluoroethylene usually codimerizes with nonfluorinated olefins more readily than it dimerizes with itself. Four-membered rings are favored over six-membered rings. Illustrations of these cyclization reactions of tetrafluoroethylene were reported (192) for the general equation:

$$C_2F_4 + CH_2{=}CR_1R_2 \longrightarrow \begin{array}{c} R_1R_2C-CH_2 \\ | \quad\quad | \\ F_2C-CF_2 \end{array}$$

where R_1 = H, CH_3, CH_2Cl, Cl

R_2 = H, CH_3, Cl, CN, $COOCH_3$, CHO

An aryl derivative is formed at 175° (193):

$$C_6H_5CH{=}CH_2 \xrightarrow[175^\circ]{13\ hr} \begin{array}{l}(C_6H_5)HC{-}CF_2\\ \quad\ \ |\qquad |\\ \quad\ H_2C{-}CF_2\end{array} + C_2F_4$$

When tetrafluoroethylene reacts with cyclopentadiene, a Diels Alder adduct is obtained as well as the cyclobutane. This adduct is rearranged thermally and hydrolyzed to form tropolone in a 20 percent overall yield (194):

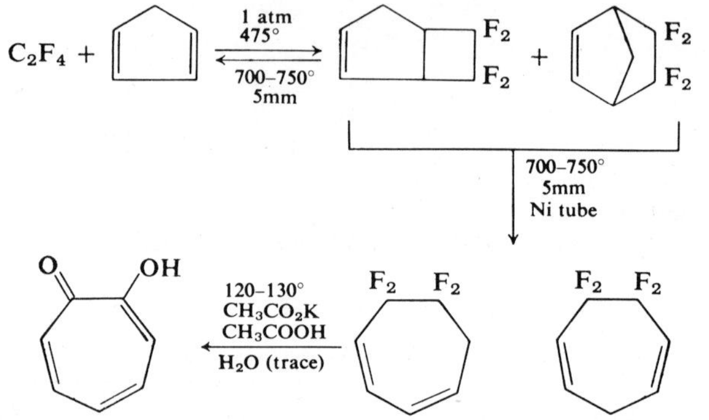

Acetylenes cyclize with tetrafluoroethylene, and subsequent pyrolytic ring openings produce dienes (195):

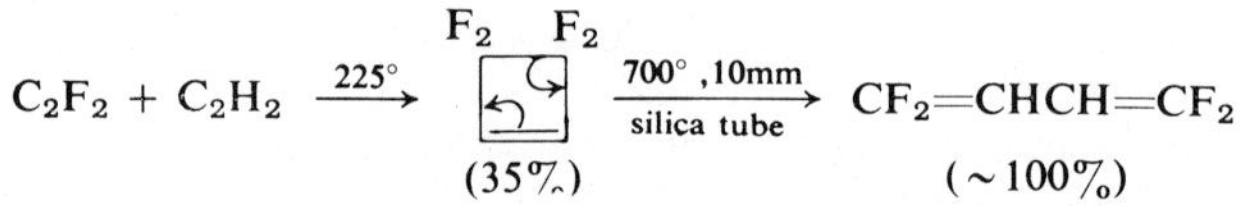

The cyclization reactions must be run with caution, since tetrafluoroethylene can polymerize explosively. In many cases an inhibitor, such as *d*-limonene or hydroquinone, is used.

Ultraviolet radiation initiates the addition of acetaldehyde to tetrafluoroethylene to give low yields of methyl tetrafluoroethylketone and, by codimerization through the double bonds, the oxetanes (196):

$$CH_3CHO + C_2F_4 \xrightarrow{h\nu} \underset{(7.2\%)}{CH_3COCF_2CHF_2} + \underset{(2.8\%)}{\begin{array}{l}F_2C{-}CF_2\\ \ \ |\qquad |\\ \ \ O{-}CHCH_3\end{array}}$$

Azobisformates have cyclized with tetrafluoroethylene (197):

$$C_2H_5OOCN{=}NCOOC_2H_5 + C_2F_4 \xrightarrow[150^\circ]{Ag} \begin{array}{l}C_2H_5O_2CN{-}NCO_2C_2H_5\\ \qquad\qquad |\qquad |\\ \qquad\quad\ F_2C{-}CF_2\end{array}$$

The nitronitroso compound obtained by the reaction of NO with tetra-fluoroethylene forms a cyclic structure with more tetrafluoroethylene (198):

$$O_2NCF_2CF_2NO + C_2F_4 \longrightarrow O_2NCF_2CF_2N\text{---}O$$
$$F_2C\text{---}CF_2$$

Kinetic data are available for the dimerization of tetrafluoroethylene (199) and its codimerization with chlorotrifluoroethylene (200). Thermal cyclization of tetrafluoroethylene is not the only reaction that occurs with heat. Perfluoropropylene is obtained in 72 percent conversion and 81.5 percent yield by heating tetrafluoroethylene at 40 mm to 750 to 810°C, followed by condensing the effluent gas with liquid nitrogen (201). The thermal stability of tetrafluoroethylene is such that the carbon-carbon bond ruptures thermally before the carbon-fluorine bonds when the monomer is heated on a platinum wire to 1000 to 1450°C (202).

Reduction of tetrafluoroethylene over palladium produces only tetra-fluoroethane, but over nickel some trifluoroethane is also obtained (203,204):

$$H_2 + C_2F_4 \xrightarrow[90°C]{Pd\text{--}Al_2O_3} CHF_2CHF_2$$
$$(97.3\% \text{ yield})$$

$$H_2 + C_2F_4 \xrightarrow[90°C]{Ni} CHF_2CHF_2 + CHF_2CH_2F$$
$$(66\% \text{ yield}) \quad (14\% \text{ yield})$$

When tetrafluoroethylene was hydrogenated over palladium on charcoal at 240°, HF was one of the products (205).

Diborane reduces tetrafluoroethylene to ethylene and then adds to form mono- and diethylboron fluorides. Vinylidene fluoride, trifluoroethylene, and 1,2-difluoroethylene were also isolated (206). Arsine adds to tetra-fluoroethylene in the absence of oxygen to give $CHF_2CF_2AsH_2$ and $(HCF_2CF_2)_2AsH$ (207).

Oxidation of tetrafluoroethylene with ozone at 100° in the gas phase produced COF_2 in 57.4 percent conversion and 70.8 percent yield with CF_3COF as the principal by-product (208). This formation of COF_2 rather than the ozonide has been confirmed in another laboratory (209). Oxygen introduced into a steel vessel containing tetrafluoroethylene destabilized the monomer and led to the formation of a shock-sensitive powder and an oxygen-containing film that decomposed thermally to COF_2 and CO_2 (210). Photosensitized tetrafluoroethylene formed hexa-fluorocyclopropane in the absence of oxygen (211) but COF_2, $(CF_2)_2O$, and a liquid polyperoxide in the presence of oxygen (212). Shock-wave treatment of tetrafluoroethylene and oxygen diluted with argon produced

CO_2 and CF_4 (214). Explosive oxidation at 200 to 450°C was studied visually (215). Oxidation of tetrafluoroethylene with oxygen difluoride gave an equimolar mixture of CF_4 and COF_2 (216).

F. Polymerization

1. INDUSTRIAL

Polymerization of tetrafluoroethylene is conducted in water at moderate temperature (40 to 80°C) and pressure (50 to 400 psig) with water-soluble free-radical initiators (8). Explosive decomposition of the monomer to carbon and carbon tetrafluoride is prevented by using the water to remove the high heat of polymerization (-41 kcal per mole). Initiators used include persulfates (217), organic peroxides (218) such as disuccinic acid peroxide (219), peroxides from ketones and aldehydes (220), amine oxides (221), and reduction-oxidation systems (222). High average molecular weight, greater than 10^6, is required to obtain a polymer that is self-supporting at sintering temperatures of 380°C. Consequently chain transfer, or inhibition due to organic additives, must be avoided. The problem is decreased as the water solubility of the organic dispersant decreases (223). Perfluorinated alkyl phosphates (224) have been used as dispersants. Nonfluorinated dispersing agents usually inhibit polymerization unless they are added after polymerization has been initiated (225).

Two general polymerization systems are used, dispersion and granular. In a batch polymerization, a stirred autoclave is used that is evacuated and charged with oxygen-free water containing the initiator, buffers, and dispersants (where used). Tetrafluoroethylene is then added. As the reaction proceeds, additional monomer is introduced to maintain the pressure. Controlled agitation is maintained throughout the polymerization. If agitation is discontinued, the reaction stops in dispersion polymerization and dies slowly in granular polymerization. Stability of dispersions has been improved by using a hydrocarbon with over 12 carbons in the polymerization (226). These dispersions can be concentrated by evaporation under reduced pressure to 60 percent solids. Granular polymerizations normally use an alkaline buffer (8) and produce polymer particles of varying shape and size. Granular polymer is sold as a dry powder with a mean particle size of 0.5 to 30 mils.

2. NONINDUSTRIAL

Polymerization of tetrafluoroethylene should not be attempted in the laboratory unless the hazards discussed earlier are fully appreciated. Several lives were lost in early work (227).

Fluorinated solvents, such as perfluorodimethylcyclobutane, with a fluorinated peroxy catalyst have been used in place of water as a heat-transfer and polymerization medium (228). Carbon tetrachloride was the solvent in a continuous polymerization process, producing a telomer wax (229). A variety of initiation systems has been used. These include an alkyl aluminum with a Friedel-Crafts compound (230) or with a titanate orthoester (231), triisobutylborane (232), chlorodiisobutylaluminum on silica (233), and activated SiO_2–Al_2O_3 (234). Irradiation of tetrafluoroethylene with light of wavelength 2536 Å produces some poly(tetrafluoroethylene), although hexafluoropropylene is the main product (235). Photoactivation, however, produces poly(tetrafluoroethylene) preferentially in the presence of light-absorbing free-radical generators, such as perfluoroazoalkanes (236), ketoacids (for example, ketopimelic acid) (237), and the chlorine atom from phosgene irradiation (238). Gamma irradiation from ^{60}Co has initiated polymerization of tetrafluoroethylene in solution and bulk (239) at 15 to 100°C. The kinetics of gamma-irradiated tetrafluoroethylene polymerization was determined at 4.2°K (240).

G. Applications

Poly(tetrafluoroethylene) is available in several forms: (1) granular polymers, 10 to 700 μ in particle size, (2) dried dispersion polymer with 0.1 μ primary particle agglomerates of 450 μ, (3) liquid dispersions with 0.16 μ particles at 60 percent solids, and (4) filled polymer containing fiber glass, powdered metals, asbestos, and molybdenum sulfide. Special techniques have been developed for the fabrication of poly(tetrafluoroethylene). Granular polymer is used for molding and extrusion. In molding, it is preformed by compression into the desired form and then sintered at 380°C. Carefully programmed heat cycles are required to obtain structurally sound moldings. Extrusion of granular polymer is accomplished by compressing presintered powder with a ram or a screw and sintering the compressed powder in a long die, 90:1 L/D ratio. The product is either machined to the finished dimensions or reheated and compressed in a finishing mold. Machining is more generally applicable, since reheating is limited to fairly small items. Skived tapes are machined from "logs" prepared by sintering granular polymer. Table 8 lists typical properties of 2-mil film.

Dispersion polymers are mixed with hydrocarbon lubricants (25 percent) and compressed into a preform. The lubricant is removed by evaporation or extraction, and then the preform is sintered. The removal of the lubricant limits the thickness of cross sections to about $\frac{1}{4}$ in. Wire coating and tapes are formed by this technique with final thicknesses as low as 2 mils as a practical minimum, although thinner ones have been made.

TABLE 8

Typical Properties of 2-mil Poly(tetrafluoroethylene) Films Cast from Dispersion Resin

Property	Units	ASTM method	Value
Specific gravity, 77°F	g/cc	D792-50	2.1–2.3
Tensile strength, 77°F	psi	D412-41 modified	2,700–3,700[a]
Elongation, 77°F	percent	D412-41 modified	500–600
Tear strength	lb/in.	D624-48	900–1,000
Stiffness, 77°F	psi	D747-48T	40,000–90,000
Moisture permeability, 0.001 in.	g/100 sq in./day	...	0.2
Dielectric constant, 1,000 cycles	...	D150-54T	2.0–2.2
Power factor, 1,000 cycles	...	D150-54T	<0.0003
Volume resistivity	ohm-cm	D149-44	>10^{15}
Dielectric strength	v/mil	D149-44	3,600–4,000
Specific heat, 100–260°F	Btu/lb/°F	...	0.25
Water absorption	percent	D570-42	<0.01
Brittleness temperature	°F	...	<−100

[a] Tensile strength in oriented film may be as high as 15,000 psi. Data shown are average values and should not be used for specifications.

Liquid dispersions are used for coating glass fibers and metal surfaces (easy-to-clean cookware, easy-shaving razor blades), the impregnation of packing materials (asbestos), glass cloth, and porous metal. The coating must be dried, using infrared or forced-air heating, heated at 300°C to remove the dispersants, and finally sintered at 380°C. Films of poly-(tetrafluoroethylene) can be formed by casting the dispersion onto a metal surface, sintering, and stripping. Fibers have been formed using thickening agents that are subsequently removed to support the fiber.

An unusual combination of valuable properties is available in poly-(tetrafluoroethylene):

1. Chemical inertness to acids, bases, and common solvents. The polymer is attacked only by alkali metals, fluorine, and strong fluorinating agents at elevated temperatures.

2. Stability through a broad temperature range, from 250° down to at least −196°C.

3. Mechanical strength.

4. Exceptional weathering resistance.

5. Low dielectric constant (1.8 to 2.0) and loss factor (8).

6. Nonflammability.

7. Low coefficient of friction.

8. Nonstick (abhesive) surface.

About half of the poly(tetrafluoroethylene) produced is used in electrical applications as insulation. Major nonelectrical uses include industrial bearings, seals, mechanical seals and packings, fluid-handling systems, skived tape for wrapping rods and tubing, and molded parts. One-fifth is used in the form of filled resins. Resistance to creep can be improved by blending in up to 25 percent glass fiber or asbestos. Silica can be added to give good dimensional stability and electrical properties. Molybdenum disulfide and graphite improve dimensional stability without affecting the low coefficient of friction in bearings, and barium ferrite–loaded poly-(tetrafluoroethylene) can be magnetized. Titanium dioxide increases the dielectric constant, and boron compounds increase the neutron resistance.

Trihydrofluoroalcohols obtained by telomerization of tetrafluoroethylene with methanol have been used to prepare the corresponding acrylates and methacrylates (241). These can be polymerized, using emulsion techniques, to plastic or rubbery materials with good chemical and thermal stability (242). The corresponding acids produced by oxidation of $H(C_2F_4)_nCH_2OH$ to $H(C_2F_4)_nCOOH$ and their salts have been used as surface-active agents (243) and in formulations for dirt and oil repellents. Various esters of the alcohols are used as textile lubricants, cutting oils for metals, foam inhibitors, cosmetic fluids, and waxes.

The copolymerization of tetrafluoroethylene with hexafluoropropylene produces a perfluorinated, melt-extrudable polymer that is designated Teflon FEP resin by du Pont. Films and extruded products of this material are available with good transparency up to 10 mils. The coefficient of friction is low but slightly higher than that of poly(tetrafluoroethylene). Its maximum use temperature is about 50°C lower than that of poly-(tetrafluoroethylene) under equivalent conditions.

II. VINYL FLUORIDE

A. Introduction

Vinyl fluoride, $CH_2{=}CHF$, was first prepared in 1901 by F. Swarts (244) by the dehydrohalogenation of 1,1-difluoro-2-bromoethane with zinc:

$$CHF_2CH_2Br \xrightarrow{Zn} CH_2{=}CHF + HBr$$

He later prepared the monomer by dehalogenation of 1-fluoro-1,2-dibromoethane (245):

$$BrCHFCH_2Br \xrightarrow{Zn} CH_2{=}CHF + Br_2$$

The conversion of vinyl fluoride to high-molecular-weight polymer was reported more than 40 years later (246,247). The polymer was introduced commercially as a film by E. I. du Pont de Nemours and Company in 1960 under the name of Tedlar PVF film. Poly(vinyl fluoride) is the lowest-priced fluorinated polymer available on a unit basis, since its density is 1.38, compared with 1.76 for poly(vinylidene fluoride) and about 2.2 for poly(tetrafluoroethylene).

B. Preparation of Monomer

Vinyl fluoride is prepared from acetylene and hydrogen fluoride. The primary addition product is 1,1-difluoroethane (248), which is then pyrolyzed to vinyl fluoride (249):

$$HC\equiv CH + 2HF \longrightarrow CH_3CHF_2 \underset{\longleftarrow}{\overset{\Delta}{\longrightarrow}} CH_2=CHF + HF$$

Direct synthesis is best accomplished by means of mercury catalysis (250,251):

$$HC\equiv CH + HF \xrightarrow{Hg} CH_2=CHF$$

In the two-step process, acetylene and a selected catalyst are introduced into the bottom of a column of liquid hydrogen fluoride. The gaseous reaction products are distilled and condensed. Quantitative conversions of acetylene to 1,1-difluoroethane have been obtained. Pyrolysis to vinyl fluoride can be accomplished using various combinations of catalysts and temperatures. An undesirable by-product, acetylene, may be reduced by pyrolyzing at subatmospheric pressures, at high linear gas velocity, and with selected catalysts. Catalysts that have been used for the pyrolysis of 1,1-difluoroethane include 6- to 8-mesh alumina at 228° (252), chromium fluoride at 400 to 850° (253), aluminum sulfate at 400° (254), and charcoal with oxides or salts of metals in groups I, II, Vb, or VIII of the periodic table (249). The preferred catalyst systems for the one-step process are based on mercury, metal salts, and oxides. These include mercury compounds on saturated organic polymers (255), basic mercuric acetate on charcoal (256), mercuric fluoride (257), double metal chromites of zinc–mercury or nickel–mercury (258), porous alumina–aluminum fluoride (259), mercuric nitrate on alumina (260), aluminum fluoride on graphite (261), cuprous cyanide on carbon (262), aluminum fluoride pellets (263, 264), cadmium sulfate or acetate or nitrate on activated wood charcoal (265), mercuric oxide in oil (this catalyst is reported to be effective down to 0°C) (266), reduced CrO_3 on charcoal (267), and mercuric nitrate on charcoal (268).

The following example of the two-step process is taken from a patent (263). Acetylene and hydrogen fluoride, in a molar ratio of 1:2.02, were passed at 30 to 32°C into 3.8 liters of FSO_3H containing 1 to 3 percent

$SnCl_4$. The rate of addition was 82 liters of acetylene per hr per liter of FSO_3H. The product, 1,1-difluoroethane containing 0.02 percent acetylene, was mixed with more acetylene to give a ratio of 1.6:1 to 1.8:1, and the gases were passed at 250 to 300°C over AlF_3 pellets at a rate of 200 to 300 volume parts per hr to yield a mixture of $CH_2\!\!=\!\!CHF$ (64 percent), CH_3CHF_2 (33 percent), C_2H_2 (1.6 percent), C_2H_4 (0.3 percent), HF (1.1 percent), and a trace of hydrogen. The unsaturated hydrocarbons were hydrogenated to ethane over Pd–C, and the vinyl fluoride was isolated by fractional distillation (263).

In an example of the one-step process, the cadmium sulfate catalyst was heated on activated wood charcoal in a mild steel tube to 300° in a current of nitrogen and hydrogen fluoride. The nitrogen was then replaced by acetylene at a mole ratio of 1:1.5 HF/C_2H_2. The flow rate was 16 liters of C_2H_2 per hr. The product, after washing in aqueous sodium carbonate, was mainly vinyl fluoride with about 0.5 percent acetylene and less than 0.1 percent 1,1-difluoroethane. The yield based on acetylene was greater than 98 percent (265).

Gamma irradiation of the catalysts prepared from transition metal oxides, such as TiO_2 or ZrO_2, on charcoal or alumina has been used for both processes. Complete conversion of 1,1-difluoroethane has been reported (269) with the irradiated catalyst.

Direct addition of HF + C_2H_2 using a Lewis acid, BF_3, produced vinyl fluoride at 25°C (270). A fluoroacid, FSO_3H, catalyzed the direct addition of hydrogen fluoride to acetylene to give a 50 percent yield of vinyl fluoride (271).

Pyrolysis of 1-chloro-1-fluoroethane, in place of difluoroethane, over powdered copper at 700°C, with a contact time of 0.9 sec, resulted in a 97.7 percent conversion to vinyl fluoride of 97.9 percent purity (272). Operation at reduced pressure favors the production of vinyl fluoride, and atmospheric-pressure pyrolysis resulted in 27 to 52 percent vinyl chloride. The effect of time, pressure, and temperature on yield is shown in Table 9 (273,274). The pyrolysis tube was packed with noble metals, carbon, copper-nickel alloys, or stainless steel.

Laboratory methods for preparing vinyl fluoride, other than the pyrolytic techniques used industrially, are the reaction of ethylene oxide and carbonyl fluoride to produce a 70 percent yield of vinyl fluoride (275), and a photolytic technique (276). When 1,3-difluoroacetone is photolyzed, CFH_2 radicals are formed. These couple to give FCH_2CH_2F, which then loses HF, yielding vinyl fluoride. The ratio of the rates of these two reactions is a function of both the pressure and the temperature.

Vinyl fluoride may be purified by passing it through soda lime towers to remove traces of hydrogen fluoride. Acetylene is removed by scrubbing

TABLE 9

Pyrolysis of 1-Chloro-1-fluoroethane (CFE)

			Percentage yield vinyl fluoride	
Temperature, °C	Pressure, mm	Contact time, sec	Based on CFE used	Based on CFE unrecovered
500	120	1.25	2.0	95
600	60	1.25	47.5	93
600	60	0.55	25.2	92
600	60	0.30	16.7	97
600	120	2.30	78.0	91
600	120	1.25	44.2	93
600	120	0.55	36.4	93
600	180	1.70	58.8	92
650	60	1.20	90.2	93
650	60	0.55	58.3	97
650	60	0.30	48.8	96
650	120	3.30	91	93
650	240	2.15	84	94
700	120	0.55	79	91
750	120	0.50	93.6	89

with ammoniacal cuprous chloride. Fractionation at 40 to 100 psia and between −50 and −25°C removes oxygen which is a contaminant (277). Extractive distillation has also been used to separate vinyl fluoride from acetylene. Solvents used were dimethyl sulfoxide and dimethylformamide or mixtures of the two (278) and N-methylpyrrolidone (279). Bromination at 0° of the pyrolysis products of difluoroethane made the separation by distillation simpler. The resulting bromides were reacted with zinc dust to give vinyl fluoride (280). Molecular sieves have been used to separate vinyl fluoride from vinylidene fluoride (281).

C. Properties

1. PHYSICAL PROPERTIES

Vinyl fluoride is a colorless gas with a mild odor. Its physical properties are listed in Table 10 (277).

Various physical properties of vinyl fluoride have been reported, based on spectral studies or calculations from these studies. These include the assignment of bond frequencies (282), the wagging fundamental and twisting frequencies (283), theoretical calculations of vibrational frequencies (284), nonplanar vibration (285), assignment of bending-force

TABLE 10

Physical Properties of Vinyl Fluoride

Property	Value
Boiling point, °C	−72.2
Critical temperature, °C	54.7
Critical pressure, kg/cm^2	55.44
Critical density, g/ml	0.32
Liquid density, g/ml:	
At −30°C	0.7753
At 10°C	0.6808
Heat of formation, kcal/g mole	−34.4
Heat of vaporization:	
At 25°C	50
At −20°C	70
Solubility in water at 80°C, g/100 g H$_2$O:	
At 500 psia (35.16 kg/cm^2)	0.94
At 1,000 psia (70.31 kg/cm^2)	1.54

constants to these out-of-plane bending motions (286), and rotational structure measurements (287). The bond lengths and angles for vinyl fluoride were assigned as shown in Table 11 (288). The nuclear-magnetic resonance spectrum (289) and mass spectrum (290) have been reported.

TABLE 11

Bond Lengths and Angles of Vinyl
Fluoride [288]

Bond		Bond length, Å
	CC	1.329 ± 0.0006
	CF	1.347 ± 0.009
	CH	1.082 ± 0.004
(cis to F)	CH	1.077 ± 0.003
(trans to F)	CH	1.087 ± 0.003

Atoms		Bond angle, deg
	FCC	120.8 ± 0.3
	FCH	110 ± 1
cis	HCC	119.0 ± 0.3
trans	HCC	120.9 ± 0.03

2. TOXICOLOGY

Vinyl fluoride has a low toxicity. The volume percent for an LC$_{50}$ for mice was 66.6 to 71. At this concentration the oxygen content was less than 6.2 percent. No narcotic effects were observed (291).

3. ANALYTICAL METHODS

Gas-liquid chromatography is used for the analysis of vinyl fluoride. For example, on a 50-ft by $\frac{1}{4}$-in. copper-tubing column packed with $33\frac{1}{2}$ percent Dow Corning silicone oil 200 on water-washed 30/60 mesh Chromosorb P, vinyl fluoride has a retention time of 16.5 min, nitrogen 7.30 min, oxygen 7.60 min, and trifluorochloromethane 11.0 min (292).

D. Storage and Handling

The major hazard of vinyl fluoride is from fire and explosions of mixtures with air. The limits of flammability in air are 2.6 to 21.7 percent by volume; the ignition temperature is $460 \pm 5°C$ (277). Storage of cylinders should be in a well-ventilated area separate from the polymerization facility. The polymerization area should also be well ventilated to ensure against buildup of explosive mixtures. Vinyl fluoride is not known to polymerize except in the presence of free radicals. It is shipped with 0.2 percent of an inhibitor such as d-limonene. The Interstate Commerce Commission accepts inhibited monomer as a red-label material. Vinyl fluoride is not shock-sensitive.

E. Chemical Reactions

Free-radical addition to vinyl fluoride has been shown to be highly specific to the methylene group. The ratio of the reaction-rate constants for addition of the CF_3 radical to the CH_2 and CHF groups is 11:1. This orientation in the propagating step was attributed to the radical stability and not to polarization effects (293,294). Other examples of this type of addition of free radicals to vinyl fluoride are gas-phase addition of CF_3 from photolysis of $(CF_3N)_2$ (295,296), photochemical addition of CF_2=CFI (297) to vinyl fluoride, benzoyl peroxide–catalyzed addition of $CBrF_2CBrClF$ (298), and CBr_2F_2 (299) to vinyl fluoride. The activation energy for the photochemical addition of HBr to vinyl fluoride to form CH_2BrCH_2F, the main product, is 1.16 kcal per mole lower than that for the formation of the alternate product CH_3CHFBr (300). The relative reactivity of the double bonds with respect to the addition of methylene from the photolysis of ketene decreases as the degree of fluorination increases. The relative activities are shown in Table 12 (301). Thiolacetic acid adds to vinyl fluoride under free-radical conditions with either benzoyl peroxide or ultraviolet light (302):

$$CH_3COSH + CH_2{=}CHF \longrightarrow CH_3CHFSCOCH_3$$
$$\downarrow \begin{smallmatrix} dry \\ HCl \end{smallmatrix}$$
$$CH_3CHFSH$$

TABLE 12
Relative Reactivity of Fluoroolefins to Methylene [301]

Olefin	Relative reactivity of double bond	Relative reactivity of CH_2
Ethylene	1	0.028
Vinyl fluoride	0.66	0.019
Vinylidene fluoride	0.33	0.009
Trifluoroethylene	0.16	0.007
Tetrafluoroethylene	0.10	

Nitrosyl chloride adds as would be expected from bond polarities, the positive NO group going to the negative C atom (303):

$$CH_2{=}CHF + NOCl \longrightarrow CHFClCH_2NO$$

Further reaction can occur to give $CHFClCCl{:}NOH$ (304).
Trifluoronitrosomethane forms the oxazetidine

$$\begin{array}{c} CF_3N{-}O \\ | \quad | \\ F_2C{-}CF_2 \end{array}$$

with tetrafluoroethylene (153), but with vinyl fluoride only an unidentified solid has been obtained (305). Clathrate compounds of phenol and vinyl fluoride have been prepared (306). The hydrate was prepared, and its structure and lattice constants determined (307). Mercury-photosensitized decomposition of vinyl fluoride at 23° in the 44 to 580 torr range yielded only acetylene and HF (308). HCN copolymerizes with vinyl fluoride in the presence of a free-radical initiator to produce a polymer containing ketimino groups (309).

F. Polymerization

Vinyl fluoride polymerization can be initiated with peroxides, such as benzoyl peroxide, diethyl peroxide, or ammonium peroxysulfate, and azo compounds, such as 2,2′-azobisisobutyramidine hydrochloride or 2,2′-azobisisobutyronitrile. Reaction times of 0.5 to 19 hr, reaction temperatures of 35 to 145°C, and pressures of 70 to 1,000 atm have been used (310). Polymerizations are normally carried out in stainless-steel, stirred autoclaves, although shaker and rocker bombs are suitable for batch polymerizations. Aqueous systems (277) and organic solvents (311) have been used for continuous-polymerization systems.

In a typical polymerization, deoxygenated water is pumped into an evacuated, stirred autoclave. The water is heated to the reaction temperature, and vinyl fluoride and the initiator are charged. Agitation is maintained throughout the process. Additional vinyl fluoride is added as required to adjust for the decrease in pressure as the monomer is polymerized. After the reaction is completed, the reactor is vented and cooled, and the poly(vinyl fluoride) is obtained after filtering and drying. If a water-soluble initiator is used, the polymer is obtained as a dispersion. If the initiator is not water-soluble, the polymer forms above the surface of the water as a fine web. A two-stage polymerization has been used for better control of the reaction in which 25 percent of the polymerization occurs in a first vessel, with polymerization being completed in a second vessel (316).

Initiators that are active below 125°C produce less branched polymer of higher molecular weight. The molecular weight is increased if lower initiator concentrations are used with higher pressures. Chain-transfer agents, such as methanol, isopropyl alcohol, and 1,3-dioxolane, are useful in controlling molecular weight (310).

The preferred initiators are thermally generated free radicals. These include azo compounds (313) and peroxides (247,314). Redox systems (315), photochemical activation (247,316), and photolysis of acetic acid (317) have been used for laboratory polymerizations but are not used industrially. Vinyl fluoride has been polymerized at −20°C with tributylborane (318) and at −60°C with triethylborane (319). The use of the reaction product of triethylborane and compounds such as ammonia, pyridine, and hydrazine as initiators for vinyl fluoride polymerization yielded polymers of high intrinsic viscosities (up to 3.5) (320). Gamma irradiation gave a maximum yield at 0°C (318). Complete conversion could be obtained by adding some benzoyl peroxide prior to irradiation (321). Maximum crystallinity of the polymer was obtained by polymerizing a vinyl fluoride–acetaldehyde equimolar mixture. This was attributed to the formation of a cyclic complex that oriented the monomer during the polymerization (322). Copolymers have been prepared (246,323–325) with a wide variety of comonomers, but none has attained commercial importance.

G. Applications

Poly(vinyl fluoride) is a high-melting, chemically inert polymer that is generally insoluble below 100°C. Above 100°C it is soluble in N-substituted amides, dinitriles, ketones, tetramethylene sulfone, and tetramethylurea. The physical and chemical properties of poly(vinyl fluoride) film have been described (277,326). The polymer is transparent to both visible and ultraviolet radiation. This property, in combination with its resistance to

TABLE 13

Electrical Properties of Poly(vinyl fluoride) Film

Property	Transparent types 20 and 30	Pigmented type	Test method
Dielectric constant, at 22.2°C	8.5	9.3	ASTM D-150
1 kc dissipation factor, percent:			
1,000 cps, at 23°C	1.6	1.5	ASTM D-150
1,000 cps, at 60°C	1.8	1.7	
100 kc, at 23°C	8.8	7.6	
100 kc, at 60°C	3.0	3.0	
Volume resistivity, Ω-cm:			
At 22.2°C	3×10^{13}	4×10^{14}	ASTM D-257
At 132°C	1×10^{10}		
Dielectric strength, 60 cps, kv/mil:			
0.5 mil	4.5		ASTM D-150
1 mil	4.1		
2 mil	3.0		
Corona endurance, hr at 60 cps, 1,000 v/mil	3	7	ASTM suggested T method

oxidation, hydrolysis, and other kinds of degradation, contributes to its excellent resistance to weathering (327). The polymer has a high dielectric constant, good dielectric strength, and intermediate electrical resistivity and dissipation factor. These are listed in Table 13 (277). It has been used as a wrap for wires and cables, in capacitors, and as insulation in hermetically sealed electrical systems. The physical properties of poly(vinyl fluoride) film are given in Table 14 (277).

Poly(vinyl fluoride) film is manufactured with an adherable surface that is used with adhesives for lamination to various substrates, such as metals, cellulosics, and plastics. Flat laminates may be postformed into various shapes without rupturing the tough film surface. Poly(vinyl fluoride) film is also produced with an abhesive (nonsticking) surface that is used as a mold release film for epoxy resins, polyester resins, and other molding materials. When one side is adherable and the other side nonadherable, laminates can be made that have a nonsticking surface, for example, on snow shovels.

It has been used in solar stills for the purification of water and in glazing applications where a tough, transparent, weatherable film is required.

TABLE 14
Physical-Thermal Properties of Poly(vinyl fluoride) Film

Description	0.5-mil transparent type 20	1.5-mil low-gloss white type 30	Test method
Designation	50 SG20TR	15 AL30WH	
Physical properties:			
Area factor, sq ft/lb	277	77	
Bursting strength, psi/mil	70	27	Mullen ASTM D-774
Coefficient of friction, film/metal	0.16	0.30	
Density at 72°F, g/ml	1.38	1.55	ASTM D-1505 density gradient tube
Impact strength, kg-cm/mil	5.3	3.1	Du Pont pneumatic tester
Moisture absorption, percent	0.5 for all types		Water immersion
Moisture vapor transmission, g/(hr)(100 sq m) for 1 mil thickness	157	200	Du Pont test method
Refractive index, n_D^{30}	1.46	...	ASTM D-542 (Abbé)
Tear strength (Elmendorf), g/mil	12	31	Elmendorf
Tear strength initial (Graves), g/mil	450	...	ASTM 1004
Tensile modulus, psi	260,000	250,000	Instron ASTM D-882 method A-100%/min
Ultimate elongation, percent	115	120	Instron ASTM D-882 method A-100%/min
Ultimate tensile strength, psi	18,000	8,500	Instron ASTM D-882 method A-100%/min
Ultimate yield strength, psi	6,000	4,800	Instron ASTM D-882 method A-100%/min
Thermal properties:			
Aging (hours to embrittlement)	3,000	3,000	Oven at 300°F
Flammability		Slow burning to self-extinguishing	Some varieties[a]
Heat sealability			
Linear coefficient of expansion, in./in. per °F	2.8×10^{-5}		
Shrinkage, percent	4 at 266°F	4 at 338°F	Oven, 30 min
Zero strength[b] at °C	300	300	Du Pont method

[a] See *Du Pont Technical Information Heat-Sealing Bulletin*, March, 1963.
[b] Temperature at which plastic film supports 1.4 kg/cm² of cross-sectional area for 50.4 sec.

The gas and vapor permeabilities are listed in Table 15 (277).

TABLE 15

Gas and Vapor Permeabilities of Poly(vinyl fluoride) Film

Gases	ml/(24 hr)(100 in.^{2}) for 1 mil thickness at 23.5°C	Method
Oxygen	3.2	
Nitrogen	0.25	
Carbon dioxide	11.0	ASTM D-1434
Hydrogen	58.1	
Helium	150	

Vapors	g/(24 hr)(100 in.^{2}) for 1 mil thickness at 23.5°C[a]	
Water, at 39.5°C (103°F)	2.8	Du Pont permeability
Ethyl alcohol	0.54	
Benzene	1.4	
Acetic acid	0.7	
Ethyl acetate	15	Modified ASTM E-96
Acetone	310	
Hexane	0.85	
Carbon tetrachloride	0.77	

[a] Except where otherwise indicated.

IV. VINYLIDENE FLUORIDE

A. Introduction

Vinylidene fluoride, $CH_2=CF_2$, was first prepared by F. Swarts (244) at the turn of the century. It was not until the late forties that an industrial process was developed for its synthesis (328). This monomer, which was then produced at the Jackson Laboratory of E. I. du Pont de Nemours and Company, has been utilized as a comonomer in various elastomers by du Pont and M. W. Kellogg Co. since the 1950s under the names of Viton and Fluorel (30:70 hexafluoropropylene–vinylidene fluoride) and Kel F elastomers (50:50 to 30:70 chlorotrifluoroethylene–vinylidene fluoride). The Three M Company subsequently purchased the manufacturing rights for the Kel F copolymers from the M. W. Kellogg Co. In 1961, Pennsalt Chemicals Corp. began offering the homopolymer on a commercial basis. E. I. du Pont de Nemours and Company, Three M, and Pennsalt produce vinylidene fluoride in the United States.

B. Preparation of Monomer

Commercial processes for the synthesis of vinylidene fluoride are based on pyrolysis. Several alternate starting materials may be used.

The monomer was produced by dehydrofluorination of 1,1,1-trifluoroethane by heating at 820°C in a platinum tube (328). Substitution of alumina for platinum and operation at 300°C gave vinylidene fluoride in a 12.7 percent yield (329), and heating to 1,200°C with a contact time of 0.006 sec over platinum produced a 56 percent conversion with a 99 percent yield of vinylidene fluoride of 99.5 percent purity (330).

$$CF_3CH_3 \xrightarrow{\text{heat}} CF_2{=}CH_2 + HF$$

Dehydrochlorination of 1,1-difluoro-1-chloroethane has been used with various catalyst systems. At 630°C in a copper tube, a 46 percent conversion to vinylidene fluoride of 96 to 98 percent purity was obtained (331). The use of a silica tube at 870°C and 1.9 sec contact time yielded 67 percent vinylidene fluoride (332). Pyrolysis in nickel tubes at 550°C gave high yields (333) that could be improved by the addition of 1.3 percent free chlorine (334):

$$CF_2ClCH_3 \xrightarrow{-HCl} CF_2{=}CH_2$$

The removal of chlorine from 1,2-dichloro-1,1-difluoroethane yields vinylidene fluoride:

$$CF_2ClCH_2Cl \xrightarrow{-Cl_2} CF_2{=}CH_2$$

This reaction has been run at 420°C (335) and 500°C (336). At the latter temperature, with some hydrogen and over nickel, a 43 percent conversion was obtained.

A combination chlorination-dehydrochlorination has been used with 1,1-difluoroethane to yield 56 percent vinylidene fluoride over silica at 600°C (337). Dichlorodifluoromethane over nickel at 650°C yielded over 96 percent vinylidene fluoride:

$$CH_3CHF_2 + Cl_2 + CCl_2F_2 \xrightarrow[650°C]{Ni}$$
$$CF_2{=}CH_2 + CF_2{=}CHCl + CF_2{=}CCl_2 + CF_2ClCCl_3$$
$$(96.5\%)$$

Other industrial methods are listed in Table 16.

Convenient laboratory methods are the original Swarts synthesis by dehydrobromination of 1,1-difluoro-2-bromoethane (244):

$$CHF_2CH_2Br + KOH \xrightarrow{\text{ethanol}} CF_2{=}CH_2 + KBr + H_2O$$

TABLE 16
Preparation of Vinylidene Fluoride

	Conditions				
Starting material	Temperature, °C	Time, sec	Tube material	Percentage of VF_2	Reference
$CHClF_2 + CH_3Cl$	900	0.72	...	60.7	339
$CCl_2F_2 + CH_3Cl$	500–1100	...	...	2.1	340
$CF_3Cl + CH_4$	600–1000	0.5	Cu	11.8	341
$CF_3Br + CH_4$	600–1000	0.5	Cu	11.8	341
$CF_3Cl + CH_4 + H_2O$	950	...	...	44.0	342
$CCl_2F_2 + CH_4$	720–760	3.8	...	6.8	343

and dehalogenations of tetrahaloethanes with zinc (344) or magnesium (345):

$$BrF_2CCH_2Cl + Zn \xrightarrow{ethanol} CF_2{=}CH_2 + ZnBrCl$$

$$CF_3CH_2I + Mg \longrightarrow CF_2{=}CH_2 + MgFI$$

A 90 percent yield is obtained in the latter reaction. Treatment of CF_3CClBr_2 with sodium amalgam and sulfuric acid at 60 to 65°C gave a 75 percent yield of vinylidene fluoride (346). A gas-phase reaction at 140 to 170°C of di-*tert*-butyl peroxide with $(CF_3N)_2$ formed vinylidene fluoride (347). The peroxide yields methyl radicals and hexafluoroazomethane forms trifluoromethyl radicals, which combine to yield vinylidene fluoride:

$$\cdot CF_3 + \cdot CH_3 \longrightarrow CF_2{=}CH_2 + HF$$

The monomer may be purified by scrubbing with water, passing over soda lime and calcium chloride, and distilling the condensate. Molecular sieves have been used for purification.

C. Properties

1. PHYSICAL PROPERTIES

Vinylidene fluoride is a colorless gas with a slight odor. The physical properties are listed in Table 17. Thermodynamic measurements and calculations on vinylidene fluoride, including ideal-gas thermodynamic functions, liquid enthalpies, and saturated-vapor enthalpies, have been reported (351). Sources for other physical properties are as follows: interatomic distances and vibrational amplitudes (94,95), infrared spectrum (352,353), Raman spectrum (353,354), nuclear-magnetic resonance spectrum (355), and microwave spectrum (356).

TABLE 17

Properties of Vinylidene Fluoride

Property	Value	Reference
Melting point, °C	-144	348
Boiling point, °C, 1 atm	-82	348
Critical temperature, °C	30.1	348
Critical pressure, psia	643	348
Critical density, g/ml	0.417	348
Heat of formation, kcal/mole	-77.5	349
Ionization potential, ev	10.30	88
Dipole moment, D	1.96	350
Liquid density, °C, g/ml	$-45.35, 1.001$	351
Liquid density, °C, g/ml	$+3.76, 0.795$	351

2. TOXICITY

Vinylidene fluoride is comparatively nontoxic. It has an LC_{50} of 128,000 ppm for rats exposed 4 hr and is less toxic than acetic acid, which has a LC_{50} of 64,000 ppm in the same test (357).

3. ANALYTICAL METHODS

Analysis of vinylidene fluoride is by gas-liquid chromatography. For example, a retention time of 25.6 min was obtained using two columns in series. The first contained silver nitrate in diethylene glycol on firebrick, the second silica gel. Hydrogen was used as carrier gas (108). Dow Corning silicone 200 oil on Chromosorb P has also been used as the stationary liquid phase (292), with a retention time of 12.4 min for vinylidene fluoride, 7.30 min for N_2, 7.60 min for oxygen, and 11.0 min for CF_3Cl.

D. Storage and Handling

Vinylidene fluoride forms explosive mixtures with air in the concentration range of 5.8 to 20.3 percent (358); hence proper ventilation must be provided. Uninhibited monomer forms peroxides on exposure to oxygen. Inhibitors, such as phenols or terpenes, may be added to prevent premature polymerization during storage or shipping, although it is routinely shipped uninhibited. Spontaneous polymerization of vinylidene fluoride has not been reported.

E. Chemical Reactions

The ease of electrophilic and free-radical addition reactions of HX decreases as the degree of fluorine substitution increases in going from ethylene to vinyl fluoride to vinylidene fluoride to tetrafluoroethylene and, finally, to chlorotrifluoroethylene. Unsubstituted ethylene reacts readily with electrophilic reagents, such as HX, to give simple addition products. Replacement of hydrogen by fluorine slows down or prevents electrophilic addition by deactivation of the double bond. The deactivation is due to the decrease in the electron density by the electronegative fluorine or by the steric influence of the fluorine. The inductive effect of the fluorine atom should facilitate nucleophilic attack on the carbon atom to which it is bonded. It increases the positive character of the carbon atom, thus facilitating the attack of a negative ion. This effect can be offset when, in the final stage of the bimolecular reaction, the ease with which X^- can be lost is decreased because of the inductive and hyperconjugative effects of fluorine and, in the attack step, because of the shielding of the carbon by the fluorine's steric and field effects.

The electrophilic additions proceeded, as would be anticipated, by ionization of the intermediate (359–362):

$$
\left. \begin{array}{c} CH_2{=}C{-}F \\[6pt] -CH_2{-}C{=}F^+ \end{array} \right\} + H^+ \longrightarrow \left\{ \begin{array}{c} CH_3{-}C{=}F^+ \\[6pt] CH_3C^+F \end{array} \right\} \xrightarrow{\ X^-\ } CH_3CFX
$$

$$HBr + CF_2{=}CH_2 \longrightarrow CF_2BrCH_3$$

$$HI + CF_2{=}CH_2 \longrightarrow CF_2I\,CH_3$$

$$ICl + CF_2{=}CH_2 \longrightarrow CF_2Cl\,CH_2I$$

$$BrCl + CF_2{=}CH_2 \xrightarrow{\ 0°\ } CF_2Cl\,CH_2Br$$

$$IBr + CF_2{=}CH_2 \longrightarrow CF_2Br\,CH_2I$$

$$HF + CF_2{=}CH_2 \xrightarrow[N_2,\,300°C]{AlF_3} CF_3CH_3$$

The activation energy E_a required for a trichloromethyl radical to attack at each end of the C=C bond in various olefins was determined. The results are shown in Table 18 (162). The addition of the methylene free

TABLE 18

Energy Requirements for the Addition of Trichloromethyl Radical to Fluoroethylenes

Attack site	E_a, kcal/mole
CH_2 of $CH_2 = CH_2$	3.2
CH_2 of $CH_2 = CHF$	3.3
CH_2 of $CH_2 = CF_2$	4.6
CHF of $CH_2 = CHF$	5.6
CF_2 of $CH_2 = CF_2$	8.3
CF_2 of $CF_2 = CF_2$	6.1

radical gave similar results (301). The dominant mode of free-radical addition of HX is reversed relative to that for electrophilic addition; that is, the principal product from $CH_2{=}CF_2$ and HX by a radical process is CH_2XCHF_2 (300,359).

Under free-radical conditions, CBr_2F_2 adds to vinylidene fluoride to form 1:1 and higher adducts (299):

$$CBr_2F_2 + CH_2{=}CF_2 \longrightarrow CBrF_2CH_2CBrF_2 \qquad CBrF_2(CH_2CF_2)_2Br \qquad etc.$$

$CBrF_2CBrClF$ forms only the 1:1 adduct, $CBrF_2CClFCH_2CBrF_2$ (298). $CBrClF_2$ adds similarly to form $CClF_2CH_2CBrF_2$ (363). Trichlorosilane forms the monoadduct $CHF_2CH_2SiCl_3$ (364). Photochemical chlorination of vinylidene fluoride has a velocity proportional to the product of the chlorine pressure and the square root of the intensity of the absorbed light (365). Photochemical reaction of hexafluoroazomethane with vinylidene fluoride resulted in attack on the CH_2 (366), as would be expected from Table 18. Photolysis of carbonyl sulfide and vinylidene fluoride formed 1,1-difluoroethylene sulfide and 2,2-difluorovinyl mercaptan (367). Sulfur chlorides formed sulfides (368):

$$CF_2{=}CH_2 + SCl_2 \longrightarrow (ClF_2CCH_2S)_2$$
$$CF_2{=}CH_2 + S_2Cl_2 \longrightarrow (ClF_2CCH_2)_2S_3$$
$$CF_2{=}CH_2 + C_2H_5SCl \longrightarrow ClF_2CCH_2SC_2H_5$$

Trifluoronitrosomethane formed a 1:1 copolymer with vinylidene fluoride considered to be $[N(CF_3)OCH_2CF_2]_n$, which eliminates HF readily (305). Nitrogen tetroxide and iodine form two products with vinylidene fluoride (369):

$$CH_2{=}CF_2 + N_2O_4 + I_2 \longrightarrow IF_2CCH_2NO_2 \quad (mainly) + (NO_2)F_2CCH_2I$$

An acid is formed by the addition of HF with chromium oxide, probably because of the formation of F_2CrO_2, which attacks the double bond electrophilically, after which the adduct is hydrolyzed to the acid, CF_3COOH (370). A fluorosulfonate is formed by adding fluorosulfonic acid (371):

$$FSO_3H + CH_2{=}CF_2 \longrightarrow CH_3CF_2OSO_2F$$

The heat of hydrogenation of vinylidene fluoride has been measured at 125°C (205).

F. Polymerization

Vinylidene fluoride is polymerized commercially using free-radical initiator systems, such as benzoyl peroxide (372,373), *in situ*-formed peroxides from ketones or aldehydes (220), di-*tert*-butyl peroxide (374) and acetyl peroxide (375), ammonium persulfate, or a redox system based on ammonium persulfate and sodium bisulfite (372). Thermal activation is adequate to initiate polymerization (376). A high-pressure vessel with an efficient stirrer is used to ensure the removal of the heat of polymerization. Water dispersions are used with a fluorinated dispersing agent to prevent chain transfer, with a resulting decrease in molecular weight and loss of properties. Other inert dispersants include pentachlorobenzoic acid and its salts (377). In a typical polymerization, there was charged into a shaker or rocker tube of 300 ml capacity 100 ml deionized and deoxygenated water, 0.8 g di-*tert*-butyl peroxide, and 35 g of vinylidene fluoride. This mixture was shaken for 18.5 hr at 122 to 124°C with a maximum pressure of 800 psig. The polymer formed, after vacuum filtering, washing with methanol and water, and vacuum drying at 102°, had a melting point of 160 to 165°C. An 83 percent conversion to polymer was obtained (374). A stirred autoclave is also used for continuous polymerizations, unreacted monomer being recycled.

Other polymerization initiators that have been reported are gamma radiation from ^{60}Co (318,378–380), trialkylboranes (318), and $TiCl_4$–organic aluminum complex (381).

Redox initiator systems are normally used for vinylidene fluoride copolymerizations (382).

G. Applications

Vinylidene fluoride is copolymerized with 30 percent hexafluoropropylene (Viton, Fluorel) and 30 to 50 percent chlorotrifluoroethylene (Kel F) to form elastomers of high heat and chemical resistance. The homopolymer is used for solid and lined piping, coatings and linings for pumps, tanks, and valves, and instrument and transfer hoses. Electrical and electronic uses include primary and jacket insulation, bobbins, connector blocks, insulating sleeves, and solder sleeves. The electrical properties are shown in Table 19 (348).

TABLE 19

Dielectric Properties of Poly(vinylidene fluoride)

	cps			
	60	10^3	10^6	10^9
Dielectric constant	8.40	7.72	6.43	2.98
Dissipation factor	0.0497	0.0191	0.159	0.11
Volume resistivity, Ω-cm	...	...	...	2×10^{14}
Dielectric strength, v/mil:				
125 mil	...	...	...	260
8 mil, short time	...	...	...	1,280

Other physical properties are shown in Table 20 (348).

TABLE 20

Physical Properties of Poly(vinylidene fluoride)

Property	Value
Crystalline melting point, °C	170
Specific gravity, g/cc	1.76
Refractive index n_D^{25}	1.42
Tensile strength, psi	7,000
Elongation, percent	300
Tensile modulus, psi	1.2×10^5

The polymer can be coated from solution or as an organosol onto metal, cellulosics, and plastics in flat or coil form that is subsequently formed into a finished shape. The excellent weather resistance of these finishes makes them useful for outdoor coatings (348).

V. CHLOROTRIFLUOROETHYLENE

A. Introduction

Chlorotrifluoroethylene, CF_2=CFCl, was first prepared in 1933 (383). It was polymerized in 1934 at I. G. Farbenindustrie (384), where it was later manufactured commercially. It is now sold in the United States by Minnesota Mining and Manufacturing Co. (3M) as Kel F and by Allied Chemical Corporation as Plaskon. European versions include the German Hostaflon, French Voltalef, and Russian Fluoroplast. It is higher in price, slightly inferior in chemical resistance, and has a lower softening point compared with poly(tetrafluoroethylene). Its electrical insulation properties are poorer because of the presence of the chlorine atom, which destroys its electrical symmetry. Poly(chlorotrifluoroethylene) can be formed into thin, transparent, flexible films with greater hardness and tensile strength than poly(tetrafluoroethylene). High-molecular-weight polymer is used in moldings, coatings, and films, and lower-molecular-weight polymers function as oils, grease, and waxes.

B. Preparation of Monomer

The monomer can be prepared from hexachloroethane and hydrogen fluoride to produce trifluorotrichloroethane, which is then dehalogenated:

$$CCl_3CCl_3 \xrightarrow{\ HF\ } CClF_2CCl_2F \xrightarrow{\ -Cl_2\ } CF_2=CClF$$

These consecutive reactions can be effected at 800°C in a quartz tube with a 1.5-sec residence time. The decomposition products are cooled to 70°C and purified (386). The use of a trace of hydrogen and a catalyst based on 30 percent Cu and 15 percent Cr_2O_3 on calcium fluoride gave a 43 percent conversion to chlorotrifluoroethylene at 400° (387). This was increased to 60 percent by the addition of barium (388). Using increased quantities of hydrogen resulted in yields of over 90 percent (389). Other catalysts used with hydrogen for the synthesis of chlorotrifluoroethylene include copper on magnesia (31), cobalt chloride on carbon (390), and nickel (391). Other pyrolytic syntheses are shown in Table 21.

TABLE 21

Other Pyrolytic Preparations of Chlorotrifluoroethylene

Starting materials	Temperature, °C	Reference
$CHClF_2$ and CH_3Cl	900	339
$CHClF_2$ and $CHFCl_2$	900–1,200	14
$C_2H_3ClF_2$ and $CHFCl_2$	900–1,200	392
CF_2ClCF_2Cl and H_2	375–430	29
$CF_2CF_2CClFCClF$	650–700	32

The current preferred synthesis uses zinc for dehalogenation. Zinc in alcohol was used to dehalogenate $CClF_2CCl_2F$ in the original synthesis of chlorotrifluoroethylene (383). This convenient preparation can be varied by using solvents such as dioxane, acetone, and pyridine (28). The use of hydrochloric acid with zinc in anhydrous (393) and aqueous methanol (394) has produced chlorotrifluoroethylene in up to 90 percent yields. Activating the zinc by washing in glacial acetic acid causes the reaction to be initiated immediately at room temperature (395). When zinc is used in an aqueous system, detergents such as polyglycol ether stearate (27), perfluorooctanoic acid, and Kel F acid 8114 (396) are helpful.

Other dechlorinating agents for $CClF_2CCl_2F$ include aqueous sodium or potassium hydroxide (397,398), sodium amalgam (399), zinc amalgam (400), iron powder alone (401) and in combination with traces of zinc (402), and magnesium alloys containing zinc, aluminum, and manganese (403).

A convenient laboratory preparation of chlorotrifluoroethylene is the dehalogenation of CF_3CClFI with zinc in dioxane at 15 to 20°C (404).

Purification of the monomer is critical to obtain acceptable polymers. In one case, silica gel was used to purify the distilled monomer. The resulting polymer, when subjected to stress testing at elevated temperatures, survived only half as long as a polymer prepared from distilled monomer purified with sulfuric acid (405). Anhydrous calcium chloride (406) was effective in purifying monomer that had been fractionated at 5 atm.

C. Properties

1. PHYSICAL PROPERTIES

The reported physical properties of chlorotrifluoroethylene are listed in Table 22.

TABLE 22

Physical Properties of Chlorotrifluoroethylene

Property	Value	Reference
Boiling point, °C, 1 atm	−28.36	407
Freezing point, °C	−168.11	407
Vapor pressure, mm Hg, from −67 to −11°C and up to 2 atm	$\log_{10} P = \dfrac{6.90199 - 850.649}{(t + 239.91)}$	407
Bubble point pressure, mm Hg, from 25 to 105.8°C	$\log_{10} P = \dfrac{7.75412 - 1{,}392.82}{(t + 319.70)}$	407
Critical temperature, °C	105.8	407
Critical pressure, °C	40.1	407
Critical density, g/ml	0.55	407
Molal heat capacity at −28.36°C, cal/deg	29.26	407
Heat of fusion at −168.11, cal/mole	1,327.1	407
Entropy values at −28.36°C:		
Liquid state, cal/deg-mole	52.74	407
Ideal gas state, cal/deg-mole	73.18	407
ΔH of vaporization, cal/mole	5,400	383
Heat of formation at 298.15°K, kcal/mole	−126	408
Refractive index at 20°C $(n - 1) \times 10^6$	926	409
Molar refraction, cc/mole:		
Observed	15.77	409
Calculated	15.90	409
Dielectric constant at 9,400 Mc/sec, debyes	0.61	410
Electric moment, debyes	0.38	411

The bond lengths and angles have been determined by electron diffraction (412). The results are given in Table 23. The solubility of chlorotrifluoroethylene in various solvents has been reported (413). Spectral

TABLE 23

Molecular Structure of Chlorotrifluoroethylene

Bond length, Å	C—Cl	1.72 ± 0.02
	C—F	1.31 ± 0.02
	C—C	1.31 ± 0.06
Bond angle, deg	F—C—F	114 ± 2
	F—C—Cl	114 ± 2
	C—C—Cl	123 ± 2

studies of chlorotrifluoroethylene include ultraviolet (106), infrared (414), Raman (415), and nuclear-magnetic resonance (416).

2. Toxicology

Chlorotrifluoroethylene is the most toxic of the four fluorovinyl monomers discussed. It has an LC_{50} for rats of 4,000 ppm for a 4-hr exposure (357) and some anaesthetic activity (417,418) at low concentrations. Continuous exposure to chlorotrifluoroethylene should be avoided, since substantial injury to laboratory workers has been observed under these conditions (19).

3. Analytical Methods

Chromatographic analysis of chlorotrifluoroethylene is the accepted technique. Care must be taken to avoid hydrolysis of the chlorine, since this causes a decrease in the measured retention volume with resulting inaccuracies (419). Impurities in chlorotrifluoroethylene have been determined with a sensitivity of 0.01 percent by gas-liquid chromatography at 20°C in a column packed with silicone oil and paraffin oil on alumina (420). The concentration of chlorotrifluoroethylene in air has been measured with an accuracy of ± 10 percent by adsorption on silica gel, followed by desorption and cracking to yield chlorine, which is absorbed in an aqueous solution of sodium carbonate and sodium formate. Titration of this solution by the Volhard method completes the analysis.

D. Storage and Handling

Chlorotrifluoroethylene explosion surges, like those of tetrafluoroethylene, move at such a high rate of speed that normal relief devices cannot handle them. Explosive limits for chlorotrifluoroethylene in air are 28.5 to 35.2 percent (358); its flammability limits are 16 to 34 volume percent in air (422). Thus, proper ventilation must be provided to prevent accumulation of vapors, and sources of ignition must be eliminated. The monomer must be shipped with inhibitor, for example, terpenes. The danger of spontaneous polymerization due to the formation of peroxides from the unintentional absorption and reaction of oxygen is thus eliminated. The exothermic dimerization of chlorotrifluoroethylene proceeds similarly to that of the previously discussed tetrafluoroethylene. This reaction is not subject to inhibition but can be prevented by maintaining the monomer below the reaction temperature. The monomer should be stored separately from the polymerization area in a protected, ventilated, barricaded area. Laboratory work should be carried out with the same precautions, using barricaded pressure reactors. Personnel should be aware of the hazards involved before working with tetrafluoroethylene or chlorotrifluoroethylene.

E. Chemical Reactions

The chemical reactions of chlorotrifluoroethylene are variations of those of tetrafluoroethylene modified by the presence of chlorine. The substitution of the bulkier chlorine for one fluorine results in a steric effect on the reactions and their products. The chlorine is less electronegative than fluorine, therefore electrophilic reagents attack chlorotrifluoroethylene more readily than tetrafluoroethylene. The electron density of the double bond, however, is still relatively low because of the four halogen substituents, so that nucleophilic attack is favored. The bulky chlorine tends to shield the carbon to which it is attached very effectively. Chlorine also has a tendency to hydrolyze with base thus providing another mode of reaction.

The mechanism of the nucleophilic addition to tetrafluoroethylene was discussed previously. Chlorotrifluoroethylene behaves similarly. Examples of nucleophilic addition are given in Table 24.

The nucleophilic attack of primary amines, such as butyl- or ethylamine, occurs at both ends of the $C=C$ bond, so that the two possible products are obtained. Piperidine, however, forms only the expected

TABLE 24

Nucleophilic Addition Reactions of Chlorotrifluoroethylene

Starting compound	Conditions	Product	Reference
CH_3OH	0°C, Na	CH_3OCF_2CHClF	115
ROH (R=Me, Et, i-Pr, t-Bu)	Na	$ROCF_2CHClF$	423
RSH (R=Me, Et, i-Pr)	115°C, 6 hr, KOH	$RSCF_2CHClF$	118
ArSH	KOH	$ArSCF_2CHClF$	424
ArOH	KOH	$ArOCF_2CHClF$	424,425

N-(2-chloro-1,1,2-trifluoroethyl)piperidine (426). Kinetics of this kind of reaction have been reported (120).

Formaldehyde adds to chlorotrifluoroethylene, as it does to tetrafluoroethylene (see Sec. IIE). Since chlorotrifluoroethylene is unsymmetrical, however, there are two possible modes of addition, and both are observed. The reaction of paraformaldehyde and chlorotrifluoroethylene in chlorosulfonic acid at 100°C, with subsequent esterification of the resulting hydroxyacids, gave $HOCH_2CF_2COOC_2H_5$ and $HOCH_2CFClCOOC_2H_5$ in yields of 30.5 and 19.8 percent respectively (427).

Aliphatic and aromatic Grignard reagents add across the double bond in a manner similar to the previously discussed bases. The resulting adduct loses MgX_2 to yield a new olefin with a longer carbon chain (428):

$$C_6H_5MgBr + CF_2{=}CFCl \longrightarrow C_6H_5CF{=}CFCl$$

Phenyl- and butyllithium react similarly (429):

$$C_6H_5Li + CF_2{=}CFCl \longrightarrow C_6H_5CF{=}CFCl$$

Electrophilic additions take place less readily with chlorotrifluoroethylene than with vinylidene fluoride because of the deactivation of the double bond by substitution with an additional fluorine and chlorine. In comparison with tetrafluoroethylene, however, chlorotrifluoroethylene is more reactive, because one chlorine and three fluorines deactivate the double bond less than four fluorines. Chlorine bulkiness is less important than electronegativity. The ease of electrophilic addition decreases in the order $CFH{=}CH_2 > CF_2{=}CH_2 > CF_2{=}CFCl > CF_2{=}CF_2$ (359). The addition of ICl, HBr, and HI were considered in this study. Chlorine can add in the dark, alone, or when mixed with oxygen (430). Fluorine adds to chlorotrifluoroethylene, even at $-150°$ (431).

Ionic mechanisms have been suggested for other additions to chlorotrifluoroethylene. Nitrosyl chloride (NOCl) adds to chlorotrifluoroethylene in the presence of ferric chloride to form $CClF_2CFCl(NO)$ in 82 percent yield. Other products were $CF_2ClCFCl_2$ (3.3 percent yield) and $CF_2(NO_2)CFCl_2$ (7.5 percent yield) (432). This reaction starts with nitric oxide and ferric chloride forming nitrosyl chloride *in situ* (432,139). Nitric oxide (NO) reacts with a free-radical mechanism in the presence of actinic light to form $CF_2(NO)CFCl(NO)$ (138). Nitrogen dioxide (N_2O_4) nitrates chlorotrifluoroethylene readily (144):

$$CF_2{=}CFCl + N_2O_4 \longrightarrow CF_2(NO_2)CFCl(ONO)$$

Trifluoronitrosomethane (F_3CNO) yields an oxazetidine, $CF_3NOCFClCF_2$, or an elastomeric or viscous copolymer $(CF_3NOCF_2CFCl)_n$. The direction of the addition was different in the two products (433). The copolymer was attributed to the formation of a paramagnetic intermediate, $CF_3N(O\cdot)R^-$ (434). Trifluoronitrosoethylene formed similar structures, that is, the oxazetidine and the polymer (155).

Lewis acids catalyze the addition of halogenated hydrocarbons to chlorotrifluoroethylene. Examples are summarized in Table 25.

Iodine and fluorine (as a mixture of I_2 and IF_5) can be added across the double bond to form CF_2ClCF_2I (55 percent) and CF_3CFClI (45 percent) (441). ICl and chlorotrifluoroethylene give $CClF_2CClFI$ (442). HBr with cupric sulfate catalysis forms $CF_2BrCHClF$ (443). An ionic route to an acid from chlorotrifluoroethylene involves adding NaOCN, followed by hydrolysis to 3-chloro-2,3,3-trifluoropropionic acid (444). Mercuric fluoride adds without dehalogenation to yield $CF_3CFCl(HgF)$ (455).

Complexes of chlorotrifluoroethylene are produced by an ionic route. Examples of this are the chlorotris(triphenylphosphine)rhodium(I), $RhCl[(C_6H_5)_3P]_2C_2F_3Cl$ (150), and the dicyclopentadienylnickel complexes (446).

Free-radical reactions of chlorotrifluoroethylene with HBr, Br_2, Cl_2, and CF_3I occur with more difficulty than those of tetrafluoroethylene and vinylidene fluoride. The ease of the reaction decreases in the following order: $CF_2{=}CH_2 > CF_2{=}CF_2 \gg CF_2{=}CHCl > CF_2{=}CFCl$ (359). The comparative stability of the intermediate radicals determines the site of the attack. Thus, for chlorotrifluoroethylene, we find the major products resulting from the intermediate radical $RCF_2CFCl\cdot$ (447). Free-radical reactions include the addition of trichlorosilane (448), HBr to form $F_2BrCCHFCl$ (449) or telomers (450), CF_3I and CF_3Br (451), $CF_2ClCFClI$ (452), bromine (453), and chlorine (454).

Telomers in which the degree of polymerization was 4 to 6 were prepared

TABLE 25

Lewis Acid–catalyzed Addition to Chlorotrifluoroethylene

Additive	Catalyst	Products	Reference
CCl_4	$AlCl_3$	$CFCl_2CF_2CCl_3$	435
CCl_4	$AlCl_3$	$CFCl_2CF_2CCl_3$ (25%),	436
		$CF_2ClCFClCCl_3$ (75%)	436
CCl_4	$AlCl_3$	$CF_2ClCFCCl_3$ and dispropor-	437
		tionation products $C_3F_4Cl_6$	
		and $C_3F_2Cl_6$	
$HCCl_3$	$AlCl_3$	$C_3Cl_4F_3H$, $C_3Cl_5F_2H$	438
$FCCl_3$	$AlCl_3$, $FeCl_3$,	$CF_2ClCF_2CCl_3$,	439
	$TiCl_4$	$CFCl(CCl_3)CF_3$	439
$i\text{-}C_3H_7Cl$	$AlCl_3$	$CF_2ClCFClCH(CH_3)_2$,	440
		$CF_2ClCCl_2CH(CH_3)_2$	440

by reaction with CF_3CCl_2I (455), and with CF_3I the degree of polymerization was 1 to 5 (456).

Silanes add to chlorotrifluoroethylene in the presence of chloroplatinic acid (H_2PtCl_6) (177). If heated in a glass tube over 470°C, however, the product dehydrohalogenates, forming an olefin (457):

$$HSiCl_3 + CF_2{=}CFCl \longrightarrow F_2C{=}CFSiCl_3 + HCl$$

Fluoroform behaves similarly, yielding $CF_2{=}CFCF_3$ (458).

The large number of cyclic addition products of tetrafluoroethylene are not duplicated with chlorotrifluoroethylene because of the substitution of fluorine by a chlorine and the resulting change in steric, inductive, and hyperconjugative effects. Tetrafluoroethylene cyclizes readily. Chlorotrifluoroethylene dimerizes in a cyclic and a linear form (459–462):

$$CClF{=}CF_2 \xrightarrow[\text{8 hr}]{200°} \begin{array}{c} F_2C{-}CF_2 \\ | \quad | \\ ClFC{-}CFCl \end{array}$$

$$CClF{=}CF_2 \xrightarrow[\text{Pyrex}]{550°} \text{cyclic product} + CClF_2CClFCF{=}CF_2$$

In the cyclic structure a head-to-head and tail-to-tail addition (460) takes place. Of the two possible isomers the ratio is 5 parts of cis to 1 part of trans when run in glassware at 300 to 500°C (200), but 56 to 44 cis to trans when cyclized at 150 to 300°C in stainless-steel cylinders at 1,500 to 2,800 psi (463). A comparison was made of the dimerizations and inter-dimerization of tetrafluoroethylene and chlorotrifluoroethylene by measuring the second-order velocity constants. This is shown in Table 26 (200).

Although chlorotrifluoroethylene has less tendency to dimerize and does so at a lower velocity than tetrafluoroethylene, a dimer is formed with 1-ethynylcyclohexene (464). With acetaldehyde and ultraviolet light

TABLE 26

Second-order Velocity Constants for Dimerization and Interdimerization of Tetra-fluoroethylene and Chlorotrifluoroethylene

Reactants	Second-order velocity constant, cc/mole-sec	Normalized reaction constants based on chlorotrifluoroethylene
Tetrafluoroethylene	16.5×10^{10}	4.70
Tetrafluoroethylene + chlorotrifluoroethylene	8.54×10^{10}	2.42
Chlorotrifluoroethylene	3.53×10^{10}	1

chlorotrifluoroethylene forms mainly the ketone, CH_3COCF_2CHFCl, but it also cyclizes to the oxetane (196):

$$
\begin{array}{ccc}
F_2C & \!\!\!-\!\!\! & CFCl \\
| & & | \\
O & \!\!\!-\!\!\! & \underset{\underset{H}{|}}{C}CH_3 \\
\end{array}
$$

Butadiene and tetrafluoroethylene form a cyclic ethynyl compound. Chlorotrifluoroethylene also reacts with butadiene to give an unidentified product that does not appear to be cyclic (465). Chlorotrifluoroethylene and acrylonitrile yield a cyclobutane derivative that, on hydrolysis and cleavage, yields α,α-difluoroglutaric acid (466).

Hydrogenation of chlorotrifluoroethylene proceeds at room temperature with a palladium-alumina catalyst to yield 75 percent CHF_2CHF_2 and low concentrations of $CF_2{=}CHF$ (203). With palladium alone, a 60 percent yield of $CF_2{=}CHF$ and a 25 percent yield of CHF_2CHF_2 were obtained. Thermal measurements of the reaction were reported (205).

Oxidation of chlorotrifluoroethylene occurs at 30° to give a liquid polymeric peroxide $(C_2F_3ClOO)_n$ and ClF_2CCOF as the main product, with COF_2 and $COFCl$ in lesser amounts. The reaction is believed to be a 1:1 copolymerization of oxygen and chlorotrifluoroethylene, with the other products resulting from the decomposition of the copolymer (467). In the absence of a catalyst, good yields of ClF_2CCOF are obtained at 100 to 300 psi and 25 to 50°C (468). When chlorotrifluoroethylene is oxidized in the dark in the presence of water and in glass, a 77 percent yield of ClF_2CCOF and small amounts of CO_2, SiF_4, and COF_2 are formed (469). With copper catalysis, the products are ClF_2CCOF, $COClF$, COF_2, and CF_3COF (470).

F. Polymerization

Polymerization of chlorotrifluoroethylene is conducted much the same as the polymerization of tetrafluoroethylene. High-pressure, stirred autoclaves are used for the reaction, which is conducted in a barricaded area. Water is the preferred polymerization medium, although solvents such as perfluorodimethylcyclobutane (228) and trichlorofluoromethane (471) have been used. Oxygen-free deionized water is added to the autoclave, which is evacuated and flushed with nitrogen. After addition of initiators, buffers, and dispersants, in an emulsion polymerization, the chlorotrifluoroethylene is introduced. Additional monomer is charged during the course of the reaction in order to maintain constant pressure. Agitation must be adequate to remove the heat of polymerization and maintain the dispersion or suspension uniformity. Suspension polymerizations are preferred, since chain transfer, telomerization, and inhibition due to the dispersant are avoided under these conditions. Perfluorooctanoic acid has been used successfully, however, as an emulsifier for a redox emulsion polymerization (472).

Chlorotrifluoroethylene was first polymerized in suspension (384). In a recent example (473), an autoclave was completely filled (no gas-free space) with 600 parts of an aqueous solution of 1.3 parts of potassium persulfate, 1 part of sodium bisulfite, and enough sodium acetate to bring the pH to 4.5. Chlorotrifluoroethylene (130 parts) was then added, displacing liquid to a second vessel. Polymerization took place at 30°C to yield 106 parts of poly(chlorotrifluoroethylene). Salts of iron (474,475), silver (476,477), cobalt, and nickel (478) have been used to increase the polymerization rate in the redox system. Treatment of the monomer with 0.05 to 1 percent terpene prior to polymerization reportedly improves the thermal stability of the resulting polymer (479). Hydrogen peroxide (480) has also been used to initiate chlorotrifluoroethylene polymerizations, as has ozone (471,481).

Various organic peroxides are initiators for chlorotrifluoroethylene polymerization. Trichloroacetyl peroxide is effective at temperatures as low as −20° (482) and, in combination with zinc, gives up to 25 percent conversion at flow rates up to 4 ft per sec (483). Trifluoropropionyl peroxide (484), benzoyl peroxide (485), acetyl peroxide, and azobisisobutyronitrile (486) have been thermally activated as initiators. Ultraviolet radiation was used to decompose ketoacids such as 3-ketopimelic acid to polymerize chlorotrifluoroethylene (237).

Organometallics initiate polymerization of chlorotrifluoroethylene. These include trialkylboranes, which function from −60° to 40°C (319, 487), borane, which must be free of oxygen and water (488), Ziegler systems (489), and chlorodiisobutylaluminum–titanium tetrafluoride catalysts on silica (233).

Gamma irradiation–initiated polymerization has been reported at liquid nitrogen temperature (379), 0° (318), and up to 100° (239). Within the limits of 30 to 1,000 r per hr, the polymerization rate is directly related to the square root of the radiation intensity (490). An electrolytic polymerization using chlorotrifluoroethylene, HF and KF at −300° and HF and KF at −30°, and 10^{-3} amp per sq cm for 7 days yielded 55 percent of the polymer (491).

G. Applications

Chlorotrifluoroethylene is used in the form of its polymers, copolymers, and telomers. The homopolymer, because of its excellent chemical resistance, is valuable to the chemical industry as seals, gaskets, valve seats and liners, and coatings. Electrical applications are wire insulation, electrical insulation of other kinds, and flexible printed circuits. In film form it is used for protective packaging, outdoor coverings, and windows as well as covers and windows for instruments. The oils, greases, and waxes formed by telomerization are used in corrosive and oxidative environments where other lubricants are ineffective or hazardous. The copolymer of chlorotrifluoroethylene with vinylidene fluoride is sold by the Three M Company as two different compositions: a 50:50 chlorotrifluoroethylene–vinylidene fluoride product designated Kel F 5500, and a 30:70 chlorotrifluoroethylene–vinylidene fluoride product designated Kel F 3700. These fluorinated elastomers were developed originally for military purposes where oil resistance and low-temperature utility were required.

References

1. *Oil, Paint Drug Reptr.*, **189**(Jan. 3):33 (1966).
2. O. Ruff and O. Bretschneider, *Z. Anorg. Allgem. Chem.*, **210**:173 (1933); *Chem. Abstr.*, **27**:2131 (1933).
3. F. Swarts, *Bull. Acad. Roy. Belg.*, **24**(3):474 (1892); *ibid.*, **29**(3):874 (1895).
4. T. Midgley, Jr. and A. L. Henne, *Ind. Eng. Chem.*, **22**:542 (1930).
5. R. J. Plunkett (to Kinetic Chemicals, Inc.), U.S. Patent 2,230,654, Feb. 4, 1941; *Chem. Abstr.*, **35**:3365 (1941).
6. M. M. Brubaker (to E. I. du Pont de Nemours & Co.), U.S. Patent 2,393,967, Feb. 5, 1946; *Chem. Abstr.*, **40**:3648 (1946).
7. R. B. Richards, Polyethylene, in "Encyclopedia of Chemical Technology," vol. 10, p. 938, Interscience, New York, 1953.
8. S. Sherratt, Polytetrafluoroethylene, in "Encyclopedia of Chemical Technology," 2d ed., vol. 9, pp. 805–831, Interscience, New York, 1966.
9. J. D. Park, A. F. Benning, F. B. Downing, J. F. Laucius, and R. C. McHarness, *Ind. Eng. Chem.*, **39**:354 (1947).
10. J. W. Edwards and P. A. Small, *Ind. Eng. Chem. Fundamentals*, **4**(4):396 (1965).
11. British Patent 917,093 (to E. I. du Pont de Nemours & Co.), Jan. 30, 1963; *Chem. Abstr.*, **59**:6255 (1963).
12. French Patent 1,397,032 (to Daikin Kogyo Co. Ltd.), Apr. 23, 1965; *Chem. Abstr.*, **63**:8197 (1965).
13. R. Müller, H. Fischer, and G. Seitz, German (East) Patent 13,224, May 13, 1957; *Chem. Abstr.*, **53**:2088 (1959).
14. V. U. Shevchuk, M. B. Fagarash, S. S. Abadzhev, A. L. Bel'ferman, and I. D. Kushina, U.S.S.R. Patent 170,962, May 11, 1965; *Chem. Abstr.*, **63**:9807 (1965).
15. F. B. Downing, A. F. Benning, and R. C. McHarness (to E. I. du Pont de Nemours & Co.), U.S. Patent 2,551,573, May 8, 1951; *Chem. Abstr.*, **45**:9072 (1951).
16. H. Madai, German (East) Patent 42,730, Jan. 5, 1966; *Chem. Abstr.*, **64**:17421 (1966).
17. A. C. Knight (to E. I. du Pont de Nemours & Co.), French Patent 1,333,489, Jul. 26, 1963; *Chem. Abstr.*, **60**:2753 (1964).

18. Netherlands Appl. 6,512,899 (to Pennsalt Chemicals Corp.), Apr. 6, 1966; *Chem. Abstr.*, **65**:5366 (1966).
19. J. Serpinet, *Chim. Anal. (Paris)*, **41**:146 (1959).
20. J. M. Hamilton, Jr., The Organic Fluorochemicals Industry, in M. Stacey, J. C. Tatlow, and A. G. Sharpe (eds.), "Advances in Fluorine Chemistry," vol. 3, p. 146, Butterworth, London, 1963.
21. O. Scherer, A. Steinmetz, H. Kühn, W. Wetzel, and K. Grafen (to Farbwerke Hoechst A.-G.), German Patent 1,073,475, Jan. 21, 1960; *Chem. Abstr.*, **55**: 17498 (1961).
22. M. Hisazumi and H. Niimiya, Japanese Patent 15,353, Oct. 14, 1960; *Chem. Abstr.*, **55**:5345 (1961).
23. French Patent 1,354,341 (to Osaka Kinzoku Kogyo Co. Ltd.), Mar. 6, 1964; *Chem. Abstr.*, **61**:4213 (1964).
24. French Patent 1,362,042 (to Osaka Kinzoku Kogyo Co. Ltd.), May 29, 1964; *Chem. Abstr.*, **61**:10589 (1964).
25. Netherlands Appl. 302,391 (to Farbwerke Hoechst A.-G.), Jun. 25, 1964; *Chem. Abstr.*, **62**:7636 (1965).
26. K. S. Revell (to Imperial Chemical Industries Ltd.), British Patent 983,222, Feb. 10, 1965; *Chem. Abstr.*, **62**:16052 (1965).
27. Belgian Patent 636,186 (to Imperial Chemical Industries Ltd.), Feb. 14, 1964; *Chem. Abstr.*, **61**:14527 (1964).
28. H. R. Davis and S. H. K. Chiang (to M. W. Kellogg Co.), U.S. Patent 2,774,798, Dec. 18, 1956; *Chem. Abstr.*, **51**:12954 (1957).
29. A. F. Benning, F. B. Downing, and R. J. Plunkett (to Kinetic Chemicals, Inc.), U.S. Patent 2,401,897, Jun. 11, 1946; *Chem. Abstr.*, **40**:5066 (1946).
30. Netherlands Appl. 6,411,278 (to Allied Chemical Corp.), Sept. 28, 1964; *Chem. Abstr.*, **63**:13074 (1965).
31. E. E. Aynsley and R. H. Watson, *J. Chem. Soc.*, **1955**:576.
32. R. M. Mantell (to M. W. Kellogg Co.), U.S. Patent 2,697,127, Dec. 14, 1954; *Chem. Abstr.*, **50**:2650 (1956).
33. A. F. Benning and E. G. Young (to E. I. du Pont de Nemours & Co.), U.S. Patent 2,615,926, Oct. 28, 1952; *Chem. Abstr.*, **47**:8770 (1953).
34. British Patent 732,269 (to E. I. du Pont de Nemours & Co.), Jun. 22, 1955; *Chem. Abstr.*, **50**:10122 (1956).
35. R. E. Burk (to E. I. du Pont de Nemours & Co.), German Patent 1,038,535, Sept. 11, 1958; *Chem. Abstr.*, **55**:3249 (1961).
36. M. W. Farlow (to E. I. du Pont de Nemours & Co.), U.S. Patent 3,081,245, Mar. 12, 1963; *Chem. Abstr.*, **59**:2645 (1963).
37. W. R. Von Tress (to The Dow Chemical Co.), U.S. Patent 3,133,871, May 19, 1964; *Chem. Abstr.*, **61**:5516 (1964).
38. R. F. Baddour and B. R. Bronfin, *Ind. Eng. Chem.*, *Process Design Develop.*, **4**(2):162 (1965).
39. M. W. Farlow and E. T. Muetterties (to E. I. du Pont de Nemours & Co.), U.S. Patent 2,732,410, Jan. 24, 1956; *Chem. Abstr.*, **50**:15574 (1956).
40. M. W. Farlow and E. T. Muetterties (to E. I. du Pont de Nemours & Co.), U.S. Patent 2,732,411, Jan. 24, 1956; *Chem. Abstr.*, **50**:15574 (1956).
41. C. S. Cleaver and M. W. Farlow (to E. I. du Pont de Nemours & Co.), U.S. Patent 2,902,521, Sept. 1, 1959; *Chem. Abstr.*, **54**:2166 (1960).
42. J. P. Landis and C. H. Manwiller (to E. I. du Pont de Nemours & Co.), U.S. Patent 2,929,771, Mar. 22, 1960; *Chem. Abstr.*, **54**:10601 (1960).

43. M. W. Farlow (to E. I. du Pont de Nemours & Co.), U.S. Patent 2,981,761, Apr. 25, 1961; *Chem. Abstr.*, **55**:20955 (1961).

44. M. W. Farlow and R. M. Joyce, Jr. (to E. I. du Pont de Nemours & Co.), U.S. Patent 2,709,183, May 24, 1955; *Chem. Abstr.*, **50**:6499 (1956).

45. M. W. Farlow (to E. I. du Pont de Nemours & Co.), U.S. Patent 2,980,739, Apr. 18, 1961; *Chem. Abstr.*, **55**:20955 (1961).

46. M. W. Farlow and E. L. Muetterties (to E. I. du Pont de Nemours & Co.), U.S. Patent 2,725,410, Nov. 29, 1955; *Chem. Abstr.*, **52**:2885 (1958).

47. W. O. Forshey, Jr. (to E. I. du Pont de Nemours & Co.), U.S. Patent 2,941,012, Jun. 14, 1960; *Chem. Abstr.*, **54**:19481 (1960).

48. M. W. Farlow and E. L. Muetterties (to E. I. du Pont de Nemours & Co.), U.S. Patent 2,709,187, May 24, 1955; *Chem. Abstr.*, **50**:6499 (1956).

49. M. W. Farlow and E. L. Muetterties (to E. I. du Pont de Nemours & Co.), U.S. Patent 2,709,189, May 24, 1955; *Chem. Abstr.*, **50**:6499 (1956).

50. E. E. Lewis and M. A. Naylor, *J. Am. Chem. Soc.*, **69**:1968 (1947).

51. E. E. Lewis (to E. I. du Pont de Nemours & Co.), U.S. Patent 2,406,153, Aug. 20, 1946; *Chem. Abstr.*, **40**:7703 (1946).

52. A. F. Benning, F. B. Downing, and J. D. Park (to Kinetic Chemicals, Inc.), U.S. Patent 2,394,581, Feb. 12, 1946; *Chem. Abstr.*, **40**:3460 (1946).

53. L. J. Hals, T. S. Reid, and G. H. Smith, Jr., *J. Am. Chem. Soc.*, **73**:4054 (1951).

54. L. J. Hals, T. S. Reid, and G. H. Smith (to Minnesota Mining & Manufacturing Co.), U.S. Patent 2,668,864, Feb. 9, 1954; *Chem. Abstr.*, **49**:2478 (1955).

55. A. D. Kirshenbaum, A. G. Streng, and M. Hauptschein, *J. Am. Chem. Soc.*, **75**:3141 (1953).

56. J. D. La Zerte, L. J. Hals, T. S. Reid, and G. H. Smith, *ibid.*, **75**:4525 (1953).

57. J. D. La Zerte (to Minnesota Mining & Manufacturing Co.), U.S. Patent 2,601,536, Jun. 24, 1952; *Chem. Abstr.*, **47**:1725 (1953).

58. R. E. Brooks (to E. I. du Pont de Nemours & Co.), U.S. Patent 2,926,203, Feb. 23, 1960; *Chem. Abstr.*, **54**:15244 (1960).

59. R. A. Mitsch, *J. Heterocyclic Chem.*, **1**(1):59 (1964).

60. G. H. Miller and J. R. Dacey, *J. Phys. Chem.*, **69**(4):1434 (1965).

61. J. N. Butler, *J. Am. Chem. Soc.*, **84**:1393 (1962).

62. J. L. Anderson (to E. I. du Pont de Nemours & Co.), U.S. Patent 2,733,278, Jan. 31, 1956; *Chem. Abstr.*, **50**:15575 (1956).

63. S. V. R. Mastrangelo (to E. I. du Pont de Nemours & Co.), U.S. Patent 3,196,114, Jul. 20, 1965; *Chem. Abstr.*, **63**:9807 (1965).

64. British Patent 868,020 (to The Dow Chemical Co.), May 17, 1961; *Chem. Abstr.*, **55**:24332 (1961).

65. J. Heicklen and V. Knight, *J. Phys. Chem.*, **69**(10):3600 (1965).

66. D. Phillips, *ibid.*, **70**(4):1235 (1966).

67. M. Lenzi and A. Mele, *J. Chem. Phys.*, **43**(6):1974 (1965).

68. F. S. Fawcett and R. D. Lipscomb, *J. Am. Chem. Soc.*, **86**:2576 (1964).

69. C. B. Miller and C. Woolf (to Allied Chemical & Dye Corp.), U.S. Patent 2,741,634, Apr. 10, 1956; *Chem. Abstr.*, **51**:460 (1957).

70. E. Tschuikow-Roux and J. E. Marte, *J. Chem. Phys.*, **42**(6):2049 (1965).

71. J. W. Hodgins and R. L. Haines, *Can. J. Chem.*, **30**:473 (1952).

72. H. Grüss, German Patent 887,648, Feb. 1, 1954; *Chem. Abstr.*, **52**:14649 (1958).

73. D. E. M. Evans (to Imperial Chemical Industries Ltd.), British Patent 922,832, Apr. 3, 1963; *Chem. Abstr.*, **59**:9789 (1963).

74. A. Wheeler (to E. I. du Pont de Nemours & Co.), U.S. Patent 2,847,391, Aug. 12, 1958; *Chem. Abstr.*, **53**:2691 (1959).
75. Netherlands Appl. 6,400,982 (to E. I. du Pont de Nemours & Co.), Aug. 7, 1964; *Chem. Abstr.*, **62**:2707 (1965).
76. French Patent 1,357,773 (to Osaka Kinzoku Kogyo Co. Ltd.), Apr. 10, 1964; *Chem. Abstr.*, **61**:4213 (1964).
77. K. Okamura, Y. Kometani, and M. Tatemoto (to Thiokol Chemical Corp.), U.S. Patent 3,221,070, Nov. 30, 1965; *Chem. Abstr.*, **64**:4936 (1966).
78. M. Mitani, M. Tatemoto, T. Fujii, and Y. Furukawa (to Daikin Kogyo Co. Ltd.), Japanese Patent 22,574, Oct. 6, 1965; *Chem. Abstr.*, **64**:4938 (1966).
79. A. H. Fainberg, D. S. Fetterman, and M. Hauptschein (to Pennsalt Chemicals Corp.), French Patent 1,365,315, Jul. 3, 1964; *Chem. Abstr.*, **61**:14527 (1964).
80. W. R. Von Tress (to The Dow Chemical Co.), U.S. Patent 3,236,030, Feb. 22, 1966; *Chem. Abstr.*, **64**:12547 (1966).
81. French Patent 1,346,901 (to Osaka Kinzoku Kogyo Co. Ltd.), Dec. 20, 1963; *Chem. Abstr.*, **60**:10821 (1964).
82. Y. Kometani, T. Sueyoshi, and M. Tatemoto (to Thiokol Chemical Corp.), U.S. Patent 3,218,364, Nov. 16, 1965; *Chem. Abstr.*, **64**:4939 (1966).
83. O. Scherer, H. Hahn, and G. Schneider (to Farbwerke Hoechst A.-G.), German Patent 1,068,247, Nov. 5, 1959; *Chem. Abstr.*, **55**:12295 (1961).
84. U. Onken and A. Pebler (to Farbwerke Hoechst A.-G.), German Patent 1,206,424, Dec. 9, 1965; *Chem. Abstr.*, **64**:8030 (1966).
85. A. H. Fainberg and M. Hauptschein (to Pennsalt Chemicals Corp.), U.S. Patent 3,215,747, Nov. 2, 1965; *Chem. Abstr.*, **64**:590 (1966).
86. M. M. Renfrew and E. E. Lewis, *Ind. Eng. Chem.*, **38**:870 (1946).
87. W. M. D. Bryant, *J. Polymer Sci.*, **56**:277 (1962).
88. G. T. Furukawa, R. E. McCoskey, and M. L. Reilly, *J. Res. Natl. Bur. Standards*, **51**:69 (1953).
89. R. Bralsford, P. V. Harris, and W. C. Price, *Proc. Roy. Soc. (London)*, **A258**:459 (1960).
90. L. C. Nelson and E. F. Obert, *Chem. Eng.*, **61**(7):203 (1954).
91. D. E. Mann, N. Acquista, and E. K. Plyer, *J. Res. Natl. Bur. Standards*, **52**:67 (1954).
92. W. F. Beckwith and R. W. Fahien, *Chem. Eng. Progr.*, *Symp. Ser.*, **59**(44):75 (1963).
93. J. H. Simons, "Fluorine Chemistry," vol. 2, p. 476, Academic, New York, 1954.
94. C. R. Patrick, *Tetrahedron*, **4**:26 (1958).
95. J. Karle and I. L. Karle, *J. Chem. Phys.*, **18**:957, 963 (1950).
96. H. A. Bent, *ibid.*, **32**:1582 (1960).
97. J. R. Lacher, R. E. Scruby, and J. D. Park, *J. Am. Chem. Soc.*, **71**:1797 (1949).
98. P. G. T. Fogg and J. D. Lambert, *Proc. Roy. Soc. (London)*, **A232**:537 (1955).
99. W. P. Slichter, *J. Polymer Sci.*, **24**:173 (1957).
100. R. E. Naylor, Jr. and S. W. Lasoski, Jr., *ibid.*, **44**:1 (1960).
101. I. E. Volokhonovich, E. F. Nosov, and L. B. Zorina, *Zh. Fiz. Khim.*, **40**(1):268 (1966); *Chem. Abstr.*, **64**:11940 (1966).
102. A. I. Zhemerdeï, *Tr. Leningr. Sanit. Gigien. Med. Inst.*, **44**:164 (1958); *Chem. Abstr.*, **55**:2896 (1961).
103. D. K. Harris, *Brit. J. Ind. Med.*, **10**:255 (1953).
104. H. E. Stokinger et al., *Arch. Environmental Health*, **9**(4):545 (1964).

105. J. R. Lacher, L. E. Hummel, E. F. Bohmfalk, and J. D. Park, *J. Am. Chem. Soc.*, **72**:5486 (1950).

106. T. D. Coyle, S. L. Stafford, and F. G. A. Stone, *Spectrochim. Acta*, **17**:968 (1961).

107. R. H. Campbell and B. J. Gudzinowicz, *Anal. Chem.*, **33**:842 (1961).

108. H. Rotzsche, *Z. Anal. Chem.*, **175**:338 (1960); *Chem. Abstr.*, **55**:4233 (1961).

109. N. L. Arthur and T. N. Bell, *J. Chromatog.*, **15**:250 (1964).

110. K. L. Berry and J. H. Petersen, *J. Am. Chem. Soc.*, **73**:5195 (1951).

111. L. V. Chepel and F. V. Shemarov, *Dokl. Akad. Nauk SSSR*, **158**(3):682 (1964); *Chem. Abstr.*, **62**:637 (1965).

112. British Patent 620,296 (to E. I. du Pont de Nemours & Co.), Mar. 23, 1949; *Chem. Abstr.*, **43**:6218 (1949).

113. D. C. England, L. R. Melby, M. A. Dietrich, and R. V. Lindsey, Jr., *J. Am. Chem. Soc.*, **82**:5116 (1960).

114. D. D. Coffman, M. S. Raasch, G. W. Rigby, P. L. Barrick, and W. E. Hanford, *J. Org. Chem.*, **14**:747 (1949).

115. W. T. Miller, Jr., E. W. Fager, and P. H. Griswold, *J. Am. Chem. Soc.*, **70**:431 (1948).

116. J. D. Park, M. L. Sharrah, W. H. Breen, and J. R. Lacher, *ibid.*, **73**:1329 (1951).

117. W. E. Hanford (to E. I. du Pont de Nemours & Co.), U.S. Patent 2,443,003, Jun. 8, 1948; *Chem. Abstr.*, **42**:6841 (1948).

118. I. L. Knunyants and A. V. Fokin, *Izv. Akad. Nauk SSSR, Otd. Khim. Nauk*, **1952**: 261 *Chem. Abstr.*, **47**:3221 (1953).

119. D. W. Wiley (to E. I. du Pont de Nemours & Co.), U.S. Patent 2,988,537, Jun. 13, 1961; *Chem. Abstr.*, **56**:330 (1962).

120. R. N. Sterlin, V. E. Bogachev, R. D. Yatsenko, and I. L. Knunyants, *Izv. Akad. Nauk SSSR, Otd. Khim. Nauk*, **1959**:2151; *Chem. Abstr.*, **54**:10848 (1960).

121. M. A. Raksha and Yu. V. Popov, *Zh. Obshch. Khim.*, **34**(10):3465 (1964); *Chem. Abstr.*, **62**:3929 (1965).

122. British Patent 583,264 (to E. I. du Pont de Nemours & Co.), Dec. 13, 1946; *Chem. Abstr.*, **41**:3479 (1947).

123. G. W. Rigby (to E. I. du Pont de Nemours & Co.), British Patent 607,103, Aug. 25, 1948; *Chem. Abstr.*, **43**:1444 (1949).

124. A. L. Henne and R. L. Pelley, *J. Am. Chem. Soc.*, **74**:1426 (1952).

125. R. A. Carboni and R. V. Lindsey, Jr., *ibid.*, **80**:5793 (1958).

126. A. V. Fokin, Yu. N. Studnev, and N. A. Proshin, U.S.S.R. Patent 172,822, Jul. 7, 1965; *Chem. Abstr.*, **64**:591 (1966).

127. T. F. McGrath and R. Levine, *J. Am. Chem. Soc.*, **77**:4168 (1955).

128. S. Dixon, *J. Org. Chem.*, **21**:400 (1956).

129. Ref. 114.

130. R. N. Haszeldine and J. E. Osborne, *J. Chem. Soc.*, **1956**:61.

131. S. Andreades, *J. Org. Chem.*, **27**:4163 (1962).

132. D. A. Barr and R. N. Haszeldine, *J. Chem. Soc.*, **1960**:1151.

133. F. S. Fawcett, C. W. Tullock, and D. D. Coffman, *J. Am. Chem. Soc.*, **84**:4275 (1962).

134. R. T. Holzmann and M. S. Cohen, *Inorg. Chem.*, **1**:972 (1962).

135. O. P. Petrii, A. A. Makhina, T. V. Talalaeva, and K. A. Kocheshkov, *Dokl. Akad. Nauk SSSR*, **167**(3):594 (1966); *Chem. Abstr.*, **64**:19462 (1966).

136. G. H. Crawford, *J. Polymer Sci.*, **45**:259 (1960).

137. J. D. Park, A. P. Stefani, G. H. Crawford, and J. R. Lacher, *J. Org. Chem.*, **26**:3316 (1961).
138. British Patent 983,486 (to Minnesota Mining & Manufacturing Co.), Feb. 17, 1965; *Chem. Abstr.*, **62**:11932 (1965).
139. J. D. Park, A. P. Stefani, and J. R. Lacher, *J. Org. Chem.*, **26**:3319 (1961).
140. R. A. Rhein and G. H. Cady, *Inorg. Chem.*, **3**(11):1644 (1964).
141. S. M. Williamson and G. H. Cady, *ibid.*, **1**:673 (1962).
142. W. P. Gilbreath and G. H. Cady, *ibid.*, **2**:496 (1963).
143. B. Tittle and G. H. Cady, *ibid.*, **4**(2):259 (1965).
144. I. L. Knunyants and A. V. Fokin, *Dokl. Akad. Nauk SSSR*, **111**:1035 (1956); *Chem. Abstr.*, **51**:9472 (1957).
145. A. I. Titov, *Dokl. Akad. Nauk SSSR*, **149**(2):330 (1963); *Chem. Abstr.*, **59**:6215 (1963).
146. H. C. Clark and J. H. Tsai, *Chem. Commun.* (*London*), **1965**(6):111.
147. *ibid.*, **1965**(6):123.
148. R. Mason and D. R. Russell, *ibid.*, **1965**(10):182.
149. G. W. Parshall and F. N. Jones, *J. Am. Chem. Soc.*, **87**:5356 (1965).
150. M. J. Mays and G. Wilkinson, *J. Chem. Soc.*, **1965**:6629.
151. R. Cramer and G. W. Parshall, *J. Am. Chem. Soc.*, **87**:1392 (1965).
152. H. H. Hoehn, L. Pratt, K. F. Watterson, and G. Wilkinson, *J. Chem. Soc.*, **1961**:2738.
153. V. A. Ginsburg, L. L. Martynova, S. S. Dubov, B. I. Tetebaum, and A. Ya. Yakubovich, *Zh. Obshch. Khim.*, **35**(5):851 (1967).
154. D. A. Barr and R. N. Haszeldine, *Nature*, **175**:991 (1955).
155. C. E. Griffin and R. N. Haszeldine, *J. Chem. Soc.*, **1960**:1398.
156. D. P. Graham and V. Weinmayr, *J. Org. Chem.*, **31**:957 (1966).
157. P. L. Barrick (to E. I. du Pont de Nemours & Co.), U.S. Patent 2,403,207, Jul. 2, 1946; *Chem. Abstr.*, **40**:6497 (1946).
158. J. M. Shreeve and G. H. Cady, *J. Am. Chem. Soc.*, **83**:4521 (1961).
159. C. G. Krespan (to E. I. du Pont de Nemours & Co.), U.S. Patent 3,119,836, Jan. 28, 1964; *Chem. Abstr.*, **60**:15834 (1964).
160. R. P. Buckley and M. Szwarc, *J. Am. Chem. Soc.*, **78**:5696 (1956).
161. J. M. Tedder and J. C. Walton, *Proc. Chem. Soc.*, **1964**(Dec.):420.
162. A. P. Stefani, L. Herk, and M. Szwarc, *J. Am. Chem. Soc.*, **83**:4732 (1961).
163. B. A. Grzybowska, J. H. Knox, and A. F. Trotman-Dickenson, *J. Chem. Soc.*, **1963**:746.
164. N. L. Arthur and T. N. Bell, *ibid.*, **1962**:4866.
165. R. N. Haszeldine and B. R. Steele, *ibid.*, **1954**:3747.
166. A. F. Benning, F. B. Downing, and R. J. Plunkett (to Kinetic Chemicals, Inc.), U.S. Patent 2,393,304, Jan. 22, 1946; *Chem. Abstr.*, **40**:2453 (1946).
167. R. E. Banks, "Fluorocarbons and Their Derivatives," p. 33, Oldbourne, London, 1965.
168. E. Castellano, N. R. Bergamin, and H. J. Schumacher, *Z. Physik. Chem.* (*Frankfurt*), **27**:112 (1961); *Chem. Abstr.*, **55**:14028 (1961).
169. R. N. Haszeldine and K. Leedham, *J. Chem. Soc.*, **1953**:1548.
170. M. S. Raasch (to E. I. du Pont de Nemours & Co.), U.S. Patent 2,451,411, Oct. 12, 1948; *Chem. Abstr.*, **43**:6645 (1949).
171. M. A. A. Beg and H. C. Clark, *Chem. Ind.* (*London*), **1962**:140.
172. C. G. Krespan and V. A. Engelhardt, *J. Org. Chem.*, **23**:1565 (1958).
173. C. Barnetson, H. C. Clark, and J. T. Kwon, *Chem. Ind.* (*London*), **1964**:458.

174. R. N. Haszeldine and R. J. Marklow, *J. Chem. Soc.*, **1956**:962.
175. A. M. Geyer and R. N. Haszeldine, *Nature*, **178**:808 (1956).
176. V. A. Ponomarenko, V. G. Cherkaev, A. D. Petrov, and N. A. Zadorozhnyĭ, *Izv. Akad. Nauk SSSR, Otd. Khim. Nauk*, **1958**:247; *Chem. Abstr.*, **52**:12751 (1958).
177. V. A. Ponomarenko, V. G. Cherkaev, and N. A. Zadorozhnyĭ, *ibid.*, **1960**: 1610; *Chem. Abstr.*, **55**:9261 (1961).
178. R. N. Haszeldine, *J. Chem. Soc.*, **1949**:2856.
179. R. N. Haszeldine, *ibid.*, **1953**:3761.
180. J. A. Bittles, Jr., and R. M. Joyce, Jr. (to E. I. du Pont de Nemours & Co.), U.S. Patent 2,559,754, Jul. 10, 1951; *Chem. Abstr.*, **46**:1026 (1952).
181. P. L. Barrick and R. E. Christ (to E. I. du Pont de Nemours & Co.), U.S. Patent 2,436,135, Feb. 17, 1948; *Chem. Abstr.*, **42**:3771 (1948).
182. W. E. Hanford (to E. I. du Pont de Nemours & Co.), U.S. Patent 2,411,159, Nov. 19, 1946; *Chem. Abstr.*, **41**:856 (1947).
183. A. A. Beer, P. A. Zagorets, V. F. Inozemtsev, G. S. Povkh, and A. I. Popov, *Neftekhimiya*, **2**:617 (1962); *Chem. Abstr.*, **58**:8885 (1963).
184. H. L. Roberts (to Imperial Chemical Industries Ltd.), British Patent 898,309, Jun. 6, 1962; *Chem. Abstr.*, **57**:15360 (1962).
185. R. D. Chambers, J. Hutchinson, R. H. Mobbs, and W. K. R. Musgrave, *Tetrahedron*, **20**(3):497 (1964).
186. B. Tittle and A. E. Platt (to Imperial Chemical Industries Ltd.), British Patent 1,007,542, Oct. 13, 1965; *Chem. Abstr.*, **64**:1956 (1966).
187. G. C. Jeffrey (to The Dow Chemical Co.), U.S. Patent 3,235,611, Feb. 15, 1966; *Chem. Abstr.*, **64**:17421 (1966).
188. A. F. Benning, F. B. Downing, and J. D. Park (to Kinetic Chemicals, Inc.), U.S. Patent 2,394,581, Feb. 12, 1946; *Chem. Abstr.*, **40**:3460 (1946).
189. E. E. Lewis and M. A. Naylor, *J. Am. Chem. Soc.*, **69**:1968 (1947).
190. B. Atkinson and V. A. Atkinson, *J. Chem. Soc.*, **1957**:2086.
191. J. D. Roberts and C. M. Sharts, Cyclobutane Derivatives from Thermal Cyclo-addition Reactions, in R. Adams et al. (eds.), "Organic Reactions," vol. 12, pp. 1–56, Wiley, New York, 1962.
192. D. D. Coffman, P. L. Barrick, R. D. Cramer, and M. S. Raasch, *J. Am. Chem. Soc.*, **71**:490 (1949).
193. P. L. Barrick (to E. I. du Pont de Nemours & Co.), U.S. Patent 2,462,346, Feb. 22, 1949; *Chem. Abstr.*, **43**:4294 (1949).
194. J. J. Drysdale, W. W. Gilbert, H. K. Sinclair, and W. H. Sharkey, *J. Am. Chem. Soc.*, **80**:3672 (1958).
195. J. L. Anderson, R. E. Putnam, and W. H. Sharkey, *ibid.*, **83**:382 (1961).
196. E. R. Bissell and D. B. Fields, *J. Org. Chem.*, **29**:249 (1964).
197. R. D. Cramer (to E. I. du Pont de Nemours & Co.), U.S. Patent 2,456,176, Dec. 14, 1948; *Chem. Abstr.*, **43**:3445 (1949).
198. J. M. Birchall, A. J. Bloom, R. N. Haszeldine, and C. J. Willis, *J. Chem. Soc.*, **1962**:3021.
199. B. Atkinson and A. B. Trenwith, *J. Chem. Phys.*, **20**:754 (1952).
200. J. R. Lacher, G. W. Tompkin, and J. D. Park, *J. Am. Chem. Soc.*, **74**:1693 (1952).
201. D. A. Nelson (to E. I. du Pont de Nemours & Co.), U.S. Patent 2,758,138, Aug. 7, 1956; *Chem. Abstr.*, **51**:3654 (1957).
202. R. K. Steunenberg and G. H. Cady, *J. Am. Chem. Soc.*, **74**:4165 (1952).

203. I. L. Knunyants, M. P. Krasuskaya, and E. I. Mysov, *Izv. Akad. Nauk SSSR, Otd. Khim. Nauk*, 1960:1412; *Chem. Abstr.*, 55:349 (1961).
204. I. L. Knunyants, E. I. Mysov, and M. P. Krasuskaya, *ibid.*, 1958:906; *Chem. Abstr.*, 53:1102 (1959).
205. J. R. Lacher, A. Kianpour, F. Oetting, and J. D. Park, *Trans. Faraday Soc.*, 52:1500 (1956).
206. F. G. A. Stone and W. A. G. Graham, *Chem. Ind. (London)*, 1955:1181.
207. Kh. R. Raver, A. B. Bruker, and L. Z. Saborovskii, *Zh. Obshch. Khim.*, 35(7):1162 (1965).
208. K. L. Cordes, *Chem. Ind. (London)*, 1966:340.
209. J. Heicklen, *J. Phys. Chem.*, 70:477 (1966).
210. A. Pajaczkowski and J. W. Spoors, *Chem. Ind. (London)*, 1964:659.
211. J. Heicklen, V. Knight, and S. A. Greene, Aerospace Corp., Los Angeles, Calif., AD 603344 (1964) from *U.S. Govt. Res. Rept.*, 39(18):18 (1964).
212. N. Cohen and J. Heicklen, *J. Chem. Phys.*, 43(3):871 (1965).
213. V. Caglioti, M. Lenzi, and A. Mele, *Nature*, 201:610 (1964).
214. A. P. Modica and J. E. La Graff, NASA Accession No. N65-18141, Rept. No. AD 456292; *Chem. Abstr.*, 63:16162 (1965).
215. R. Kiyama, J. Osugi, and S. Kushuhara, *Rev. Phys. Chem. Japan*, 27:22, 24 (1957).
216. J. K. Ruff and R. F. Merritt, *J. Org. Chem.*, 30:3968 (1965).
217. M. M. Brubaker (to E. I. du Pont de Nemours & Co.), U.S. Patent 2,393,967, Feb. 5, 1946; *Chem. Abstr.*, 40:3648 (1946).
218. R. M. Joyce, Jr. (to E. I. du Pont de Nemours & Co.), U.S. Patent 2,394,243, Feb. 5, 1946; *Chem. Abstr.*, 40:3648 (1946).
219. M. M. Renfrew (to E. I. du Pont de Nemours & Co.), U.S. Patent 2,534,058, Dec. 12, 1950; *Chem. Abstr.*, 45:2262 (1951).
220. British Patent 604,580 (to E. I. du Pont de Nemours & Co.), Jul. 6, 1948; *Chem. Abstr.*, 43:905 (1949).
221. G. L. Dorough (to E. I. du Pont de Nemours & Co.), U.S. Patent 2,398,926, Apr. 23, 1946; *Chem. Abstr.*, 40:4389 (1946).
222. M. I. Bro and R. C. Schreyer (to E. I. du Pont de Nemours & Co.), U.S. Patent 3,032,543, May 1, 1962; *Chem. Abstr.*, 57:2426 (1962).
223. K. L. Berry (to E. I. du Pont de Nemours & Co.), U.S. Patent 2,559,752, Jul. 10, 1951; *Chem. Abstr.*, 46:3064 (1952).
224. A. F. Benning (to E. I. du Pont de Nemours & Co.), U.S. Patent 2,559,749, Jul. 10, 1951; *Chem. Abstr.*, 46:3066 (1952).
225. J. E. Duddington and S. Sherratt (to Imperial Chemical Industries Ltd.), British Patent 821,353, Oct. 7, 1959; *Chem. Abstr.*, 55:1084 (1961).
226. S. G. Bankoff (to E. I. du Pont de Nemours & Co.), U.S. Patent 2,612,484, Sept. 30, 1952; *Chem. Abstr.*, 47:3618 (1953).
227. C. E. Schildknecht, "Vinyl and Related Polymers," p. 485, Wiley, New York, 1952.
228. M. I. Bro, R. J. Convery, and R. C. Schreyer (to E. I. du Pont de Nemours & Co.), British Patent 840,080; Jul. 6, 1960; *Chem. Abstr.*, 55:1071 (1961).
229. E. Fischer and G. Bier (to Farbwerke Hoechst A.-G.), German Patent 1,086,433, Aug. 4, 1960; *Chem. Abstr.*, 55:19344 (1961).
230. O. W. Burke, Jr., British Patent 821,971, Oct. 14, 1959; *Chem. Abstr.*, 55:1088 (1961).
231. D. Sianesi and G. Caporiccio (to "Montecatini" Società Generale per l'Industria

1346   VINYL AND DIENE MONOMERS

Mineraria e Chimica), Belgian Patent 618,320, Sept. 17, 1962; *Chem. Abstr.*, **58**:9247 (1963).
232. T. Satokawa, M. Yoneya, and T. Yoshimura (to Osaka Kinzoku Kogyo Co. Ltd.), Japanese Patent 15,291, 1961; *Chem. Abstr.*, **57**:1075 (1962).
233. J. C. MacKenzie (to Cabot Corp.), French Patent 1,375,985, Oct. 23, 1964; *Chem. Abstr.*, **62**:11940 (1965).
234. British Patent 996,273 (to Allied Chemical Corp.), Jun. 23, 1965; *Chem. Abstr.*, **63**:18308 (1965).
235. B. Atkinson, *J. Chem. Soc.*, **1952**:2684.
236. D. D. Coffman (to E. I. du Pont de Nemours & Co.), U.S. Patent 3,047,553, Jul. 31, 1962; *Chem. Abstr.*, **57**:13990 (1962).
237. N. G. Carlson (to Minnesota Mining & Manufacturing Co.), U.S. Patent 2,912,373, Nov. 10, 1959; *Chem. Abstr.*, **54**:4047 (1960).
238. D. G. Marsh and J. Heicklen, *J. Am. Chem. Soc.*, **88**:269 (1966).
239. P. V. Zimakov, E. V. Volkova, A. V. Fokin, A. D. Sorokin, and V. M. Belikov, *Radioaktivn. Izotopy i Yadernye Izlucheniya v Nar. Khoz. SSSR, Tr. Vses. Soveshch. Riga,* **1**:219 (1960); *Chem. Abstr.*, **56**:6161 (1962).
240. M. A. Bruk, V. F. Gromov, I. V. Chernyak, P. M. Khomikovskii, and A. D. Abkin, *Vysokomolekul. Soedin.*, **8**(5):961 (1966); *Chem. Abstr.*, **65**:5540 (1966).
241. J. A. Bittles, Jr. (to E. I. du Pont de Nemours & Co.), U.S. Patent 2,628,958, Feb. 17, 1953; *Chem. Abstr.*, **47**:5728 (1953).
242. F. A. Bovey, J. F. Abere, G. B. Rathmann, and C. L. Sandberg, *J. Polymer Sci.*, **15**:520 (1955).
243. N. L. Jarvis and W. A. Zisman, Surface Chemistry of Fluorochemicals, in "Encyclopedia of Chemical Technology," 2d ed., vol. 9, p. 718, Interscience, New York, 1966.
244. F. Swarts, *Bull. Acad. Roy. Belg.*, **1901**:383; *J. Chem. Soc. Abstr.*, **1902**: 129.
245. F. Swarts, *Bull. Acad. Roy. Belg.*, **1909**:728; *Chem. Abstr.*, **5**:883 (1911).
246. D. D. Coffman and T. A. Ford (to E. I. du Pont de Nemours & Co.), U.S. Patent 2,419,009, Apr. 15, 1947; *Chem. Abstr.*, **41**:4964 (1947).
247. A. E. Newkirk, *J. Am. Chem. Soc.*, **68**:2467 (1946).
248. R. E. Burk, D. D. Coffman, and G. H. Kalb (to E. I. du Pont de Nemours & Co.), U.S. Patent 2,425,991, Aug. 19, 1947; *Chem. Abstr.*, **42**:198 (1948).
249. J. Harmon (to E. I. du Pont de Nemours & Co.), U.S. Patent 2,599,631, Jun. 10, 1952; *Chem. Abstr.*, **47**:1725 (1953).
250. D. D. Coffman and T. A. Ford (to E. I. du Pont de Nemours & Co.), U.S. Patent 2,419,010, Apr. 15, 1947; *Chem. Abstr.*, **41**:4964 (1947).
251. J. Söll (to I. G. Farbenindustrie A.-G.), U.S. Patent 2,118,901, May 31, 1938; *Chem. Abstr.*, **32**:5409 (1938).
252. R. M. Hedrick (to Monsanto Chemical Co.), U.S. Patent 2,695,320, Nov. 23, 1954; *Chem. Abstr.*, **50**:2652 (1956).
253. D. D. Coffman and R. D. Cramer (to E. I. du Pont de Nemours & Co.), U.S. Patent 2,461,523, Feb. 15, 1949; *Chem. Abstr.*, **43**:3437 (1949).
254. B. F. Skiles (to E. I. du Pont de Nemours & Co.), U.S. Patent 2,674,632, Apr. 6, 1954; *Chem. Abstr.*, **49**:4007 (1955).
255. L. F. Salisbury (to E. I. du Pont de Nemours & Co.), U.S. Patent 2,469,848, May 10, 1949; *Chem. Abstr.*, **43**:7035 (1949).
256. L. F. Salisbury (to E. I. du Pont de Nemours & Co.), U.S. Patent 2,519,199, Aug. 15, 1950; *Chem. Abstr.*, **45**:2496 (1951).

257. F. Nerdel, *Naturwissenschaften*, **39**:209 (1952); *Chem. Abstr.*, **47**:8632 (1953).
258. G. M. Whitman (to E. I. du Pont de Nemours & Co.), U.S. Patent 2,401,850, Jun. 11, 1946; *Chem. Abstr.*, **40**:6091 (1946).
259. R. Petit, C. Kaziz, and G. Wetroff (to Produits Chimiques Pechiney-Saint-Gobain), French Patent 1,325,750, May 3, 1963; *Chem. Abstr.*, **59**:11248 (1963).
260. R. Petit, C. Kaziz, and G. Wetroff (to Produits Chimiques Pechiney-Saint-Gobain), French Patent 1,324,408, Apr. 19, 1963; *Chem. Abstr.*, **59**:11247 (1963).
261. Kh. U. Usmanov, A. A. Yul'chibaev, G. S. Dordzhin, A. A. Patenko, and M. K. Asamov, U.S.S.R. Patent 174,622, Sept. 7, 1965; *Chem. Abstr.*, **64**:1956 (1966).
262. J. W. Clark (to Union Carbide & Carbon Corp.), U.S. Patent 2,626,963, Jan. 27, 1953; *Chem. Abstr.*, **48**:1407 (1954).
263. Netherlands Appl. 6,405,140 (to Diamond Alkali Co.), Nov. 11, 1964; *Chem. Abstr.*, **62**:10334 (1965).
264. F. J. Christoph, Jr., and G. Teufer (to E. I. du Pont de Nemours & Co.), French Patent 1,383,927, Jan. 4, 1965; *Chem. Abstr.*, **62**:11686 (1965).
265. L. Foulletier (to Société d'Electrochimie, d'Electrometallurgie et des Acieries Electriques d'Ugine), French Patent 1,410,317, Sept. 10, 1965; *Chem. Abstr.*, **64**:589 (1966).
266. I. M. Dolgopol'skii, K. Vaistariene, and J. Kriauciunas, *Lietuvos TSR Mokslu Akad. Darbai, Ser. B.*, no. 3, 95 (1965); through *Chem. Abstr.*, **64**:6474 (1966).
267. B. F. Skiles (to E. I. du Pont de Nemours & Co.), U.S. Patent 2,892,000, Jun. 23, 1959; *Chem. Abstr.*, **54**:1296 (1960).
268. British Patent 580,910 (to E. I. du Pont de Nemours & Co.), Sept. 24, 1946; *Chem. Abstr.*, **41**:2428 (1947).
269. P. J. Manno and W. H. Snaveley (to Continental Oil Co.), German Patent 1,210,799, Feb. 17, 1966; *Chem. Abstr.*, **64**:15740 (1966).
270. C. B. Linn (to Universal Oil Products Co.), U.S. Patent 2,762,849, Sept. 11, 1956; *Chem. Abstr.*, **51**:10556 (1957).
271. J. D. Calfee and F. H. Bratton (to Allied Chemical & Dye Corp.), U.S. Patent 2,462,359, Feb. 22, 1949; *Chem. Abstr.*, **43**:3834 (1949).
272. French Patent 1,399,051 (to Daikin Kogyo Co. Ltd.), May 14, 1965; *Chem. Abstr.*, **63**:5529 (1965).
273. D. Sianesi and G. Nelli (to "Montecatini" Società Generale per l'Industria Mineraria e Chimica), Belgian Patent 633,525, Dec. 12, 1963; *Chem. Abstr.*, **61**:1755 (1964).
274. British Patent 1,020,716 (to "Montecatini" Società Generale per l'Industria Mineraria e Chimica), Feb. 23, 1966; *Chem. Abstr.*, **64**:15740 (1966).
275. K. O. Christe and A. E. Pavlath, *J. Org. Chem.*, **30**:1639 (1965).
276. G. O. Pritchard, M. Venugopalan, and T. F. Graham, *J. Phys. Chem.*, **68**:1786 (1964).
277. L. E. Wolinski, Poly(vinyl Fluoride), in "Encyclopedia of Chemical Technology," 2d ed., vol. 9, pp. 835–840, Interscience, New York, 1966.
278. Belgian Patent 626,014 (to Kali-Chemie A.-G.), Mar. 29, 1963; *Chem. Abstr.*, **59**:9789 (1963).
279. K. H. Mieglitz, H. Heinze, and H. Linke (to Kali-Chemie A.-G.), German Patent 1,191,804, Apr. 29, 1965; *Chem. Abstr.*, **63**:6859 (1965).
280. K. Vaistariene and J. Kriauciunas, *Lietuvos TSR Mokslu Akad. Darbai, Ser. B*, no. 3, 103 (1965); through *Chem. Abstr.*, **64**:6474 (1966).

281. W. C. Percival (to E. I. du Pont de Nemours & Co.), U.S. Patent 2,917,556, Dec. 15, 1959; *Chem. Abstr.*, **54**:5461 (1960).
282. H. J. Bernstein, *J. Chem. Phys.*, **24**:910 (1956).
283. J. R. Scherer and W. J. Potts, *ibid.*, **31**:1691 (1959).
284. L. M. Sverdlov, Yu. V. Klochkovskiĭ, U. S. Kukina, and T. P. Mezhueva, *Opt. i Spektroskopiya*, **9**:728 (1960); *Chem. Abstr.*, **55**:17215 (1961).
285. P. Torkington, *Proc. Roy. Soc. (London)*, **A206**:17 (1951).
286. K. S. Pitzer and N. K. Freeman, *J. Chem. Phys.*, **14**:586 (1946).
287. A. R. H. Cole and H. W. Thompson, *Proc. Roy. Soc. (London)*, **A200**:10 (1949).
288. D. R. Lide, Jr., and D. Christensen, *Spectrochim. Acta*, **17**:665 (1961).
289. C. N. Banwell and N. Sheppard, *Proc. Roy. Soc. (London)*, **A263**:136 (1961).
290. C. Lifshitz and F. A. Long, *J. Phys. Chem.*, **67**(11):2463 (1963).
291. J. Kopecny, N. Lucanska, F. Sipka, E. Cerny, and D. Ambros, *Pracovni Lekar.*, **16**(7):301 (1964); *Chem. Abstr.*, **61**:16690 (1964).
292. H. T. Rein, H. E. Milville, and A. H. Fainberg, *Anal. Chem.*, **35**(10):1536 (1963).
293. T. J. Dougherty, *J. Am. Chem. Soc.*, **86**:460 (1964).
294. R. N. Haszeldine and B. R. Steele, *J. Chem. Soc.*, **1953**:1199.
295. J. M. Pearson and M. Szwarc, *Trans. Faraday Soc.*, **60**:553 (1964).
296. P. S. Dixon and M. Szwarc, *ibid.*, **59**:112 (1963).
297. J. D. Park, R. F. Seffl, and J. R. Lacher, *J. Am. Chem. Soc.*, **78**:59 (1956).
298. P. Tarrant and M. R. Lilyquist, *ibid.*, **77**:3640 (1955).
299. P. Tarrant, A. M. Lovelace, and M. R. Lilyquist, *ibid.*, **77**:2783 (1955).
300. P. I. Abell, *Trans. Faraday Soc.*, **60**(12):2214 (1964).
301. F. Casas, J. A. Kerr, and A. F. Trotman-Dickenson, *J. Chem. Soc.*, **1965**:1141.
302. E. K. Ellingboe (to E. I. du Pont de Nemours & Co.), U.S. Patent 2,439,203, Apr. 6, 1948; *Chem. Abstr.*, **42**:5046 (1949).
303. A. Ya. Yakubovich, V. A. Shpanskiĭ, and A. Lemke, *Dokl. Akad. Nauk SSSR*, **96**:773 (1954); *Chem. Abstr.*, **49**:8785 (1955).
304. A. Ya. Yakubovich, V. A. Shpanskiĭ, and A. Lemke, *Zh. Obshch. Khim.*, **24**:2257 (1954); *Chem. Abstr.*, **50**:206 (1956).
305. R. E. Banks, R. N. Haszeldine, H. Sutcliffe, and C. J. Willis, *J. Chem. Soc.*, **1965**:2506.
306. M. v. Stackelberg, *Rec. Trav. Chim.*, **75**:902 (1956).
307. M. v. Stackelberg, *Z. Elektrochem.*, **58**:162 (1954).
308. A. R. Trobridge and K. R. Jennings, *Trans. Faraday Soc.*, **61**:2168 (1965).
309. D. W. Woodward (to E. I. du Pont de Nemours & Co.), U.S. Patent 2,579,061, Dec. 18, 1951; *Chem. Abstr.*, **46**:3327 (1952).
310. G. H. Kalb, D. D. Coffman, T. A. Ford, and F. L. Johnston, *J. Appl. Polymer Sci.*, **4**(10):55 (1960).
311. E. Fischer and G. Bier (to Farbwerke Hoechst A.-G.), German Patent 1,086,433, Aug. 4, 1960; *Chem. Abstr.*, **55**:19344 (1961).
312. Netherlands Appl. 6,511,117 (to E. I. du Pont de Nemours & Co.), Feb. 28, 1966; *Chem. Abstr.*, **65**:3994 (1966).
313. F. L. Johnston and D. C. Pease (to E. I. du Pont de Nemours & Co.), U.S. Patent 2,510,783, Jun. 6, 1950; *Chem. Abstr.*, **46**:1299 (1952).
314. D. D. Coffman and T. A. Ford (to E. I. du Pont de Nemours & Co.), U.S. Patent 2,419,008, Apr. 15, 1947; *Chem. Abstr.*, **41**:4963 (1947).
315. J. M. Hamilton (to E. I. du Pont de Nemours & Co.), U.S. Patent 2,569,524, Oct. 2, 1951; *Chem. Abstr.*, **46**:1299 (1952).

316. W. Moschel, W. Müller, and H. Knopf (to Farbenfabriken Bayer A.-G.), German Patent 850,668, Sept. 25, 1952; *Chem. Abstr.*, **52**:10644 (1958).

317. F. A. Raal and C. J. Danby, *J. Chem. Soc.*, **1950**:1596.

318. P. J. Manno, *Am. Chem. Soc. Div. Polymer Chem. Preprints*, **4**(1):79 (1963).

319. Belgian Patent 562,433 (to SOLVIC Soc.), May 16, 1958; *Chem. Abstr.*, **53**:11887 (1959).

320. D. Sianesi, G. Caporiccio, and E. Strepparola (to "Montecatini" Società Generale per l'Industria Mineraria e Chimica), Belgian Patent 635,081, Nov. 18, 1963; *Chem. Abstr.*, **62**:663 (1965).

321. Kh. U. Usmanov, A. A. Yul'chibaev, R. Mukhamedzhanov, A. A. Gordienko, A. A. Patenko, G. S. Dordzhin, and A. Valiev, *Fiz. i Khim. Prirodn. i Sintetich. Polimerov, Akad. Nauk Uz. SSR, Inst. Khim. Polimerov*, no. 1, 205 (1962); through *Chem. Abstr.*, **59**:11666 (1963).

322. Yu. A. Sangalov, K. S. Minsker, G. A. Razuvaev, and A. S. Shevlyakov, *Plaste Kautschuk*, **10**(8):464 (1963); *Chem. Abstr.*, **60**:1844 (1964).

323. B. W. Howk and L. J. Plambeck (to E. I. du Pont de Nemours & Co.), U.S. Patent 2,499,097, Feb. 28, 1950; *Chem. Abstr.*, **44**:4723 (1950).

324. C. A. Thomas (to Monsanto Chemical Co.), U.S. Patent 2,406,717, Aug. 27, 1946; *Chem. Abstr.*, **40**:7702 (1946).

325. C. A. Thomas (to Monsanto Chemical Co.), U.S. Patent 2,362,960, Nov. 14, 1944; *Chem. Abstr.*, **39**:3179 (1945).

326. V. L. Simril and B. A. Curry, *J. Appl. Polymer Sci.*, **4**(10):62 (1960).

327. V. L. Simril and B. A. Curry, *Mod. Plastics*, **36**:121 (1959).

328. F. B. Downing, A. F. Benning, and R. C. McHarness (to Kinetic Chemicals, Inc.), U.S. Patent 2,480,560, Aug. 30, 1949; *Chem. Abstr.*, **44**:4922 (1950).

329. A. E. Pavlath and F. H. Walker (to Stauffer Chemical Co.), French Patent 1,330,146, Jun. 21, 1963; *Chem. Abstr.*, **60**:406 (1964).

330. M. Hauptschein and A. H. Fainberg (to Pennsalt Chemicals Corp.), U.S. Patent 3,188,356, Jun. 8, 1965; *Chem. Abstr.*, **63**:6859 (1965).

331. R. M. Mantell and W. S. Bernhart (to M. W. Kellogg Co.), U.S. Patent 2,774,799, Dec. 18, 1956; *Chem. Abstr.*, **51**:12955 (1957).

332. C. F. Feasly and W. A. Stover (to Socony-Vacuum Oil Co., Inc.), U.S. Patent 2,627,529, Feb. 3, 1953; *Chem. Abstr.*, **48**:1406 (1954).

333. C. B. Miller (to Allied Chemical & Dye Corp.), U.S. Patent 2,628,989, Feb. 17, 1953; *Chem. Abstr.*, **48**:1406 (1954).

334. M. E. Milville and J. J. Earley (to Pennsalt Chemicals Corp.), U.S. Patent 3,246,041, Apr. 12, 1966; *Chem. Abstr.*, **64**:19410 (1966).

335. E. T. McBee (to Purdue Research Foundation), U.S. Patent 2,644,845, Jul. 7, 1953; *Chem. Abstr.*, **48**:7044 (1954).

336. J. D. Calfee and C. B. Miller (to Allied Chemical & Dye Corp.), U.S. Patent 2,734,090, Feb. 7, 1956; *Chem. Abstr.*, **50**:9441 (1956).

337. H. Johnston (to The Dow Chemical Co.), U.S. Patent 2,722,558, Nov. 1, 1955; *Chem. Abstr.*, **50**:10758 (1956).

338. I. Litant and C. B. Miller (to Allied Chemical & Dye Corp.), U.S. Patent 2,723,296, Nov. 8, 1955; *Chem. Abstr.*, **50**:10758 (1956).

339. D. M. Marquis (to E. I. du Pont de Nemours & Co.), U.S. Patent 3,073,870, Jan. 15, 1963; *Chem. Abstr.*, **59**:446 (1963).

340. Ref. 16.

341. F. Olstowski and J. D. Watson (to The Dow Chemical Co.), U.S. Patent 3,089,910, May 14, 1963; *Chem. Abstr.*, **60**:1586 (1964).

342. S. Okazaki and N. Sakauchi (to Kureha Chemical Industry Co. Ltd.), Japanese Patent 22,453, Oct. 5, 1965; *Chem. Abstr.*, **64**:3349 (1966).
343. F. C. McGrew and E. H. Price (to E. I. du Pont de Nemours & Co.), U.S. Patent 2,687,440, Aug. 24, 1954; *Chem. Abstr.*, **49**:12527 (1955).
344. R. N. Haszeldine and B. R. Steele, *J. Chem. Soc.*, **1957**:2193.
345. H. Gilman and R. G. Jones, *J. Am. Chem. Soc.*, **65**:2037 (1943).
346. D. Goerrig, W. Moschel, and H. Jonas (to Farbenfabriken Bayer A.-G.), German Patent 952,713, Nov. 22, 1956; *Chem. Abstr.*, **53**:5129 (1959).
347. L. Batt and J. M. Pearson, *Chem. Commun.*, **1965**(22):575.
348. W. S. Barnhart and N. T. Hall, Poly(vinylidene Fluoride), in "Encyclopedia of Chemical Technology," 2d ed., vol. 9, pp. 840–847, Interscience, New York, 1966.
349. C. A. Neugebauer and J. L. Margrave, *J. Phys. Chem.*, **60**:1318 (1956).
350. V. N. Vasil'eva, N. A. Kocheshkov, T. V. Talalaeva, E. M. Panov, G. V. Kazennikova, R. S. Sorokina, and O. P. Petrii, *Dokl. Akad. Nauk SSSR*, **143**:844 (1962); *Chem. Abstr.*, **57**:3328 (1962).
351. W. H. Mears, R. F. Stahl, S. R. Orfeo, R. C. Shair, L. F. Kells, W. Thompson, and H. McCann, *Ind. Eng. Chem.*, **47**:1449 (1955).
352. I. P. Torkington and H. W. Thompson, *Trans. Faraday Soc.*, **41**:236 (1945).
353. W. F. Edgell and C. J. Ultee, *J. Chem. Phys.*, **22**:1983 (1954).
354. W. F. Edgell and W. E. Byrd, *ibid.*, **17**:740 (1949).
355. H. M. McConnell, A. D. McLean, and C. A. Reilly, *ibid.*, **23**:1152 (1955).
356. A. Roberts and W. F. Edgell, *ibid.*, **17**:742 (1949).
357. C. P. Carpenter, H. F. Smyth, Jr., and U. C. Pozzani, *J. Ind. Hyg. Toxicol.*, **31**:343 (1949).
358. A. N. Baratov and X. M. Kucher, *Zh. Prikl. Khim.*, **38**(5):1068 (1965).
359. R. N. Haszeldine and J. E. Osborn, *J. Chem. Soc.*, **1956**:61.
360. E. T. McBee, H. B. Hass, W. A. Bittenbender, W. E. Weesner, W. G. Toland, Jr., W. R. Hausch, and L. W. Frost, *Ind. Eng. Chem.*, **39**:409 (1947).
361. R. N. Haszeldine and B. R. Steele, *J. Chem. Soc.*, **1954**:923.
362. C. B. Miller and L. B. Smith (to Allied Chemical & Dye Corp.), U.S. Patent 2,669,590, Feb. 16, 1954; *Chem. Abstr.*, **49**:2477 (1955).
363. P. Tarrant and A. M. Lovelace, *J. Am. Chem. Soc.*, **77**:768 (1955).
364. G. H. Wagner (to Union Carbide & Carbon Corp.), U.S. Patent 2,637,738, May 5, 1953; *Chem. Abstr.*, **48**:8254 (1954).
365. N. R. Bergamin, E. Castellano, and H. J. Schumacher, *Z. Physik. Chem. (Frankfurt)*, **40**(34):246 (1964); *Chem. Abstr.*, **60**:14357 (1964).
366. V. A. Ginsburg, E. S. Vlasova, N. M. Vasil'eva, N. S. Mirzabekova, S. P. Makarov, A. I. Shchekotikhin, and A. Ya. Yakubovich, *Dokl. Akad. Nauk SSSR*, **149**:97 (1963); *Chem. Abstr.*, **59**:5008 (1963).
367. H. A. Wiebe, A. R. Knight, O. P. Strausz, and H. E. Gunning, *J. Am. Chem. Soc.*, **87**:1443 (1965).
368. I. L. Knunyants and E. G. Bykhovskaya, *Izv. Akad. Nauk SSSR, Otd. Khim. Nauk*, **1955**:852 *Chem. Abstr.*, **50**:9322 (1956).
369. M. Hauptschein, R. E. Oestetling, M. Braid, E. A. Tyczkowski, and D. M. Gardner, *J. Org. Chem.*, **28**:1281 (1963).
370. L. S. German and I. L. Knunyants, *Dokl. Akad. Nauk SSSR*, **166**(3):602 (1966); *Chem. Abstr.*, **64**:12534 (1966).
371. J. D. Calfee and P. A. Florio (to Allied Chemical & Dye Corp.), U.S. Patent 2,628,972, Feb. 17, 1953; *Chem. Abstr.*, **48**:1413 (1954).

372. T. A. Ford and W. E. Hanford (to E. I. du Pont de Nemours & Co.), U.S. Patent 2,435,537, Feb. 3, 1948; *Chem. Abstr.*, **42**:3215 (1948).
373. T. A. Ford and W. E. Hanford (to E. I. du Pont de Nemours & Co.), British Patent 590,817, Jul. 29, 1947; *Chem. Abstr.*, **42**:794 (1948).
374. British Patent 941,913 (to Pennsalt Chemicals Corp.), Nov. 13, 1963; *Chem. Abstr.*, **61**:7137 (1964).
375. E. T. McBee, H. M. Hill, and G. B. Bachman, *Ind. Eng. Chem.*, **41**:70 (1949).
376. C. B. Miller and J. D. Calfee (to Allied Chemical & Dye Corp.), U.S. Patent 2,635,093, Apr. 14, 1963; *Chem. Abstr.*, **47**:6701 (1953).
377. H. Iserson (to Pennsalt Chemicals Corp.), U.S. Patent 3,031,437, Apr. 24, 1962; *Chem. Abstr.*, **57**:3638 (1962).
378. J. W. Borland, C. B. Miller, and J. H. Pearson (to Allied Chemical Corp.), U.S. Patent 2,865,824, Dec. 23, 1958; *Chem. Abstr.*, **53**:5749 (1959).
379. R. Roberts, F. L. Dalton, P. Hayden, and P. R. Hills, *Proc. U. N. Intern. Conf. Peaceful Uses At. Energy*, 2nd, Geneva, **29**:408 (1958), published 1959.
380. E. V. Volkova, A. V. Fokin, A. D. Sorokin, and L. A. Bulygina, *Zh. Vses. Khim. Obshchestva im. D. I. Mendeleeva*, **7**:593 (1962).
381. British Patent 856,469 (to Minnesota Mining & Manufacturing Co.), Dec. 14, 1960; *Chem. Abstr.*, **55**:14990 (1961).
382. T. A. Ford (to E. I. du Pont de Nemours & Co.), U.S. Patent 2,468,054, Apr. 26, 1949; *Chem. Abstr.*, **43**:5638 (1949).
383. H. S. Booth, P. E. Burchfield, E. M. Bixby, and J. B. McKelvey, *J. Am. Chem. Soc.*, **55**:2231 (1933).
384. F. Schloffer and O. Scherer (to I. G. Farbenindustrie A.-G.), German Patent 677,071, Jun. 17, 1939; *Chem. Abstr.*, **33**:6999 (1939).
385. R. Bringer, Poly(chlorotrifluoroethylene), in "Encyclopedia of Chemical Technology," 2d ed., vol. 9, pp. 832–833, Interscience, New York, 1966.
386. D. Goerrig and H. Jonas (to Farbenfabriken Bayer A.-G.), German Patent 1,014,985, Sept. 5, 1957; *Chem. Abstr.*, **54**:294 (1960).
387. C. B. Miller and L. B. Smith (to Allied Chemical & Dye Corp.), U.S. Patent 2,864,873, Dec. 16, 1958; *Chem. Abstr.*, **53**:10035 (1959).
388. Netherlands Appl. 6,411,279 (to Allied Chemical Corp.), Apr. 2, 1965; *Chem. Abstr.*, **63**:9809 (1965).
389. J. T. Rucker and D. B. Stormon (to Hooker Electrochemical Co.), U.S. Patent 2,760,997, Aug. 28, 1956; *Chem. Abstr.*, **51**:3653 (1957).
390. J. W. Clark (to Union Carbide & Carbon Corp.), U.S. Patent 2,704,777, Mar. 22, 1955; *Chem. Abstr.*, **50**:15574 (1956).
391. J. W. Clark (to Union Carbide & Carbon Corp.), U.S. Patent 2,685,606, Aug. 3, 1954; *Chem. Abstr.*, **48**:12788 (1954).
392. V. U. Shevchuk and M. B. Fagarash, U.S.S.R. Patent 166,674, Dec. 1, 1964; *Chem. Abstr.*, **62**:10335 (1965).
393. O. A. Blum (to M. W. Kellogg Co.), U.S. Patent 2,590,433, Mar. 25, 1952; *Chem. Abstr.*, **47**:1180 (1953).
394. L. B. Smith and C. B. Miller (to Allied Chemical & Dye Corp.), U.S. Patent 2,635,121, Apr. 14, 1953; *Chem. Abstr.*, **48**:2756 (1954).
395. J. G. Abramo and R. H. Reinhard (to Monsanto Chemical Co.), U.S. Patent 2,903,489, Sept. 8, 1959; *Chem. Abstr.*, **54**:5462 (1960).
396. E. M. Ilgenfritz and R. P. Ruh (to The Dow Chemical Co.), U.S. Patent 2,996,556, Jul. 11, 1961; *Chem. Abstr.*, **56**:1344 (1962).

397. A. L. Dittman and J. M. Wrightson (to M. W. Kellogg Co.), U.S. Patent 2,690,459, Sept. 28, 1954; *Chem. Abstr.*, **49**:11681 (1955).
398. J. M. Wrightson and A. L. Dittman (to M. W. Kellogg Co.), U.S. Patent 2,667,518, Jan. 26, 1954; *Chem. Abstr.*, **49**:2478 (1955).
399. British Patent 681,067 (to Farbenfabriken Bayer A.-G.), Oct. 15, 1952; *Chem. Abstr.*, **48**:1406 (1954).
400. T. Zdichynec, Czechoslovakian Patent 107,919, Jul. 15, 1963; *Chem. Abstr.*, **60**:5335 (1964).
401. British Patent 756,027 (to Farbwerke Hoechst A.-G.), Aug. 29, 1956; *Chem. Abstr.*, **51**:9671 (1957).
402. O. Scherer and H. Kühn (to Farbwerke Hoechst A.-G.), German Patent 907,173, Mar. 22, 1954; *Chem. Abstr.*, **52**:10141 (1958).
403. H. Madai, German (East) Patent 12,182, Oct. 6, 1956; *Chem. Abstr.*, **53**:2089 (1959).
404. W. T. Miller, Jr., E. Bergman, and A. H. Fainberg, *J. Am. Chem. Soc.*, **79**:4159 (1957).
405. B. F. Landrum, K. H. Kahrs, H. Kuhn, and R. Schaff (to Minnesota Mining & Manufacturing Co.), U.S. Patent 2,909,571, Oct. 20, 1959; *Chem. Abstr.*, **54**:7558 (1960).
406. O. Scherer and H. Kühn (to Farbwerke Hoechst A.-G.), German Patent 953,972, Dec. 13, 1956; *Chem. Abstr.*, **53**:11222 (1959).
407. G. D. Oliver, J. W. Grisard, and C. W. Cunningham, *J. Am. Chem. Soc.*, **73**:5719 (1951).
408. F. W. Kirkride and F. G. Davidson, *Nature*, **174**:79 (1954).
409. M. T. Rogers, J. G. Malik, and J. L. Speirs, *J. Am. Chem. Soc.*, **78**:46 (1956).
410. J. E. Boggs, C. M. Crain, and J. E. Whiteford, *J. Phys. Chem.*, **61**:482 (1957).
411. M. T. Rogers and R. D. Pruett, *J. Am. Chem. Soc.*, **77**:3686 (1955).
412. T. Kawai, H. Sekine, and M. Igarashi, *Bull. Chem. Soc. Japan*, **34**:1472 (1961); *Chem. Abstr.*, **56**:11030 (1962).
413. A. G. Pokorny, *Chem. Zvesti*, **10**:135 (1956); *Chem. Abstr.*, **50**:9108 (1956).
414. D. E. Mann, N. Acquista, and E. K. Plyler, *J. Chem. Phys.*, **21**:1949 (1953).
415. J. A. Rolfe and L. A. Woodward, *Trans. Faraday Soc.*, **50**:1030 (1954).
416. T. D. Coyle, S. L. Stafford, and F. G. A. Stone, *Spectrochim. Acta*, **17**:968 (1961).
417. T. H. S. Burns, J. M. Hall, A. Bracken, and G. Gouldstone, *Anaesthesia*, **17**:337 (1962).
418. J. A. Zapp, Jr., *Am. Ind. Hyg. Assoc. J.*, **20**:350 (1959).
419. J. Janák and M. Rusek, *Chem. Listy*, **48**:207 (1954).
420. V. A. Balandina and M. S. Kleshcheva, *Gaz. Khromatogr. Moscow, Sb.*, **1964**(2):61; *Chem. Abstr.*, **64**:15987 (1966).
421. J. E. Peterson, H. R. Hoyle, and E. J. Schneider, *Am. Ind. Hyg. Assoc. J.*, **17**:429 (1956); *Chem. Abstr.*, **54**:9611 (1960).
422. Ref. 19, p. 142.
423. P. Tarrant and H. C. Brown, *J. Am. Chem. Soc.*, **73**:1781 (1951).
424. J. Lichtenberger and R. E. Rusch, *Bull. Soc. Chim. France*, **1962**:254.
425. J. Lichtenberger and A. M. Geyer, *ibid.*, **1957**:581.
426. K. E. Rapp, R. L. Pruett, J. T. Barr, C. T. Bahner, J. D. Gibson, and R. H. Lafferty, Jr., *J. Am. Chem. Soc.*, **72**:3642 (1950).
427. B. L. Dyatkin, E. P. Mochalina, and I. L. Knunyants, *Dokl. Akad. Nauk SSSR*, **139**:106 (1961); *Chem. Abstr.*, **56**:311 (1962).

428. P. Tarrant and D. A. Warner, *J. Am. Chem. Soc.*, 76:1624 (1954).

429. Ref. 128.

430. J. R. Lacher, A. S. Rodgers, and J. D. Park, *Univ. of Colo. Studies, Ser. Chem. Pharm.* (*N.S.*), no. 5, (1963); *Chem. Abstr.*, 59:9771 (1963).

431. W. T. Miller, Jr., J. O. Stoffer, G. Fuller, and A. C. Currie, *J. Am. Chem. Soc.*, 86:51 (1964).

432. J. D. Park, A. P. Stefani, and J. R. Lacher, *J. Org. Chem.*, 26:4017 (1961).

433. D. A. Barr, R. N. Haszeldine, and C. J. Willis, *J. Chem. Soc.*, 1961:1351.

434. V. A. Ginsburg, A. N. Medvedev, L. L. Martynova, N. N. Vasileva, M. F. Lebedeva, S. S. Dubov, and A. Ya. Yakubovich, *Zh. Obshch. Khim.*, 35(11):1924 (1965).

435. D. D. Coffman, R. Cramer, and G. W. Rigby, *J. Am. Chem. Soc.*, 71:979 (1949).

436. O. Paleta, *Tech. Publ. Stredisko Tech. Inform. Potravinar. Prumyslu*, no. 161, 65 (1959–61) (Publ. 1962); *Chem. Abstr.*, 60:5315 (1964).

437. A. L. Henne and D. W. Kraus, *J. Am. Chem. Soc.*, 73:5303 (1951).

438. P. R. Austin (to E. I. du Pont de Nemours & Co.), U.S. Patent 2,449,360, Sept. 14, 1948; *Chem. Abstr.*, 43:661 (1949).

439. A. Posta and O. Paleta, Czechoslovakian Patent 105,966, Dec. 15, 1962; *Chem. Abstr.*, 60:2753 (1964).

440. J. T. Barr, J. D. Gibson, and R. H. Lafferty, Jr., *J. Am. Chem. Soc.*, 73:1352 (1951).

441. M. Hauptschein and M. Braid (to Pennsalt Chemicals Corp.), British Patent 930,758, Jul. 10, 1963; *Chem. Abstr.*, 60:2753 (1964).

442. Ref. 440.

443. M. Hudlicky, Czechoslovakian Patent 107,661, Jun. 15, 1963; *Chem. Abstr.*, 60:5334 (1964).

444. A. Ya. Yakubovich and A. P. Sergeev, *Khim. Nauka i Prom.*, 4:682 (1959); *Chem. Abstr.*, 54:8615 (1960).

445. H. Goldwhite, R. N. Haszeldine, and R. N. Mukherjee, *J. Chem. Soc.*, 1961: 3825.

446. D. W. McBride, R. L. Pruett, E. Pitcher, and F. G. A. Stone, *J. Am. Chem. Soc.*, 84:497 (1962).

447. R. N. Haszeldine, *J. Chem. Soc.*, 1954:4026.

448. R. N. Haszeldine and J. C. Young, *ibid.*, 1960:4503.

449. M. Hudlicky and I. Lejhancova, *Collection Czech. Chem. Commun.*, 28:2455 (1963); *Chem. Abstr.*, 59:11227 (1963).

450. Ref. 165.

451. A. L. Henne and D. W. Kraus, *J. Am. Chem. Soc.*, 73:1791 (1951).

452. R. N. Haszeldine, *J. Chem. Soc.*, 1955:4291.

453. J. D. Park, R. W. Lamb, and J. R. Lacher, *Bull. Chem. Soc. Japan*, 37(7):946 (1964); *Chem. Abstr.*, 61:8173 (1964).

454. D. L. Bunbury, J. R. Lacher, and J. D. Park, *J. Am. Chem. Soc.*, 80:5104 (1958).

455. M. Hauptschein and M. Braid (to Pennsalt Chemicals Corp.), U.S. Patent 3,219,712, Nov. 23, 1965; *Chem. Abstr.*, 64:8031 (1966).

456. R. N. Haszeldine and B. R. Steele, *J. Chem. Soc.*, 1953:1592.

457. R. Mueller and M. Dressler, *J. Prakt. Chem.*, (4), 22(1–2):29 (1963); *Chem. Abstr.*, 60:13265 (1964).

458. French Patent 1,399,414 (to Imperial Chemical Industries Ltd.), May 14, 1965; *Chem. Abstr.*, 63:9864 (1965).

459. J. Harmon (to E. I. du Pont de Nemours & Co.), U.S. Patent 2,436,142, Feb. 17, 1948; *Chem. Abstr.*, **42**:3776 (1948).

460. A. L. Henne and R. P. Ruh, *J. Am. Chem. Soc.*, **69**:279 (1947).

461. W. T. Miller (to Allied Chemical & Dye Corp.), U.S. Patent 2,668,182, Feb. 2, 1954; *Chem. Abstr.*, **49**:2478 (1955).

462. F. N. Teumac and L. W. Harriman (to The Dow Chemical Co.), U.S. Patent 3,214,479, Oct. 26, 1965; *Chem. Abstr.*, **64**:590 (1966).

463. W. C. Soloman and L. A. Dee, *J. Org. Chem.*, **29**:2790 (1964).

464. C. M. Sharts and J. D. Roberts, *J. Am. Chem. Soc.*, **83**:871 (1961).

465. P. L. Barrick (to E. I. du Pont de Nemours & Co.), U.S. Patent 2,462,345, Feb. 22, 1949; *Chem. Abstr.*, **43**:4293 (1949).

466. A. L. Barney and T. L. Cairns, *J. Am. Chem. Soc.*, **72**:3193 (1950).

467. D. Ambros, *Chem. Prum.*, **11**:60 (1960); *Chem. Abstr.*, **55**:18556 (1961).

468. V. R. Hurka (to E. I. du Pont de Nemours & Co.), U.S. Patent 2,676,983, Apr. 27, 1954; *Chem. Abstr.*, **49**:5510 (1955).

469. R. N. Haszeldine and F. Nyman, *J. Chem. Soc.*, **1959**:1084.

470. W. T. Miller (to Allied Chemical & Dye Corp.), U.S. Patent 2,712,555, Jul. 5, 1955; *Chem. Abstr.*, **50**:6505 (1956).

471. British Patent 729,010 (to Farbenfabriken Bayer A.-G.), Apr. 27, 1955; *Chem. Abstr.*, **49**:13693 (1955).

472. British Patent 805,103 (to Minnesota Mining & Manufacturing Co.), Nov. 26, 1958; *Chem. Abstr.*, **53**:11890 (1959).

473. D. Bachmann, H. Fritz, W. Grundmann, and R. Schäff (to Farbwerke Hoechst A.-G.), German Patent 938,037, Jan. 19, 1956; *Chem. Abstr.*, **53**:6683 (1959).

474. A. L. Dittman, H. J. Passino, and J. M. Wrightson (to M. W. Kellogg Co.), U.S. Patent 2,689,241, Sept. 14, 1954; *Chem. Abstr.*, **49**:11681 (1955).

475. French Patent 1,155,143 (to Société d'Electrochimie d'Electrometallurgie et des Acieries Electriques d'Ugine), Apr. 23, 1958; *Chem. Abstr.*, **54**:12659 (1960).

476. H. J. Passino, A. L. Dittman, and J. M. Wrightson (to M. W. Kellogg Co.), U.S. Patent 2,783,219, Feb. 26, 1957; *Chem. Abstr.*, **51**:7758 (1957).

477. J. M. Hamilton, Jr., *Ind. Eng. Chem.*, **45**:1347 (1953).

478. G. F. Roedel (to General Electric Co.), U.S. Patent 2,613,202, Oct. 7, 1952; *Chem. Abstr.*, **47**:1979 (1953).

479. B. F. Landrum and R. L. Herbst, Jr. (to M. W. Kellogg Co.), U.S. Patent 2,753,379, Jul. 3, 1956; *Chem. Abstr.*, **50**:14266 (1956).

480. R. L. Herbst, Jr., and B. F. Landrum (to Minnesota Mining & Manufacturing Co.), U.S. Patent 2,888,446, May 26, 1959; *Chem. Abstr.*, **53**:16589 (1959).

481. E. Huss (to Farbenfabriken Bayer A.-G.), German Patent 949,082, Sept. 13, 1956; *Chem. Abstr.*, **52**:19251 (1958).

482. H. J. Passino, A. L. Dittman, and J. M. Wrightson (to Minnesota Mining & Manufacturing Co.), U.S. Patent 2,820,026, Jan. 14, 1958; *Chem. Abstr.*, **52**:5884 (1958).

483. J. W. Jewell (to Minnesota Mining & Manufacturing Co.), U.S. Patent 3,014,015, Dec. 19, 1961; *Chem. Abstr.*, **58**:6945 (1963).

484. A. L. Dittman and J. M. Wrightson (to M. W. Kellogg Co.), U.S. Patent 2,783,219, Feb. 26, 1957; *Chem. Abstr.*, **51**:7758 (1957).

485. F. G. Pearson (to American Viscose Corp.), British Patent 578,168, Jun. 18, 1946; *Chem. Abstr.*, **41**:1881 (1947).

486. W. M. Thomas and M. T. O'Shaughnessy, *J. Polymer Sci.*, **11**:455 (1953).

487. D. R. Schultz, *J. Polymer Sci., Pt. B*, **1**(1):613 (1963).

488. R. G. Heiligmann and F. Benington (to Borden Co.), U.S. Patent 2,685,575, Aug. 3, 1954; *Chem. Abstr.*, **48**:14295 (1948).
489. G. H. Crawford (to Minnesota Mining & Manufacturing Co.), U.S. Patent 3,089,866, May 14, 1963; *Chem. Abstr.*, **59**:1777 (1963).
490. M. Lazar, R. Rado, and N. Kliman, *Chem. Zvesti*, **10**:585 (1956).
491. D. Goerrig, H. Jonas, and W. Moschel (to Farbenfabriken Bayer A.-G.), German Patent 935,867, Dec. 1, 1955; *Chem. Abstr.*, **53**:3949 (1959).

3. MISCELLANEOUS VINYL MONOMERS

J. P. SCHROEDER AND DOROTHY C. SCHROEDER, *University of North Carolina, Greensboro*

Contents

Many vinyl compounds have been synthesized that were not dealt with in other chapters. All are, in principle, within the scope of this chapter. The majority, however, have not been studied in depth as monomers and have not provided, or shown promise of providing, useful polymers. Only miscellaneous vinyl monomers that meet these practical criteria will be discussed here. Even with such a limitation, the volume of literature remains awesome. In an effort to hold the presentation to a reasonable length, the general policy of discussing only the early references to a given discovery has been followed unless subsequent publications describe significant modifications.

The monomers in each of the other chapters are closely related chemically and, therefore, are amenable to uniform treatment. In order to follow a similar convenient format in this chapter, the miscellaneous monomers are presented by chemical type also. They are named as indexed in *Chemical Abstracts*, but if there are other commonly used designations, they are given in parentheses.

I. *N*-VINYL COMPOUNDS

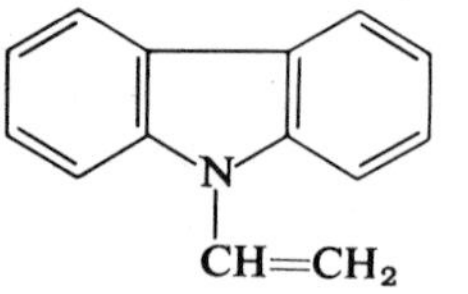

$$-(CH_2)_n-$$
$$N-\!\!\!\!-\!\!\!\!-C=\!\!O$$
$$CH=CH_2$$

1

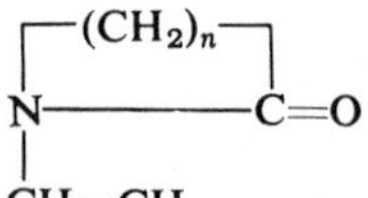

2a, $n = 3$
b, $n = 4$
c, $n = 5$

A. Introduction

9-Vinylcarbazole (*N*-vinylcarbazole) (**1**) was first prepared in 1924 by *N*-chloroethylation of carbazole, followed by dehydrochlorination (1). Commercial interest in **1** began when a manufacturing process involving vinylation of carbazole with acetylene was developed by Reppe and his associates at I. G. Farbenindustrie (2).

The *N*-vinyl lactams, 1-vinyl-2-pyrrolidinone (*N*-vinylpyrrolidone) (**2a**), 1-vinyl-2-piperidone (*N*-vinylpiperidone) (**2b**), and 1-vinylhexahydro-(2H)azepin-2-one (*N*-vinyl-ε-caprolactam) (**2c**), were a direct outgrowth of the same research program of Reppe and coworkers, the synthetic method again being vinylation of the corresponding —NH— compound with acetylene (2b,3). Other *N*-vinyl lactams have been prepared, but only **2a, b**, and **c** have received much attention, and **2a** is by far the most important.

Soon after development of the vinylation manufacturing process, the polymerization of these monomers was investigated. A process for polymerizing **1** was patented in 1936 (4), a reference to copolymerization of **1** and **2a** appeared in 1941 (5), and polymerization of *N*-vinyl lactams was patented in 1942 (6). Commercially available homopolymers of **1** and **2a** were described in Germany in 1937 and 1943, respectively. Polyvinylcarbazole, with the trade name Luvican, was stated to have excellent chemical resistance, mechanical strength, and electrical insulating properties and to be easy to fabricate (7). A buffered aqueous solution of polyvinylpyrrolidone (8), with the trade name Periston, proved to be a highly effective blood-plasma substitute with no apparent toxicity to man (9). This solution was used extensively in Germany during the Second World War, reportedly for over 100,000 cases of acute blood loss and shock in the German army (10).

Substantial quantities of polyvinylcarbazole were produced during the Second World War both in Germany and the United States (trade name Polectron) (11,12) for use as electrical insulation and as a capacitor dielectric. Because of its high cost, dark color, and brittleness, however, this plastic did not fare well in competition with the many polymers that appeared after the war. Although still considered a good capacitor dielectric and promising as an organic semiconductor, it is no longer produced in large volume. Polyvinylpyrrolidone, on the other hand, has become a well-established industrial polymer with a variety of applications because of its excellent, unusual properties and relatively low cost.

There does not appear to be any plant-scale production of **1** today. **2a** is manufactured by two former units of the dismantled I. G. Farben. industrial empire—General Aniline and Film Corp. (United States) and

Badische Anilin- und Soda-Fabrik A.-G. (West Germany). None of the other *N*-vinyl lactams has achieved commercial status in the western world. Judging from the number of Russian publications on the subject, there is great interest in *N*-vinyl lactams there, but whether or not they are being made in quantity in the U.S.S.R. or other Communist countries is uncertain.

In addition to use as a monomer, **2a** is a good pigment dispersant and varnish solvent and undergoes addition reactions with phenols to give products that are useful as dye intermediates, softeners for polyamides, and hydrophilic modifiers of phenolic resins, for example, to improve their adhesion to glass (13). The presence of a minor quantity of **2a** in Ziegler-Natta propylene polymerizations is claimed to reduce the amount of atactic product significantly (14). A novel property of **2a**, which has been studied extensively in the U.S.S.R., is its ability to prevent radiation damage to yeast and other living cells, apparently by scavenging harmful free radicals (15). Acaricidal activity of **1** against chiggers and other pests of the order Acarina has been demonstrated (16).

B. Synthetic Methods

1. 9-VINYLCARBAZOLE (1)

The manufacturing process is the reaction of carbazole (**3**) from coal tar with acetylene in an inert diluent at elevated temperature and pressure in the presence of an alkaline catalyst:

$$+ \ CH{\equiv}CH \ \longrightarrow \ \mathbf{1} \tag{1}$$

As diluents, cyclohexane (17–19), methylcyclohexane (**2b**), dimethylcyclohexane (**2b**,20), naphtha (21–23), decalin (24), 1-methylpyrrolidinone (25), ethyl vinyl ether (26), and liquid ammonia (27) have been used. The acetylene may be mixed with nitrogen (**2b**,17,22,25) or a volatile liquid diluent (19) before charging to the autoclave in order to minimize the risk of explosion.

Temperatures of 100 to 200° and pressures of 12 to 56 atm have been claimed to be effective, but 180° (2,18,28) and 25 atm (18,25) appear to be the preferred conditions. The catalyst may be an alkali metal, its hydroxide or alcoholate (28), but ammonia or a tertiary amine (**2b**,29) and zinc or cadmium compounds (for example, ZnO, ZnS) (**2b**,18–20,30) have been used in conjunction with the base, which may also be an alkaline earth metal hydroxide. A catalyst mixture of sodium amide and calcium carbide was used in one instance (22). A simple catalyst (potassium carbazole) at a relatively low temperature (140°), however, was found to give excellent results with short reaction times (23). Yields of 92 to 99 percent of **1** based on consumed **3** have been reported (19,21,23).

A somewhat related process is the reaction of moist potassium carbazole with vinyl chloride in dimethylcyclohexane and nitrogen at 180 to 190° and > 20 atm (24). Recently, a general vinylation procedure based on ethylene was developed that can be applied to the preparation of **1** (31). In this process, ethylene and a compound with an OH or NH moiety react in the presence of a catalyst containing a group VIII metal, such as palladium, to give a product in which the active hydrogen is replaced by a vinyl group. For example, ethylene and **3** yield **1**:

$$\begin{array}{ccc} \diagdown \diagup \\ N & + \; CH_2{=}CH_2 \longrightarrow \\ H \\ \textbf{3} \end{array} \qquad \begin{array}{c} \diagdown \diagup \\ N \\ | \\ CH{=}CH_2 \\ \textbf{1} \end{array} \; + \; 2H \qquad (2)$$

Typical conditions are 93°, 20 psi, and $PdCl_2$ as catalyst. The catalyst is reduced in this dehydrogenation process. It can be regenerated by heating with O_2 and HCl.

In the acetylene process, crude **1** is sometimes recovered by filtration of the reaction mixture to remove insolubles, distillation of the diluent, and extraction of the residue with methanol. In another modification, it is recovered by vacuum distillation. The product is purified by fractional distillation or by recrystallization from methanol, cyclohexane, or benzene but may still be contaminated with sulfur compounds (from the coal tar source of the starting material, **3,** that interfere with free-radical polymerization. These can be removed by treatment of the molten monomer with a free-radical–producing azo compound at temperatures below 120°, followed by recrystallization from methanol (32). Column chromatography on alumina with benzene as solvent has also been used to separate **1** from unreacted **3** (26).

Several synthetic methods for **1** which do not involve acetylene under pressure, and thus are more suitable for laboratory preparations, have been described. The first synthesis of **1** was accomplished by chloroethylation of **3** with 2-chloroethyl *p*-toluenesulfonate, followed by dehydrochlorination with ethanolic KOH (1):

$$\begin{array}{c} \diagdown \diagup \\ N \\ H \\ \textbf{3} \end{array} + CH_3{-}\!\!\left\langle \bigcirc \right\rangle\!\!{-}SO_3CH_2CH_2Cl \xrightarrow{\text{NaOH}} \begin{array}{c} \diagdown \diagup \\ N \\ | \\ CH_2CH_2Cl \end{array} \xrightarrow{\text{KOH}} \begin{array}{c} \diagdown \diagup \\ N \\ | \\ CH{=}CH_2 \\ \textbf{1} \end{array} \quad (3)$$

Treatment of **3** with ethylene oxide or 2-chloroethanol and dehydration of the resulting N-(2-hydroxyethyl)carbazole (**4**) is another method (33,34). Alternatively, **4** may be acylated and the ester (**5**) pyrolyzed to yield **1** (35,36):

$$
\underset{\substack{\text{3}}}{\underset{\substack{| \\ H}}{N}} + \underset{\substack{\text{(or ClCH}_2\text{CH}_2\text{OH)}}}{\underset{\substack{\diagdown \\ O}}{CH_2\!\!-\!\!CH_2}} \xrightarrow{\text{base}} \underset{\substack{\text{4}}}{\underset{\substack{| \\ CH_2CH_2OH}}{N}} \xrightarrow[\text{heat}]{\substack{\text{base} \\ -H_2O}} \underset{\substack{\text{1}}}{\underset{\substack{| \\ CH=CH_2}}{N}}
$$

$$(RCO)_2O \downarrow$$

$$
\underset{\substack{\text{5}}}{\underset{\substack{| \\ CH_2CH_2OCOR}}{N}} \xrightarrow[-\,RCOOH]{\text{heat}}
$$

(4)

2. N-Vinyl Lactams (2a, b, c)

The manufacturing process for the N-vinyl lactams is vinylation of the corresponding unsubstituted lactam with acetylene under conditions similar to those described earlier for vinylation of carbazole. Reppe and coworkers developed a process for **2a** starting with acetylene, formaldehyde, and ammonia (2b,10,37) that is the basis for the processes now used at General Aniline (38,39) and BASF. Acetylene decomposes spontaneously at elevated pressures or at moderate pressures if energy is introduced. This may create a 200-fold rise in pressure and, consequently, destructive explosions. Therefore, special handling techniques were devised to eliminate explosive wave propagation and to limit the hazards to slower decompositions, resulting in only about a 12-fold rise in pressure. All equipment is designed to withstand such pressures. If a high partial pressure of acetylene is required, free space is reduced to a minimum by using small bore lines and packing unavoidably large spaces with fillers such as Raschig rings. If a low partial pressure is satisfactory, an inert gaseous diluent is mixed with the acetylene.

Acetylene and formalin solution with added $NaHCO_3$ react in the presence of a copper acetylide catalyst to give 1,4-dihydroxy-2-butyne (**6**), with propargyl alcohol as a by-product. The effluent liquor is stripped to remove the latter and unreacted acetylene and formaldehyde, leaving an aqueous solution of **6**. This is hydrogenated at 2,000 to 3,000 psi over a Ni–Cu–Mn catalyst to 1,4-butanediol (**7**), which can be freed of water by fractional distillation or used as obtained (in aqueous solution) in the next step, simultaneous dehydration and dehydrogenation of **7** to γ-butyrolactone (**8**) over a metallic copper catalyst. The pure lactone is isolated by fractional distillation, and reaction at 200° with liquid ammonia (40) yields 2-pyrrolidinone (9). Vinylation of 9 with acetylene in the presence of a diluent and a basic catalyst gives 2a. The monomer is purified by vacuum distillation:

$$CH{\equiv}CH + CH_2O \longrightarrow HOCH_2C{\equiv}CCH_2OH \xrightarrow[\text{cat.}]{H_2} HOCH_2CH_2CH_2CH_2OH \xrightarrow[\text{cat.}]{}$$
$$\qquad\qquad\qquad\qquad\qquad \textbf{6} \qquad\qquad\qquad\qquad\qquad\qquad \textbf{7}$$

$$\text{(5)}$$

$$\textbf{8} \qquad\qquad\qquad\qquad \textbf{9} \qquad\qquad\qquad\qquad \textbf{2a}$$

In the original Reppe vinylation process (2b,3,10,24,37), a mixture of 9 and powdered KOH was allowed to react to form the potassium salt of 9. Part of the mixture was then distilled off to remove water of reaction, the residual solution heated to 150 to 160°, and an acetylene-nitrogen (2:1) mixture introduced at 22 to 25 atm. A feed of pure 9 was supplied to the reactor, and crude 2a taken off overhead continuously. This was condensed, and the acetylene-nitrogen mixture in the stream enriched and recycled to the reactor. The optimum conversion of 9 to 2a is about 70 percent; a drastic drop in rate occurs in attempts to go beyond this point (17). The yield based on converted 9 is approximately 90 percent.

Inert liquid diluents, such as toluene (2b,41), butanol (42), dioxane (43a), and tetrahydrofuran (44), have been used. A recent development is continuous vinylation in a tube reactor in the complete absence of a gas phase (45). Conversions of 50 to 60 percent and yields of 90 to 95 percent (based on 9) are obtained using KOH as catalyst at a temperature of 210 to 240°. The high solvent power of 9 toward acetylene is a critical feature of this process. Another recent modification is the use of less basic catalysts, such as potassium aluminate, stannate, vanadate, and titanate on activated carbon at higher temperatures (290 to 310°). This is reported (46) to provide yields of 2a in excess of 95 percent. Aside from this reference, the catalyst is invariably a strong base, for example, an alkali metal hydroxide or alcoholate. The temperature and pressure ranges said to be effective with these strongly basic catalysts are 130 to 300° and 10 to 120 atm, respectively. Pretreatment of 9 with a strong base and distillation before vinylation is claimed to eliminate the formation of polymeric by-products (47).

Many of the references to the preparation of 2a cited above state that similar, and often identical, conditions are also suitable for vinylation of

2-piperidone and ε-caprolactam to **2b** and **2c**, respectively. The use of 130 to 150° petroleum ether as a diluent in the synthesis of **2c** (43b) and vinylations at atmospheric pressure to give **2b** and **2c** (48) are the only significant process variations found in the literature referring specifically to these compounds and not to **2a**.

N-Vinyl lactams have also been prepared by treatment of the corresponding unsubstituted lactam with vinyl chloride in the presence of NaOH, a copper chromite catalyst, and the salt of an iron group metal in its highest oxidation state (49) or with ethylene by the general vinylation process (31) described earlier—Eq. (2).

Neither **2b** nor **2c** is produced commercially today, although the starting material for the latter, ε-caprolactam, is a large-volume industrial chemical as the precursor of nylon 6 (50). Apparently, polymers of **2b** and **2c** have demonstrated no advantages over polyvinylpyrrolidone.

As is true for **1**, synthetic routes to the *N*-vinyl lactams have been described that do not involve high pressures. Several of these start with 1-(2-hydroxyethyl)-2-pyrrolidinone (**10**), which can be obtained from γ-butyrolactone (**8**) and ethanolamine:

$$\text{(8)} + \text{HOCH}_2\text{CH}_2\text{NH}_2 \longrightarrow \text{(10)} + \text{H}_2\text{O} \qquad (6)$$

Dehydration of **10** over Al_2O_3 produces **2a** directly (51). In other procedures, **10** is converted to the *N*-(2-chloroethyl) derivative, which is dehydrochlorinated with KOH or by Hofmann elimination (52). Alternatively, the acetate of **10** is prepared and pyrolyzed to **2a** (53).

N-(1-Alkoxyethyl) derivatives of lactams (**11**) have been pyrolyzed to *N*-vinyl lactams (43c,54). The starting materials are made by reaction of the appropriate unsubstituted lactam and a 1-chloroethyl alkyl ether:

$$\begin{array}{c}\text{(CH}_2)_n\\ \text{NH} \quad \text{C=O}\end{array} + \begin{array}{c}\text{CH}_3\text{CHOR}\\ |\\ \text{Cl}\end{array} \longrightarrow \begin{array}{c}\text{(CH}_2)_n\\ \text{N} \quad \text{C=O}\\ |\\ \text{CHOR}\\ |\\ \text{CH}_3\end{array} \xrightarrow{\text{heat}} \begin{array}{c}\text{(CH}_2)_n\\ \text{N} \quad \text{C=O}\\ |\\ \text{CH=CH}_2\end{array} \qquad (7)$$

11 **2**

N-vinylation of lactams with ethyl vinyl ether using a phenylmercuric acetate–cupferron catalyst has also been reported (55).

C. Properties

1. PHYSICAL

Physical constants of the N-vinyl monomers are listed in Table 1. 9-Vinylcarbazole is insoluble in water but readily soluble in nonpolar solvents, such as benzene, ether, and CCl_4. It is sparingly soluble in methanol and ethanol at room temperature but dissolves on heating. The N-vinyl lactams are water-soluble and are also dissolved by most organic solvents. Phase equilibria, heat of mixing, and viscosity for the system **2a**–H_2O have been determined (62). Although the results showed strong interaction between the components, there was no evidence of hydrate formation. The viscosity of **2a** at 25° is 2.07 cP (13).

2. PHYSIOLOGICAL

9-Vinylcarbazole may cause dermatitis (2b,24,63,64) and sensitizes the skins of guinea pigs and rabbits (64,65). When administered orally, it produces death in mice at 0.05 g per kg and in guinea pigs and rabbits at 0.5 g per kg (64).

Patch tests on human subjects showed that **2a** is neither a sensitizer nor a primary skin irritant and has very little fatiguing action. The acute oral toxicity values toward rats in cubic centimeters per kilogram were found to be LD_0 1.0, LD_{50} 1.5, and LD_{100} 2.5 (13). The homologous **2b** and **2c** would be expected to have similar toxicological properties.

3. ANALYTICAL METHODS AND SPECTRA

The determination of monomeric **1** in homopolymer and in copolymers with methyl methacrylate has been accomplished (66) by the reaction of mercury(II) acetate and methanol with the vinyl group (67):

$$RCH{=}CH_2 + Hg(OAc)_2 + CH_3OH \longrightarrow \underset{\underset{OCH_3}{|}}{RCHCH_2HgOAc} + HOAc \qquad (8)$$

The liberated acetic acid is titrated with standardized base solution.

N-Vinyl lactams have been determined by hydrolytic removal of the vinyl group as acetaldehyde with 10 percent aqueous H_2SO_4 in the presence of $NaHSO_3$. The excess $NaHSO_3$, after reaction with the liberated aldehyde, is titrated with iodine (43d). A related method is the subjection of N-vinyl lactams to the conditions of the iodoform reaction, followed by acidification and titration of unreacted iodine with sodium thiosulfate (43e). The **2a** content in homopolymers was measured using the monomer's absorption maximum at 233 to 234 mμ in the UV (68). The mercury(II) acetate method (67)—Eq. (8)—has been applied to **2a** also (69). In this reference, both the acetic acid and the addition compound produced in

TABLE 1

Physical Properties of N-Vinyl Monomers **1**, **2a**, **2b**, and **2c**

	Compound							
	1	Reference	**2a**	Reference	**2b**	Reference	**2c**	Reference
Boiling point, °C (mm)	70(0.02)	24	65(1)	43a	72(2)	42	95(4)	56,57
	140(1)	23	76(3)	43a	94.5(6.5)	43a	105.5(7.5)	43b
	178(15)	24	93(11)	54	111(12)	46	119(10)	24
	180(15)	2a	96(14)	13	126(25)	58	130(20)	54
			123(50)	13			132(22)	57
			148(100)	13				
			170(200)	13				
			193(400)	13				
Melting point, °C	67	28	17	52a	48	58	38	55
	66	1,35b,36	15	52b	45	43a	35	43a,54,56,57
	64	2a,21,23, 24,27, 59,60,61	13.5	13	42.5	46		
					42	24		
Refractive index, n_D (°C)			1.5113(20)	46			1.5133(20)	43a
			1.5117(20)	43a			1.5135(20)	54
			1.511(25)	13			1.5138(20)	43b
			1.5120(25)	3c			1.5051(40)	43a,56
			1.5019(28.5)	52b				
Density, d_4 (°C)			1.0458(20)	43a			1.0287(20)	43a
			1.04(25)	13			1.0084(40)	43a,56

the reaction were determined by base titration and polarographic reduction, respectively. An assay method for **2a** has been described in which the sample is iodinated in aqueous sodium acetate solution, and the excess iodine back-titrated with sodium thiosulfate (13).

Gas-liquid partition chromatography of **2a**, with particular emphasis on the use of esters as stationary liquid phases, has been studied (70).

The IR spectra of **1** (71), **2a** (13,71,72), and **2c** (71), the NMR spectra of **1** (73), **2a** (72–75), **2b** (73), and **2c** (73), and the UV spectrum of **2a** (68,72) have been determined. In another publication (76), the UV and Raman spectra of **2a, b,** and **c** are presented.

D. Storage and Handling

1 is a crystalline solid at ordinary temperatures, but can usually be handled as a liquid at 70 to 80° without danger of polymerization, because it contains "natural" inhibitors—sulfur compounds carried over from the coal tar–based starting material, carbazole (23). If necessary, inhibitors can be added. Nitrogen bases, such as morpholine, inhibit polymerization at 100° for at least 24 hr (77). An inorganic base, anhydrous sodium acetate, has been used as an inhibitor to prevent polymerization of **1** during its preparation and distillation (23). Formamides and acetamides (78) or stearyl alcohol (79) provide stability at 70°. Anthracene and phenanthrene also prevent rapid polymerization of molten **1** (80). The monomer is extremely sensitive to aqueous acids, hydrolyzing readily to acetaldehyde and carbazole (24). For this reason, ammonia is often added when **1** is in contact with water. Since it is toxic and may cause dermatitis, inhalation of fines and contact of **1** with the skin should be carefully avoided.

The *N*-vinyl lactams polymerize readily on exposure to light or heat. They are stable, however, in the presence of basic substances. Thus, alkali metal hydroxides, alkoxides, sulfides, and carbonates are suitable inhibitors (81). Hydrazone and azine stabilizers have also been patented (82). General Aniline uses 0.1 percent flake caustic in its commercial **2a** but suggests that, in special cases, other insoluble inhibitors (sodium methylate, sulfide, or carbonate) and soluble inhibitors, such as thioureas, pyrogallol, and sulfur, may be useful. Stabilized samples of **2a** have been stored for more than 13 months without polymerization and, after storage at 60° in contact with mild steel (simulated tank-car conditions) for 2 months, neither the **2a** nor the steel was affected. The flake-caustic inhibitor is easily removed from the monomer before use by decantation or filtration. Vacuum distillation of **2a** is feasible if the temperature is kept below 120° (13).

Like **1**, the *N*-vinyl lactams are hydrolyzed by aqueous acids to the parent lactam and acetaldehyde (43a,d), but the rate is very low at room temperature above pH 4. Because of its low toxicity, **2a** can be handled with only the normal precautions customarily accorded any organic liquid. Care should be exercised to avoid contact with the small amount of corrosive NaOH inhibitor, however. The flash point (open cup) and fire point of **2a** are 98.4 and 100.5°, respectively (13). Protection of all the monomers from light and heat is desirable.

E. Chemical Reactions

Addition to the vinyl double bond occurs readily in all the compounds. **1** has been hydrogenated catalytically to 9-ethylcarbazole (24) and **2a** to 1-ethyl-2-pyrrolidinone (2b,83). Hydrogen sulfide, mercaptans, thiophenols (24), and methanol (84,85) add to **1**; phenols (24), amides (2b), and alcohols (86,87) to *N*-vinyl lactams. The mode of addition to the *N*-vinyl monomers is illustrated in Eq. (9):

$$\begin{array}{c} \diagdown \\ \diagup \end{array}\text{N—CH=CH}_2 \qquad \diagdown \qquad $$

$$\updownarrow \quad + \text{ HA} \longrightarrow \quad \begin{array}{c}\diagdown\\\diagup\end{array}\text{NCHACH}_3 \qquad (9)$$

$$\begin{array}{c} \diagdown \\ \diagup \end{array}\overset{+}{\text{N}}\text{=CH—}\overset{-}{\text{CH}}_2$$

$$(\text{A} = \text{OR, OAr, SH, SR, SAr})$$

The reaction of **2a** with dry HCl at 5° in CCl$_4$ results in a viscous mass of adducts (87). As indicated earlier, the *N*-vinyl monomers are hydrolyzed by aqueous acids to acetaldehyde and the parent —NH— compound— Eq. (10):

$$\begin{array}{c}\diagdown\\\diagup\end{array}\text{NCH=CH}_2 + \text{H}_2\text{O} \xrightarrow{\text{H}^+} \begin{array}{c}\diagdown\\\diagup\end{array}\text{NH} + \text{CH}_3\text{CHO} \qquad (10)$$

The *N*-vinyl lactams give a positive iodoform test (43e).

2c undergoes the Diels-Alder reaction with tetraphenylcyclopentadienone (88) but not with cyclopentadiene (89). After reaction with the dienone, carbon monoxide and ε-caprolactam are eliminated from the adduct, leaving 1,2,3,4-tetraphenylbenzene as the final product.

The picrate of **1** melts at 185° (35b,36).

A crystalline dimer, variously reported to melt at 193.5, 194, and 196.5°, has been obtained from **1** by electrolysis in acetonitrile with mercury(II) cyanide as electrolyte (90), by treatment with chloranil and visible light in methanol (85), and by reaction with iron(III) nitrate in methanol (84,91). The dimer is known to be a bis(9-carbazyl)cyclobutane, but there is disagreement as to its exact structure. On the basis of NMR data, Ellinger et al. (92) and Wang (91) conclude that it is the *trans*-1,2 isomer (12), but Breitenbach and coworkers (90) favor the *trans*-1,3 configuration (13):

| 12 | 13 |

Two mechanisms for the dimerization with $Fe(NO_3)_3$ have been proposed —Eq. (11). Both start with abstraction of an electron from 1 by the Fe^{3+} ion to give a cation radical (14) but diverge in the path by which this intermediate undergoes reductive coupling to produce the dimer. Route A (84) could yield either the 1,2 or 1,3 isomer, route B (91) only the 1,2:

$$(11)$$

Similar mechanisms are plausible for the other methods of preparation with the electrodes and the chloranil, respectively, serving as the means of electron transfer. Higher polymers of 1 are also obtained by these experimental procedures (see Sec. F. Polymerization).

There is some hydrolysis of **2a** by 5 percent aqueous HCl, but 1,1-bis(2-pyrrolidinon-1-yl)ethane (**15**), mp 88.5°, is produced, too, from addition of the hydrolysis product **9** to **2a** (93):

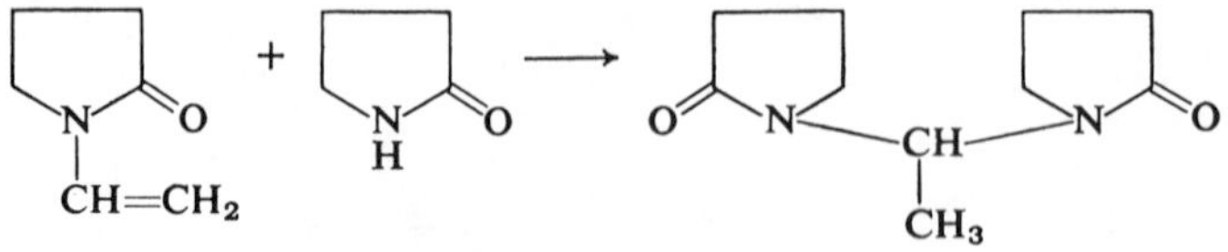

$$
\begin{array}{ccccc}
\text{2a} & & \text{9} & & \text{15}
\end{array}
\tag{12}
$$

Compound **15** is also obtained (43c), along with **11** ($n = 3$, R = butyl), in the reaction of **9** and 1-chloroethyl butyl ether—Eq. (7)—probably by way of the addition of **9** to **2a**—Eq. (12)—the **2a** being formed by elimination of butanol from **11**—Eq. (7). A true dimer (**16**), mp 75°, results when **2a** is treated with dry HCl (93):

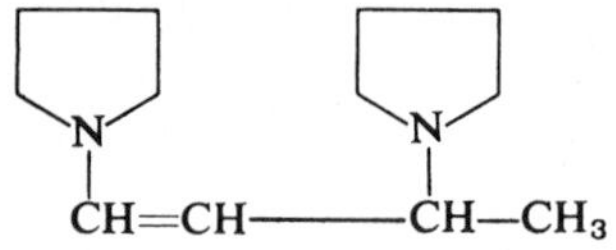

16

The same reagent in ether solution dimerizes **2c** in 93 percent yield (89). The high melting point (143.5°) of the dimer suggests a symmetrical structure like **13** rather than one analogous to **16**. Small amounts of concentrated H_2SO_4 together with acetic acid or a pyrrolidinone in an inert diluent convert **2a** into water-soluble oligomers, particularly trimers (94). Telomers of **2a** have been obtained using mercaptans as telogens and 2,2'-azobisisobutyronitrile as initiator (95).

F. Polymerization

1. 9-VINYLCARBAZOLE (1)

Pure **1** polymerizes readily by free-radical initiation; the heat of polymerization has been found to be 15.2 ± 0.3 kcal per mole (96). The commercial grade of monomer, however, is less reactive, even if not deliberately inhibited, because it contains sulfur impurities. Preparation of **1** by vinylation of synthetic **3** gave a product which was low in sulfur content and which polymerized very readily by free-radical initiation (23).

Pretreatment of the commercial monomer with air (97) or an azonitrile (98) makes it polymerize much more readily on heating, presumably because of scavenging of the sulfur-containing inhibitors. On the other hand, rapid polymerization has been effected simply by heating molten **1** in the presence of di-*tert*-butyl peroxide or 2,2′-azobisisobutyronitrile (99). When used as a capacitor dielectric, polyvinylcarbazole is produced *in situ* by impregnation of the capacitor assembly with the molten monomer at 70 to 85°, followed by free-radical–initiated polymerization at 120 to 140° (11). Suspension (20,100,101) and emulsion (99) polymerizations of **1** have also been described.

1 is polymerized by ionizing radiation (102–104) and more rapidly in the solid than in the molten state (103b). For irradiation of solid **1**, the polymer yield and molecular weight are greater for large crystals, the polymer chains grow linearly, mainly during the postirradiation period, and the chains end abruptly on reaching the crystal face (104). Apparently, the vinyl groups are held in a spatial relationship to one another in the crystal lattice which is highly favorable to chain propagation.

Cationic initiation (11,35b,83,105–107) tends to give low-molecular-weight polymers. Stereoregular products from the polymerization of **1** with a butyllithium–TiCl$_4$ (108) or triisobutylaluminum–TiCl$_4$ (109) catalyst have been reported, but other investigators found that, although **1** undergoes facile cationic polymerization with these and similar catalyst systems, the polymers are completely amorphous and atactic (110). The polymerization of **1** with nitroalkanes is retarded by water and amines and, therefore, appears to proceed by a cationic mechanism (111).

There are several examples of data concerning the polymerization of **1** that are difficult to explain by analogy with the behavior of other vinyl compounds. Although sulfur compounds usually act toward **1** as inhibitors, 2-naphthalenethiol and thiourea are potent initiators (112). Aryl phosphates and phosphites are also active initiators, probably by a mechanism related to previous exposure of the esters to air (102a,113).

The first commercial method of polymerizing **1**, with Na$_2$CrO$_4$ in hot aqueous dispersion (2b,10,20) was discovered accidentally by Reppe and coworkers while attempting to oxidize **1** to carbazole *N*-carboxylic acid (2b). Halogens were later found to be initiators (99). More recently, it was claimed (103a,114) that CCl$_4$ polymerizes **1**, but another research group found no evidence for this (115). Although agreeing that **1** sometimes polymerizes in CCl$_4$ solution, these workers ascribed the behavior to traces of an oxidant, such as chlorine, in the solvent and demonstrated that molecular chlorine is indeed an initiator.

In 1963, Ellinger (116) and Scott, Miller, and Labes (117) independently reported that electrophilic compounds, such as chloranil, *sym*-trinitrobenzene, tetracyanoethylene, tetranitromethane, and maleic anhydride, are effective initiators for the polymerization of **1**. Ellinger later described his work in greater detail and showed that even weak electron acceptors, for example, ethyl cyanoacetate, cyclopentadiene, acrylonitrile, and methyl methacrylate, act as initiators (85). The reaction does not appear to be either free-radical or ionic in the usual sense, and the other oxidative polymerizations with Na_2CrO_4 and halogens, which are also difficult to explain in such terms, probably occur by a similar mechanism. There is little doubt that the first step is formation of a charge-transfer complex between 1 and the electron acceptor. Neither Ellinger nor Scott et al. suggest complete charge separation in the complex to form the cation-radical 14. The latter propose, however, that the complex is in equilibrium with such a species and the anion-radical from the electron acceptor, polymerization occurring by a cationic type mechanism—Eq. (13):

$$M + O \rightleftharpoons M...O \rightleftharpoons M\overset{+}{\cdot} + O\overset{-}{\cdot}$$
$$\text{complex} \tag{13}$$

$$M\overset{+}{\cdot} + nM \longrightarrow M^+_{n+1} \rightleftharpoons M^+_{n+1}...O\overset{-}{\cdot}$$
$$+ \qquad\qquad +$$
$$O\overset{-}{\cdot} \qquad\qquad O\overset{-}{\cdot}$$

$$M = \text{monomer} = 1$$
$$O = \text{oxidant}$$

Takakura and coworkers (118) take the position that 14 is produced, the odd electron in it is delocalized over the carbazole ring system, and cationic polymerization ensues with the anion radical acting as counterion.

A related initiation by electron acceptors is the polymerization of 1 with Fe(III), Cu(II), and Ce(IV) salts (119). In these cases, there is undoubtedly complete transfer of an electron from 1 to the metal cation, resulting in reduction of the latter and leaving the cation-radical 14, which then initiates polymerization. Still another related process is the formation of polyvinylcarbazole on electrolysis of 1 in acetonitrile solution (90). Here again 14 would be expected to form by electron transfer from 1 to the anode.

Although two of the three wartime grades of Luvican produced in Germany were 1–styrene copolymers and other copolymers have been mentioned in the literature (102,120), 1 is not a reactive comonomer. Its reluctance to copolymerize has been ascribed to steric hindrance because of the large size of the carbazyl group (10), but this explanation is dubious, since it would seem to predict difficult homopolymerization also and this, of course, is not true. Published copolymerization parameters are presented in Tables 2 and 3.

Grafting of polyvinylcarbazole chains onto other polymers, primarily polyolefins, has been studied extensively (102). This is a means of modifying the properties, for example, dyeability, of the parent polymer by introduction of controlled amounts of the relatively polar —NH— group.

There are two reviews on the polymerization and polymers of 1 written

TABLE 2

Relative Reactivity Ratios for Free-radical Copolymerization of N-Vinylcarbazole and N-Vinyl Lactams (M_1)

M_1	Comonomer (M_2)	r_1	r_2	Reference
1	Allyl chloride	∞	0	131
	1,2-Dichloropropene-2	∞	0	131
	2,5-Dichlorostyrene	0.016 ± 0.002	8.0 ± 0.5	131
	Methyl acrylate	0.050	0.50	132
	Methyl methacrylate	0.20 ± 0.03	2.0 ± 0.3	133
		0.040	2.0	132
	Styrene	0.012 ± 0.002	5.5 ± 0.8	133
		0.035	5.7	132
	Vinyl acetate	2.68	0.126	134
	Vinyl butyrate	1.28	0.059	134
	Vinyl formate	4.22	0.196	134
	Vinyl propionate	1.68	0.076	134
	Vinylidene chloride	3.7	0.020	132
2a	Acrylic acid	0.15 ± 0.1	1.3 ± 0.2	135
	Acrylonitrile	0.06 ± 0.07	0.18 ± 0.07	13
	Allyl acetate	1.6	0.17	13
	Allyl alcohol	1.0	0.0	13
	Dioctyl fumarate	0.03 ± 0.03	0.041 ± 0.007	136
	Maleic anhydride	0.16 ± 0.03	0.08 ± 0.03	13
	Methyl acrylate	0.04 ± 0.02	0.3 ± 0.2	137
	Methyl methacrylate	0.02 ± 0.02	5.0	13
		0.005 ± 0.05	4.7 ± 0.5	138
	Styrene	0.045 ± 0.05	15.7 ± 0.5	138
		0.11	9.0	137
	Trichloroethylene	0.54 ± 0.04	<0.01	13
	Tris(trimethylsiloxy)vinylsilane	4.0 ± 0.6	0.10 ± 0.02	139
	Vinyl acetate	2.0	0.24	13
		3.30 ± 0.15	0.205 ± 0.015	138
		0.44[a]	0.38	140
		2.3 ± 0.2	0.24 ± 0.04	136
	Vinyl benzoate	2.45 ± 0.1	0.44 ± 0.09	141
	Vinyl chloride	0.38	0.53	142
	Vinyl cyclohexyl ether	3.84	0	143
	Vinylene carbonate	0.7	0.4	140
	Vinyl ethyl ether	1.68	0	144
	Vinyl isopropyl ether	2.97 ± 0.01	0	144
	Vinyl phenyl ether	4.43	0.22	143
	N-Vinylurethane	2	0.42	145
2c	Vinyl ethyl ether	1.88 ± 0.02	0	144
	Vinyl isopropyl ether	1.39 ± 0.01	0	144
	Vinyl phenyl ether	2.53	0.39	143

[a] Bork and Coleman (138) recalculated this r_1, applying their correction factor for Kjeldahl nitrogen determinations on 2a copolymers, and obtained a value near 3.3.

TABLE 3
Q and e Values for N-Vinylcarbazole and N-Vinyl Lactams

Monomer	Q	e	Reference
1	0.44	-1.34	137
	0.41	-1.40	146
	0.30	-1.2	132
	-2.3 kcal/mole(q)[a]	-0.39×10^{-10}esu(ϵ)[a]	132
	-1.6 kcal/mole(q)[a]	-0.33×10^{-10}esu(ϵ)[a]	147
2a	0.14	-1.14	137,146
	0.11	-1.64	140
	0.10	-1.4	132
	0.093	-1.17	138
	0.048	-1.07	140
2c	0.081	-1.55	143

[a] Schwan-Price resonance q and electrical ϵ factors (147).

by Cornish (102) in 1962 and 1963. Ellinger (85) has reviewed polymerization methods in the introduction to a 1964 paper.

2. N-Vinyl Lactams

2a polymerizes by free-radical initiation with peroxides, azonitriles, or irradiation. Bulk, emulsion, suspension, and solution processes have been used, but polymerization in aqueous solution is the preferred method. The basic process, which was developed at I. G. Farben., is carried out at 50° and a monomer concentration of 30 to 60 percent with H_2O_2 as initiator in the presence of NH_3 or an amine. The nitrogen base acts both as a polymerization activator and to prevent acid hydrolysis of 2a to acetaldehyde and 2-pyrrolidinone. Unreacted monomer is removed by extraction with methylene chloride (2b,10,121). Residual monomer may also be removed by distillation (122) or by a second polymerization step (123). 2,2′-Azobisisobutyronitrile in combination with ultraviolet light (124) and in ethanolic solution with a hydroperoxide as chain-transfer agent (125) has also been used as initiator. Gel formation during continuous polymerization in aqueous solution is said to be prevented by the addition of certain water soluble compounds, for example, isopropanol, mercaptoacetic acid, or dimethylformamide (126).

Cationic initiators, such as BF_3 etherate (13,83), polymerize 2a as do Hg(II), Sb(III), and Bi(III) salts (127). The earliest reported method of polymerizing 2a is in aqueous solution using sodium bisulfite as catalyst (6). Initiation in this reaction, which was discovered accidentally by Reppe et al. (2b), is probably by a bisulfite-oxygen or bisulfite-monomer redox system (128).

The other *N*-vinyl lactams, **2b** and **2c**, have not been studied so thoroughly, but the available data indicate that their polymerization characteristics are practically identical to those of **2a**. This similarity is specifically noted in several of the references to the polymerization of **2a**, which are cited above (124–127).

The *N*-vinyl lactams copolymerize readily by a free-radical mechanism with a large number of other monomers and under a variety of conditions. Bulk, solution, emulsion, and suspension processes have been used. 2,2′-Azobisisobutyronitrile is the preferred initiator (13). Recently, several patents (129,130) have been issued on ethylene copolymers that are purported to be useful as dispersants for water-in-oil emulsions. Published copolymerization parameters are presented in Tables 2 and 3.

For reviews of the literature on the polymerization and polymers of **2a**, see references 2a, 10, 13, and 148–154.

G. Applications of Polymers

The properties of polyvinylpyrrolidone (PVP) resemble in many ways those of proteins, but, being a synthetic resin, these can be modified deliberately to suit a particular application, whereas this is difficult to achieve for a natural protein. As a result, PVP has become an important commercial polymer with a large number of uses. It is readily soluble in water and in many organic solvents, forms clear, hard, glossy films on casting from a variety of solvents, and is compatible with many plasticizers, inorganic salts, and other resins. Small amounts stabilize emulsions and suspensions by a protective colloid action, and it forms complexes with many substances, for example, iodine and phenols. Solid PVP is hygroscopic but otherwise highly stable, and aqueous solutions are also stable if protected from molds by standard preservatives. The polymer exhibits very low toxicity.

The most important applications of PVP are in pharmaceuticals and cosmetics. The first large use, that is, as a blood-plasma expander or substitute, is still the principal one in Europe, but the conservative attitude of the United States government toward drugs has prevented a similar development here except in veterinary medicine. Two other major pharmaceutical uses are based on the complexing ability of PVP. The first is detoxification and is illustrated by the PVP-iodine complex so familiar as an antiseptic for cuts and bruises (Isodine). The complex retains the full germicidal activity of elemental iodine while drastically reducing irritation to the skin and oral toxicity. The second is attenuation of drug effects by slow liberation of the active constituent from the complex. The action of the drug is prolonged and in some cases increased by the presence of

PVP, making possible a reduction in frequency of administration and dosage level. PVP is also used pharmaceutically as a tablet binder, coating and granulating agent, emulsion stabilizer, solubilizer, and in ophthalmic preparations. In cosmetics, PVP is widely used as the resinous component in hair sprays and as a stabilizer in creams and lotions.

In other fields, PVP finds applications in textile fibers, beverage clarification, adhesives, agricultural sprays, coatings, detergents and soaps, lithography, photography, and paper. The industrial literature (153,154) should be consulted for a detailed discussion of uses.

Because of its excellent electrical properties and good heat resistance, polyvinylcarbazole performs well as a dielectric in capacitors (11,12,102, 155). Recently, its photoconductive properties have been applied to duplicating processes (156–159). Polymers and copolymers of **1** have been patented as dye-receptive additives to polyacrylonitrile textile fibers (160). To our knowledge, however, neither **1** nor its polymers are produced in the United States. Present interest in polyvinylcarbazole appears to be centered in western Europe, but there is no evidence of large-scale production anywhere.

II. VINYLPYRIDINES AND VINYLQUINOLINES

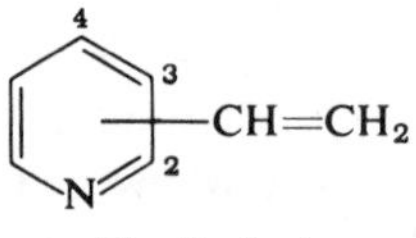

17a, 2-vinyl
 b, 3-vinyl
 c, 4-vinyl

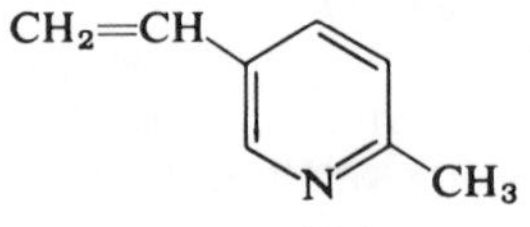

18

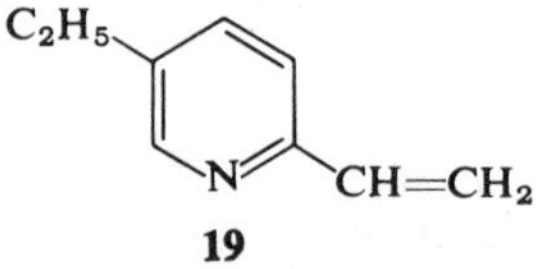

19

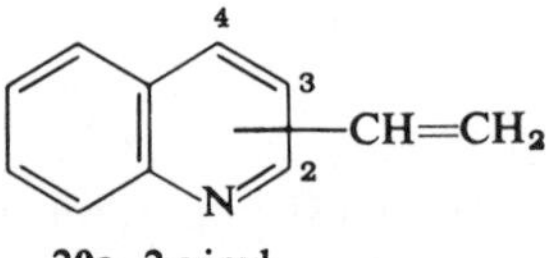

20a, 2-vinyl
 b, 4-vinyl

A. Introduction

In 1887, Ladenburg (161a) prepared 2-vinylpyridine (**17a**) by passing ethylene and pyridine through a hot tube. He later synthesized **17a** in better yield by dehydration of 2-(2-hydroxyethyl)pyridine (161b,c). Einhorn and Lehnkering (162) prepared 2-vinylquinoline (**20a**) in 1888 by dehydrobromination and decarboxylation of 3-bromo-3-(2-quinolyl)-propionic acid (**21**) in boiling aqueous $CaCO_3$ solution—Eq. (14):

$$\text{21} \qquad \xrightarrow[-\text{CO}_2]{-\text{HBr}} \qquad \text{20a} \qquad (14)$$

The superior Ladenburg method of dehydrating the appropriate 2-hydroxyethyl derivative was soon adapted to the syntheses of **20a** (163) and 5-ethyl-2-vinylpyridine (**19**) (164). In 1920, **17c** and **20b** were obtained by dehydrohalogenation of the corresponding 2-haloethyl compounds (165) and, in 1937, **17b** was first prepared from 3-(1-chloroethyl)pyridine by treatment with alcoholic KOH (166).

Commercial interest in the vinylpyridines as monomers began just before the Second World War and, as in the case of the *N*-vinyl monomers, was first evinced by I. G. Farbenindustrie. Patents covering the polymerization and copolymerization of vinylpyridines were issued to this company in 1939 and 1940 (167). In the United States, vinylpyridine research commenced during the war in connection with the government's synthetic-rubber program. This began as a crash project stimulated by the imminent cutting off of natural-rubber supplies from Asia by the advancing Japanese armies. The workhorse synthetic elastomer that emerged was the now familiar SBR (styrene-butadiene rubber) known then as GR-S. There was a wholesale investigation of many other comonomers, however, and it is not surprising that among these were the vinylpyridines, which resemble styrene in molecular structure. **17a** was the front runner, because it was the most readily available (from coal-tar 2-picoline and formaldehyde) and its copolymers with butadiene were found to have excellent properties. There was concern, however, about the limited and unpredictable nature of the coal-tar source, and so emphasis was shifted to **19** and 2-methyl-5-vinylpyridine (**18**), both of which can be synthesized from paraldehyde and ammonia by way of 5-ethyl-2-methylpyridine (aldehyde collidine) (**22**). In 1952, **18** became commercially available from Phillips Petroleum Company (168). The development of vinylpyridine rubbers in the United States to 1957 is described in an excellent review by Haws (169).

At present, the most important vinylpyridines are **17a** and **18**. **17a** is produced from 2-picoline by Reilly Tar and Chemical Corp. in the United States and by Midland Tar Distillers Ltd., in England. These companies also have the facilities to supply **17c**. Research samples of **17a**, **17c**, **18**, and **19** have been made available by the chemical industry in Japan. The properties and uses of **17a** are discussed by Wallsgrove (170) in a 1959 review article. At one time, both Phillips Petroleum and Union Carbide manufactured **18**, but the former is now the only producer in the United States. Carbide still makes the precursor, **22**. There have been a number of publications from the U.S.S.R. and Communist China on the manufacture of **18** and its use in synthetic rubber. Prices (United States, tank-car lots, f.o.b. works) in February, 1968, were $0.95 to 1.10 per lb for **17a** and $1.34 per lb for **18**. There are no known commercial applications for the vinylpyridines other than as monomers.

B. Synthetic Methods

The industrial process for making **18** starts with the condensation of paraldehyde and ammonia under heat and pressure in the presence of an ammonium salt or salts to give **22** as the main product—Eq. (15):

$$4(CH_3CHO)_3 + 3NH_3 \longrightarrow 3\underset{\textbf{22}}{\left[\begin{array}{c} C_2H_5 \\ \\ \\ N \quad CH_3 \end{array}\right]} + 12H_2O \qquad (15)$$

This is an example of the Chichibabin pyridine synthesis (171). The yield of **22** is raised from 30 to 70 percent by increasing the ammonia concentration from 1.17 to 10.7 times the theoretical amount (172). 2-Picoline and other by-products are also formed (173–175).

Catalytic dehydrogenation of **22** yields **18**—Eq. (16):

$$\underset{\textbf{22}}{\left[\begin{array}{c} CH_3CH_2 \\ \\ \\ N \quad CH_3 \end{array}\right]} \xrightarrow[\text{catalyst}]{\text{heat}} \underset{\textbf{18}}{\left[\begin{array}{c} CH_2{=}CH \\ \\ \\ N \quad CH_3 \end{array}\right]} + H_2 \qquad (16)$$

The preferred conditions are similar to those for converting ethylbenzene to styrene, that is, a temperature of 600 to 700°, reduced pressure or an inert diluent, and a mixture of metal oxides as catalyst. The diluent is usually steam (176–178), which is claimed to obviate regeneration of the catalyst (177) and to minimize tar formation (178). Steam can be replaced by CO_2, N_2, or flue gas (179a). Operation without diluent at reduced pressure and relatively low temperatures (427 to 649°) has been described (180).

Although catalysts containing as many as six components have been used, relatively simple systems are reported to be effective also. Examples are Cr_2O_3–Al_2O_3 in the first patent reference to this synthesis (181), Cr_2O_3 or MoO_3 alone or on alumina or bauxite, ThO_2–Al_2O_3, Cr_2O_3 plus an alkaline oxide or hydroxide (180), and CaO–MgO (182a). In addition, oxides of iron, zinc and copper, boron phosphate, K_2SO_4, K_2CrO_4, K_2CO_3, and compounds of cerium, tungsten, and vanadium have been used in catalysts. Chromium(III) oxide is the component cited most often. Reported catalysts that are chemically unrelated to the metal oxide type are sulfur or a sulfur compound (183a), triethanolamine (183b), and iodine and oxygen (184).

The yields of **18** based on converted **22** are 70 to 80 percent. By-products include 3-ethylpyridine, 3-vinylpyridine, 2,5-lutidine, 2- and 3-picoline, and pyridine (174,175). The most troublesome feature of the process is the isolation of pure **18** from the crude reaction product. Of 44 patent references to the dehydrogenation, 23 are concerned exclusively with this problem. The major difficulty lies in the similar boiling points of **18** and unreacted **22**. Their separation has been effected by fractional distillation (179a,185,186), but the operation requires extreme care and special equipment (186). Just as ethylbenzene is lower-boiling than styrene, **22** is lower-boiling than **18**.

Because of the difficulties associated with simple fractional distillation, modifications have been used. Codistillations with steam (187), CO_2 (188), or air (189) are reported to improve separation and reduce polymer formation. In all these fractionations, even if carried out at reduced pressure, it is necessary to add a polymerization inhibitor, such as sulfur. Polymers already present in the crude reaction product before fractionation can be removed by precipitation with n-pentane (190) or by flash distillation after adding a high-boiling polymer solvent, for example, diethylene glycol, to maintain fluidity (191). Depolymerization of the polymers by heating at reduced pressure results in a 50 percent recovery of **18** (192).

Again by analogy with the styrene-ethylbenzene system, **18** is higher-melting than **22**, and this has been utilized in separating them by fractional crystallization. A continuous process has been described (193).

The electron-withdrawing vinyl group in **18** in place of the electron-donating ethyl group in **22** results in a lower electron density at the nitrogen atom, and, hence, **18** is the weaker base. Treatment of **18–22** mixtures with an aqueous acid, for example, $NaHSO_3$ (194), H_2CO_3 (195), or H_3PO_4 (196), results in preferential solubility of **22** in the aqueous phase. A water-immiscible solvent, such as pentane, may be used to aid phase separation. Conversely, **18** has been extracted from **22** with aqueous ammonia (197).

Finally, **18** has been isolated from the crude reaction product as its crystalline oxalate (198) or hydrochloride (199) and then sprung with strong base.

The vinylpyridines **17a, b**, and **c** have also been prepared by dehydrogenation of the corresponding ethylpyridines. The reaction conditions are similar to those described above, but the work-up procedure is less demanding, because the boiling points of the starting materials and products are more widely separated than those of **22** and **18**. Thus, **17a** is obtained in 75 to 90 percent yield by passing a mixture of 2-ethylpyridine and CO_2 over a cerium-containing catalyst at 700°. The boiling points of **17a** and 2-ethylpyridine are 11° apart, and so the mixture can be separated by fractional distillation. 4-Ethylpyridine is converted to **17c** in an analogous manner (179b). Processes for making **17b** (200) and vinylquinolines (201,202) by dehydrogenation have been described also.

The classical Ladenburg synthesis is the best method for commercial preparation of **17a**. 2-Picoline (**23**), available from coal tar, is reacted with formaldehyde to give 2-(2-hydroxyethyl)pyridine (**24**) and this is dehydrated to **17a**—Eq. (17):

$$\text{23} \qquad\qquad\qquad\qquad \text{24} \qquad\qquad\qquad\qquad \text{17a} \tag{17}$$

Formaldehyde adds to **23** (also to 2-methylquinoline and the vinylogous 4-picoline (**25**) and 4-methylquinoline), because the methyl group, attached to a carbon atom multiply bonded to nitrogen, is reactive in the same sense as the methyl groups in acetonitrile (**26**) and crotononitrile (**27**). The analogy is apparent from a comparison of the structures of these compounds:

$$\text{23} \qquad\qquad \text{26} \qquad\qquad \text{25} \qquad\qquad \text{27}$$

3-Methylpyridines do not exhibit this reactivity and, therefore, cannot be used to prepare 3-vinylpyridines by this route.

The first step has been carried out successfully without catalysis by heating **23** with paraformaldehyde (161c,203) or formalin (204) under pressure in an atmosphere of hydrogen or nitrogen. Acid catalysts have been used, but they seem to result in lower yields (205,206). Good results are obtained with persulfate catalysis, provided that oxidation inhibitors are also present (207). Some **17a** is often produced in this step by dehydration of **24**, particularly when catalysts are used.

Dehydration of **24** to **17a** proceeds readily at 190 to 200° (80 to 100 percent yields). The preferred dehydrating agent is KOH, although Al_2O_3 is also effective. Others have been tried but offer no advantage. For example, H_2SO_4 and KOH gave yields of 65 and 78 percent **17a**, respectively, in a comparative study. The same work showed that addition of a polymerization inhibitor is advantageous. Thus, with hydroquinone, the yields of **17a** were 76 percent with $KHSO_4$ and 91 percent with KOH as dehydrating agents (208).

The dehydration is the subject of several patents. In two of these, **24** is heated with a solid alkali metal hydroxide at reduced pressure to a temperature of 190 to 200°. The **17a** distils under these conditions as it is formed (207,209). Using aqueous alkali at 150 to 190°, the **17a** steam distils as it is produced in yields of > 95 percent (210). Even better results (98 to 99 percent yields) are reported to be obtained with NaOH in aqueous polyglycol or triethanolamine at 180 to 200° (211). Use of Al_2O_3 at 290 to 325° gives an 84 percent yield of **17a** and very little polymer (212). In two of the processes, the intermediate **24** is not isolated. The overall yields of **17a** are 76 percent (based on CH_2O) with Al_2O_3 as dehydrating agent (204) and 80 percent (based on **23**) with KOH (207). The product is purified by fractional distillation in the presence of a polymerization inhibitor.

Some **17a** is produced in the preparation of 2-ethylpyridine by reaction of 2-picoline with formaldehyde at 450° over alumina. At lower temperatures, and particularly when zinc fluoride is added to the alumina, **17a** becomes the main product (213). Undoubtedly, **24** is the intermediate in this synthesis also, being reduced by formaldehyde to 2-ethylpyridine at high temperatures and dehydrated to **17a** under the milder conditions.

The same sequence of reactions—Eq. (17)—has been used to prepare **17c, 19, 20a,** and **20b,** starting with the appropriate methylpyridine or methylquinoline under conditions similar to those described above for **17a.** For example, **19** has been made by reaction of **22** and formaldehyde in the presence of a persulfate catalyst and an oxidation inhibitor at 220 to 250° (172,214). Paraformaldehyde was found to be superior to formalin, and the use of ethanol as solvent improved the yield. The hydroxyethyl derivative, which was also formed, was dehydrated to **19** with either Al_2O_3 or KOH (172).

Small-scale preparations of 2- and 4-vinylpyridines and quinolines are generally carried out by the Ladenburg synthesis—Eq. (17)—for which several laboratory procedures are available (172,215). Other preparative methods are dehydrohalogenation of the 2-haloethyl derivative with base (**17c** and **20b**) (165) and pyrolysis of the 1-acetoxyethyl derivative (**17a** and c) (216). **17a** is also obtained in low yield by heating butadiene and acrylonitrile with a Cr_2O_3–Al_2O_3 catalyst, apparently by Diels-Alder reaction of the diene with the C≡N group followed by aromatization of the adduct through loss of hydrogen (217). Although not a preparative method, the conversion of 2,3-dihydro-4H-oxazino(2,3-a)pyridinium bromide (28) to 17a in 85 percent yield by treatment with aqueous base (218) should be noted here:

$$\text{28} \xrightarrow{\text{OH}^-} \text{17a} + H_2C{=}O \qquad (18)$$

It was pointed out earlier that 3-vinylpyridines cannot be prepared by the Ladenburg method, because 3-methylpyridines are not reactive toward formaldehyde. Dehydrohalogenation of the corresponding 1-haloethyl compound has been used to make 17b (156) and 18 (219). 17b (156,220, 221a) and 18 (215a,219,222) have also been obtained by dehydration of the -(1-hydroxyethyl) compound and 18 by pyrolysis of the analogous acetate (222). The best reported yield (65 percent) of 17b is from 3-(1-chloroethyl)pyridinium chloride by way of its quaternary ammonium salt with trimethylamine and treatment of the latter with NaOH (221). Pyrolysis of nicotine produces 17b among other products (220).

C. Properties

1. PHYSICAL

Some of the physical constants of the vinylpyridines are presented in Table 4. The effects of temperature on the vapor pressure of 17c (233) and

TA▮
Physical Properties o▮

	Monomer					
	17a		17b		17c	
		Reference		Reference		Refer▮
Boiling point, °C (mm)	51(11)	216	68(18)	215a	52(7)	216▮
	60(17)	224	72(21)	174	54(7)	21▮
	65(20)	226			65(15)	21▮
	71(30)	215a				
	80(50)	208				
	110(150)	229				
	160	230				
n_D (°C)	1.5476(20)	216	1.5530(20)	174	1.5485(20)	216▮
	1.5495(20)	215a		215a	1.5499(20)	21▮
	1.5497(20)	226			1.5490–	21▮
					1.5495(25)	
	1.5509(20)	229				
	1.5518(20)	232a				
	1.5442(31)	224				
d_4 (°C)	0.9746(20)	229	0.9879(20)	174	0.9863(20)	303▮
	0.9757(20)	226			0.979(25)	23▮
	0.9770(20)	232a				
	0.9661(31)	224				

of **18** (231) and on the density and refractive index of **17c** (233) have been determined. The pK_a for the equilibrium $BH^+ \rightleftharpoons B + H^+$ (B = vinyl-pyridine) was calculated from spectrophotometric data to be 4.8 for **17a** and **17b** and 5.5 for **17c** (234). The dipole moments of **17a, b,** and **c** are 1.86, 2.36, and 2.46 D, respectively (234). A pK_a of 4.98 (235) and a dipole moment of 2.08 D (226) for **17a** have also been reported. The heat of combustion of **17a** is 8,883.2 cal per g (236).

The vinylpyridines are readily soluble in most organic solvents and, as nitrogen bases, are soluble in aqueous acids. They are sparingly soluble in water but can display a strong affinity toward it. Thus, the solubility of **17a** in water is only 2.75 g per 100 g, but that of water in **17a** is 18.8 g per 100 g (229).

2. PHYSIOLOGICAL

Like other pyridines, **17a** has a pungent, unpleasant odor and is irritating to the skin, eyes, and respiratory tract (229). It is highly toxic both by skin contact with the liquid and inhalation of the vapor. A maximum permissible concentration of 0.005 mg per liter in industrial atmospheres has been suggested (237). The single dose oral LD_{50} of **19** for rats is 1.23 g per kg (238). Comparable data for the other monomers in this section were not found, but it seems safe to assume that they are relatively toxic substances also.

·lpyridines and 2-Vinylquinoline

		Monomer			
18		**19**		**20a**	
	Reference		Reference		Reference
7)	223	65(3)	172	92(0.8)	215d
15)	174	89.5(15)	225	104(3)	226
	215a	102(19)	164	125(7)	215b
45)	179a				
	227a				
calcd)	228				
5415(20)	179a	1.5371(20)	225	1.6439(20)	215b
5450(20)	231	1.5383(20)	172	1.6461(20)	215d
5454(20)	174			1.6485(20)	226
	215a				
5459(20)	227b				
5420(25)	223				
5411(25)	228				
961(16)	228	0.9432(20)	303	1.0705(20)	226
9521(20)	223			1.0661(25)	215b
956(20)	215a				
9597(20)	231				
9646(20)	174				

The melting point of **18** is reported to be $-7°$ (**181**) and $-12°$ (174,215a).

3. ANALYTICAL METHODS AND SPECTRA

Polarography (239,240) and near infrared spectra (241) have been used for the quantitative determination of vinylpyridines. The composition of 2-methyl-5-vinylpyridine (18) –2-methyl-5-ethylpyridine (22) mixtures can be measured by refractive index or bromine titration (231).

Picrates are important derivatives in pyridine chemistry, and the vinylpyridines are not exceptional in this respect. Their picrates are frequently used for identification and isolation purposes. The melting points of the vinylpyridine picrates that have been published are listed in Table 5.

The ultraviolet spectra of 17a (224), 17b (244,245), and 19 (246) and the NMR spectra of 17a, 17c, and 18 (73,247) have been determined.

TABLE 5

Vinylpyridine Picrates

Compound	Picrate, mp, °C	Reference
2-Vinylpyridine (17a)[a]	159	218,242
3-Vinylpyridine (17b)	148	174
	147.5	221
	144	166,215a,220,221
4-Vinylpyridine (17c)	157[b]	165
2-Methyl-5-vinylpyridine (18)	161	181
	159	174
	158	215a
5-Ethyl-2-vinylpyridine (19)	130.5	172
4-Vinylquinoline (20b)	189	165

[a] The β-resorcylic acid salt, mp 113°, has also been used for identification of 17a (243).

[b] Resolidified after melting at 157° and remelted at 199°.

D. Storage and Handling

The vinylpyridines and vinylquinolines are liquids, and the former have relatively high vapor pressures at room temperature. Because of their irritant properties, contact with the skin and inadequate ventilation while handling should be avoided. They are also flammable; the flash point (open cup) of 18 is 74° (228).

Vinylpyridines polymerize readily, especially when subjected to heat and light. For example, although commercial 17a and 18 contain polymerization inhibitors, the manufacturers recommend storage under refrigeration (228,229). That undesired polymerization of vinylpyridines during production and storage is a major problem is obvious from the number of references to inhibitors for them. Phillips Petroleum Company alone has 21 United States patents on the subject. These cover naphthylamine and naphthol sulfonic acids; aromatic nitro compounds, including nitro-substituted amines, phenols, and their chloro derivatives; cupferron and related compounds; organic nitrites; azo- and azoxybenzene; phenyl-hydrazine and hydrazides of aliphatic acids; azine, phenoxazine, indigo, and thioindigo dyes; organic and inorganic polysulfides; thiocyanates and dithiocarbamates; iron halides; and furnace black (248). Other companies have patented alkyl nitrites (249), diarylamines and their nitro derivatives (250a), aryl thiols and polysulfides (182b), dimethyl sulfoxide (251), and dyes such as methylene blue (250b,252), malachite green, methyl violet (250b), sulfonaphthaleins (250c), phenothiazines (253), and 1-nitroso-2-naphthol (252). Other inhibitors that have been used are sulfur, methylaminophenol sulfate (methol), picric acid (254), picrocarmine, trinitrobenzene, phenol, and pyrogallic acid (227c).

It should be emphasized that there are two types of inhibitors, one for long-term storage at ambient temperatures, the other to minimize polymerization at high temperatures during synthesis and purification. As examples of the former, methylene blue and 1-nitroso-2-naphthol were patented specifically as storage inhibitors for vinylpyridines (252). As an example of the latter, Russian workers reported (254) that sulfur is the better inhibitor during separation of **18** and **22** by fractional distillation, although methylaminophenol sulfate is preferred for storage.

On heating during distillation, for example, crude **18** displays a marked tendency to form a popcorn polymer. Picric acid or trinitrobenzene prevents this in the distillation autoclave but not in the fractionating column. Maleic anhydride reduces but does not eliminate the difficulty. Pure **18** shows no such behavior, and so it was concluded that an impurity was responsible (215a), possibly 2,5-divinylpyridine (255). Four of the Phillips inhibitor patents (248) deal specifically with prevention of popcorn formation.

For shipment and storage of **18**, Phillips Petroleum Company uses 0.1 percent *tert*-butylcatechol and 0.05 percent *o*-aminophenol as inhibitors under an air cover or, if color is critical and a small amount of polymer permissible, 0.15 percent of only the former compound (228). Procedures for determination of these inhibitors in **18** by spectrophotometry (256) and for their removal by treatment with Al_2O_3 or an alkali metal aluminate (257) have been described. Formaldehyde, formaldehyde polymers, and methylolureas are claimed to be color stabilizers for vinylpyridines (258).

E. Chemical Reactions

Vinylpyridines and vinylquinolines react both as tertiary amines and as olefins. The 2- and 4-vinyl compounds have an additional structural feature in the conjugation of the olefinic C=C bond with the C=N bond in the ring. As a result, they exhibit many of the reactions of acrylonitrile.

As amines, the vinylpyridines form salts and complexes, for example, hydrochlorides (163,166,221a,259), chloroaurates (162,163,166,230), chloroplatinates (161b,162,163,165,166), and picrates (see Sec. II-C-3). Coordination complexes with $HgCl_2$ (163,166,260), Zn(II) salts (260), and $CuCl_2$ (261,262) are known. The reactions of the $CuCl_2$–17a complex indicate that both the nitrogen atom and the vinyl double bond in 17a act as electron donors to the copper(II) cation (261).

2-(2-Chloroethyl)pyridine (29a) rearranges to the hydrochloride of 17a (30a), sometimes suddenly as a result of mechanical shock (259):

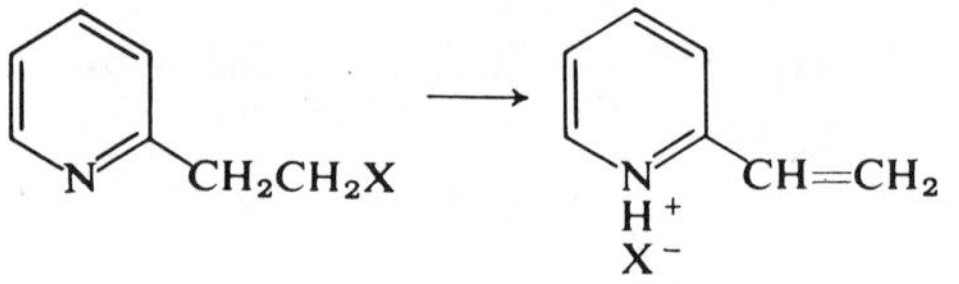

$$(19)$$

29a, X = Cl
b, X = Br

30a, X = Cl
b, X = Br

At 160°, 30b is converted to the dimeric pyridinium salt 31. The same compound is formed from the hydrobromide of 29b on neutralization with aqueous Na_2CO_3 and extraction with ether (218):

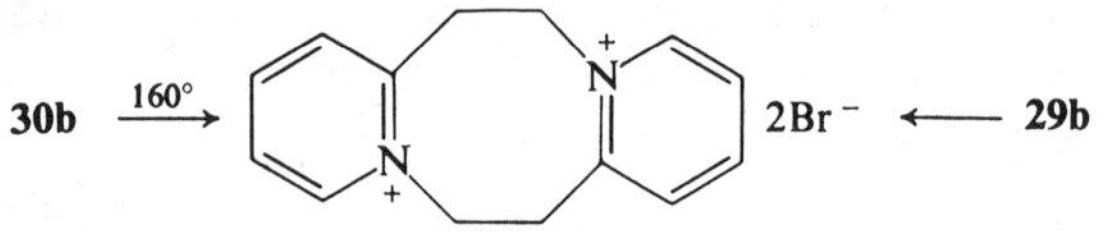

$$(20)$$

31

It is not known if 29b is an actual intermediate in the 30b to 31 reaction. A dimer of 20b was also obtained by heating its hydrochloride or picrate, followed by treatment with base (165). The structure was not determined, but the compound's molecular weight, facile solubility in benzene, and ability to form a hydrochloride indicate a true dimer rather than a pyridinium salt analogous to 31. Its failure to reduce permanganate further suggests a cyclobutane rather than a linear unsaturated dimer.

N-Oxides have been obtained both directly from vinylpyridines and by indirect means. 2-Methyl-5-vinylpyridine N-oxide was prepared by oxidation of 18 with peracetic acid (263), 2-vinylpyridine N-oxide by dehydration of 2-(2-hydroxyethyl)pyridine N-oxide with NaOH (179c) or $KHSO_4$

(264,265), as in the Ladenburg synthesis—Eq. (17). The N-oxide of **19** was also prepared by the Ladenburg route (265). The N-oxide of **17a** reacts readily with nucleophiles (264) in the manner of **17a** itself (see below).

Oxidation of **17a**, **17c**, and **20b** with $KMnO_4$ produces the corresponding pyridine or quinoline carboxylic acid (161a,165,232b). **17c** is oxidized much more readily by CrO_3 than is **17a** (266), and electrolytic oxidation of **17a** requires high potentials (267).

17a is reduced by sodium and ethanol to 2-ethylpyridine (161a). Electrolysis of mixtures of a vinylpyridine and an α,β-unsaturated nitrile, ester, or ketone results in reductive coupling products at the cathode. For example, 5-(2-pyridyl)valeronitrile (**32**) was obtained from **17a** and acrylonitrile (268). Heating liquid **18** in the presence of atomic hydrogen gives **22**, the homopolymer of **18** and 1,4-bis-5-(2-methylpyridyl)butane (**33**) (269):

32	**33**

Reference was made earlier to the structural resemblance of **17a** to acrylonitrile and the resulting similarity of behavior toward nucleophilic reagents. By way of example, the reactions with diethyl malonate in the presence of base are shown in Eqs. (21) and (22):

$$CCH{=}CH_2 + \ ^-CH(COOC_2H_5)_2 \xrightarrow{\text{base}} NC\bar{C}HCH_2CH(COOC_2H_5)_2 \xrightarrow{H^+}$$

Acrylonitrile

$$NCCH_2CH_2CH(COOC_2H_5)_2 \quad (21)$$

$$(22)$$

Because **17c** is a vinylog of **17a**, it reacts in an analogous manner to give 4- rather than 2-pyridylethyl derivatives.

The Michael type of condensation of acrylonitrile with "active hydrogen" compounds (cyanoethylation) has been studied extensively, particularly by Bruson (270). Not surprisingly, the comparable reaction of **17a** and **17c** has been given the name "pyridylethylation" (271a) and has also been investigated thoroughly. Doering and Weil (221a) showed that **17a** reacts in this way with diethyl malonate, ethyl acetoacetate, HCN, diethylamine, piperidine, ethanol, and $NaHSO_3$, that **17c** gives similar products with $NaHSO_3$ and HCN, but that **17b**, in which the vinyl group is not conjugated with the C=N bond, does not react with $NaHSO_3$. Independently, Boekelheide and Rothchild (272a) discovered the same reaction of **17a** and diethyl malonate. Following these early publications, three groups in particular investigated pyridylethylation in greater detail: Boekelheide (272), Levine (271), Profft (206,225,273) and their associates. This work explored ketones, esters, ketoesters, and amines as nucleophiles and showed, among other things, that acids as well as bases can be used as catalysts (271d,272b,273c). Further examples of these reactions have been reported by others (208,259,274,275).

Additional nucleophilic reagents that have been used include mercaptans and thioacids (276), sodium azide (277), dialkyl phosphites (278), trichlorosilane (179d), and thiourea (279). The literature on pyridylethylations with **17a** and **17c** to 1961 has been summarized (279). Comparable additions of amines and thiols to **20a** have been studied (215d). A complete coverage of pyridylethylation was not attempted, since it is beyond the scope of this chapter; but we believe that the examples above demonstrate the broad applicability of the reaction.

Vinylpyridines undergo the Diels-Alder reaction with 1,3-dienes. The main product from **17a** and 2-substituted butadienes is the para rather than the alternative meta adduct—Eq. (23) (R = C_6H_5 or CH_3) (280):

$$\tag{23}$$

17a

17a is not appreciably more reactive as a dienophile toward butadiene than **17b** or styrene (221b) despite its more readily polarizable double bond. The same reference describes the adduct from **17a** and 2,3-dimethyl-1,3-butadiene. Diels-Alder reactions of **17a** and **17c** with a variety of dienes have been studied in the U.S.S.R. (232).

Like other vinyl aromatic compounds, **17a** yields the corresponding arylacetamide under the conditions of the Willgerodt reaction—Eq. (24) (281):

$$+ (NH_4)_2S_x + (NH_4)_2S_2O_3 \xrightarrow[\text{pyridine}]{150°} \tag{24}$$

17a

F. Polymerization

The vinylpyridines and vinylquinolines homopolymerize and co-polymerize with monomers, such as butadiene, isoprene, methyl metha-crylate, and styrene, by a free-radical mechanism using heat, light, or peroxides to initiate the process (167,215a,b,233,282,283). Since they are nitrogen bases, they can also be polymerized as salts in aqueous solution with a water-soluble initiator; neutralization with base precipitates the polyamine (284). Polymerization of vinylpyridine coordination complexes has also been studied (260).

Vinylpyridines are polymerized by catalytic amounts of copper(II) acetate (285). The reaction rate is higher in polar solvents than in bulk and **17c** polymerizes more rapidly than either **17a** or **18**. It has been suggested (286) that a one-electron transfer from monomer to Cu^{++} occurs in the $(17c)_2-Cu^{++}$ coordination complex and that polymerization is initiated by the resulting radical cation in a manner analogous to the polymerization of 9-vinylcarbazole by Cu(II), Fe(III), and Ce(IV) salts (see Sec. I-F-1).

Like acrylonitrile, which they resemble structurally (see above), **17a** and **17c** undergo anionic polymerization very readily with such initiators as sodium naphthalene and butyllithium to produce stable polymeric anions ("living polymers") that may be used to prepare block copolymers (287–290). A soluble graft copolymer was obtained by anionic polymeriza-tion of **17c** with poly(vinylbenzophenone sodium ketyl) as initiator (291).

Natta and coworkers (292) obtained crystallizable, presumably stereo-regular polymers from **17a**, by anionic initiation with alkyls, aryls, hydrides, and alkyl- and arylamino compounds of group I, II, and III metals or the corresponding halometal derivatives, for example, C_6H_5MgBr, $(C_2H_5)_2Mg$, $(C_2H_5)_2NMgX$. The products are insoluble in boiling aliphatic hydrocarbons, ether, and methyl ethyl ketone and melt at 185 to 215°, depending on the degree of crystallinity. The infrared spectra of **17a** polymers prepared with aryl or arylamino magnesium compounds indicate aryl terminal groups in agreement with the postulated anionic polymerization mechanism. It appears that both the nitrogen atom and the vinyl group of **17a** are coordinated with the initiator, since **17c**, in which these groups are too far apart for such simultaneous co-ordination, gives noncrystallizable polymers under the same conditions.

Acetone-insoluble polymers are obtained from **17a** using C_6H_5MgBr, 2,2'-azobisisobutyronitrile, and ammonium persulfate (in $2M$ HCl) as initiators (293). The authors claim, on the basis of the NMR spectra of the products, that they are isotactic, a mixture of isotactic-syndiotactic, and syndiotactic, respectively. There is no evidence of stereoregularity in polymers of **17c** prepared in a similar manner.

Treatment of **17a** and **17c** polymers in *tert*-butanol with optically active acids yields optically active salts of the polymers. The salts of D-tartaric acid and isotactic poly(2-vinylpyridine) have higher optical rotation values than those from atactic polymer (294).

Fuoss and Strauss (295) prepared high-molecular-weight poly(4-vinyl-pyridine) by polymerization of **17c** in toluene with benzoyl peroxide initiation. The polymer forms a water-soluble hydrochloride (**34**) that behaves like the salt of an extremely weak base ($K_b = 0.011 \times 10^{-9}$):

$$(-CH_2-CH-)_n \qquad (-CH_2-CH-)_n$$

$$H^+Cl^- \qquad \qquad + \ Br^- \\ C_4H_9$$

34   **35**

Addition of bromobutane to the polyvinylpyridine gives a quaternary ammonium salt (**35**) that is also water-soluble. Fuoss and coworkers later made a systematic study of **17c** polymerization (bulk, bead, and emulsion methods) (296) and extended the investigation to **17a** polymers (297). Strauss and Jackson (298) also prepared poly(2-vinylpyridine) and quaternized it with bromododecane in nitroethane solution. Aqueous solutions of the product exhibit soaplike properties—solubilization of hydrocarbons, viscosities indicating compact molecules resembling micelles—and so the authors named this type of polymer a *polysoap*. Both Fuoss and Strauss, with their associates, have continued to investigate polyelectrolytes based on vinylpyridines. Quaternary salts of **17a** and **20a** polymers are the subject of a patent (299).

Polymers of vinylpyridine *N*-oxides may be prepared from the monomers (179c) or by oxidation of vinylpyridine polymers (300).

The polymerization kinetics of **17c** (233) and **18** (223) have been studied. Heats of polymerization and activation energies for polymerization are listed in Table 6, relative reactivity ratios in Table 7, and Q and e values in Table 8. Styrene and **17a** have been copolymerized in the presence of

TABLE 6

Heats of Polymerization and Activation Energies for Polymerization of Vinylpyridines

Monomer	$\Delta H_{polymn,}$ kcal/mole	Reference	$E_{act,}$ kcal/mole	Reference
17a	17.1 ± 0.1	96	20 ± 1	226
17c	18.7 ± 0.3	301	26.9	302
20a			18 ± 1	226

TABLE 7

Relative Reactivity Ratios for Free-radical Copolymerization of Vinylpyridines and their N-Oxides (M_1)

M_1	Comonomer (M_2)	r_1	r_2	Reference
17a	Acrylonitrile	0.47 ± 0.03	0.113 ± 0.002	303
		22 ± 6	0.05 ± 0.01	226b,c
	Butyl acrylate	2.51 ± 0.05	0.10 ± 0.04	304
	Chloroprene	0.06 ± 0.01	5.19 ± 0.03	226b
	2,5-Dichlorostyrene	1.1	0.9	305
	Ethyl acrylate	0.23 ± 0.05	0.19 ± 0.06	135
	Isoprene	0.47 ± 0.07	0.59 ± 0.05	226
	Isopropenylacetylene	1.65 ± 0.05	0.55 ± 0.1	306
	Methacrylic acid	1.55 ± 0.1	0.58 ± 0.05	307
	Methyl acrylate	2.0 ± 0.5	0.20 ± 0.09	308
		1.58 ± 0.05	0.168 ± 0.003	309
	Methyl methacrylate	0.86 ± 0.06	0.395 ± 0.025	310
		0.77 ± 0.02	0.439 ± 0.002	309
		0.7 ± 0.1	0.33 ± 0.05	311
	Phenylacetylene	4.0 ± 0.7	0.20 ± 0.05	283
	Styrene	1.81 ± 0.05	0.55 ± 0.03	226
		1.33 ± 0.08	0.57 ± 0.05	312
		1.135 ± 0.08	0.55 ± 0.025	310
		0.9 ± 0.2	0.56 ± 0.02	309
	Vinyl acetate	30 ± 4.5	0.0	311
		10	0.3	313
	1-Vinyl-2-ethylacetylene	1.5 ± 0.5	0.6 ± 0.1	306
17c	Acrylonitrile	0.41 ± 0.09	0.113 ± 0.005	303
	Butyl acrylate	5.15 ± 0.09	0.46 ± 0.09	304
	Methyl acrylate	1.7 ± 0.2	0.22 ± 0.01	309
	Methyl methacrylate	0.79 ± 0.05	0.574 ± 0.004	309
	Styrene	0.7 ± 0.1	0.54 ± 0.03	309
		0.52 ± 0.06	0.62 ± 0.02	314
	Vinyl chloride	23.4	0.02	302
18	Acrylamide	0.01 ± 0.09	0.56 ± 0.09	315
	Acrylonitrile	1.1 ± 0.2	0.10 ± 0.05	316
		0.27 ± 0.04	0.116 ± 0.003	317
	Butadiene	0.72 ± 0.03	1.32 ± 0.01	318
	Diallyl ether	80	0	319
	Methyl acrylate	0.9 ± 0.1	0.172 ± 0.007	317
	Methyl methacrylate	0.61 ± 0.08	0.46 ± 0.02	317
	Styrene	1.2 ± 0.1	0.9 ± 0.2	223
		0.91 ± 0.02	0.812 ± 0.005	317
		0.7 ± 0.1	0.6 ± 0.1	320
19	Acrylonitrile	0.43 ± 0.05	0.02 ± 0.02	303
	Methyl acrylate	1.16 ± 0.08	0.179 ± 0.006	309
	Methyl methacrylate	0.69 ± 0.03	0.395 ± 0.003	309
	Styrene	1.2 ± 0.2	0.79 ± 0.03	309

continued

TABLE 7 (*continued*)

M_1	Comonomer (M_2)	r_1	r_2	Reference
20a	Chloroprene	0.38 ± 0.03	2.1 ± 0.1	226a,c
	Isoprene	1.88 ± 0.02	0.53 ± 0.01	226b,c
	Styrene	2.1 ± 0.55	0.5 ± 0.1	226b,c
17a	Methyl methacrylate	3.9 ± 0.03	0.13 ± 0.03	265
N-Oxide	Styrene	2.1 ± 0.6	0.11 ± 0.01	265
19	Methyl methacrylate	5.5 ± 0.5[a]	0.11 ± 0.01[a]	265
N-Oxide		4.7 ± 0.6	0.12 ± 0.02	265
	Styrene	2.6 ± 0.3	0.10 ± 0.01	265

[a] These values are for bulk copolymerization; the others from reference 265 are for copolymerization in ethanolic solution.

TABLE 8

Q and e Values for Vinylpyridines

Monomer	Q	e	Reference
17a	1.30	−0.50	137,146
	1.2	−0.6	309
	1.1	−0.6	309
	1.09	−0.6	321
	1.07	−0.1	321
	0.94	0.0	309
	0.50	−0.5	303
	−3.2 kcal/mole(q)[a]	−0.07 × 10^{-10}esu(ϵ)[a]	147
17c	1.91	−0.51	302
	1.0	−0.4	309
	0.90	−0.5	309
	0.83	−0.2	309
	0.82	−0.20	137,146
	0.49	−0.6	303
	−2.9 kcal/mole(q)[a]	0.09 × 10^{-10}esu(ϵ)[a]	147
18	1.22	−0.55	137
	1.05	−0.8	317
	1.04	−0.7	317
	0.99	−0.58	146
	0.76	−0.2	317
	0.38	−0.7	317
19	1.6	−1.0	303
	1.37	−0.74	137,146
	1.2	−0.7	309
	1.1	−0.7	309
	1.1	−0.6	309
20a	3.79	−0.82	137,146
17a N-Oxide	3.77	−0.01	146
19 N-Oxide	4.52	−0.10	146

[a] Schwan-Price resonance q and electrical ϵ factors (147).

additives of varying acid strength (ethanol, methanol, phenol, acetic acid). With increasing acidity, the monomer versus copolymer composition curve acquires an S shape and the $r_1 \cdot r_2$ product decreases, indicating an increased degree of alternation. The authors suggested that, in an acidic medium, the basic **17a** acquires a more positive charge, so that it tends to alternate with the relatively negative styrene (312). Along with many other monomers, **17a** and **17c** exhibit a correlation between rate of reaction with methyl radicals and copolymerization reactivity ratio (322).

G. Applications of Polymers

Vinylpyridines impart to their homo- and copolymers much the same mechanical properties as styrene, which is considerably less expensive (323). Therefore, they cannot compete with styrene on this basis but are used in specialty applications where incorporation of the basic, polar tertiary amino group into a polymer is desirable.

The two major areas of use for vinylpyridines are synthetic rubbers and acrylic fibers. Butadiene-vinylpyridine copolymers are excellent elastomers, resembling SBR in mechanical properties. They have high vulcanization rates and good freeze resistance, and can be made in conventional SBR production equipment by only slightly modified procedures. The copolymer latexes are particularly effective as cord dips for the tire industry to improve adhesion between rubber and nylon cord. Butadiene-vinylpyridine rubbers can be made oil-resistant by quaternization with organic halides during vulcanization (169). Addition of vinylpyridines to acrylic fibers, either as polymers (by blending) or as comonomers, improves dye receptivity, especially toward acidic dyestuffs (170).

Other industrial uses of vinylpyridine polymers that have been studied are as emulsifiers for polymerizations in acidic media, anion exchange resins, polyelectrolytes, and components of photographic emulsions (169, 170,324).

A possible use in medicine is suggested by the observation that injections of polyvinylpyridine N-oxides are effective against silicosis caused by inhalation of quartz dust (325). Activity decreases and finally vanishes as the molecular weight of the polymer is lowered. It is believed that the N—O moiety forms hydrogen bonds with silanol groups on the surface of the silica particles that would otherwise hydrogen-bond with lipids and proteins in cell membranes. This reaction would aid in elimination of the particles from the body and neutralizing, or at least attenuating, their pathogenic effects on living tissues (326–328).

III. VINYLSILANES

A. Introduction

Triethylvinylsilane, reported in 1937, was the first vinylsilane to appear in the literature. Ushakov and Itenburg (329) prepared this compound by chlorinating tetraethylsilane with phosphorus pentachloride and then dehydrohalogenating with alcoholic sodium hydroxide. Since this silane would not polymerize, even in the presence of peroxides, it caused little stir in the world of polymers.

Eight years later, Hurd (330) at General Electric prepared trichloro-vinylsilane (36) and dichlorodivinylsilane by the "direct method" previously used in synthesizing chloromethylsilanes—Eq. (25) (331,332):

$$CH_2{=}CHCl \xrightarrow[\substack{300-350° \\ N_2}]{Si-Cu} CH_2{=}CHSiCl_3 + (CH_2{=}CH)_2SiCl_2 \qquad (25)$$
$$\phantom{CH_2{=}CHCl \xrightarrow[\substack{300-350° \\ N_2}]{Si-Cu} CH_2{=}CHSi} \mathbf{36}$$

Yields were poor, but he also explored a second and more successful approach. Trichloroethylsilane was chlorinated with sulfuryl chloride in the presence of benzoyl peroxide (333) to give trichloro(1-chloroethyl)- and trichloro(2-chloroethyl)silane. Distillation of this mixture with dry quinoline led to a 50 percent overall yield of 36—Eq. (26):

$$C_2H_5SiCl_3 \xrightarrow[Bz_2O_2]{SO_2Cl_2} Cl(C_2H_4)SiCl_3 \xrightarrow[quinoline]{dist.} \mathbf{36} \qquad (26)$$

Dichloromethylvinylsilane (37) was also prepared, both by the dehydro-halogenation of dichloro(chloroethyl)methylsilane and by treating 36 with methylmagnesium bromide.

Hurd noted that hydrolysis of these compounds yielded polyvinyl-siloxanes that were capable of further polymerization through their C=C unsaturation. A patent in 1947 (334) claimed that these compounds are useful as monomers, chemical intermediates, and water repellants. This attracted the attention of many workers in the organosilicon field, and soon the literature was flooded with improved syntheses and further applications for these monomers, as well as a variety of new vinylsilanes.

Since 36 is a useful starting material for many of these vinyl compounds, a great deal of effort has been directed toward finding commercially feasible processes for its manufacture. In 1956, a British patent (335) claimed good yields of 36 when vinyl chloride and trichlorosilane were heated at 600°. This method, with only minor modifications, is still used. With the development of safer techniques for handling acetylene, the addition of trichlorosilane to this alkyne rapidly became the most important commercial synthesis. Platinum catalysts (336,337) are preferred for the reaction.

During the middle fifties, the preparation of vinyl Grignard reagents in tetrahydrofuran was developed (338–341). This made possible the direct synthesis of organosilanes with more than one vinyl group.

Although many vinylsilanes have been prepared and many applications claimed, only six of the monomers are now commercially available (see Table 9). With the exception of **41** these silanes are ambifunctional, that is, able to undergo both condensation (after hydrolysis) and vinyl polymerization. This characteristic is responsible for the majority of their applications, since it enables them to serve as coupling agents between hydroxyl-containing substrates and organic resins (342).

This property has been utilized extensively in glass-reinforced plastics. When glass fabrics are treated with one of these silanes before incorporation into polyester- or epoxy-bonded laminates, strength, color, water resistance, and electrical properties are greatly improved (343–354). Glass fibers treated in a similar manner show greater adhesion when used as fillers in resins (355,356) and are also easier to fabricate into yarns and cloth (344).

The nature of the bonding between glass and resins has been studied in detail (357–359). It has been found that the silane-resin bond is stronger than that between silane and glass. This is believed to be advantageous in laminates, since a weaker bond permits some slippage on the glass surface, relieving local stresses set up during cooling or mechanical working. Vinylsilanes with three hydrolyzable substituents give optimum bonding (343). Since **36**, **38**, **39**, and **40** all give vinylsilanetriol upon hydrolysis, a

TABLE 9

Commercially Available Vinylsilanes

No.	Compound	Structure	Source[a]
36	Trichlorovinylsilane (vinyltrichlorosilane)	$CH_2{=}CHSiCl_3$	a, b, c, d
37	Dichloromethylvinylsilane (methylvinyldichlorosilane)	$CH_2{=}CH(CH_3)SiCl_2$	b, c
38	Triethoxyvinylsilane (vinyltriethoxysilane)	$CH_2{=}CHSi(OC_2H_5)_3$	a, b, c
39	Triacetoxyvinylsilane (vinyltriacetoxysilane)	$CH_2{=}CHSi(OCOCH_3)_3$	c
40	Tris(2-methoxyethoxy)vinylsilane (vinyltris(2-methoxyethoxy)silane)	$CH_2{=}CHSi(OCH_2CH_2OCH_3)_3$	b, d
41	Diphenyldivinylsilane (divinyldiphenylsilane)	$(C_6H_5)_2Si(CH{=}CH_2)_2$	a

[a] (a) Borden Chemical Company, (b) Union Carbide Corp., Silicones Division, (c) Dow Corning, (d) General Electric.

choice must be made based on cost and ease of handling. Although **36** is the least expensive, hydrogen chloride is evolved when it is hydrolyzed. This is hazardous and could interfere with the subsequent vinyl polymerization.

The coupling action of these vinylsilanes also promotes resin-resin and resin-metal bonding (342,360–362). They can be used with a variety of other inorganic and organic fillers to improve the properties of reinforced plastics (363–367).

In protective coatings, vinylsilanes with hydrolyzable groups function in several ways. They facilitate the wetting and dispersion of pigments in organic vehicles (368). Pigments treated with these compounds are resistant to water and oxidation, making them especially useful in roof coatings (369) and marking compositions for mine fields (370). Copolymerization with unsaturated polyesters gives surface coatings with short bake cycles that are hard, flexible, and resistant to acids and alkalies (371–373). Thermoplastic articles coated with these silanes are heat-, stain-, and scratch-resistant (374). Alone, or as an adduct with other resinous materials, they impart water repellence to leather, wood, and a variety of textiles (375–377). Pyrolytic films suitable for semiconducting devices result when alkoxyvinylsilanes are heated at temperatures ranging from 700 to 1100° (378,379). Glass surfaces treated with alkaline solutions of vinylsilanols are made hydrophobic (380). Hydrophilic surfaces, useful in lithography, can be formed on metals by first applying trialkoxyvinylsilanes, followed by acrylic acid (381). It is also claimed that optic glass can be protected against fungi by treating with an alkoxysilane, mercuric acetate, and ethanol (382).

A closely related use of alkoxyvinylsilanes is encapsulation. Oxidants, for example, NH_4NO_3 and NH_4ClO_4, show excellent storage stability when coated with these compounds (383,384). Particles of sodium bicarbonate, used in fire-extinguishing compositions, do not cake when treated in a similar manner (385).

There are many instances in which the addition of a very small quantity of a vinylsilane with hydrolyzable groups to a composition gives greatly improved properties. Elastomers with good resistance to heat and solvents result when alkyl acrylates are copolymerized with 0.5 to 5.0 percent of an alkoxyvinylsilane (386). Copolymers of olefins and chlorovinylsilanes, subjected to hydrolysis and crosslinking, are claimed (387,388) to have exceptional thermal stability, although the silane content is very small. The addition of 0.1 to 10.0 percent of **40** to a diorganopolysiloxane improves adhesion to smooth, solid surfaces (389). Less than 1 percent of **37** in a silicone rubber gives low compression set and increased hardness (390). A similar amount of alkoxyvinylsilane is claimed to improve the properties of concrete (391). In urethane foams, 0.05 to 3.0 percent of **38** gives more uniform fine cells (392).

Similar amounts of homopolymers of vinylalkoxysilanes as plasticizers in synthetic resins, and especially poly(vinyl alcohol), improve toughness and solvent resistance (393). They also serve as antifoaming agents in hydrocarbon lubricating oils (394). Copolymers of **38** with *N*-vinylcarbamates are useful as emulsifying agents for silicone oil-water mixtures (395). Butyl rubbers show better tensile strength, modulus, and elasticity when a vinylorganopolysiloxane is added (396).

Thus, though the applications of monomers **36** to **40** are varied, the quantities used are often relatively small, and their production volume low.

No specific applications for **41** were found in the literature. With two vinyl groups, **41** should be an effective cross-linking agent, but its high cost ($260 per 500 g) (397) precludes large-scale commercial development at present.

B. Synthetic Methods

1. TRICHLOROVINYLSILANE (36)

a. Direct Method

Although initial attempts to combine vinyl chloride with elemental silicon in the presence of a copper catalyst—see Eq. (25)—were disappointing (330), workers have continued to investigate this approach. Major difficulties are the tendency of vinyl chloride to pyrolyze and a low conversion to **36**. Operation at lower temperatures (375,398) and shorter contact times (399–401) seem to minimize the former. Nitrogen is generally used to dilute the vinyl chloride, since it aids in temperature control and reduces the contact time (330). Catalysts, such as $SnCl_2$ or $CuCl$ (402) and Ni (403), are reported to improve the yield of **36**. The best results to date have been obtained using a mixture of silicon powder and copper oxides bonded with water glass and 5 percent ferric chloride as an activator (401). With this catalyst (particle size 3 to 4 mm) and introduction of vinyl chloride at 0.1 ml per min and a temperature of 360°, a 35 percent conversion is claimed. $SiHCl_3$, $SiCl_4$, $(CH_2=CH)_2SiCl_2$, and $CH_2=CHSiHCl_2$, which are useful intermediates, are the by-products of this reaction in addition to some polymer. The method lends itself well to continuous operation.

b. From Olefins and Chlorosilanes

Vinyl chloride and trichlorosilane react at high temperatures to give **36**—Eq. (27):

$$CH_2=CHCl + HSiCl_3 \xrightarrow{550–700°} 36 + HCl \qquad (27)$$

The reactants may be introduced separately, but simultaneously, or as a mixture. Careful control of the flow rate and stoichiometry of the reactants have boosted the yield of **36** from initial reports of 56 percent (335) to 80 percent (404). A catalyst, Si–CuO–Mg, is said to permit operation at lower temperatures (405,406), but it appears to lower the yield as well.

Although moderately successful results have been reported using SiCl$_4$ (406) and Cl$_3$SiSiCl$_3$ (407,408), HSiCl$_3$ is the preferred starting material (407). Ethylene can be substituted for vinyl chloride but is far less effective (407,409,410).

Dichloromethylvinylsilane (**37**) can also be prepared in this manner if trichlorosilane is replaced by dichloromethylsilane (400,404,407,411). Yields of **37** are somewhat lower, however, than those of **36**. It has, in fact, been generally observed that, as an increasing number of chlorines on trichlorosilane are replaced by alkyl groups, yields of the corresponding vinylsilanes decrease (407,412). In the extreme situation, trimethylsilane gives no trimethylvinylsilane at all; only ethylene and chlorotrimethylsilane are isolated. Recent studies (412) indicate that the thermal stability of both the initial chlorosilanes and the final products are responsible for this trend. There is reason to believe that yields in all cases would be almost quantitative were it not for the accompanying decomposition.

It is commonly agreed that this reaction has a free-radical mechanism. However, it is uncertain whether HCl splits from vinyl chloride, yielding acetylene, which then reacts with the chlorosilane, or whether addition occurs first, followed by HCl elimination (413,414). The latter is in keeping with the fact that trichloro(chloroethyl)silanes can be dehydrohalogenated by pyrolysis (see Sec. III-B-1-d). Furthermore, the formation of ethylene and chlorotrimethylsilane from trimethylsilane and vinyl chloride could readily arise from addition followed by β-elimination—Eqs. (28) to (31) (415):

$$(CH_3)_3SiH + CH_2{=}CHCl \longrightarrow (CH_3)_3SiCH_2CH_2Cl \qquad (28)$$

$$(CH_3)_3SiCH_2CH_2Cl \xrightarrow{\text{heat}} (CH_3)_3SiCH_2CH_2{}^+ + Cl^- \qquad (29)$$

$$(CH_3)_3Si\overset{\frown}{CH_2}CH_2{}^+ \longrightarrow (CH_3)_3Si^+ + CH_2{=}CH_2 \qquad (30)$$

$$(CH_3)_3Si^+ + Cl^- \longrightarrow (CH_3)_3SiCl \qquad (31)$$

The good yields, easy accessibility of starting materials, and simplicity of operation make this a good laboratory or industrial synthesis of **36**. Both vinyl chloride and trichlorosilane (416) are commercially available.

c. From Acetylene and Chlorosilanes

It has been known for some time that 2 moles of chlorosilane add readily to acetylene at low temperatures in the presence of such catalysts as metal chlorides, metal oxychlorides (417), peroxides, or ultraviolet light (418) in two steps—Eqs. (32) and (33):

$$CH{\equiv}CH + HSiCl_3 \longrightarrow \underset{\textbf{36}}{CH_2{=}CHSiCl_3} \qquad (32)$$

$$CH_2{=}CHSiCl_3 + HSiCl_3 \longrightarrow Cl_3SiCH_2CH_2SiCl_3 \qquad (33)$$

With the rising interest in **36** as a monomer, and improved techniques in handling acetylene, the use of the first step as a synthetic method was investigated and is now the most important laboratory and industrial synthesis.

Platinum (336,337) and palladium (419,420) catalysts on a variety of supports, that is, silica gel, carbon, asbestos, and alumina, were found to improve the yield of **36** greatly. Excellent results (80 percent of theory) were obtained using $Pt-Al_2O_3$ at temperatures just over $100°$ and pressures of less than 260 psi (337). With an aromatic solvent, such as 1,2-dichlorobenzene and finely divided platinum, an 87 percent yield of **36** is claimed (421). A basic catalyst, triphenylphosphine, when used in conjunction with ethyl acetate as solvent, is reported to reduce pressure requirements and also improve yields (422).

d. Dehydrohalogenation

Trichloro(chloroethyl)silane can be dehydrohalogenated in a variety of ways. Hurd's original procedure (330,334) using quinoline—see Eq. (26)—has become a standard laboratory preparation for **36** (423). This is also adaptable to commercial synthesis, since the quinoline can be recovered and used again. Other organic bases, such as diethylaniline (424), diethylamine (425), isoquinoline (426), and piperidine (427,428), may be used instead of quinoline. Some aliphatic dinitriles are also claimed to be effective (429). With all these reagents, as with quinoline (334), trichloro-(2-chloroethyl)silane (**42**) is far more reactive than trichloro(1-chloroethyl)silane (**43**). Trichloro(bromoethyl)silanes give similar results (424):

$$\underset{\textbf{42}}{ClCH_2CH_2SiCl_3} \qquad \underset{\textbf{43}}{CH_3CHClSiCl_3}$$

Dehydrohalogenation can also be achieved with aluminum chloride or bromide (430,431), but this is a reversible reaction—Eq. (34):

$$Cl_3SiCH_2CH_2Cl \underset{\longleftarrow}{\overset{AlCl_3}{\longrightarrow}} \underset{\textbf{36}}{Cl_3SiCH{=}CH_2} + HCl \qquad (34)$$

A low concentration of aluminum chloride and low temperature are needed to shift the equilibrium to the right (431). Both **42** and **43** undergo this reaction, but the latter requires a higher temperature, 125 versus 90° (430).

Pyrolysis of trichloro(chloroethyl)silanes at 520 to 610° (413,432) gives small quantities of **36** along with tetrachlorosilane, ethylene, and other by-products. Somewhat lower temperatures (300 to 500°) can be used if the reaction is carried out in the presence of Si or Si–Cu (433,434). With ferrosilicon (435–439), yields of over 75 percent are claimed at 150°. Very little difference is noted here between **42** and **43** as starting materials, and it has been postulated (433) that **43** rearranges to **42** under these conditions. A mechanism has been proposed (440) for the gas-phase reaction of **42** at 356 to 417°.

A mixture of **42** and **43** results from the chlorination of trichloroethylsilane with sulfuryl chloride and benzoyl peroxide (330,333). Trichloroethylsilane can be prepared either from ethyl Grignard reagent and tetrachlorosilane (441) or by direct synthesis (331).

2. DICHLOROMETHYLVINYLSILANE (37)

a. Grignard Reaction

Two approaches are possible in synthesizing **37** by this method—Eqs. (35) and (36):

$$36 + CH_3MgBr \longrightarrow 37 \qquad (35)$$

$$CH_2{=}CHMgBr + CH_3SiCl_3 \longrightarrow 37 \qquad (36)$$

Hurd (330,334) used the first of these in preparing **37** in 1945. The development of vinyl Grignards in tetrahydrofuran in the 1950s gave impetus to the process in Eq. (36).

b. From Olefins and Dichloromethylsilane

(See Sec. III-B-1-b.) The equimolar condensation of vinyl chloride and dichloromethylsilane at 530 to 540° is reported (404) to give 75 to 80 percent of **37**. Yields are considerably lower when vinyl chloride is replaced by ethylene (407).

c. From Acetylene and Dichloromethylsilane

The nature of the catalyst determines the optimum temperature for this reaction. Thus, with Pd–Al$_2$O$_3$ (419,442) higher temperatures (260 to 300°) favor higher yields. Excellent results are obtained with platinum catalysts at temperatures as low as 20 to 60° (443,444) and 110° (421):

$$CH_3SiHCl_2 + HC\equiv CH \xrightarrow[\substack{iso\text{-}C_3H_7OH \\ 20\text{-}60°}]{H_2PtCl_6} 81\% \; 37 \qquad (37)$$

$$CH_3SiHCl_2 + HC\equiv CH \xrightarrow[110°]{Pt;\, 1,2\text{-dichlorobenzene}} 86\% \; 37 \qquad (38)$$

Addition of a solvent, such as chloroform, in Eq. (37) is said to improve the yield (445). The 1,2-dichlorobenzene in Eq. (38) may be replaced by other aromatic compounds.

d. Dehydrohalogenation

Dichloro(chloroethyl)methylsilane can be dehydrohalogenated in the same manner as trichloro(chloroethyl)silane, but yields of **37** are somewhat lower than those of **36**—Eq. (39) (439). Quinoline (430,434,446,447), piperidine (427), and ferrosilicon (435,436,439) are commonly used to promote this reaction:

$$ClC_2H_4(CH_3)SiCl_2 \longrightarrow 37 + HCl \qquad (39)$$

Dichloro(chloroethyl)methylsilane can be prepared by treating dichloro-ethylmethylsilane with sulfuryl chloride in the presence of benzoyl peroxide (330,334,446,448). The methyl group is not chlorinated, and the product is a mixture of the 1- and 2-chloroethyl isomers. Dichloroethyl-methylsilane is readily synthesized from trichloromethylsilane and ethyl Grignard reagent (446,448).

3. TRIETHOXYVINYLSILANE (**38**)

Alcoholysis of **36** with anhydrous ethanol is the usual method of preparing **38**—Eq. (40):

$$36 + 3C_2H_5OH \longrightarrow 38 + 3HCl \qquad (40)$$

It has been claimed that the addition of pyridine (449) improves the yield from 50 to 60 percent. Excellent results (66 percent), however, have been achieved without it (450). It is also possible to prepare **38** from triethoxy-silane and acetylene, using a platinum catalyst (336) or by treating ethyl orthosilicate with vinyl Grignard (451).

4. TRIACETOXYVINYLSILANE (**39**)

Anhydrous sodium acetate reacts with trichlorovinylsilane to give **39** (452). Better results are obtained using acetic anhydride and a trace of triethanolamine (453).

5. TRIS(2-METHOXYETHOXY)VINYLSILANE (**40**)

Treatment of either **36** (454) or **38** (354) with 2-methoxyethanol produces **40**. With the former, a yield of 46 percent has been reported.

6. DIPHENYLDIVINYLSILANE (41)

An excellent yield (80 percent) of 41 is obtained by mixing vinyl Grignard in tetrahydrofuran with dichlorodiphenylsilane in pentane (340). The dichlorodiphenylsilane can be prepared by direct synthesis (331), by the high-temperature condensation of chlorobenzene and dichlorophenylsilane (410), or from tetrachlorosilane and phenylmagnesium bromide (455).

An alternate, though less successful, route is the preparation of dichlorodivinylsilane, followed by treatment with phenylmagnesium bromide. Difficulties arise in the first step, since yields by direct synthesis (330,334) or vinyl Grignard and tetrachloroethane (456) are poor. There is little improvement when dichlorodivinylsilane is prepared by a double dehydrohalogenation of dichlorobis(chloroethyl)silane (330,334), and it involves two extra steps.

C. Properties

1. PHYSICAL

Monomers 36 to 41 are colorless liquids. Their physical properties are listed in Table 10.

2. PHYSIOLOGICAL

In general, liquid vinylsilanes are irritating to the skin, eyes, and respiratory tract (457). The individual monomers vary a great deal, however, in their toxicities. Chlorosilanes, 36 and 37, are extremely corrosive, since they hydrolyze readily to give hydrogen chloride (342,458). Thus, they are capable of causing severe burns on contact or inhalation. Hydrolysis of 39 yields acetic acid, which, although less toxic than hydrogen chloride, can still cause burns and irritation. Alkoxysilanes, 38 and 40, are not corrosive but are irritating to the eyes and upper respiratory tract and have a drying effect on the skin (458). The single oral dose LD_{50} in rats has been found to be 22.5 g per kg for 38 and 1.28 g per kg for 36 (238).

3. ANALYTICAL METHODS AND SPECTRA

Chemically, bromine addition can be used to determine the vinyl groups in vinylsilanes. With 38 this is a rapid quantitative reaction, but with 36 it is slow and incomplete. Good results can be achieved if 36 is first treated with methanol and then titrated with a standard methanolic solution of bromine and sodium bromide (431).

Mixtures of 36 and 37 have been separated by gas-liquid partition chromatography (467). The best results were obtained using kieselguhr (Svetlofiltr)–nitrobenzene (10:3) at 25°.

Russian workers have studied the Raman and infrared spectra of vinylsilanes in great detail (468–478). From this work they have determined the electronegativities of moieties attached to silicon (469) and of a number of silyl groups (470). The intensity of absorption of the vinyl group at 1,600 cm^{-1} for compounds of the type, $Cl_n(CH_3)_{3-n}SiCH{=}CH_2$, was found to increase strongly as n increased (472). Quantitative infrared methods for determining vinyl groups in silanes have been worked out that are faster and more accurate than chemical procedures (473,474,477).

Mixer and Bailey at Linde (463) found typical vinyl absorption bands in the infrared at 963, 1,010, 1,277, 1,408, 1,603, and 3,053 cm^{-1} for **38** that were completely missing for its homopolymer. The infrared spectrum of **41** at 3,500 to 680 cm^{-1} has been published (479).

Ultraviolet spectra at 1,700 to 2,300 Å are reported for a group of vinylsilanes (480). A bathochromic shift is induced by the silyl moieties.

Electron diffraction studies have been used to determine the geometrical parameters of gaseous **36** (481). The Si—Cl bond was found to be 0.04 Å longer than in $HSiCl_3$, $SiCl_4$, or CH_3SiCl_3.

A number of groups have investigated the NMR spectra of vinylsilanes (478,482–485). Chemical shifts and coupling constants have been determined for a variety of these compounds, and, for the most part, the results are consistent with the idea of a variable $d\pi$-$p\pi$ resonance effect between silicon and the vinyl group—Eq. (41):

$$\overset{\diagdown}{\underset{\diagup}{-}}SiCH{=}CH_2 \longleftrightarrow \overset{\diagdown}{\underset{\diagup}{-}}\overset{\ominus}{Si}{=}CH{-}\overset{\oplus}{C}H_2 \qquad (41)$$

Studies of acid-base strengths (486), halide additions (487,488), dipole measurements (489), and infrared spectra (71,475,476,490) also support this theory.

D. Storage and Handling

Monomers **36** to **40** are readily hydrolyzed and so should be stored in closed containers to exclude moisture (458). Under anhydrous conditions, mild steel is satisfactory for tanks, pipes, and valves in contact with **36** and **37**. Wet chlorosilanes, however, require materials that are passive to hydrogen chloride, for example, glass or chlorinated polyethers (491). All these monomers are best stored in cool, well-ventilated areas away from open flames (342). This is especially true of **38**, which yields volatile, flammable ethyl alcohol on hydrolysis or heating, and **41**, which shows a tendency to polymerize when heated (340). There is no need for inhibitors with **36** to **40** because of their reluctance to undergo polymerization.

Although the toxicity of these monomers is not fully known, it is

	Compound			
	36		37	
		Reference		Reference
Boiling point, °C (mm)	−10.7(10)	457,459	38(100)	342
	1.3(20)	459	91(742)	419,442
	14.5(40)	459	92.5(743.2)	444
	18(50)	458	88–92(760)	419,442
	22.9(60)	459	90–4(760)	462
	34.2(100)	457,459	92(760)	342
	35(100)	342	92–3(760)	330,334
	51.3(200)	459	92.4(760)	421
	70.5(400)	457,459	92.5–3(760)	411
	92(742)	419	92.9(760)	457
	90.5(752)	460	93(760)	435,436,45
	92.5–4.0(752)	399		
	88–91(760)	397		
	89–91(760)	421		
	90(760)	336		
	90–2(760)	401		
	90–3(760)	434		
	90.5(760)	342		
	90.6(760)	457,458,459, 461a		
	91–2(760)	436,439		
	91.5–2.5(760)	334		
	92(760)	330,423		
	92–3(760)	438		
	92–5(760)	437		
Melting point, °C	−94.6	464		
	−95	457		
Refractive index, n_D (°C)	1.4349(20)	460	1.4270(20)	419,442
	1.4350(20)	419	1.4295(20)	411
	1.436(20)	461a	1.4270(25)	342
	1.432(25)	457		
	1.4320(25)	458		
	1.4330(25)	342		
Density, d_4 (°C)	1.2650(20)	399,419	1.0868(20)	419,442
	1.2710(20)	401	1.0870(20)	442
	1.2717(20)	439	1.080(25)	342
	1.28(20)	461a	1.081(25)	330,334
	1.26(25)	458	1.085(25)	457
	1.260(25)	457		
	1.270(25)	342		
	1.260(27)	330,334,423		
Flash point, °C	16.4	342	29.5	457
	19	458		
	21.1(coc)	457		
Ignition temp, °C (ASTM)	263	465		
Coefficient of expansion per °C	0.0016	457	0.0014	457
Viscosity, cSt at 25°C	0.50	457		
Specific heat, 30–60°C, cal/g/°C	0.20	457		
Heat of vaporization, Btu/lb	88	457		
Thermal conductivity (avg $T = 32°C$), cal/(sec)(cm²)(°C/cm)	3.1×10^{-4}	466		
Solubilities:[a]				
Acetone	SR	342,458	SR	342
	R	457	R	457
Benzene	S	342,457,458	S	342,457
Ethyl ether	S	342,457,458	S	342,457
Carbon tetrachloride	S	457,458	S	457
Ethyl acetate	S	458		
Hexane	S	458		
Heptane	S	342	S	342
Methanol	R	342,458	R	342
Water	R	342,457,458	R	342,457

[a] S = soluble, I = insoluble, R = reacts, SR = soluble and reacts.

	Compound							
	38		**39**		**40**		**41**	
	Reference		Reference		Reference		Reference	
.4(10)	457,459	115(10)	342	136.2(5.5)	454	130–1(0.05)	340	
.6(20)	459	126–8(25)	453	144.5–5.5(11)	354			
.5–3(20)	449	164(100)	342	153(50)	458			
(30)	393,463	223(760)	342	285(760)	457,458			
.2(40)	459							
(50)	458							
.4(60)	459							
.9(100)	457,459							
.6(200)	459							
.6(400)	457,459							
–8(760)	336							
–9(760)	397							
.5(760)	457,459							
–1(760)	461b							
.398(20)	461b	1.4200(25)	342	1.4268(25)	458	1.5350(25)	340	
.3960(25)	449			1.4270(25)	354			
.3966(25)	393,463			1.427(25)	457			
.397(25)	458			1.4271(25)	454			
.90(20)	461b	1.167(25)	342	1.0336(25)	454	1.0092(25)	340	
.897(25)	458			1.04(25)	354,457,			
.9027(25)	449				458			
.9036	393,463							
	458	4.4	342	66	458			
.4	457							
	465	>400	465					
.0011	457			0.0011	457			
.70	457			1–2	457			
.25	457							
	457							
.7 × 10^{-4}	466							
S	457,458	S	342	S	457,458			
S	457,458	S	342	S	457,458			
S	457,458	S	342	S	457,458			
S	457,458			S	457,458			
S	458			S	458			
S	458			S	458			
		S	342					
SR	458	R	342	SR	458			
R	457	R	342	R	457			
SR	458			SR	458			

recommended that they be handled in well-ventilated areas and that inhalation be avoided (461). Goggles or face shields, rubber gloves, and protective clothing should be worn, especially when handling chlorosilanes (457,458).

E. Chemical Reactions

The reactions of monomers **36** to **40** may be divided into two categories: those which involve the hydrolyzable substituents (silicon-functional) and those which occur because of the vinyl group (organo-functional). Obviously, **41** can only engage in the latter.

1. SILICON-FUNCTIONAL REACTIONS

a. Hydrolysis

Compounds **36** to **40** are easily hydrolyzed, especially in the presence of strong mineral acids (342,451,457,458). Under similar conditions, their order of vulnerability is chloro > acetoxy > ethoxy > 2-methoxyethoxy (458,492). Silanols are the initial products, but these are unstable and condense to polysiloxanes—Eqs. (42) and (43) (342,431,458,492):

$$CH_2{=}CHSiY_3 + 3H_2O \longrightarrow CH_2{=}CHSi(OH)_3 + 3HY \qquad (42)$$
$$(Y = Cl, OAc, OC_2H_5, or OCH_2CH_2OCH_3)$$

$$CH_2{=}CHSi(OH)_3 \longrightarrow CH_2{=}CHSiO^{3/2} + \tfrac{3}{2}H_2O \qquad (43)$$

Neutral conditions and low temperatures enhance the stability of the silanols, and, with care, silanediols may be isolated—Eqs. (44) and (45) (453,493,494):

$$CH_2{=}CH(CH_3)Si(OAc)_2 \xrightarrow[\text{sat. aq. NaCl}]{\sim 25^\circ} (CH_2{=}CH(CH_3)(OH)Si)_2O \qquad (44)$$

$$CH_2{=}CH(CH_3)Si(OC_2H_5)_2 \xrightarrow{C_2H_5OH{-}H_2O} CH_2{=}CH(CH_3)Si(OH)_2 \qquad (45)$$

This tendency toward siloxane formation, however, plays an important role in the usefulness of these monomers.

b. Alcoholysis

It has already been observed (see Secs. III-B-3 and 5) that **36** reacts readily with alcohols to give the corresponding alkoxy compounds. This is true of chlorosilanes in general and can be extended to include phenols as well—Eq. (46) (454):

$$36 + 3C_6H_5OH \xrightarrow[\text{8 hr}]{\text{reflux}} 62\% \ CH_2{=}CHSi(OC_6H_5)_3 + 3HCl \qquad (46)$$

Usually an organic base, such as pyridine, is added to remove the hydrogen chloride that is evolved. Excellent yields of phenoxysilanes have been obtained by preparing first a cyclic or polysilazane and treating this with a phenol—Eq. (47) (495):

$$2CH_2{=}CH(CH_3)SiCl_2 + 2NH_3 \xrightarrow[\text{cooling}]{\text{benzene}}$$
$$\underset{37}{}$$

$$CH_2{=}CH(CH_3)Si{-}NH{-}Si(CH{=}CH_2)(CH_3)NH$$

$$\Big\downarrow 4C_6H_5OH$$

$$2CH_2{=}CH(CH_3)Si(OC_6H_5)_2 \longleftarrow$$

$$(47)$$

The ability of alkoxyvinylsilanes to undergo ester interchange with high-boiling alcohols was illustrated in the synthesis of **40** from **38** (see Sec. III-B-5). Reactivity decreases with increasing size of the alkoxy group (342,354). Similar exchanges occur with phenols (458).

c. Grignard Coupling Reactions

The usefulness of this reaction in synthesizing **36, 37,** and **41** has already been demonstrated (see Secs. III-B-1-d, III-B-2-a, and III-B-6). Under carefully controlled conditions, one or all of the chlorines attached to silicon may be replaced by alkyl (330,334,487,492,496,497), aryl (498–500), or aralkyl (498,501) groups. The ease of replacement tends to decrease with an increase in the number of organic substituents (502a). Improved yields have been claimed (503) when diethyl ether was replaced by dibutyl ether as solvent.

Alkoxyvinylsilanes also react with Grignard reagents but somewhat less readily than the corresponding chloro compounds (502b). Thus, reactions of **38** are more easily controlled to give the desired degree of substitution than those of **36** (494,504–506). Stepwise use of Grignards with **38** makes possible the preparation of optically active vinylsilanes—Eq. (48) (507):

$$38 + C_6H_5MgBr \xrightarrow[\text{Et}_2\text{O–xylene}]{4.5 \text{ hr}} 35\% \ CH_2{=}CH(C_6H_5)Si(OC_2H_5)_2$$

$$\Big\downarrow \overset{\text{1-C}_{10}\text{H}_7\text{MgBr}}{\underset{\text{heat}}{\text{THF}}} \quad (48)$$

$$73\% \ CH_2{=}CH(C_6H_5)Si(1{-}C_{10}H_7)OC_2H_5 \longleftarrow$$
$$\underset{44}{}$$

When **44** is treated with *l*-menthol and sodium methoxide in toluene, two diastereoisomers are obtained. One melts at 89 to 95°, $[\alpha]_D^{20} -43.5°$, and the other is an oil, $[\alpha]_D^{20} -34°$.

d. Reduction

Chlorovinylsilanes are readily reduced to the corresponding vinylsilanes with lithium aluminum hydride (508,509), lithium hydride (460), or sodium hydride (510)—Eq. (49):

$$36 + LiAlH_4 \xrightarrow[N_2]{dioxane} CH_2{=}CHSiH_3 \tag{49}$$

Alkoxyvinylsilanes are reduced in a similar manner (458).

e. Miscellaneous

The hydrolyzable substituents of vinylsilanes can be replaced by a variety of other groups. Chloro- and alkoxyvinylsilanes are converted in good yield to the corresponding fluorovinyl compounds by treatment with hydrogen fluoride (458,511,512), zinc fluoride (513), or antimony trifluoride (456):

$$37 + SbF_3 \xrightarrow{vac.} 86\% \ CH_2{=}CH(CH_3)SiF_2 \tag{50}$$

The reaction of **37** with ammonia to form a vinylsilazane was illustrated in Eq. (47). Chlorosilanes also react readily with primary and secondary amines—Eqs. (51) and (52) (514):

$$37 + (C_2H_5)_2NH \xrightarrow[15-20°]{benzene} 51.6\% \ CH_2{=}CH(CH_3)Si(N(C_2H_5)_2)_2 \tag{51}$$

$$37 + CH_3NH_2 \longrightarrow$$
$$CH_2{=}CH(CH_3)Si(NHCH_3)_2 + (CH_2{=}CH(CH_3)SiNHCH_3)_2NCH_3 \tag{52}$$
Main product   Small amount

Alkoxy- and acyloxyvinylsilanes, that is, **38** and **39**, give polymeric products when treated with phosphoric acid (or its anhydride) or boric acid (515,516):

$$38 + H_3BO_3 \xrightarrow{heat} 95\% \ (CH_2{=}CHSiO_3B)_n \tag{53}$$

$$38 + P_2O_5 \longrightarrow 96\% \ [(CH_2{=}CHSi(OC_2H_5)_2PO_{2.5})O]_n \tag{54}$$

The reaction of **36** with sodium isothiocyanate yields vinylsilyl triisothiocyanate (517):

$$36 + 3NaCNS \xrightarrow[1.5\,hr]{reflux} 82\% \ CH_2{=}CHSi(CNS)_3 \tag{55}$$

When dimethylcyclosiloxane is heated with **37**, telomers of the type $CH_2{=}CHSi(CH_3)Cl(OSi(CH_3)_2)_nCl$ are formed (518).

2. ORGANO-FUNCTIONAL REACTIONS

The vinylsilanes undergo many reactions characteristic of olefins. However, the vulnerability of the other groups on silicon to a variety of reagents and heat must be kept in mind, since this can cause complications in some instances. Free-radical reactions generally proceed smoothly.

Irregular results in ionic reactions suggest that $d\pi$-$p\pi$ bonding exists between silicon and the vinyl group—see Sec. III-C-3 and Eq. (41). This would markedly lower the basicity of the double bond and reverse the polarity of the vinyl carbon atoms.

a. Hydrogenation

The double bond in vinylsilanes can be hydrogenated using Raney nickel (519) or PtO_2 (488) as catalyst. Conditions must be chosen carefully so as not to affect hydrolyzable substituents. In practice, vinylsilanes are rarely hydrogenated, since the resulting ethylsilanes are more readily prepared by other means (520).

b. Addition of Halides

At 25°, bromine adds to **38** more readily than to **36** (see Sec. III-C-3). In the addition to **36**, the yield of 1,2-dibromoethyltrichlorosilane can be improved by raising the reaction temperature to 50° and irradiating with a 150-watt lamp (521). **37** has been brominated with cooling and illumination (522).

Chlorine adds less readily to **36** than does bromine. This has been demonstrated by comparing yields under similar conditions (413,521) and by competitive reactions where chlorine and bromine were introduced simultaneously (523). Chlorination of **36** has been studied in both the liquid and vapor phase at a variety of temperatures (431). In the liquid phase at 25 to 50°, a 4:1 ratio of trichloro(1,2-dichloroethyl)silane and 1,4-dichloro-1,3-trichlorosilylbutane is obtained. The use of $SiCl_4$ as diluent eliminates the latter product. The course of vapor-phase chlorination of **36** depends on the temperature—Eq. (56):

$$
\begin{array}{ccc}
& \xrightarrow{100\text{--}250°} & ClCH_2CHClSiCl_3 \\
& & \mathbf{45} \\
\mathbf{36} + Cl_2 & \xleftarrow{200\text{--}350°} & CH_2{=}CClSiCl_3 \qquad\qquad (56) \\
& & \mathbf{46} \\
& \xrightarrow{300\text{--}400°} & SiCl_4 + \text{decomposition}
\end{array}
$$

Since **45** is thermally stable up to 400°, **46** must result from direct substitution of Cl for H rather than by addition followed by elimination of hydrogen chloride.

c. Addition of Hydrogen Halides

Hydrogen chloride adds to **36** in the presence of aluminum chloride to give the 2-chloro isomer **42**. This is a reversible reaction—Eq. (34)—for which higher temperatures and excess aluminum chloride favor the formation of **42**. The anti-Markownikoff addition in this case could be attributed to the inductive effect of the strongly electronegative chlorines attached to silicon. If these are replaced by electron-releasing methyl groups, however, addition of hydrogen halides (HBr and HI) occurs in the same way (487,488). This is convincing evidence for $d\pi$-$p\pi$ bonding in vinylsilanes and indicates that it is the dominant influence in ionic additions to the double bond.

d. Friedel-Crafts Reactions

Benzene is readily alkylated with **36** in the presence of aluminum chloride—Eq. (57) (431,524):

$$36 + C_6H_6 \xrightarrow[\substack{75-80° \\ 4.5\ hr}]{AlCl_3} C_6H_5CH_2CH_2SiCl_3 \qquad (57)$$
$$(66.8\%)$$

Polyalkylation is suppressed by using an excess of benzene. Reactivity decreases as the chlorines on silicon are replaced by methyl or alkoxy groups. Thus, **37** gives only a 59 percent yield of dichloromethyl(2-phenylethyl)silane, and trimethylvinylsilane and **38** do not react at all. If toluene or chlorobenzene replaces benzene in the reaction with **36** or **37**, para-substituted derivatives are formed (525).

e. Addition of Alkyl and Alkenyl Halides

Certain halo compounds add to vinylsilanes in the presence of peroxides (500,526) or ultraviolet light (527)—Eqs. (58) and (59):

$$36 + Cl_3CBr \xrightarrow[137°,\,20\,hr]{Bz_2O_2} Cl_3CCH_2CHBrSiCl_3 \qquad (58)$$

$$36 + F_2C{=}CFI \xrightarrow[12\,days]{UV} F_2C{=}CFCH_2CHISiCl_3 \qquad (59)$$

At higher temperatures, fluoroalkenes codimerize with **36**, **37**, or **38** to give silyl-substituted cyclobutanes—Eq. (60) (527,528):

$$F_2C{=}CX_2 + CH_2{=}CHSiR_3 \xrightarrow[autoclave]{200-220°} \overline{X_2C{-}CF_2CH_2CHSiR_3} \qquad (60)$$
$$(X = F,\ Cl,\ or\ I;\ R = C_2H_5O,\ Cl,\ or\ CH_3)$$

Addition of a small amount of hydroquinone inhibits the dimerization of the fluoroalkene that would otherwise occur.

f. Reactions with Dienes

Diels-Alder reactions occur in varying degrees, depending on the nature of the vinylsilane and the diene. Thus, **36** reacts more readily with cyclopentadiene than with butadiene, and **38** reacts less readily with cyclo-pentadiene than does **36** (431):

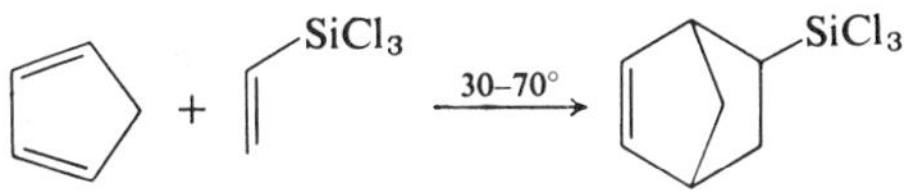

(61)

36

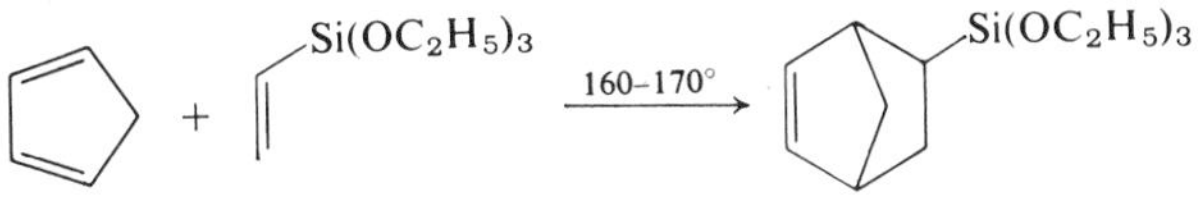

(62)

(Dimer)   **38**

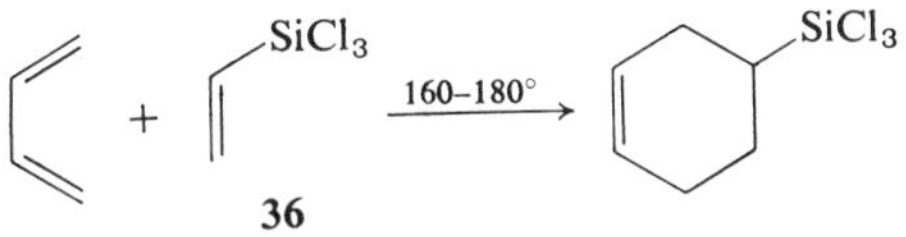

(63)

36

With hexachlorocyclopentadiene at 190° the corresponding hexachloro-bicycloheptenylsilanes are obtained (529). Furan does not form adducts with **36** or **38** even when heated under pressure to 200° (431).

g. Addition of Hydrides

The tendency of silicon hydrides to add to vinylsilanes has already been mentioned—see Sec. III-B-1-c and Eq. (33). Using chloroplatinic acid as catalyst, it has been demonstrated that an accumulation of chlorine atoms on the silicon of the vinylsilane hinders this addition—Eqs. (64) and (65) (530,531):

$$2 \ \mathbf{36} + H_2SiCl_2 \ \xrightarrow{H_2PtCl_6} \ 27\% \ Cl_2Si(CH_2CH_2SiCl_3)_2 \qquad (64)$$

$$2 \ \mathbf{37} + H_2SiCl_2 \ \xrightarrow{H_2PtCl_6} \ 52\% \ Cl_2Si(CH_2CH_2Si(CH_3)Cl_2)_2 \qquad (65)$$

Phenyltin hydrides add readily to vinylsilanes without a catalyst (532):

$$\mathbf{41} + 2(C_6H_5)_3SnH \ \xrightarrow[4.5 \ hr]{80°, \ vac.} \ 86\% \ [(C_6H_5)_3SnCH_2CH_2]_2Si(C_6H_5)_2 \qquad (66)$$

h. Miscellaneous Additions

Organic compounds containing active hydrogen atoms add to vinyl-silanes in the presence of a base—Eq. (67) (533):

$$\mathbf{38} + YH \ \xrightarrow{NaOC_2H_5} \ YCH_2CH_2Si(OC_2H_5)_3 \qquad (67)$$
$$(Y = C_2H_5O, \ (C_3H_7)_2N, \ C_4H_9S)$$

Diethylphosphine and **37** yield dichloro(2-diethylphosphinoethyl)vinyl-silane when irradiated with ultraviolet light (534). 4-Alkenylsilanes can be prepared by heating vinylsilanes with certain olefins under pressure in the presence of a polymerization inhibitor—Eq. (68) (535):

$$38 + (CH_3)_2C{=}CH_2 \xrightarrow[240°,\,2\,hr,\,100\,atm]{p\text{-}C_6H_4(OH)_2} CH_2{=}C(CH_3)CH_2CH_2CH_2Si(OC_2H_5)_3 \quad (68)$$

Pyrolysis of trichloro(trichlorosilyl)methane yields dichlorocarbene—Eq. (69)—which adds to vinylchlorosilanes—Eq. (70) (536):

$$Cl_3CSiCl_3 \xrightarrow{\text{heat}} Cl_2C{:} + SiCl_4 \quad (69)$$

$$Cl_2C{:} + 36 \xrightarrow[250°]{2\ hr} 50\% \quad \text{(structure)} \quad (70)$$

Hydrolysis of the ozonide from **36** with dilute hydrogen peroxide cleaves the C—Si bond, implying that carbonyl groups attached to silicon are unstable (431).

F. Polymerization

1. HOMOPOLYMERIZATION

In general, vinylsilanes substituted as in **47** are reluctant to undergo homopolymerization and form, at best, polymers of relatively low molecular weight (431,463):

$$CH_2{=}CHSiR_nR'_{3-n} \quad (R = CH_3, C_2H_5, \text{ or } C_6H_5; R' = OC_2H_5 \text{ or } Cl; n = 0{-}3)$$
47

The case of **47** where R is ethyl and $n = 3$ has already been discussed (see Sec. III-A). Equally disappointing results are obtained when R = methyl and $n = 3$ (488,537).

At 290° under moderate pressures, **36** gives mainly tetramer (431). The reaction time and temperature can be decreased by raising the pressure to 50,000 psi, but there is little effect on the molecular weight of the viscous liquid product. At 95,000 psi, solid thermoplastic polymers of 800 to 1,000 molecular weight are formed. The use of peroxide or acid catalysts does not increase the molecular weight of the products. It has been observed, however, that aluminum, ferric, and zinc chlorides do increase the rate of reaction.

The monomer giving the greatest degree of polymerization is **38** (431, 463). In contrast to **36** homopolymerization of **38** is enhanced by per-oxides. A study of 13 catalysts (393,463) showed di-*tert*-butyl peroxide to be superior. Using the optimum amount, 0.1 wt percent, the best yields are obtained at 145° and the highest molecular weight at 135°. The clear, colorless viscous polymer produced at 130° is soluble in most organic solvents and has a molecular weight of 5,000 to 8,000 (431). It has been claimed that 0.0001 percent of an alkylchlorosilane has a synergistic effect on the polymerization of **38** with di-*tert*-butyl peroxide (538). Thus, amyltrichlorosilane added in this amount raises the viscosity of the product from 19,000 to 47,000 centistokes. Air, oxygen, and water retard the polymerization (393,463).

The use of γ rays at dose rates of 1.8 × 10^4 r per hr and temperatures of −78° to 50° to induce polymerization of **38** has also been described (539). The resulting polymers are soft, rubbery gums or powders, depending on the radiation dose. These products swell in solvents but are not soluble. Unlike the polymers formed with peroxides (463), their infrared spectra show that a considerable number of vinyl groups are retained.

Mechanisms for the free-radical polymerization of **38** have been proposed (463,539). The low molecular weight of the polymers is attributed to chain transfer involving abstraction of an α hydrogen (or a methyl group) from an ethoxysilyl moiety of either the polymer or monomer—Eqs. (71) and (72):

$$\sim\!\!CH_2\overset{\bullet}{C}HSi(OC_2H_5)_3 + CH_3CH_2OSi(OC_2H_5)_2R \longrightarrow$$
$$\sim\!\!CH_2CH_2Si(OC_2H_5)_3 + CH_3\overset{\bullet}{C}HOSi(OC_2H_5)_2R \quad (71)$$
$$\mathbf{48}$$

$$\mathbf{48 + 38} \longrightarrow R(C_2H_5O)_2SiOCH(CH_3)CH_2\overset{\bullet}{C}HSi(OC_2H_5)_3 \quad (72)$$

The ratio of chain transfer with polymer to that with monomer for high conversion is 4.5.

40 is less readily polymerized than **38**. Using di-*tert*-butyl peroxide, only a very small amount of polymer is formed (393). Little information is available on the polymerization of **39** or **41**, but it has been stated that the latter must be distilled at extremely low pressures in order to avoid heat-induced polymerization (340). Monomers **36** to **40** are converted by hydrolysis to polyvinylsiloxanes, which can be cross-linked by peroxides to brittle materials (431,540,541).

2. COPOLYMERIZATION

Sluggish copolymerization of **36** with other vinyl monomers has been observed. Using BF$_3$ as catalyst (431), copolymers with styrene contained only 2 percent of **36**, and with aliphatic olefins no more than 10 percent. Ziegler catalysts have also been used (387) in the copolymerization of chlorovinylsilanes with olefins. Copolymers of **36** are formed with vinyl acetate, methyl methacrylate, and acrylonitrile in the presence of peroxide catalysts (431), but, again, the percentage of silane in these products is very small. No reaction was observed between **36** and maleic anhydride or ethyl maleate.

In contrast, **38** reacts extensively (1:1) with maleic anhydride or ethyl maleate when peroxide catalysts are present (431). The resulting copolymers are hard, brittle resins that are soluble in a variety of organic solvents and have molecular weights up to 3,000. Copolymerization of **38** and ethylene with benzoyl peroxide has been reported (542).

Thompson (543) prepared copolymers of **38** with vinyl chloride and acrylonitrile using diacetyl peroxide as initiator at 50°. He noted that the polymerization rate decreases rapidly as the percentage of **38** is increased. There is no homopolymerization of **38** nor copolymerization with chlorostyrene under these conditions. Assuming $r_1 = 0$ ($M_1 = 38$), Thompson calculated the relative reactivity ratio r_2 to be 0.9 for vinyl chloride and 4.5 for acrylonitrile. Pike and Bailey (139) noted that the reactivity ratios that they determined for the copolymerizations of **49**, **50**, and **51** with a variety of vinyl monomers were not affected by the structure of the vinylsiloxane:

$$[(CH_3)_3SiO]_3SiCH=CH_2 \qquad [(CH_3)_3SiO]_2Si(CH_3)CH=CH_2$$
$$\textbf{49} \qquad\qquad\qquad\qquad \textbf{50}$$
$$(CH_3)_3SiOSi(CH_3)_2CH=CH_2$$
$$\textbf{51}$$

Furthermore, their r_1 and r_2 values for copolymerizations with vinyl chloride and acrylonitrile were similar to those reported by Thompson, and the reactivity of triphenylvinylsilane toward acrylonitrile was about the same as that shown by **49**, **50**, and **51**. Therefore, they postulated that their relative reactivity ratios are valid for any vinylsilicon compound under similar conditions. Later work by Scott and Price (450) in copolymerization studies of **38** and trimethylvinylsilane with styrene and acrylonitrile, using 2,2′-bisazoisobutyronitrile as catalyst, supports this theory. They found r_1 to be 0 ($M_1 = 38$), $r_2 = 22 \pm 5$ ($M_2 =$ styrene), and $r_2 = 5 \pm 1$ ($M_2 =$ acrylonitrile). With trimethylvinylsilane as M_1, r_1 is 0 with styrene and 0.07 ± 0.03 with acrylonitrile ($r_2 = 26 \pm 8$ and 3.9 ± 3, respectively). Their Q values for both monomers (0.03 to 0.035) are in good agreement with the Q values calculated from Thompson's data (0.02). The more positive e of **38** (0.10) compared with -0.14 to -0.10 for trimethylvinylsilane is ascribed to the greater electron-withdrawing effect of the ethoxy versus the methyl groups.

Hurd and Roedel (390) studied the copolymerization of polymethylvinylsiloxane with diethyl maleate, diallyl phthalate, butyl methacrylate, styrene, dichlorostyrene, and methyl methacrylate using peroxide catalysts. Properties of the resulting products were dependent on the temperature, catalyst, and ratio of components. With methyl methacrylate, copolymers were obtained that showed good crazing resistance and electrical properties.

The easy copolymerization of polyvinylsiloxanes with many vinyl monomers plays an important role in the applications of vinylsilanes (see Sec. III-A). Graft copolymers of **38** and polyolefins (544) and **37** with polyacrylonitrile (545) have been reported.

A number of excellent books and reviews have been published that include sections on vinylsilanes (447,451,502,520,546–548). None of these is restricted to commercially available monomers.

IV. ACROLEIN, METHACROLEIN, AND VINYL KETONES

A. Introduction

Acrolein (acrylaldehyde, acraldehyde, acrylic aldehyde, 2-propenal) (**52**) has yet to become a commercially high-volume monomer. Thus, although the literature concerning this compound is staggering, coverage here will be held to a minimum. There are a number of excellent books and review articles available (549–555). In addition to **52**, methacrylaldehyde (methacrolein) (**53**), 3-buten-2-one (methyl vinyl ketone) (**54**), acrylophenone (phenyl vinyl ketone) (**55**), and 1,4-pentadien-3-one (divinyl ketone) (**56**) are discussed in this chapter:

$$CH_2{=}CHCHO \qquad CH_2{=}C(CH_3)CHO \qquad CH_3COCH{=}CH_2$$
$$\textbf{52} \qquad\qquad \textbf{53} \qquad\qquad\qquad \textbf{54}$$

$$C_6H_5COCH{=}CH_2 \qquad (CH_2{=}CH)_2CO$$
$$\textbf{55} \qquad\qquad\quad \textbf{56}$$

Acrolein was first reported by Redtenbacher (556) in 1843. He noted its formation in the dry distillation of fats and described its synthesis by dehydration of glycerol. The product he obtained polymerized spontaneously to a white, insoluble, infusible solid he called *disacryl*. Before the turn of the century, **52** had also been prepared from dihaloacetone, ethylene, and carbon monoxide, and by the decomposition of a propylene–mercuric sulfate complex (557,558). The dehydration of glycerol, however, remained the principle source of **52** for almost 100 years.

Extremely important in acrolein history was the discovery of stabilizers for this most reactive compound by Moureu and coworkers (559). Their studies revealed a variety of effective substances of which hydroquinone is the most extensively used. This made possible the investigation of a wide variety of reactions of **52** as well as prolonged storage. Consequently, a search began for new and better syntheses with commercial potential.

In the 1930s, Shell Development Co. obtained patents for the preparation of **52** by vapor-phase oxidation of allyl alcohol (560) and from epichlorohydrin (561). The former was carried out in the presence of activated copper catalyst, and the latter by heating with water and an inorganic acid.

The first commercial process for the manufacture of acrolein was developed by Degussa, a German company, around 1940 (550,562–564). They obtained yields of over 80 percent by condensing formaldehyde and acetaldehyde in the vapor phase in the presence of a suitable catalyst. Concurrently in the United States, Eastman Kodak Co. (565) pursued this same approach. Du Pont (566) and the Acrolein Corporation (567) explored the commercial feasibility of oxidizing propylene with the aid of mercuric sulfate in acid solution. This was an extension of Deniges' synthesis of **52** in 1898 (558). Imperial Chemicals Ltd., in England, obtained **52** (85 percent) and ethylene by thermal fission of dihydropyran (568), and Shell Chemical Co. (569) produced sizable quantities of acrolein by noncatalytic pyrolysis of diallyl ether, a by-product in the manufacture of allyl alcohol.

Battelle Memorial Institute (570) obtained a patent in 1945 for the catalytic oxidation of unsaturated hydrocarbons with air using metal selenites or tellurites as catalysts at elevated temperatures (175 to 450°). The catalyst can be promoted by CuO or other metallic oxides.

In 1948 a patent was issued to Shell Development Co. for the vapor-phase oxidation of propylene to acrolein with air or oxygen in the presence of a cuprous oxide catalyst (571). Following this, a wide variety of other catalysts was developed, and this rapidly became the most widely used commercial synthesis. The vapor-phase cross-condensation of acetaldehyde with formaldehyde, however, is also well suited for industrial production of **52**. The choice of method is dictated by the availability of raw materials.

Acrolein was manufactured on a small commercial scale by Degussa during the Second World War. Later, Shell Chemical Corporation and Union Carbide offered it in limited quantities as an experimental compound. Large-scale production by Union Carbide began in 1952 and has expanded greatly since that time. The largest producer of acrolein is Shell. Their plant at Norco, Louisiana, which was installed in 1959, produced 1,100 to 1,300 tons per month in 1960 (572). By comparison, Union Carbide's output was 100 tons per month and Degussa's 50 tons per month (552). Other manufacturers of acrolein are Montecatini, in Italy (pilot-plant scale) (573), and Sumitomo Chemical Company, in Japan.

Although it was anticipated that the price of acrolein would decrease in time, it has remained remarkably stable at 31 cents per lb (tank-car lots) from 1961 to 1968 in the United States. The price in Europe was $0.84 to $1.20 per kg in 1963 (552).

The thermal dimer of acrolein, 3,4-dihydro-2H-pyran-2-carboxaldehyde (**57**), is also available from Shell Chemical Co.:

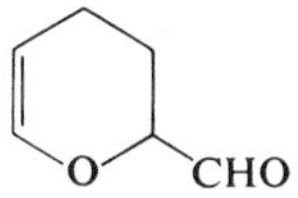

57

It readily dissociates to the monomer at approximately 400° and can thus serve as a convenient and controlled source of pure acrolein. It is not a strong lachrymator and is considered only slightly toxic. Therefore, it is less hazardous to handle than **52** (574).

The double bond in **57**, although not in conjugation with the carbonyl, is adjacent to an ether linkage. Therefore, it undergoes addition and hydrolytic reactions characteristic of vinyl ethers. These reactions have been noted for over 20 years, but only recently have detailed investigations appeared in the literature. Its potential as a chemical intermediate is shown by the number of patents issued in this area since 1950 (549,574). Of special interest is the hydrolysis of **57** to hydroxyadipaldehyde, which is easily reduced to 1,2,6-hexanetriol. These materials prepared from acrolein dimer, along with the dimer alcohol and the sodium salt of the dimer acid, have been available from Shell in research and development quantities since 1963 (574).

The largest use of acrolein at present is as an intermediate in organic syntheses. Most of the material manufactured by Shell is used in the production of glycerol (552). Relatively large amounts are also used to make 1,2,6-hexanetriol, hydroxyadipaldehyde, and glutaraldehyde (551). Another large use is in the manufacture of methionine (575). This amino acid is added to chicken feeds to improve nutritional efficiency and accelerate chick growth. It is also used to fortify dog food and is being tested in swine and ruminant feeds. Synthetic racemic (DL) methionine ($1.43 per lb) is as effective as the natural (L) form (551).

Acrolein as a sizing in cellulose fabrics (576,577) is claimed to improve crease resistance and crease recovery. Studies of the mode of interaction of **52** with cellulose (578,579) indicate that it is mainly the carbonyl, rather than the C=C double bond, of acrolein that is involved. Treatment with **52** is also said to improve the properties of polyvinyl alcohol fibers (580) and to decrease the alkali solubility of wool (581). Acrolein condensates with formaldehyde and melamine (582,583) or modified ureas (584–586) are also used to treat fabrics, making them wrinkle- and shrink-proof.

Because of its potent and aggravating odor, acrolein has been used as a warning agent for detecting leakage of refrigerants (587,588). Approximately 1 percent is said to be adequate.

Acrolein is a herbicide (589,590) and shows special promise in the control of aquatic weeds (591–595). It is a bactericide (596) and is used for this purpose in distillate fuels (597). It destroys soil fungi (598) and the eggs of the yellow-fever mosquito (599). Acrolein is also a molluscicide (600), and its piscicidal properties are useful in the control of goldfish (601). In the paper industry, it has proved to be an effective slimicide (602,603) that is safe to use in the manufacture of containers for food products. In contrast to this variety of "-cidal" properties, 52, in small doses, stimulates growth in animals (604). Addition of 50 mg per day to the feed or drinking water of lambs increased body-weight gains by 0.09 to 0.23 lb per day.

Acrolein is used to harden and stabilize image-receiving layers in photography (605–607), and in histology it is useful in fixing tissue stains (608–611). In vitro, 52 shows a cytoxic effect on leukemic cells (612).

Collagen (613,614) is readily cross-linked by acrolein, as are polyamides (615) and poly(oxymethylenes) (616). The products in the latter case are tough elastomers that can be fabricated by conventional methods.

Condensation products of acrolein and formaldehyde can be used as lacquers (617) or in treating fibrous materials (618,619). In combination with a polyalkeneamine, 52 is used to give bright nickel electroplating (620). Its condensation polymers with phenol are adhesives for glass and paper and display good electrical properties (621). The reaction of acrolein with unsaturated polyesters is claimed to improve their pot life (622).

Zeisel and Daniek (623) reported the preparation of methacrolein dimethyl acetal from α-bromoisobutyraldehyde dimethyl acetal and hot KOH solution in 1909. 53 itself, however, is not mentioned in the literature until 1935. At that time, a flurry of patents appeared for its synthesis by dehydrogenation or oxidation of methallyl alcohol using a variety of catalysts (560,624–627). Patents were also obtained for its preparation by dehydration of β-methylglycerol (628) and from β-methylepichlorohydrin (561,629). Further development of industrial production of 53 by cross-condensation of formaldehyde and propionaldehyde (565,566), thermal pyrolysis of bis(methallyl) ether (570), and the vapor-phase oxidation of isobutylene (571,572) has paralleled that of acrolein.

Today, methacrolein is produced commercially by the oxidation of butylene and by the condensation of formaldehyde with propionaldehyde (551). Recently, however, Russian chemists (630) have described a promising process for large-scale production of 53 by means of the Mannich reaction, and Eastman Kodak has developed a possible commercial synthesis from isobutyraldehyde.

Methacrolein was introduced on a limited commercial scale about the same time as acrolein. Both Shell Chem. Co. and Union Carbide have listed **53** among their experimental chemicals for some time. Large-scale quantities have been available from Union Carbide since 1952. In 1961, their price for methacrolein was $1 per lb (less than carload lots of drums f.o.b., Institute) (551).

53 is similar to **52** in properties and can be used for many of the same purposes. It is useful as an intermediate in a variety of organic syntheses, in the treatment of textiles and fibers (576,581), and in fixing tissue stains (610). Phenol-methacrolein resins are tanning agents (631), and meth-acrolein-formaldehyde resins display good resistance to heat, oxidation, and alkalies (632). In view of the higher cost of methacrolein, however, it is only logical that acrolein would be preferred unless **53** proved far superior. Except in syntheses where the α-methyl group is important, this is not the case.

Blaise and Maire (633) reported the preparation of methyl vinyl ketone (**54**) from 2-chloroethyl methyl ketone and boiling diethylaniline in 1908. They studied some of its reactions and noted its tendency to polymerize, especially in the presence of alkali. Shortly thereafter, Krapiwin (634) synthesized **54** from ethylene and acetyl chloride in the presence of aluminum chloride. Early patents by Bayer and Co. (635) described the preparation of **54** from 2-hydroxyethyl methyl ketone and $ZnCl_2$ and by the pyrolysis of 2-acetylacrylic acid. In 1924, Wohl and Prill (636) reported the synthesis of **54** using a modification of the Bayer method in which the aldol condensation product from acetone and formaldehyde (2-hydroxy-ethyl methyl ketone) was dehydrated with $ZnCl_2$ without isolation.

Du Pont patents in the early thirties (637–639) described the hydration of vinylacetylene in the presence of a transition metal salt and an acid to give **54**. This approach, which was also used by Russian chemists (640–642) and by I. G. Farben. (643,644), appears to be suitable for large-scale production of **54**. The hydration of vinylacetylene under certain conditions also serves as a source for 2-hydroxyethyl methyl ketone. Interest in the dehydration of this compound to **54** has continued (645).

In the forties, Shell (646) and Wingfoot Corp. (647) obtained patents for the manufacture of **54** by the pyrolysis of 1-acetoxyethyl methyl ketone. There was also interest in 2,3-butanediol, a common fermentation product of carbohydrates, as a starting material for the production of **54** (648), since dehydrogenation to acetoin and subsequent dehydration go smoothly. Catalytic oxidation of methyl vinyl carbinol was also reported to give good yields of **54** (649). Vapor-phase oxidation of 1-butene was developed at the same time as it was for propylene and isobutylene (571) and is adaptable to commercial production. The oxidation can also be effected in the liquid phase by way of a complex of the olefin with mercuric salts. Cross-condensation of acetone and formaldehyde in the vapor phase was also found to be feasible. In recent years, the possibility of large-scale production of **54** by means of the Mannich reaction has been suggested (630).

Although there has been great interest in **54**, no large-volume use has evolved and, thus, no large-scale production. Methyl vinyl ketone is available from Chas. Pfizer and Co. in 5- and 55-gal drums. Borden, Roberts Chemicals, and Polysciences, Inc., offer it in experimental quantities. As of 1967, the Monomer-Polymer Laboratories of Borden listed its price at $11 per 500 g and $55 per 3 kg (397).

Because of its reactive nature, **54** is extremely useful in organic syntheses. An important example is the commercial synthesis of vitamin A (650,651). It is also used in the preparation of perfumes (652).

Methyl vinyl ketone is a fungicide and as such is useful in soil fumigation (653,654). Like **52** and **53**, **54** is effective in textile sizing (655). Its higher cost, however, is a disadvantage.

I. G. Farben. has devoted a great deal of effort toward finding uses for methyl vinyl ketone, as well as to its synthesis and chemistry. They claim in a number of patents that condensation products of **54** with formaldehyde are excellent adhesives (656,657) and useful in coatings (658). Further reaction of these condensates with phenols, amines, or ureas and with polybasic acids or anhydrides yields resins that can be used as electrical insulation (657) and in lacquers (659), respectively.

In 1906, Schafer and Tollens (660) reported that they had obtained phenyl vinyl ketone (**55**) by steam distillation of tris(β-benzoylethyl)amine hydrochloride. Three years later, Kohler (661) synthesized **55** by heating α,β-dibromopropiophenone with alcoholic KI. Shortly thereafter, Beaufour (662) prepared it by heating 1,3-dimethoxy-1-phenylpropene with dilute H_2SO_4, and Straus and Berkow (663) by acid hydrolysis of **55** dimethyl acetal obtained by treating 1-chloro-3-methoxy-3-phenylpropene with methanol and sodium methoxide. Since then **55** has been prepared in a variety of ways, but synthesis by the Mannich reaction appears to be the most common (630,664).

The tendency of **55** to polymerize was reported by early workers, and they also noted its ability to undergo a wide variety of other chemical reactions. It is in the latter category that phenyl vinyl ketone finds its greatest use. It is, for example, an intermediate in the synthesis of chloramphenicol (665).

55 shows bacteriostatic activity toward both gram-positive and gram-negative organisms but a greater effect against the former (666,667). It has fungicidal properties (666) and inhibits cholinesterase in horse serum (668). Its carcinogenic action has been demonstrated in that subcutaneous injections in the rat and mouse produce local tumors (669). Dyes can be fixed on textile fibers by **55** produced *in situ* from sulfonium compounds of the type $(BzCH_2CH_2SR_2)^+A^-$ (670).

Divinyl ketone (**56**) is an extremely reactive substance. Thus, although derivatives of **56** have been known for many years, it was not until 1946 that its synthesis and isolation were reported by Nazarov and Torgov (671). They prepared it by a series of reactions starting with vinylacetylene. The immediate precursor, bis(2-methoxyethyl) ketone, was distilled under reduced pressure to yield **56**. On standing, the product polymerized to a yellow glass.

Other syntheses have since evolved, and it is generally agreed that the final step is best carried out in the presence of a stabilizer, such as hydroquinone (672,673). This monomer must be used promptly, and it is, in fact, usually advantageous to prepare **56** *in situ* (674,675), just prior to reaction.

Divinyl ketone is a useful intermediate in organic syntheses. It has also been used in the treatment of textiles (655).

B. Synthetic Methods

1. Acrolein (**52**)

a. Dehydration of Glycerol

52 is conveniently prepared in the laboratory, in 50 percent yield, by heating glycerol with a mixture of $KHSO_4$ and K_2SO_4 at 190° (559c,676, 677). Other less effective or less convenient dehydrating agents are $KHSO_4$ alone, $NaHSO_4$ alone or with Na_2SO_4, $MgSO_4$, H_2SO_4, H_3PO_4, "bleaching earths" (for example, franconite), and alkali metal, iron, or copper phosphates (556,559c,678–684).

b. Oxidation of Propylene with Mercuric Salts

Mercuric salts form complexes with propylene that decompose to **52**, mercurous salts, and free mercury. Thus, the addition of propylene to an aqueous suspension of mercuric sulfate in sulfuric acid at 50 to 60°C yields ~ 75 percent **52** (552). When ammonium sulfate is included and the temperature raised to 90°, an 89 percent yield is claimed (685). Another improvement is to carry out the reaction in the presence of air or oxygen and salts of transition metals that act as oxygen carriers in converting mercury(I) back to mercury(II) (686–688). Instead of the sulfate, mercuric perchlorate (689,690) or nitrate (691,692) may be used. The addition of propylene to a suspension of mercuric trifluoracetate in $C_2H_4Cl_2$ at 20° and subsequent heating with either aqueous trifluoracetic or perchloric acid rapidly liberates **52** in > 90 percent yield (693).

c. Vapor-phase Condensation of Acetaldehyde and Formaldehyde

In the process used by Degussa (550,562–564), a catalyst, such as calcium triphosphate, containing small amounts of lithium or aluminum phosphate, at temperatures from 200 to 400°C gives 70 to 85 percent **52**. The Eastman synthesis uses dehydrating catalysts of silica or alumina gel in combination with alkaline earth salts at 250 to 325° (565). Workers in Japan (694,695) have since investigated a Li_2CO_3–SiO_2 catalyst and report a maximum single-pass yield of 52 to 53 percent at 350°. Polish chemists have studied catalysts prepared by treatment of silica with alkali silicates, tungstates, titanates, or hydroxides (696–698) and the kinetics of the catalyzed reaction (699).

d. Vapor-phase Oxidation of Propylene

In the process developed by Shell in 1948 (571), propylene and molecular oxygen, diluted with steam and/or N_2, CO_2, or C_3H_8, were passed over cuprous oxide on SiC at 368° to give 65 percent **52**. For over a decade, Cu_2O on a variety of supports, for example, SiC, pumice, silica gel, and alumina (700–702), remained the preferred catalyst with operating temperatures of 300 to 375°. The addition of volatile organic halides (703), selenium (704), and palladium (705) is claimed to improve yields. Copper silicate, chromate, molybdate, sulfate, tungstate, and vanadate are also reported to be effective catalysts (704,706).

A process developed by Montecatini (573,707,708) uses no supported catalysts and no diluent. Propylene and oxygen are passed through small-bore copper coils immersed in a heat-transfer fluid. The flow of the reactants causes a thin layer of copper oxide to form on the inner surfaces of the tubes that acts as the catalyst. Hot spots do not develop, because the heat is dispersed by the heat-transfer fluid. This eliminates the necessity of a diluent and gives a higher unit capacity. Using a C_3H_6/O_2 ratio of 4:1 to 5:1, temperature of 390°, and contact time of 0.5 sec at 4 atm pressure, a yield of 78 percent **52**, based on propylene, is obtained.

In 1960, Standard Oil of Ohio (709) patented the use of bismuth molybdate and phosphomolybdate as catalysts for the oxidation of propylene with oxygen. With the former they claimed 95 percent conversion and a 73+ percent yield of **52**. Distillers Co., in England, reported silver, tin, or bismuth tungstates (710) and antimony oxide with oxides of Sn, Mo, W, Te, Cu, Ti, or Co (711) to be improved catalysts. Bayer, in West Germany (712), obtained a patent that included salts of 16 metals, mostly transition elements, and thermally stable oxyacids of B, P, Mo, W, or V. A Cu(II) phosphate gave 12 percent conversion to **52** with no CO or CO_2 in the waste gases. It now appears from the literature that almost every conceivable combination of Cu, Sn, Si, Bi, Mo, W, Te, Ti, Co, Sb, B, Fe, P, V, As, Zn, Al, Sr, Ba, Ce, Cr, U, Ni, Mn, Se, Ga, Si, Ge, Hg, Ag, and Au oxide, silicate, sulfide, selenite, chromate, molybdate, sulfate, tungstate, vanadate, antimonate, arsenate, arsenite, phosphate, phosphomolybdate, phosphotungstate, silicomolybdate, vanadomolybdate, arsenomolybdate, arsenovanadate, boromolybdate, vanadotungstate, and molybdochromate has been patented (713–734).

In sifting through this maze of catalysts, bismuth, molybdenum, and phosphorus seem to appear most frequently as components. At temperatures of 450 to 550°, bismuth molybdate and phosphomolybdate show great selectivity for the oxidation of propylene to **52** (716a,717a,727,733, 735). The addition of less than 0.5 percent each of Te and TeO_2 to the former (713e) and a trace of bromide ion to the latter (719a) are reported to give even better results. A two-stage catalyst in which the first portion contacted by the feed is Cu_2O on SiC and the second is bismuth molybdate on SiO_2 (736) is claimed to give conversions $1\frac{1}{2}$ times as great as either catalyst alone.

Conversions and yields as reported in patents tend to be confusing and, therefore, it is difficult to make direct comparisons between the various catalysts. However, the 85 percent yields claimed by Allied Chemical Co. (734) with bismuth vanadotungstate at 425° and bismuth arsenomolybdate at 450°, as well as a 99.5 percent conversion and 88 percent yield using a catalyst of Fe, Bi, P, Mo, and O on silica (in a continuous operation) reported by Knapsack (720b), are impressive. Distillers Co. has continued to claim good conversions with Sb_2O_3 in combination with other oxides, especially those of Sn and U on graphite (713b,f,h,737). These catalysts, which are generally used at 470 to 480°, can be made resistant to erosion by adding compounds of metals of atomic number 20 to 56 in small amounts, for example, ~ 2 percent $CaSO_4$ (738).

In addition to choice of catalyst and support, it is also necessary to optimize the temperature, feed concentration, space velocity, and contact time for maximum conversion of propylene to **52** with a minimum of by-products (saturated aldehydes, CO, CO_2, acrylic acid). A great deal of study has been devoted to these variables and to reaction mechanisms. For copper catalysts (Cu–Cu_2O–CuO), the Cu_2O has been shown to be the component that is most selective toward **52** formation (739). Further oxidation of **52** to CO_2 and water is suppressed by the addition of steam or other diluents to the feed (739–746) and by keeping the contact time as short as possible (747). The surface area of the catalyst is influenced by the support (748). If an excess of catalyst is used, however, it can block the pores and decrease the surface (749). Optimum conditions for this reaction are reported to be temperature, 380°; contact time, 0.3 sec; C_3H_6–O_2,

4:1 to 5.6:1; 40 volume percent steam in feed; copper content in catalyst ≤ 1.5 wt percent; carrier, carborundum (750).

The role of selenium with copper catalysts has been explored quite thoroughly (751,752). CuO alone gives mainly oxidation of C_3H_6 to CO_2 and water. Under the conditions of the reaction, selenium is oxidized to SeO_2 while CuO is reduced to Cu_2O. SeO_2 also may react with CuO to give $CuSeO_3$. These products are all selective for the formation of **52**.

The oxidation of propylene to **52** over bismuth molybdate on silica at 450 to 550° has been shown to be first-order in olefin and independent of the concentration of oxygen or products (735). Studies using tagged molecules indicate that the mechanism for formation of **52** using Bi–Mo catalysts, with or without P, is the same as with Cu_2O (753–756). It is believed that the reaction proceeds through a symmetrical allyl intermediate that results from the abstraction of a hydrogen atom from the methyl group. Further reaction can then occur at either end of the molecule.

Although the addition of Mo or W oxides to copper oxide catalysts does not change the mechanism for formation of **52**, it does change the mechanism for CO_2 formation and raises its activation energy (741,757). Rate constants for the oxidation of propylene to **52** over V_2O_5 and MoO_3 catalysts have been determined at 314 and 560° (758). It was noted that the addition of acidic materials to V_2O_5 raises the energy of activation and that alkali metal cations lower it. In a study of the catalytic activity of a number of metal oxides, it was established that the order of selectivity toward **52** is $Cu > Co > Fe > Ni > V$ (759). The selectivities and activities of Cu, V, and Mo oxides, with and without P_2O_5 or Bi_2O_3, have been compared, and the activation energies of the resulting oxidation products of propylene have been determined (760). Japanese workers (761) have made an extensive study of two- and three-component catalysts containing arsenic. A combination of iron and arsenic oxides was found to have excellent selectivity for **52**, especially if the concentration of As^{+5} was high with respect to As^{+3}.

Even under the best of conditions, the product of vapor-phase oxidation of propylene is a mixture, often containing only a small percentage of **52**. Thus, efficient methods for the purification of **52** and recovery of unchanged starting materials are essential. Generally, most of the water is removed by condensation on cooling in one or more steps to 49°. This also separates some of the water-soluble by-products. The **52** in the remaining vapor can then be selectively extracted with water in an absorption tower, and the aqueous solution fractionally distilled to give 99+ percent recovery (762). A second method is the extraction of the almost dry vapor at low temperatures with hydrocarbons, high-boiling ketones or ethers, followed by distillation (763–765). Another approach is to cool the acrolein-rich vapor to -78 to $-50°$ and distill the resulting condensate (764b,765).

Water can also be extracted from crude 52 with 1,2,6-hexane- or 1,2,4-butanetriol (766). Small amounts of saturated aldehydes in anhydrous 52 can be converted to high-boiling condensation products with aluminum alcoholates and removed by distillation (767). Acrolein can be distilled at reduced pressure without polymerization if oxygen is excluded, a pH of 5 to 8 (preferably 6 to 7) is maintained, and a stabilizer such as hydroquinone is added (765,768).

e. Miscellaneous

Acetaldehyde, formaldehyde, and diethylamine hydrochloride heated for 20 min at 80° and a pH of 7 in an autoclave give a Mannich base—Eq. (73):

$$CH_3CHO + HCHO + Et_2NH \cdot HCl \longrightarrow OCHCH_2CH_2NEt_2 \cdot HCl \qquad (73)$$

Steam distillation of this base hydrochloride gives a 49 percent overall yield of 52 (769). Acrolein can also be prepared from vinyl Grignard reagent–tetrahydrofuran complex (1 mole) and methyl formate (4 moles) (770). In recent years there has been some renewed interest in the preparation of 52 from ethylene and carbon monoxide (771)—Eq. (74)—and by oxidation of allyl alcohol (772)—Eq. (75):

$$C_2H_4 + CO \xrightarrow[150°, \ 4 \ hr, \ autoclave]{\gamma-Al_2O_3-Co-BF_3} 52 \qquad (74)$$

$$CH_2{=}CHCH_2OH + air \xrightarrow[200-530°]{\begin{subarray}{c} Cu \ spirals \\ Ag-Al_2O_3 \end{subarray}} 52 \qquad (75)$$

2. METHACROLEIN (53)

a. Dehydration of β-Methylglycerol

β-Methylglycerol dehydrates far more readily than glycerol—Eq. (76):

$$\underset{\underset{OH}{|}}{\overset{\overset{CH_3}{|}}{HOCH_2-C-CH_2OH}} \xrightarrow[120°]{aq \ H_2SO_4} 53 + 2H_2O \qquad (76)$$

Thus, distillation with 12 percent H_2SO_4 at atmospheric pressure gives a 100 percent yield of 53 (628,629). Under these same conditions, β-methylglycidol, β-methylepichlorohydrin, and the derivatives of β-methylglycerol in which one or both of the primary hydroxyl groups have been replaced by chlorine are also converted to 53 (561,629). This is a convenient laboratory preparation.

b. Oxidation of Isobutylene with Mercuric Salts

Isobutylene, like propene, forms complexes with mercuric salts at room temperature. These hydrolyze on heating with aqueous acid at 70 to 100° to yield **53**. Mercuric sulfate (773,774), perchlorate (690), nitrate (691), or trifluoroacetate (693) may be used. In general, yields are somewhat lower than in the preparation of **52**. Thus, using $HgNO_3$ at 70°, isobutylene gives 80 percent **53**, compared with 91.2 percent **52** from propene (691).

c. Mannich Reaction

The Mannich reaction provides a convenient synthesis of **53**—Eq. (77):

$$EtCHO + CH_2O + R_2NH \cdot HCl \longrightarrow MeCH(CHO)CH_2NR_2 \cdot HCl + H_2O$$

$$R_2NH \cdot HCl + 53 \xleftarrow{\quad 45-100° \quad} \qquad (77)$$

Although other secondary amines have been used successfully (775–777), diethylamine hydrochloride seems to be favored (630,769,778,779). The kinetics of this reaction have been studied, and optimum conditions for the reaction determined (630,779). Approximately 99 percent yields of **53** are obtained from 1 mole $Et_2NH \cdot HCl$: 1 mole 35 to 38 percent formalin : 1.1 mole EtCHO at pH 7 and 45° in 20 min. The product is distilled as the **53**–H_2O azeotrope. In view of its simplicity, mild conditions, and excellent yields, this process has been proposed for large-scale manufacture (630).

d. Oxidation of Isobutyraldehyde

Isobutyraldehyde is oxidized to **53** by contact with oxides of Mo, U, As, Sb, or Bi at elevated temperatures (780,781)—Eq. (78):

$$CH_3CH(CH_3)CHO \xrightarrow[100-600°]{\text{metal oxide}} 53 + H_2O + \text{reduced oxide} \qquad (78)$$

Addition of an equimolar portion of an inert gas, such as N_2, aids in the separation of the product and disperses the heat produced by the reaction. Using Sb_2O_4 at 450°, 42 percent conversion and an 84 percent yield of **53** are claimed. The catalyst must, however, be regenerated, or the conversion falls. It is also possible to feed oxygen simultaneously into the reaction and thus obtain a catalytic type of oxidation. This reaction, using oxides of As, Sb, or Bi, has been developed into a possible commercial synthesis (781).

e. Cross-condensation of Propionaldehyde and Formaldehyde

This condensation, carried out in the vapor phase, is a commercial synthesis of **53**—Eq. (79):

$$EtCHO + CH_2O \xrightarrow[\text{catalyst}]{200-400°} 53 + H_2O \qquad (79)$$

Catalysts and conditions are the same as those used in preparing **52** from acetaldehyde and formaldehyde (564,565,696,697). With a catalyst of Na_2O–SiO_2 at 275°, a 46.6 percent yield of **53** is reported (697). Considerable attention has also been given to the liquid-phase condensation of these two aldehydes accompanied by dehydration at temperatures of 100° or lower. This can be accomplished using aqueous BF_3 (782–784), Na_2CO_3 (785), mineral acids, especially H_2SO_4, or a sulfonic acid ion exchange resin (786,787). Sulfuric acid (10 percent aq) at $\sim 95°$ is reported to give a 60 percent yield of **53** (786).

f. Vapor-phase Oxidation of Isobutylene

Developments in the vapor-phase oxidation of isobutylene to **53** have followed the same path as those in the comparable oxidation of propylene to **52**. Cuprous oxide on a variety of supports, with or without added promoters, was the most commonly used catalyst prior to 1960 (571, 700–706). The addition of selenium to copper catalysts appears to be especially beneficial (704,788). Using $CuSiO_3$ on SiO_2 at 320° in the presence of Se, oxidation of isobutylene with O_2 is claimed to give 65 to 70 percent yields of **53** (788c,d). Addition of an alkali metal oxide and an alkali metal halide to a CuO–Cu_2O–Cu catalyst is reported to give 84 percent selectivity to **53** at 320° (789).

The Montecatini process (see Sec. IV-B-1-d) is also suitable for the production of **53** (573,707,708,790–792). Pilot-plant operation at 390°, 4 atm, and 0.5 sec contact time gives, in one pass, a 58 percent yield of **53** as compared with a 78 percent conversion of propylene to **52** under similar conditions (791).

Many of the catalyst patents for **52** issued after 1960 also apply to **53** (711,713g,h,714c,e,f,715a,b,c,d,716a,b,717e,718a,b,c,719a,721,727–731, 733). Others, using a variety of components, especially Mo, V, P, and Bi in various combinations, deal specifically with **53** (793–798). A catalyst containing oxides of Bi, V, and P on silica or pumice is reported to give 56 mole percent conversion at 400° (796b). With oxides of V, P, and Mo on alundum or SiC at 475 to 650°, yields up to 81 mole percent are claimed (795a). A complex catalyst of Cu, Ni, Te, P, Mo, and O_2 at 477° is said to give a 50 percent conversion and 82 percent yield of **53** (797). The two-stage catalyst of Cu_2O on SiC and $Bi_2(MoO_4)_3$ on SiO_2 (736) also gives improved results with isobutylene. Combinations of Sb and Sn are also effective catalysts (711,737).

Optimum reaction conditions are the same as those for the oxidation of propylene. Addition of water and other diluents aids in temperature control and prevents further oxidation of **53** (739,799a). The kinetics of the reaction over copper catalysts have been studied and mechanisms postulated (799,800). In the temperature range of 330 to 385°, the heats of activation for isobutylene are lower than those for propylene, which is less strongly adsorbed on the catalyst (799b).

Isolation of **53** from the reaction products and its purification are accomplished in the same manner as for **52** (762b,763,764,768,769,801). Preliminary distillation of **53** with $C_5H_{11}OH$ (8:1 by volume) in the presence of a stabilizer at 70° is a convenient method of removing low-boiling impurities, including **53**–H_2O azeotrope. Distillation of the residue is claimed to give 99.99 percent pure **53** (802).

g. Miscellaneous

Methallyl alcohol can be either dehydrogenated or oxidized to **53**. The dehydrogenation can be accomplished over a conventional catalyst, such as brass spelter at 500° (624), but this also gives 15 to 25 percent isobutyraldehyde, which cannot be removed by fractional distillation (803). Oxidation is preferred using catalysts of Cu, Ni, $V_2(SO_4)_3$, and Ag (560, 625,626,803). Silver is especially good, since it is not oxidized at the temperatures, 500 + °, required for the reaction. The best method is to use O_2 rather than air, steam to control the temperature, and a catalyst of Ag gauze. In this manner, yields of 90 percent **53** are obtained with only about 2 percent isobutyraldehyde (803). It is also possible to oxidize methallyl alcohol with SeO_2 or H_2SeO_4 (627,804). The oxidizing agent may be dissolved in an indifferent solvent, preferably dioxane. Yields up to 62 percent have been reported (627). Although the oxidation of methallyl alcohol received some attention in the late thirties, there has been little interest in it as a synthesis for **53** since then.

Bis(methallyl) ether, when subjected to thermal noncatalytic pyrolysis at 550 to 600°, yields **53** (569). The product is recovered by condensation and fractional distillation.

Methacrolein can also be prepared from methallyl Grignard reagent–THF complex and dimethylformamide at −60° (805), by dehydrohalogenation of α-chloroisobutyraldehyde with NaOAc in ethanol (806), by oxidizing isobutylene with SeO_2 in BuOH (807), from α-keto-β-methyl-γ-butyrolactone by heating at 280 to 300° (808), or by treating 5-methyl-5-nitro-1,3-dioxane with polyphosphoric acid at 160° (809). An 85 percent yield of **53** is claimed when β-methallyl chloride is oxidized with a palladium salt in aqueous solution at 50° (810).

3. METHYL VINYL KETONE (54)

a. Cross-condensation of Acetone and Formaldehyde

Liquid-phase condensation of acetone and formaldehyde gives 2-hydroxyethyl methyl ketone, which can be dehydrated to **54**—Eq. (80):

$$(CH_3)_2CO + HCHO \longrightarrow CH_3COCH_2CH_2OH \xrightarrow{-H_2O} 54 \qquad (80)$$

An alkali metal hydroxide (650,811–813) is the most common catalyst for the condensation, although a volatile tertiary amine may also be used (814). The product has been dehydrated without isolation by heating with $ZnCl_2$ (636), a base (645), or an acid (645,814a,815). A 65 percent overall yield is obtained by dehydration with H_3PO_4 under N_2 in the presence of Cu and hydroquinone (815).

Isolation of the hydroxyketone, which is difficult because it is unstable in the alkaline medium required for the aldol condensation, has been accomplished by maintaining the pH at 10.2 and removing the product from the alkaline zone as it is formed. Under these conditions the reaction proceeds satisfactorily with minimum product decomposition. Acetone is refluxed in a fractionating column into which formalin adjusted to the desired pH with NaOH is introduced from the top. The condensation occurs in the column, and the high-boiling product runs down into the boiler, where the basic catalyst is neutralized with citric acid. This prevents dehydration and polymerization, which would otherwise take place on continued heating in the presence of alkali. The yield of 2-hydroxyethyl methyl ketone is 84 percent based on formaldehyde (812). Tartaric acid has also been used to neutralize the base in a similar procedure (813). A continuous process has been described in which the pH of the reaction mixture is maintained at 9.5 to 11.2 by means of an anion-exchange resin (814a).

2-Hydroxyethyl methyl ketone, which is also prepared by hydration of vinylacetylene (see Sec. IV-B-3-d), can be dehydrated to **54** with $ZnCl_2$ (635,816), acetic anhydride (811), phosphoric acid (811,814a), or oxalic acid (813). Thermal dehydration occurs at 166° in dibutyl phthalate (650) and at 240 to 260° on Al_2O_3 using a liquid paraffin diluent (817). Yields of over 95 percent are claimed for the last method.

Acetone and formaldehyde can also be condensed in the vapor phase. The kinetics of this reaction have been studied using silica impregnated with alkali metal hydroxides at 260 to 300° (697,818). RbOH (1 percent) at 280 to 300° gave the best yield (39.1 percent). Similar results are reported using rare earth oxides, acid clays, and cadmium and calcium phosphates as catalysts. Optimum yields are obtained if the temperature is ∼400° and a 1:1 molar ratio of acetone and formaldehyde is used (819).

b. Mannich Reaction

Acetone with formalin (30 to 35 percent) and a secondary amine hydrochloride (preferably $Et_2NH \cdot HCl$) form the corresponding Mannich base, which decomposes on heating to yield **54**—Eq. (81):

$$(CH_3)_2CO + CH_2O + Et_2NH \cdot HCl \longrightarrow CH_3COCH_2CH_2NEt_2 \cdot HCl + H_2O$$

$$Et_2NH \cdot HCl + 54 \xleftarrow{\;\Delta\;} \qquad\qquad (81)$$

A lower pH (0.6 to 1.0) is required for the Mannich condensation of ketones than for aldehydes (630,820,821), and higher temperatures are needed to decompose the products. The decomposition may be accomplished in Dowtherm at 150° (822), in a hot tube filled with Raschig rings (823), by cleaving with dry HCl in diethylene glycol at 220° (824), or by heating under reduced pressure (630,820,821). All these methods give excellent results. If the condensation is carried out at 100° and a pH of 1 for 1.5 hr, followed by decomposition at 150 to 210° and 50 to 80 mm, an 87 percent overall yield of **54** is obtained (630). This method is a convenient laboratory preparation and also is suitable for large-scale production.

c. Oxidation of Butene

Vapor-phase oxidation of butene to **54** can be carried out over metal selenites or tellurites (570) or cuprous oxide (571,700,705,784) on silica, pumice, SiC, and the like, at 300 to 400°. Conditions are very much the same as those for oxidizing propylene and isobutylene, and if the need for large-scale production arose, many of the other catalysts developed for these olefins would, no doubt, prove effective. Although both 1- and 2-butene have been used, higher conversions and yields result at lower temperatures with the 1 isomer. Thus, using a Cu_2O–SiC catalyst, 2-butene gives a 13 percent conversion and 19 percent yield of **54** at 380 to 400°, and 1-butene gives a 35 percent conversion and 32.3 percent yield at 300 to 320° (700).

In the liquid phase, 1-butene is readily oxidized to **54** by mercuric salts and an oxyacid (689,691,693,825,826)—Eq. (82):

$$CH_3CH_2CH{=}CH_2 + 4HgY + H_2O \longrightarrow 54 + 2Hg_2Y + 2H_2Y \qquad (82)$$
$$(Y = SO_4, (NO_3)_2, ClO_4, \text{ or benzenesulfonate})$$

This can be a one- or two-stage process, depending on whether the reaction is run above or below the decomposition temperature of the complex. Using the two-step approach with $HgSO_4$ and H_2SO_4, a 93 percent yield of **54** is reported (826). This process is adaptable to batch, intermittent, and continuous operation on a commercial scale (825). Liquid-phase oxidation of 1-butene in an autoclave with copper and silver oxides and NaOH at 230° and 35 atm (827) or with cumene hydroperoxide at 95 to 120° and 60 atm (828) has been reported to give **54** along with other oxidation products.

d. Hydration of Vinylacetylene

Mercuric salts with aqueous sulfuric acid, at 25 to 65°, are most commonly used to hydrate vinylacetylene to **54** (639,641,642)—Eq. (83):

$$HC{\equiv}C{-}CH{=}CH_2 + H_2O \xrightarrow[H_2SO_4]{Hg^{+2}} 54 + Hg \qquad (83)$$

However, Ag, Cd, Cu, and Zn salts of phosphoric, acetic, formic, chloro-acetic, and dihydroxyfluoroboric acids are also claimed to be effective (637,638,829,830). The reaction is promoted by the addition of an oxidizing agent, such as $Fe_2(SO_4)_3$ (639,641), which gave a 93 percent yield of **54** when used with $HgSO_4$ and H_2SO_4 at 60 to 62° (641). Free mercury formed in the reaction can be recovered to the extent of 94 percent. Dilution of the vinylacetylene with nitrogen and careful control of temperature and contact time in specially designed reactors is claimed to improve both the yield and purity of **54** (831). I. G. Farben. has patented a process for continuous production using a countercurrent trickling apparatus (643,832).

Some 2-hydroxyethyl methyl ketone is usually formed as a by-product in this reaction. Its production can be encouraged by using a lower concentration of acid, that is, 0.5 to 5.0 percent (833), or catonite K U-2 in water at 80° (834). Dehydration of the hydroxyketone to **54** was discussed previously (see Sec. IV-B-3-a).

Vinylacetylene can also be hydrated at high temperatures (350 to 375°) over catalysts of Cd–Ca phosphate or Cd tungstate (835). Using the former, yields to 83 percent **54** were reported.

Considerable attention has been given to the isolation and purification of **54** prepared by this method. The crude product contains water, and colored by-products are often present. The latter can be removed by heating in the presence of an inert diluent, for example, water or toluene, until test portions give a colorless distillate (836a,b). Subsequent steam distillation gives colorless **54**. A stabilizing agent may be added prior to the distillation (836c), but the presence of steam alone inhibits polymerization (836d). Treatment of the distillate with a dehydrating agent, such as $CaCl_2$, removes the water.

54 can also be recovered from aqueous solutions by extraction with unsaturated hydrocarbons, for example, C_3H_6, iso-C_4H_8, vinylacetylene or butadiene (836e,f). The ketone-hydrocarbon layer floats on the surface and can be separated, dried, and fractionally distilled. If the pH of the ketone is adjusted to < 7.1, with acetic or formic acid, polymerization is prevented during the distillation (836g).

e. From Butanediols

Dehydrogenation and dehydration of 2,3-butanediol may be carried out stepwise (648,837) or simultaneously (838)—Eq. (84):

$$CH_3CHOHCHOHCH_3 \xrightarrow{-2H} CH_3COCHOHCH_3 \xrightarrow{-H_2O} \textbf{54} \qquad (84)$$
$$\underset{-2H,\ -H_2O}{\rule{6cm}{0.4pt}}$$

In the two-step process, a mild oxidizing agent, such as $FeCl_3$ (837), converts the diol to acetoin. Dehydration to **54** may be accomplished with $NaHSO_4$ (839) or, preferably, with oxides of W, Cr, or Mo on Al, Be, Mg, Si, Ti, or Th (837,840). With a catalyst of Al, Al_2O_3, and WO_3 at 300°, an 82 percent yield of **54** is claimed (837). Water vapor is mixed with the acetoin before passing it over the catalyst. Introduction of an amine is also advantageous. Thus, when triethylamine is used with $Al-SiO_2-WO_3$, excellent yields are obtained at 230 to 235° (840). Simultaneous oxidation and dehydration occur with a catalyst of NaH_2PO_4 on asbestos (838). Yields, however, are lower than those from the two-step process.

Good yields of **54** from 1,3-butanediol are claimed using a catalyst of ZnO_2 with silica at $\sim 400°$ (841). The addition of 6 to 12 wt percent of oxides of Zr, Ce, or Th is reported to improve and stabilize the action of the catalyst. Reaction of 1,3-butanediol monohydrogen sulfate with CrO_3 also gives **54** (842).

f. Miscellaneous

2-Chloroethyl methyl ketone can be dehydrohalogenated by pyrolysis (843) or by treatment with diethylaniline (633,844) or aqueous sodium benzoate (845)—Eq. (85):

$$ClCH_2CH_2COCH_3 \xrightarrow{-HCl} 54 \qquad (85)$$

With the latter, a 77 percent yield is reported. The chloroethyl ketone can be prepared from chloropropionyl chloride and dimethylzinc (633), acetyl chloride and ethylene in the presence of $AlCl_3$ (845), 2-hydroxyethyl methyl ketone and HCl (843), or from acetone, paraformaldehyde, and HCl (844).

Acetoin acetate can be pyrolyzed at 500 to 550° to give **54** (647,846)— Eq. (86):

$$CH_3CH(OAc)COCH_3 \xrightarrow{\Delta} 54 + HOAc \qquad (86)$$

This is preferably done in the presence of an inert gas in a glass tube containing quartz chips (646), and the product is recovered by distillation. The starting material can be obtained by treating 3-chloro-2-butanone with sodium acetate in acetic acid (647,846).

2-Methoxyethyl methyl ketone is formed when vinylacetylene (847) or 4-methoxy-2-butyne (848) is treated with an alcohol and water in the presence of $HgSO_4$—Eqs. (87) and (88):

$$CH_2{=}CHC{\equiv}CH + CH_3OH + H_2O \xrightarrow[60-65°]{HgSO_4} 94\% \ CH_3COCH_2CH_2OCH_3 \quad (87)$$

$$CH_3OCH_2C{\equiv}CCH_3 + H_2O \xrightarrow[\Delta]{\substack{HgSO_4 \\ CH_3OH}} CH_3COCH_2CH_2OCH_3 \qquad (88)$$

Distillation of the ketone with p-toluenesulfonic acid (847–849) or heating it at 250 to 400° (850) gives good yields of **54**.

Methyl vinyl carbinol can be oxidized in the vapor phase using catalysts of Cu and Zn or Ag (649,851). Oxidation with air and a pumice-supported catalyst of Cu on Ag at 530° gives over 90 percent yields of **54**. The carbinol can be prepared by hydrohalogenation of butadiene followed by hydrolysis and separation of isomers (851).

In the presence of water and cuprous chloride, 2,4-dichloro- and 2-chloro-4-hydroxy-2-butene are converted to **54** (852). Boiling water alone is claimed to achieve the same results with 1,3-dichloro-2-butene (853).

A Knoevenagel condensation of acetaldehyde with *tert*-butyl acetoacetate followed by distillation of this product with catalytic amounts of p-toluenesulfonic acid gives **54** (854)—Eq. (89):

$$CH_3COCH_2CO_2\text{-}t\text{-}Bu + CH_3CHO \xrightarrow[\substack{\text{piperidine} \\ \text{EtOH}}]{-5 \text{ to } -15°} CH_3COC(=CHCH_3)CO_2\text{-}t\text{-}Bu$$

$$46\% \; \mathbf{54} \xleftarrow{\qquad\qquad} \quad \substack{13-40° \; | \; \text{vac.}} \tag{89}$$

3-Hydroxy-1-butyne yields **54** when heated with steam or nitrogen over phosphoric acid catalysts (855,856). With H_3PO_4–NaH_2PO_4 on graphite, a 50 percent yield of **54** is obtained at 280 to 300°.

1,4-Dibromo-2-butyne and 1-bromo-4-phenoxy-2-butyne yield butatriene when treated with zinc. Further reaction of the triene with 78 percent H_2SO_4 at 0° gives 37 percent **54** (857).

54 is also obtained by pyrolysis of levulinic acid, its esters or α- or β-angelicalactone over silica, pumice, or aluminum silicate at 450 to 650° (858), methyl propyl ketone at 500 to 530° (859), or divinyl ether over alumina at 350° (860).

4. PHENYL VINYL KETONE (55)

a. Mannich Reaction

The oldest and still most widely used synthesis of **55** is the decomposition of a Mannich base formed from acetophenone, formaldehyde, and an amine hydrochloride. Ammonia (660) and primary and secondary amines have been used—Eqs. (90) to (92):

$$3C_6H_5COCH_3 + 3HCHO + NH_3 \cdot HCl \xrightarrow{-3H_2O}$$
$$(C_6H_5COCH_2CH_2)_3N \cdot HCl \longrightarrow \mathbf{55} + (C_6H_5COCH_2CH_2)_2NH \cdot HCl \tag{90}$$

$$2C_6H_5COCH_3 + 2HCHO + RNH_2 \cdot HCl \xrightarrow{-2H_2O}$$
$$(C_6H_5COCH_2CH_2)_2NR \cdot HCl \longrightarrow \mathbf{55} + C_6H_5COCH_2CH_2NHR \cdot HCl \tag{91}$$

$$C_6H_5COCH_3 + HCHO + R_2'N \cdot HCl \xrightarrow{-H_2O}$$

$$C_6H_5COCH_2CH_2NR_2' \cdot HCl \longrightarrow 55 + HNR_2' \cdot HCl \quad (92)$$

[R = CH_3 (664,861); $HOCH_2CH_2$ (862). R' = CH_3 (824,861,863,864); C_2H_5 (630,664,820,865); $R_2' = (CH_2)_5$ (866).]

Secondary amines are the most efficient, and the highest yields have been reported using diethylamine hydrochloride. Two papers (864,867) which claim that this amine salt does not undergo the Mannich reaction with acetophenone and formaldehyde appear to be in error.

A pH of 1.0 or lower is essential during the condensation. Usually, equimolar amounts of acetophenone and amine hydrochloride are used with a slight excess of formaldehyde (formalin or paraformaldehyde) (862,865). Alcohol is frequently used as a solvent (664,861), but a mixture of toluene, nitromethane, and alcohol is reported to give a better yield (54 percent) (865). A reflux period of 0.5 to 3 hr is necessary to bring the reaction to completion. The condensation can also be carried out in an autoclave without solvent (630,820). After removal of the volatiles, the Mannich base hydrochloride can be decomposed to 55 by steam distillation (660,664,824,862,863,866) or by pyrolysis under vacuum, preferably in the presence of a stabilizer (630,664,820,864,865).

Optimum conditions for this reaction on a large scale have been determined (630,820). The condensation is carried out in an autoclave at 100° and a pH of 1.0 for 1.2 hr. Volatiles are removed at 120° and 40 mm, and the base decomposed at 150 to 210° and 10 mm with simultaneous distillation of the product. Reproducible overall yields of greater than 80 percent are claimed.

b. Dehydrohalogenation of 2-Chloroethyl Phenyl Ketone

Pure 55 is prepared in good yield (~70 percent) by treating 2-chloroethyl phenyl ketone with potassium acetate in anhydrous alcohol followed by distillation (868). This reaction is also used to prepare 55 *in situ* (869,870). The starting material can be synthesized in excellent yield from 2-chloropropionyl chloride and benzene in the presence of aluminum chloride (868).

c. From Benzoyl Chloride and Ethylene

A Friedel-Crafts reaction between benzoyl chloride and ethylene gives 55 (871,872). If the reaction is carried out in tetrachloroethane at 2 to 5° for 30 hr using equimolar quantities of acid chloride and aluminum chloride, a 70 percent yield is obtained (872). An excess of aluminum chloride seriously suppresses the formation of 55 (873).

d. Miscellaneous

55 can be prepared by means of the Wittig reaction from benzoyl-methylenetriphenylphosphorane and paraformaldehyde (874)—Eq. (93):

$$(C_6H_5)_3P\!\!=\!\!CHCOC_6H_5 + HCHO \xrightarrow[78-80°,\ 3\ hr]{C_6H_5CH_3,\ 2\text{-}C_{10}H_7NHC_6H_5}$$

$$83\%\ 55 + (C_6H_5)_3PO \quad (93)$$

The starting phosphorane is readily obtained from 2-bromoacetophenone and triphenylphosphine.

When β-propiolactone is treated with phenylmagnesium bromide, phenyl vinyl ketone and 2-bromopropionic acid are formed (875,876). This lactone reacts with benzene in the presence of aluminum chloride to give small yields of **55** (877).

Phenyllithium reacts with equimolar quantities of benzoylmethyltrimethylammonium bromide to give over 40 percent **55** (878). The reaction proceeds through a Stevens rearrangement. It is also possible to prepare **55** by reacting vinylmagnesium chloride in tetrahydrofuran with benzonitrile or with benzoyl chloride and zinc chloride (770).

55 can be purified by codistillation at low temperature with toluene, xylene, or water, or it can be recrystallized from low-molecular-weight hydrocarbons (879). At atmospheric pressure, **55** tends to polymerize upon distillation even in the presence of an inhibitor and under argon. It can, however, be distilled at reduced pressure in the presence of an inhibitor (820).

5. DIVINYL KETONE (56)

a. Mannich Reaction

Both acetone and methyl vinyl ketone react with formaldehyde and diethylamine hydrochloride to form Mannich bases that can be decomposed to yield **56**—Eqs. (94) and (95):

$$(CH_3)_2CO + 2HCHO + 2Et_2NH \cdot HCl \xrightarrow[100°,\ 10\ hr]{pH\ =\ 0.48} OC(CH_2CH_2NEt_2)_2 \cdot 2HCl$$

$$\downarrow \Delta$$

$$\mathbf{56} + 2Et_2NH \cdot HCl \quad (94)$$

$$CH_2\!\!=\!\!CHCOCH_3 + HCHO + Et_2NH \cdot HCl \xrightarrow[100°,\ 2\ hr]{pH\ =\ 0.75}$$

$$CH_2\!\!=\!\!CHCOCH_2CH_2NEt_2 \cdot HCl$$

$$\downarrow \Delta$$

$$\mathbf{56} + Et_2NH \cdot HCl \quad (95)$$

The condensations are best carried out in an autoclave, and the pH is carefully adjusted by adding hydrochloric acid. Volatiles are then removed under reduced pressure (120° and 40 mm), and the bases decomposed by heating to 150 to 210° at 100 to 110 mm in the presence of hydroquinone. Yields to 55 percent are obtained starting with acetone, but those from **54** are somewhat lower. The amine hydrochloride can be recovered and used again (630,673).

b. Dehydrohalogenation of Bis(2-chloroethyl) Ketone

Distillation of bis(2-chloroethyl) ketone from anhydrous sodium carbonate gives 40 to 50 percent yields of **56** (880,881). The best results are obtained if this is done rapidly under reduced pressure, and the distillate condensed in an acetone–dry ice bath to inhibit polymerization. Attempts to dehydrohalogenate with alcoholic KOH, AlCl$_3$, or C$_6$H$_5$NEt$_2$ gave only polymers (880). Bis(2-chloroethyl) ketone is prepared from β-chloropropionyl chloride and ethylene in the presence of aluminum chloride. This can be done using an excess of the acid chloride as solvent (882), but better yields are obtained in solutions of ethylene or methylene dichloride (883).

Frequently it is desirable to prepare **56** *in situ*. This is conveniently done using bis(2-chloroethyl) ketone and sodium acetate (674).

c. Cleavage of Bis(2-methoxyethyl) Ketone

Slow distillation of bis(2-methoxyethyl) ketone under reduced pressure (90 mm) in the presence of p-toluenesulfonic acid gives ~ 35 percent **56** (671,884). The starting material can be obtained from vinylacetylene in two steps—Eqs. (96) and (97):

$$H_2C{=}CH{-}C{\equiv}CH + CH_2O \xrightarrow[\text{Et}_2\text{O, N}_2, \, <6°]{\text{EtMgBr}} 65\% \; H_2C{=}CH{-}C{\equiv}CCH_2OH \quad (96)$$

$$H_2C{=}CH{-}C{\equiv}CCH_2OH + 2CH_3OH \xrightarrow[40°]{\text{HgSO}_4, \text{H}_2\text{SO}_4} 42\% \; (CH_3OCH_2CH_2)_2CO$$
$$(97)$$

The reaction described by Eq. 96 can be carried out at room temperature (885), but yields are lower.

d. Oxidation of Divinyl Carbinol

Divinyl carbinol is oxidized to **56** using MnO$_2$ under mild conditions, 24 hr at 20° in dry chloroform (672). The type and amount of MnO$_2$ are important. Yields to 50 percent of **56** are claimed using a commercial grade of MnO$_2$ and an alcohol to MnO$_2$ ratio of 1:10 to 1:15 by weight. The most consistent results are obtained with a 1:10 ratio using a specially activated MnO$_2$. Pure **56** is isolated by distillation at reduced pressure in the presence of hydroquinone.

C. Properties

1. PHYSICAL

Table 11 summarizes the physical properties of acrolein (52) and methacrolein (53), Table 12 those of the vinyl ketones. The monomers

TABLE 11

Physical Properties of Acrolein and Methacrolein

	Acrolein		Methacrolein	
		Reference		Reference
Boiling point, °C (mm)	−36(10)	886	−25(10)	551,886
	−35(10)	551	3(50)	551
	−9(50)	551	42(300)	886
	28(300)	886	66.5(752)	887a
	52.15(751.3)	888	66.8(752)	889
	52.46(760)	888	67.9(760)	551
	52.6(760)	890	69(760)	886
	52.6(766)	891	68	775
	52.69(760)	892,893	68.4	630,803
	53(765)	889	68.5	802
Boiling point coefficient	0.0355	892	0.042	886
(dt/dp) at bp, °C/mm	0.039	886		
Melting point, °C	−87	886,888	−81	551,886
	−86.95	892,893		
Vapor pressure, mm (°C)	14.22(−30)	892	121(20)	886
	27.01(−20)	892		
	48.49(−10)	892		
	82.90(0)	892		
	135.71(10)	892		
	213.84(20)	892		
	325.70(30)	892		
	481.28(40)	892		
	692.15(50)	892		
	760.00(52.69)	892		
Vapor pressure, atm (°C)	1.13(60.0)	888,892		
	2.07(80.0)	888,892		
	4.04(100.0)	888,892		
	6.69(120.0)	888,892		
	9.68(137.2)	888,892		
Density, d_4 (°C)	0.86205(0)	888,892, 893	0.837(20)	803
	0.8506(10)	892,893	0.845(20)	630,886
	0.8447(15)	892,893	0.8138(25)	887a

continued

TABLE 11 (*continued*)

	Acrolein		Methacrolein	
		Reference		Reference
	0.8389(20)	888,892, 893		
	0.8404(20)	891		
	0.8377(25)	894		
	0.8269(30)	892,893		
	0.8179(40)	892,893		
	0.8075(50)	892,893		
Apparent specific gravity, 20°/20°C	0.8427	551,886	0.8474	551,886
Coefficient of expansion, °C^{-1} (°C)	0.00143(20)	892	0.00133(55)	886
	0.00147(55)	886		
Refractive index, n_D (°C)	1.40475(15)	888	1.4100(20)	783
	1.4048(15)	892,893	1.4144(20)	803
	1.3992(20)	891	1.4156(20)	630
	1.3998(20)	892,895	1.4169(20)	551,886
	1.4013(20)	886	1.4098(22)	783
	1.4017(20)	892,893	1.4053(25)	889
	1.4022(20)	892	1.4098(25)	887a
Viscosity, cP (°C)	1.26(−60)	892	0.49(20)	886
	0.820(−40)	892		
	0.586(−20)	892		
	0.431(0)	892		
	0.329(20)	892		
	0.260(40)	892		
Heat of vaporization at 760 mm, cal/g mole	7,260	892	6,930	886
	6,730	886		
Heat of combustion (liquid) at 25°, kcal/g	6.95	892		
Specific heat (liquid) at 17–44°, cal/(g) (°C)	0.511	892		

discussed in this section are soluble in the ordinary organic solvents, and methyl vinyl ketone (54) is completely miscible with water (896). In a saturated mixture of 52 and water at 20°, the aqueous phase contains 21 wt percent of acrolein, and the aldehyde phase 7 wt percent of water (551,886,892). Comparable figures for 53 are 5.9 and 1.7 wt percent (551,886). Above 87.5°, 52 and water are miscible in all proportions. The heat of mixing of 52 with excess water at 20° is 450 cal per mole (892).

52 (892), 53 (802,803), and 54 (811,851,903a) form minimum-boiling azeotropes with water. 54 also does so with methanol (850), and 52 with

TABLE 12
Physical Properties of Vinyl Ketones

	Methyl vinyl ketone	Reference	Phenyl vinyl ketone	Reference	Divinyl ketone	Reference
Boiling point, °C (mm)	32(120)	896	38(0.05)	868	30(16)	880
	32(121)	857	59(2)	897	38(50)	672
	34(130)	895,898	76(4)	864	38(52)	881
	35(135)	899	76(5)	874	41(62)	630,673
	36.8(145)	900	94(14)	866	44(75)	671
	79(743)	630	112(14)	630	49(100)	671
	80.8(752)	887b	115(18)	661	52(102)	901
	81.4(760)	811,896	116(18)	664		
	81	650,850,870	117(18)	872		
	81.4	868	118(20)	663		
	81.5	902	—a			
Refractive index, n_D (°C)	1.4120(15)	811,899	1.5522(20)	630	1.4440(20)	630,673
	1.4086(20)	634,895,896	1.5550(20)	864	1.4485(20)	671
	1.4091(20)	854	1.5587(20)	874	1.4497(20)	672
	1.4103(20)	845	1.5588(20)	868	1.4540(20)	901
	1.4108(20)	870			1.4561(20)	881
	1.4109(20)	850				
	1.4115(20)	630				
	1.4096(20.7)	868				
	1.4095(22)	650				
	1.4083(25)	900				
	1.4084(25)	896				
	1.4086(25)	887b				
Density, d_4 (°C)	0.8620(20)	816	1.0387(20)	864	0.8811(20)	671
	0.8636(20)	634,896	1.041(20)	874	0.8839(20)	630,673
	0.8666(20)	630	1.060(20)	630		
	0.8393(25)	896				
	0.840(25)	811				
	0.8407(25)	887b				
	0.842(25)	650				
	0.8429(25)	900				

a The melting point of phenyl vinyl ketone is −11.5° (874).

carbon disulfide (904), diethyl ether, carbon tetrachloride, and cyclo-pentane (892). Vapor-liquid equilibrium studies of **52** (892,905) and of **54** (903) in admixture with other components have been described.

Dipole moment values (in Debye units) are as follows: acrolein 2.88 (889,891), 2.90 (890), 3.04 (906), 3.11 (907); methacrolein 2.68 (906), 2.72 (887a,889); methyl vinyl ketone 2.98 (887b), 3.16 (908). These data, along with electron-diffraction results for acrolein (909) and microwave spectra for acrolein (907) and methyl vinyl ketone (908), indicate that the s-trans (**58a**) predominates over the s-cis isomer (**58b**) in all three compounds:

$$\text{58a} \qquad\qquad \text{58b}$$

(R = H or CH₃)

Such geometric isomerism can occur about the single bond in C=C—C=O systems, since it has some double-bond character. The situation is reversed, that is, the s-cis form predominates, in phenyl vinyl ketone because of the large bulk of the phenyl substituent (910).

The best single source of information on physical properties of acrolein is the excellent review by Anderson and Hood (892), which not only covers the literature through 1959 but also includes a considerable amount of previously unpublished material.

2. PHYSIOLOGICAL

Acrolein (**52**) is highly toxic by inhalation and oral ingestion and moderately toxic by absorption through the skin. Results of toxicity tests on animals are shown in Table 13.

It is believed that toxic effects from repeated exposure to **52** at low concentrations are not cumulative but severe overexposure can cause serious lung injury. Autopsies on animals killed by breathing the vapor showed hemorrhaging and edema in the lungs. Microscopic examination of tissues from other organs revealed no damage (911). Bronchial con-striction was noted in guinea pigs breathing air containing only 0.4 to 1.0 ppm of **52** (912).

The maximum average concentration of **52** vapor considered safe by the American Conference of Governmental Industrial Hygienists for 8 hr per day exposure is 0.5 ppm. The comparable value for chlorine and phosgene is 1.0 ppm (911). Fortunately, **52** is a powerful irritant and has a pene-trating odor, so that its presence can be detected at concentrations well below the danger level. The sensory responses of humans to the vapor at various concentrations are given in Table 14.

TABLE 13
Toxicity of Acrolein to Animals [911]

Method of exposure	Test animal	Lethal dose
		LC_{50}, ppm (exposure time)
Inhalation	Mouse	875 (1 min)
		175 (10 min)
	Dog	150 (30 min)
	Rat	8 (4 hr)
		LD_{50}, mg/kg
Oral	Rat	42, 46
	Mouse	28
Percutaneous	Rabbit	200 (undiluted acrolein)
		562 (undiluted acrolein)
		335 (20% in water)
		1,022 (10% in water)
		164 (20% in mineral spirits)
		238 (10% in mineral spirits)

TABLE 14
Physiological Responses of Man to Acrolein Vapor [913]

Vapor concentration, ppm	Time	Nasal irritation	Eye irritation	Other
0.25	5 min	Moderate	Moderate	
1.0	1 min	Slight		
	2–3 min	Slight	Moderate	
	4–5 min	Moderate	Practically intolerable	
1.8	30 sec			Odor apparent
	1 min		Slight	
	3–4 min		Practically intolerable	Profuse lachrymation
5.5	5 sec	Moderate	Moderate	Slight odor
	20 sec	Painful	Painful	
	1 min			Marked lachrymation
				Practically intolerable
21.8	Immediately			Intolerable

Liquid **52** can cause skin irritations ranging from simple reddening to severe blisters and scabs. As indicated in Table 13, absorption through the skin of animals in sufficient quantities results in systemic poisoning. Liquid **52** can also cause severe eye damage (911,914).

Methacrolein is less volatile than **52**, not so powerful a lachrymator and less toxic (551). The oral LD_{50} for rats is 140 mg per kg. With rabbits, the skin penetration LD_{50} (24 hr covered contact) is 430 mg per kg, and the liquid causes severe eye injury but only slight skin irritation (886). Methyl vinyl ketone is reported (896) to be toxic, a skin irritant and a strong lachrymator.

3. ANALYTICAL METHODS AND SPECTRA

The usual impurities in commercial **52** are water, other aldehydes (chiefly CH_3CHO and C_2H_5CHO), and hydroquinone, which has been added as an inhibitor. Aged samples may, in addition, contain dimer and polymers (551,915). Gas-liquid chromatography and mass spectrometry are useful in identifying the volatile components, but there are also specific tests for some of the individual constituents that supplement such data.

Water is determined by modified Karl Fischer procedures, infrared spectrometry, or the cloud point of a solution of the sample in *n*-heptane; hydroquinone by colorimetry, polarography, ultraviolet spectrometry, or infrared spectrometry (915). The colorimetric procedure is based on complex formation with Millon's reagent, a solution of mercury in nitric and nitrous acids (915), or with pyrrole in the presence of an oxidizing agent (551).

Volatile aldehyde and ketone impurities can be separated from **52** for measurement by treatment of the sample with excess sodium bisulfite, neutralization and distillation. In the first step, the impurities form bisulfite addition compounds—Eq. (98):

$$R_2C{=}O + NaHSO_3 \rightleftharpoons R_2C(OH)SO_3Na \qquad (98)$$

52 reacts similarly but combines with a second mole of bisulfite by Michael addition to the C=C double bond—Eq. (99):

$$CH_2{=}CHCHO + 2NaHSO_3 \longrightarrow NaO_3SCH_2CH_2CH(OH)SO_3Na \qquad (99)$$

On neutralization, the addition compounds in Eq. (98) revert to the original aldehydes and ketones, but the acrolein derivative does not. The volatile impurities are then distilled from the nonvolatile sodium sulfonate and determined by the hydroxylamine hydrochloride method (915). Alternatively, the total carbonyl content is measured by the latter procedure, and **52** by bromination of its C=C double bond. Then the saturated carbonylic impurities can be calculated by difference (551).

Acrolein dimer is readily determined by gas-liquid chromatography, and high polymers by evaporation to dryness with a suitable correction for hydroquinone (915).

Acrolein content can be measured by bromination (551,915,916). Corrections may be applied for bromine-consuming impurities (dimer, polymer, allyl alcohol, ketones, and aldehydes) known to be present. Polarography at pH 7 is another satisfactory assay method. Formaldehyde, acetaldehyde, propionaldehyde, and acetone do not interfere (915). Polarography in alkaline (917) and acidic (918) media has also been used.

Many methods for the determination of **52** in mixtures in which it is not the major component are described in the literature. Gas-liquid chromatography (919–924) is one of these; another is precipitation as the 2,4-dinitrophenylhydrazone (925). When other carbonyl compounds are present in the mixture, the hydrazones can often be separated by column (915), paper (926,927), or thin-layer (928) chromatography. Ultraviolet spectrometry has been used to identify acrolein absorbed in water from air samples (929).

Acrolein reacts with amines, phenols, and other compounds to give colored products. Colorimetric methods for its determination have been reported based on the following reagents: m-phenylenediamine (930), tryptophan (931–934), fuchsin (935–937), resorcinol (938,939), hexylresorcinol (934,939–941), phloroglucinol (942–944), chromotropic acid (945,946), J acid (947), anthrone (948), ferrocyanide complexes (934,949), benzidine, phenol, pyrogallol (950), and m-dinitrobenzene (951).

Peters (915) has discussed the analysis of acrolein and described the most important procedures in detail.

Many of the analytical methods used for acrolein are also applicable to methacrolein and methyl vinyl ketone. The former has been determined by gas-liquid chromatography (919,920,952), bromination (916), polarography (783), preparation of the 2,4-dinitrophenylhydrazone (783), and reaction with chromotropic acid (946); methyl vinyl ketone by gas-liquid chromatography (919,920,953–956), polarography (649,838,957–959), bromination (958,960), the hydroxylamine hydrochloride method (896, 958), and reaction with 2,4-dinitrophenylhydrazine (961,962), chromotropic acid (946), and m-phenylenediamine (930,963). Analysis for methacrolein in the presence of ketones has been accomplished by oxidation to methacrylic acid followed by titration with base (964), and methyl vinyl ketone has been assayed by catalytic hydrogenation (965). The 2,4-dinitrophenylhydrazones of phenyl vinyl ketone (966a) and divinyl ketone (672) have been prepared, and the infrared (966b) and ultraviolet (966b,967) spectra of the former determined.

The spectra of the monomers in this section are of great interest,

TABLE 15

Literature References to Spectra of Acrolein, Methacrolein, and Vinyl Ketones

Compound	Type of spectrum				
	Ultraviolet	Infrared	Raman	NMR	Mass
Acrolein	892,893,895,968–991	892,893,984b, 986,992–1000	892,1001–1003	892,1004–1009	892,1010, 1011
Methacrolein	783,976,980,984, 986	986,1012	1003	1008,1013, 1014	
Methyl vinyl ketone	887b,895,899, 900,961,984b, 986,989,1015, 1016	986,996,1012, 1016–1020	1003,1021, 1022	73,1023–1026	955
Phenyl vinyl ketone	910	910,1027		910b	
Divinyl ketone		672		73	

because they are the simplest compounds having the moiety C=C—C=O; but space limitation precludes a discussion of the subject here. The spectral properties of acrolein have been reviewed by Anderson and Hood (892). Literature references to the spectra of all five monomers are presented in Table 15. In addition to these studies, the ultrasonic absorption spectra of acrolein (1028), methacrolein and methyl vinyl ketone (1028c), the microwave absorption spectrum of methyl vinyl ketone (908), and the magneto-optical properties of acrolein (1029) have been investigated.

D. Storage and Handling

Acrolein (52) is flammable, highly reactive, toxic, and a powerful irritant, but, by exercising proper precautions, it can be handled and stored safely. The monomer is volatile, with a low flash point (Tag. open cup $< -29°C$, closed cup $-25°C$) and explosive limits in air of 2.8 to 31 volume percent. Its toxicological properties have been discussed in Sec. IV-C-2. 52 polymerizes readily on exposure to air, light, and certain chemicals (especially bases), with heat evolution. Commercial 52 generally contains 0.1 wt percent of hydroquinone as an inhibitor and is blanketed with an inert gas during shipment and storage. Under these conditions it will not polymerize appreciably in several months at ordinary temperatures. Uninhibited 52 cannot be legally shipped in interstate commerce and should never be stored.

Bases and oxidizing agents can initiate extremely rapid and exothermic polymerization. A pressure of 8,500 psi and a temperature of 400°F have been produced under confined conditions in the presence of sodium hydroxide, which is a particularly effective initiator. Equipment for handling **52** must be scrupulously clean, and great care must be taken to prevent contamination, because even traces of impurities can be dangerous in this regard.

Electrical equipment in acrolein service should be of class 1, group D explosion-proof construction. All fixed equipment should be permanently grounded, and ground wires provided to link unloading pipes with the tank car, drum, or other container being unloaded.

Persons handling **52** must be thoroughly trained in safe operating techniques and emergency procedures. Handling instructions and safety precautions should be posted conspicuously in areas where **52** is used or stored. Personnel should have rubber gloves, rubber aprons, and properly fitted safety goggles, not spectacles. Either gas masks or self-contained breathing apparati with full face pieces should be available in the event of emergencies. Conventional canister gas masks containing activated charcoal provide protection for up to 30 min at acrolein vapor levels of 2 percent or less. Because of the monomer's good warning properties (see Sec. IV-C-2), it is unlikely that workers will voluntarily or inadvertently overexpose themselves to harmful vapor concentrations. Those with a history of asthma, allergies, or other hypersensitivities, however, may experience a strong reaction at low concentrations that are tolerable and innocuous to the average person.

Because of the volatility of **52** and its irritating odor, good housekeeping is important. Spills should be "neutralized" at once with an excess of 10 to 20 percent aqueous sodium bisulfite solution. (*Caution:* this reaction is exothermic. Also, sodium bisulfite is corrosive to equipment and should never be poured on the skin.) After neutralization, the liquid is not hazardous and may be flushed to the sewer. Certain protective foams, for example, National Foam System's Airfoam 99, are effective in containing acrolein vapor from large spills. In an emergency, water sprays can also be used to suppress the vapors.

Liquid **52** on the skin should be rinsed away immediately with copious amounts of water, followed by thorough washing with soap and water. Contaminated clothing should be removed at once and laundered before reuse. The exposed person should be cautioned as to the possibility of a delayed respiratory irritation. Liquid **52** in the eye must be removed immediately by flushing with water for at least 15 min. The affected eye should then be examined at once by a physician.

In the laboratory, all work involving **52** requires a fume hood. Exploratory experiments should be done on a small scale with slow addition of **52** to the reaction mixture rather than the reverse. There should be provision for rapid cooling, since many reactions of **52** are exothermic. Particular care must be exercised when bases are present because of the possibility of violent polymerization. If such a reaction appears to be occurring, it can usually be stopped by adding an excess of aqueous acetic acid.

Before distilling **52**, the equipment must be cleaned thoroughly and purged with nitrogen or natural gas to remove oxygen. (Carbon dioxide should not be used for purging.) It is also good practice to place hydroquinone in the receiver. If inhibitor-free monomer is required, the distillate should be protected from light and used immediately. **52** containing small amounts of water may be dried over anhydrous calcium sulfate.

Steel containers are best for laboratory storage, but brown bottles with metal guard cases are also satisfactory. Storage in a well-ventilated area is advisable, and the stored monomer must contain inhibitor. A convenient method for disposing of **52** is to pour it slowly into dilute aqueous sodium bisulfite.

Drum storage away from other chemicals in cool, well-ventilated, fireproof sheds with automatic sprinkler systems is recommended. The drums should be placed on end with the bungs up and not be stacked. Application of inert-gas pressure to empty or sample drums minimizes introduction of air. A periodic check of inhibitor content is advisable when monomer is stored for an extended time.

Pressure vessels are better than atmospheric tanks for bulk storage, because air should be excluded and the irritating monomer vapor confined. Vent lines from the tanks are connected through a central collection system to either a water scrubber or flare stack. Fouling of lines, valves, and pressure-relief devices with polymer is a potential problem. For this reason, low or dead spots in lines are to be avoided, regular checks for polymer buildup made, and relief devices provided that are protected by inert-gas bleed streams just under them. Liquid-outlet lines should have check valves to prevent backflow into the storage vessel. The tank should be located outdoors in a diked area that drains to a waste-disposal unit.

Tank-car shipments of **52** are blanketed with natural gas (normal total pressure 40 psi maximum). Unloading is done through a dip tube extending from the dome to a sump in the bottom of the car by applying inert-gas pressure to the vapor space above the liquid **52**. During unloading, the car is electrically grounded, and only nonsparking tools and fittings are used.

In plant facilities, caustic solutions should not be piped through an acrolein area unless they are to be used in the process, and in that case, there must be no possibility of cross-connections with lines containing **52**. Carbon steel, stainless steel, copper, and copper alloys are suitable construction materials for handling **52**, but galvanized metals cause excessive polymerization. Glass equipment should be avoided, but, if used, adequate protection for personnel must be provided. All equipment should be washed with dilute acid, rinsed with water, and purged with inert gas before being charged with **52**. Untreated wastes must not be discharged to a sewer system or natural waterway unless it is known that the acrolein content is low enough to be harmless. In sufficient quantity, **52** upsets biological sewage-treatment processes, is highly toxic to fish, and could also cause a serious explosion. Concentrated solutions in organic solvents are best disposed of by burning. Aqueous solutions should be neutralized with sodium bisulfite before disposal (911,914).

The storage and handling techniques for **52** apply equally well to closely related methacrolein (**53**). Although somewhat less volatile, irritating, and toxic than **52**, this monomer is also highly flammable and very reactive. It has a low flash point (Tag. open cup $-14°C$) and must be kept away from oxygen (886). Hydroquinone is an effective polymerization inhibitor for **53** (1030,1031,1032a). Others that have been claimed in patents are nitric oxide (1033) and oximes (1034). One apparent difference between methacrolein and acrolein is that the condensing vapor of the former polymerizes in the presence of Fe^{2+} and Fe^{3+} ions. Therefore, carbon steel is not a satisfactory construction material for distillation equipment used in methacrolein purification. Stainless steel and Monel metal are reported to be suitable (1032).

The same general precautions and techniques in storage and handling also apply to methyl vinyl ketone. Again, hydroquinone is used as the polymerization inhibitor. Storage under refrigeration is recommended, but the inhibited monomer will not polymerize appreciably in about a month at room temperature (896). Other inhibitors for methyl vinyl ketone that have been suggested are aliphatic acids (836g,1035) and potassium iodide (1036).

E. Chemical Reactions

It would be redundant to describe the many reactions of acrolein (**52**) in detail here, since this has been done elsewhere for the literature through 1959 (549). The discussion will be confined to the reactions of **52** that, in our judgment, are most important and a more complete coverage of methacrolein (**53**) and vinyl ketone reactions, these having not been previously reviewed. If no reference is cited for **52**, it may be assumed that the information was obtained from reference 549. All the monomers contain the $C{=}C{-}C{=}O$ moiety, so that they may often be discussed together. This structure leads to a wide variety of reactions, since the two double bonds may react separately, in concert, or consecutively.

These monomers react readily with typical reagents for the carbonyl group, such as hydroxylamine, semicarbazide, and hydrazines. The oximes of **52** and **53** (864), their oxime methyl ethers (from methoxylamine) (1037a), dialkylhydrazones (1937b), semicarbazones, and nitrophenylhydrazones are formed in the conventional manner. **52, 53** (1038), **54** (633), and **55** (660,661,663,872,875,897), however, react with hydrazine and phenylhydrazine to give pyrazolines. These arise from cyclization of the initially-formed hydrazone through addition of the —NH— moiety to the olefinic double bond—Eq. (100):

$$CH_2{=}CRCOR' + R''NHNH_2 \longrightarrow CH_2{=}CRC(R'){=}NNHR'' \longrightarrow$$

RCH———CR'

$|$ $\|$

CH$_2$ N   (R = H or CH$_3$; R' = H, CH$_3$, or C$_6$H$_5$; R'' = H or C$_6$H$_5$) (100)

N

$|$

R''

Since they are useful for identification purposes, the solid derivatives of these types are listed in Table 16.

Sodium bisulfite gives a carbonyl adduct with **52** but also adds across the C=C double bond (see Sec. IV-C-3). The carbonyl monoadduct has been identified in acidic aqueous solution by spectrometric and other analytical methods, but not isolated. With phenyl vinyl ketone, addition to the olefinic bond is the only reaction (661,1050). Acrolein cyanohydrin has been prepared from **52**, KCN, and acetic acid in ether solution. An aldimine of **53** was obtained by an exchange reaction with a saturated ketimine (1051). **52** and **53** react with phosphorus ylids to yield the normal products—Eq. (101) (1052):

$$CH_2{=}CRCHO + Ar\bar{C}H\overset{+}{P}Ph_3 \longrightarrow CH_2{=}CRCH{=}CHAr + Ph_3PO \quad (101)$$
$$(R = H \text{ or } CH_3)$$

Acetalization of **52** and **53** in the usual manner with alcohol and acid— Eq. (102)—is invariably accompanied by some addition of the alcohol to the C=C double bond—Eq. (103) (1053,1054a):

$$CH_2{=}CRCHO + 2R'OH \xrightarrow{H^+} CH_2{=}CRCH(OR')_2 + H_2O \quad (102)$$

$$CH_2{=}CRCHO + 3R'OH \xrightarrow{H^+} R'OCH_2CHRCH(OR')_2 + H_2O \quad (103)$$
$$(R = H \text{ or } CH_3)$$

This is minimized if only a trace of acid catalyst, a long reaction time, and continuous removal of water by azeotropic distillation are used (1053). The desired acetal can be recovered from by-product alkoxyacetal by

TABLE 16

Solid Derivatives of Acrolein, Methacrolein, and Vinyl Ketones

Derivative	Monomer	mp, °C[a]	Reference
Semicarbazone	Acrolein	171	1039
	Methacrolein	198	807,1038
		195.5	1040
	Methyl vinyl ketone	194	842
		143	857
		141.5	847
		140	1041
	Phenyl vinyl ketone	227.5	1042
Oxime	Phenyl vinyl ketone	110	1043
Phenylhydrazone	Acrolein (P)[b]	52	1039
	Methacrolein (P)[b]	74	1038
	Phenyl vinyl ketone	86	661
	Phenyl vinyl ketone (P)[b]	158	661
		153	872,875
p-Nitrophenylhydrazone	Acrolein	160,151	1039
	Methacrolein	163	1038
		158	808
2,4-Dinitrophenylhydrazone	Acrolein	165	1039
	Methacrolein	238	805b
		207	807
		206.5	1038
		206	1040,1044
		201	1045
		198	810
	Methyl vinyl ketone	217	1046
		142	1047
		140.5	1048
	Phenyl vinyl ketone	251	1049
		196	966a
		160	1049
	Divinyl ketone	102	672

[a] The wide discrepancy among values for the same derivative may be due to isolation of different geometric isomers.

[b] The symbol P means that the pyrazoline was obtained instead of the phenylhydrazone.

pyrolysis (1054b). Another successful method is reaction of the aldehyde with an alkyl orthoformate (1055).

It is much easier to prepare cyclic acetals of these aldehydes from glycols (1053,1056–1058). The products from pentaerythritol and 2 moles of 52 or 53—Eq. (104) (1053,1059–1062):

$$2CH_2{=}CRCHO + C(CH_2OH)_4 \xrightarrow{H^+}$$
$$(R = H \text{ or } CH_3)$$

$$CH_2{=}CR-\underset{\underset{59}{}}{\overset{O-\qquad-O}{\underset{O-\qquad-O}{\bigotimes}}}-CR{=}CH_2 + 2H_2O \qquad (104)$$

which show promise as resin intermediates, are particularly noteworthy.

Carboxylic anhydrides and ketene add to the carbonyl group in **52** under acidic conditions to give 1,1-diacyloxy-2-propenes—Eq. (105)—and 3-vinylpropiolactone—Eq. (106)—respectively:

$$CH_2{=}CHCHO + (RCO)_2O \longrightarrow CH_2{=}CHCH(OCOR)_2 \qquad (105)$$

$$CH_2{=}CHCHO + CH_2{=}C{=}O \longrightarrow CH_2{=}CHCH\overset{\overset{CH_2}{\diagup}}{\underset{\underset{O}{\diagdown\,\diagup}}{\quad}}CO \qquad (106)$$

A trisphenol is obtained by reaction of **52** with phenol in the presence of an acid catalyst—Eq. (107):

$$CH_2{=}CHCHO + 3C_6H_5OH \xrightarrow{H^+} HOC_6H_4CH_2CH_2CH(C_6H_4OH)_2 + H_2O \qquad (107)$$

PCl_5 and **52** yield mainly the expected 3,3-dichloro-1-propene (allylidene dichloride) but also 1,3-dichloro-1-propene and 1,1,3-trichloropropane. The first by-product may result from allylic rearrangement of allylidene dichloride. Methyl (1063) and phenyl (663) vinyl ketone also give the rearranged product—Eq. (108):

$$RCOCH{=}CH_2 + PCl_5 \longrightarrow [RCCl_2CH{=}CH_2] \longrightarrow RCCl{=}CHCH_2Cl \quad (108)$$
$$(R = CH_3 \text{ or } C_6H_5)$$

Addition of organometallic reagents to the carbonyl group in **52** and in methyl vinyl ketone (1064–1068) has been used to prepare substituted allyl alcohols. Grignard reagents add 1,4 to the C=C—C=O system in methyl (1069) and phenyl (661) vinyl ketone. The adduct from the former may condense further with a second mole of **54**.

52 and **53** (1070–1088) are oxidized to the corresponding acids—Eq. (109)—by a number of reagents:

$$CH_2{=}CRCHO + O \longrightarrow CH_2{=}CRCOOH \qquad (109)$$
$$(R = H \text{ or } CH_3)$$

tert-Butyl hypochlorite and irradiation produce mixtures of the *tert*-butyl ester, acid chloride, and acid (1089). Peroxide oxidation of **53** in methanol yields methyl methacrylate (1077,1090). **52** and **53** (1091) are converted to nitriles by oxygen, ammonia, and an oxide catalyst (ammoxidation)—Eq. (110):

$$CH_2{=}CRCHO + \tfrac{1}{2}O_2 + NH_3 \xrightarrow{\text{catalyst}} CH_2{=}CRCN + 2H_2O \quad (110)$$
$$(R = H \text{ or } CH_3)$$

Epoxidation occurs with hydrogen peroxide at pH 8 to 8.5 (1092) or with sodium hypochlorite (1093)—Eq. (111):

$$CH_2{=}CRCHO + O \xrightarrow[\text{NaOCl}]{H_2O_2 \text{ or}} CH_2\underset{O}{\overset{}{\diagdown\diagup}}CRCHO \quad (111)$$
$$(R = H \text{ or } CH_3)$$

52 can also be oxidized to glyceraldehyde.

Methyl vinyl ketone is oxidized by mercury(II) sulfate to biacetyl (1094), by sodium hypochlorite to acrylic acid (1095), and undergoes ammoxidation to acrylonitrile (1096). It is epoxidized by *tert*-butyl hydroperoxide (1097) or H_2O_2 (1098). The kinetics of methyl vinyl ketone oxidation have been studied (902,1099). Treatment of divinyl ketone with H_2O_2 in alkaline aqueous dioxane gives a low yield of the diepoxide (884). Halohydrins are produced from α,β unsaturated aldehydes, and ketones by electrolytic oxidation in aqueous hydrogen halide solution (1100).

Catalytic hydrogenation of **52** generally attacks both the C=C and C=O double bonds, producing mainly propionaldehyde and propyl alcohol. The former is apparently formed in part by rearrangement of allyl alcohol, since this is a major product when catalysts on which the rearrangement is slow, for example, cadmium-copper, are used. The carbonyl group is selectively attacked in hydride ion reductions, such as the Meerwein-Pondorff-Verley method, vapor-phase hydrogen transfer with alcohols, and with metal hydrides. Electron transfer—Eqs. (112) and (113)—as in electrolytic and metal-acid reductions, seems to involve protonation as the first step:

$$CH_2{=}CHCHO + H^+ \longrightarrow [CH_2{=}CHCHOH]^+ \quad (112)$$

$$[CH_2{=}CHCHOH]^+ + e \longrightarrow CH_2{=}CH\overset{\cdot}{C}HOH \quad (113)$$

Coupling of the postulated free radical with itself or with a hydrogen atom accounts for the main products, divinylglycol and allyl alcohol. Cobalt hydrocarbonyl (1101) and catalytic hydrogenation on ruthenium-rhodium (1102) selectively reduce the olefinic double bond.

Reduction of the C=C or C=O group is inhibited by an alkyl substituent. Thus, with Cd–Cu catalysts, hydrogenation of **53** gives a high yield of methallyl alcohol, and **54** is converted mainly to methyl ethyl ketone (1103). Methallyl alcohol is also the principal product of meth-acrolein reduction by alcohols (1062) and cadmium–sulfuric acid (1104). Disproportionation of **52** and **53** to the corresponding alcohol and acid occurs over a silver catalyst (1105). Selective catalytic hydrogenation (1106) and hydroboration (1107) of **52**, **53**, and **54** have been described.

Reduction of **54** to methyl vinyl carbinol is accomplished by hydrogen exchange with propanol (1108), catalytic hydrogenation over oxides of cadmium and copper (1109), diphenyl- or dibutyltin dihydride (1110), and triphenyltin hydride (1111). Cobalt hydrocarbonyl (1101) and hydro-genation over nickel (1112) convert the monomer to methyl ethyl ketone. Electrolytic reduction and reactive metals with water, acids, or alcohols cause reductive coupling to the 2,7-diketone or pinacol—Eq. (114) (1113–1117):

$$CH_3COCH{=}CH_2 \xrightarrow{H} CH_3CO(CH_2)_4COCH_3 + CH_2{=}CHC\underset{\underset{OH}{|}}{\overset{\overset{CH_3}{|}}{C}}\underset{\underset{OH}{|}}{\overset{\overset{CH_3}{|}}{C}}CH{=}CH_2$$

(114)

Electrolytic cross-coupling with diethyl fumarate has also been reported (1118). Divinyl ketone undergoes catalytic hydrogenation to diethyl ketone readily (671,880). Polarographic reduction of methyl (1119) and phenyl (1120) vinyl ketone has been discussed.

The monomers in this section act as dienophiles, and all but divinyl ketone have been reported to act also as dienes in the Diels-Alder reaction. Since these compounds are able to perform both roles, dimers are formed as shown in Eq. (115):

(115)

(R = H, CH₃ or C₆H₅; R′ = H or CH₃)

There is the opportunity for another isomeric product here in which the acyl substituent is in the 3 rather than the 2 position, and this has been isolated in the case of acrolein (R = R′ = H). By the same token, isomeric products have also been observed in the addition of the monomers, as dienophiles, to substituted dienes. The Diels-Alder reactions of meth-acrolein and the vinyl ketones, including dimerization, are summarized in Table 17. Detailed information has been published on the many analogous reactions of acrolein (1121,1122).

Substituted cyclobutanes are obtained by the coupling of **52**, **53**, and **54** with tetrafluoroethylene (1155,1156) and of **52** and **53** with allene (1157).

TABLE 17
Diels-Alder Reactions of Methacrolein and Vinyl Ketones

Dienophile	Diene	Reference
Methacrolein	Acrolein	1123
	Butadiene	1062,1124
	Cyclopentadiene	1125
	Isoprene	1062
	2-Methyl-1,3-pentadiene	1126
	4-Methyl-1,3-pentadiene	1125
	Myrcene	1127
	Methacrolein	1030,1062,1123, 1128–1130
n-Butyl vinyl ether	Methacrolein	1131a,1132
Methyl vinyl ether	Methacrolein	1131a,1132
Methyl methacrylate	Methacrolein	1062,1131b,1132
α-Methylstyrene	Methacrolein	1131c,1132
Methyl vinyl ketone	Acrolein	1133
	1,1'-Bicyclohexenyl	1134
	Butadiene	1135–1139
	Chloroprene	1139
	2-Cyanobutadiene	1140
	1,3-Cyclohexadiene	1135
	2,3-Dimethylbutadiene	1135,1139
	1,3-Diphenylisobenzofuran	1141,1142
	α- and β-Eleostearic acid	1143a
	2-Ethoxybutadiene	1144
	Fluoroprene	1145
	Hexachlorocyclopentadiene	1146
	Isoprene	1135,1138,1139, 1147
	2-Methoxybutadiene	1144
	Methyl α-eleostearate	1143b
	1,3-Pentadiene	1139
	1-Phenylbutadiene	1148,1149
	Tung oil	1143c
	Methyl vinyl ketone	1133,1150a,1151
Butyl vinyl ethers	Methyl vinyl ketone	1131a,1132,1151
Eneamines	Methyl vinyl ketone	1152
Ethyl vinyl ether	Methyl vinyl ketone	1151
Methacrylate esters	Methyl vinyl ketone	1151
α-Methylstyrene	Methyl vinyl ketone	1131c,1132
Styrene	Methyl vinyl ketone	1151
Phenyl vinyl ketone	1,3-Cyclohexadiene	1153
	Cyclopentadiene	1153,1154
	1,1'-Dicyclohexenyl	1153
	1,1'-Dicyclopentenyl	1153
	2,3-Dimethylbutadiene	1154
	1,4-Diphenylbutadiene	1154
	2,3-Diphenylbutadiene	1154
	Ethyl sorbate	1154
	Isoprene	1153
	Methyleneanthrone	1154
	1-Phenyl-1,3-pentadiene	1154
	Substituted cyclopentadienones	1154
	Phenyl vinyl ketone	1150b
Divinyl ketone	Cyclopentadiene	675

Halogens add to the C=C double bond in **52**, **53** (1055,1158), and phenyl vinyl ketone (661,862) in the liquid phase. Steam distillation of the products from **52** gives α-haloacroleins in good yield, and analogous derivatives are obtained from **54** by treatment with copper(II) chloride or bromide (1159). The chlorohydrin, 2-chloro-3-hydroxypropionaldehyde, is produced by chlorination of **52** in aqueous solution or by its reaction with pure hypochlorous acid—Eq. (116):

$$CH_2{=}CHCHO + HOCl \longrightarrow HOCH_2CHClCHO \qquad (116)$$

Depending on conditions, vapor-phase chlorination of **52** can give 2- and 3-chloroacrolein (1160) or acrylyl chloride.

Addition of Grignard reagents to the olefinic double bond was mentioned earlier. Other such additions that do not involve an "active hydrogen" reagent are the reactions of **54** with diazonium salts (1161) and with a substituted diazomethane (1162) and of nitrosyl chloride with vinyl ketones (1163).

A vast amount of research has been done on additions of "active hydrogen" compounds to acrolein, methacrolein, and vinyl ketones. Whether or not the reaction is a Michael addition in the presence of base, these reagents invariably add in the sense of Eq. (117):

$$CH_2{=}CRCOR' + HA \longrightarrow ACH_2CHRCOR' \qquad (117)$$

Under alkaline (Michael) conditions, the situation is complicated by the possibility of a concurrent or consecutive aldol type of reaction of the carbonyl group as well—Eq. (118):

$$CH_2{=}CRCOR' + HA \longrightarrow CH_2{=}CR\overset{\displaystyle R'}{\underset{\displaystyle OH}{\overset{|}{\underset{|}{C}}}}A \qquad (118)$$

The attacking species (HA) may be the original reagent or the product of its addition to the double bond (ACH₂CHRCOR'). If HA is an ester, Claisen condensation with elimination of alcohol may ensue. Alcohols, phenols, ammonia, amines, and amides react with the carbonyl group as well as with the olefinic double bond. The resulting complexity of reaction routes and the large number of such reactions, which are described in the literature, preclude a thorough discussion here. A complete listing of reactant types and pertinent references, however, is given in Table 18.

There are several reactions of a miscellaneous nature that should be mentioned. **52** and **54** (657a,b,658,659,1233–1235) condense with formaldehyde to give resinous compositions. **52** and **54** form complexes with molybdenum dicarbonyl (1236), and **52** does so with copper(I) chloride (1237). The Willgerodt-Kindler reaction of phenyl vinyl ketone has been studied (1238).

TABLE 18
Addition Reactions of Active Hydrogen Compounds with Acrolein, Methacrolein, and Vinyl Ketones

Reactant[a]	Monomer	Reference
⟩CHCO—	Acrolein	1164
	Methacrolein	1165–1167
	Methyl vinyl ketone	815,1168–1178
	Phenyl vinyl ketone	869b–d,1179,1180
	Divinyl ketone	881,885
⟩CHCH=CHCHO	Acrolein	1181
	Methacrolein	1181
⟩CHCN	Acrolein	1164
	Methyl vinyl ketone	1182
2- and 4-picoline	Methyl vinyl ketone	1176b
⟩CHNO₂	Acrolein	1164
	Methacrolein	1183,1184
	Methyl vinyl ketone	870,1169a,1185–1187
	Phenyl vinyl ketone	869a,d,870,1188
ROH and ArOH	Acrolein	1053
	Methacrolein	576,1031,1054a,1189–1195
	Methyl vinyl ketone	650,1048,1196a,1197,1198a,1199, 1200
	Phenyl vinyl ketone	661
	Divinyl ketone	671
H₂O	Acrolein	1201
	Methacrolein	1202
	Methyl vinyl ketone	1196b
RCOOH and ArCOOH	Acrolein	1203
	Methyl vinyl ketone	845,1204,1205
⟩NH	Acrolein	1039
	Methacrolein	1206,1207
	Methyl vinyl ketone	636,1176b,1198a–c,1200,1208– 1211
	Phenyl vinyl ketone	664,1212–1214
—NHCO—	Acrolein	1039
	Methacrolein	1215
	Methyl vinyl ketone	1216,1217
—NHNO₂	Methyl vinyl ketone	1218
HN₃	Acrolein	1309
ArH	Acrolein	1164,1219
	Methacrolein	1219
	Methyl vinyl ketone	898,1220–1224
	Phenyl vinyl ketone	1221

continued

TABLE 18 (*continued*)

Reactant[a]	Monomer	Reference
RSH	Acrolein	1053
	Methacrolein	1225
	Methyl vinyl ketone	1226–1228
H_2S	Acrolein	1229
	Methyl vinyl ketone	1176b,1198d
RCOSH	Acrolein	1053
	Methacrolein	1053
	Methyl vinyl ketone	845
$(RO)_2PS$—SH	Methyl vinyl ketone	1230
$NaHSO_3$	Acrolein	1229
	Phenyl vinyl ketone	661,1050
HX	Acrolein	1229
	Methyl vinyl ketone	1197
	Phenyl vinyl ketone	661,863,872
HCN	Acrolein	1229
	Methyl vinyl ketone	1176b,1231
$CH_2=C=O$	Methyl vinyl ketone	1232

[a] R = alkyl, Ar = aryl, X = halogen.

F. Polymerization

The polyfunctionality and reactivity of **52** are important factors in its polymerization behavior. The monomer polymerizes readily by free-radical, cationic, or anionic initiation, but the products are usually cross-linked. Schulz, Kern, and coworkers showed that the free-radical polymers have primarily a fused polytetrahydropyran structure (**60**):

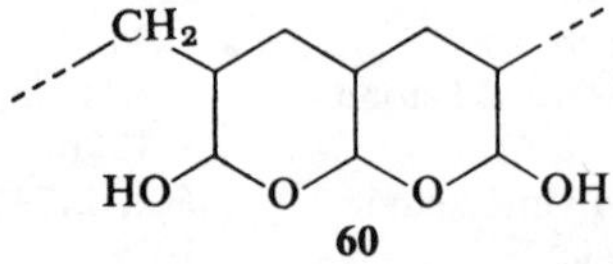

60

Sequences of two to four monomer units are joined in this manner and in between are 10 to 20 mole percent of free aldehydic acrolein units. There are also about 4 mole percent of C=C double bonds and 3 to 5 mole percent of keto groups from side reactions. Cross-links appear to be interchain acetal or hemiacetal moieties, since they are easily split by treatment with such reagents as sodium bisulfite and thiophenol to give soluble polymers (552,1239). Hot pyridine–water (65:35) (552) and dimethyl sulfoxide (1240) are also solvents for polyacrolein.

According to Schulz (552,1239), the main propagation process is head-to-tail vinyl polymerization, followed by cyclization to the pyran structure **60**—Eq. (119):

$$CH_2{=}CHCHO \longrightarrow \begin{array}{ccc} CH_2 & CH_2 & CH_2 \\ CH & CH & CH \\ | & | & | \\ CHO & CHO & CHO \end{array} \longrightarrow 60 \qquad (119)$$

The cyclization and formation of acetal cross-links are presumably initiated by traces of acid or base. Another proposal is that the product is actually a copolymer of **52** and its Diels-Alder dimer—Eq. (120) (1241):

$$3n CH_2{=}CHCHO \longrightarrow \left(\begin{array}{c} -CH_2CH- \\ | \\ CHO \end{array} \right)_n \qquad (120)$$

but the very high density of the polymer (1.32 to 1.37) favors the compact, fused tetrahydropyran structure. Diels-Alder polymerization—Eq. (121):

$$2CH_2{=}CHCHO \longrightarrow \underset{O\quad CHO}{\bigcirc} \xrightarrow{\; nCH_2{=}CHCHO \;} \qquad (121)$$

would also give this structure, and high polymers resembling free-radical polyacrolein have been obtained as by-products of acrolein dimerization (1121).

Anionic initiators, such as sodium naphthalene, are reported to cause polymerization mainly through the C=O bond with only a minor amount of vinyl addition—Eq. (122) (552):

$$(m + n)CH_2{=}CHCHO \longrightarrow \underset{\substack{| \\ CH{=}CH_2 \\ \text{Major structure}}}{(-CH-O-)_m} + \underset{\substack{| \\ CHO}}{(-CH_2-CH-)_n} \qquad (122)$$

Besides these mechanisms, 1,4-polymerization—Eq. (123):

$$nCH_2{=}CHCHO \longrightarrow (-CH_2CH{=}CH-O-)_n \qquad (123)$$

is said to occur with organometallic reagents (1242,1243) and with γ rays (1244). Using sodium as initiator, 1,4 addition is the major polymerization route (1245). Cationic initiation has been investigated less thoroughly. The aldehyde content of the polymers is low (1246), and polymerization through the C=O bond has been observed here also (1247). In general, ionic polymers of **52** are low in molecular weight or cross-linked, the carbonyl content is far below the theoretical value, and cross-links, when present, involve C—C rather than C—O bonds (1239).

Free-radical initiation is the most important means of polymerizing **52**. Although it has been done in bulk and in organic solvents, the preferred method is polymerization in water with water-soluble initiators or redox systems (552,1248–1250). An interesting modification is to use the NaHSO$_3$ (or H$_2$SO$_3$) adduct of polyacrolein as the emulsifier or the reducing component of the redox initiator (552,1239,1251–1256). Higher molecular weights are obtained in this manner. Acids are sometimes added (1248c,1257–1260) to promote hydrolysis of acetal cross-links. The heat of free-radical polymerization of **52** in bulk is 19.1 ± 4 kcal per mole (96).

High-molecular-weight **52** homopolymers are not soluble in ordinary solvents and are difficult to fabricate. They sinter without melting at 220° (552), and it is significant that compression molding is the only reported method of making test pieces (552,1261,1262). This intractability has deterred their commercial development, and, understandably in light of acrolein's otherwise attractive potential as a monomer, there have been many attempts to circumvent the problem. One approach is to "tame" **52** by modifying its reactive carbonyl group before polymerization. A great deal of work has been done on cyclic acetals, particularly the divinyl compound **59** (R = H) from **52** and pentaerythritol (1053). Other monomeric derivatives that have been studied are the diacetate (1203, 1263–1265), the diethyl acetal (1053,1263), the oxime (1264), and the sodium bisulfite adduct (prepared *in situ*) (1266).

A second technique is to modify polyacrolein chemically, so that it becomes soluble or moldable, especially by means of carbonyl reagents (552,1248a,1267–1269). Of these, sodium bisulfite (552,1248b,1254,1270, 1271) is the most important, yielding water-soluble products that are claimed to be useful in paper treatment. A study of the equilibrium between polyacrolein and its bisulfite adduct in solution has been published (1272). Alkaline formalin also gives water-soluble polymers (1273–1275). Polycarboxylic acid derivatives have been obtained by treatment with base (Cannizzaro reaction) (552) or by oxidation (552,1275).

A third approach is to take advantage of acrolein's polyfunctionality in the preparation of cross-linked polymers. Reaction with ammonia, amines, formaldehyde, phenols, and urea, or combinations of these, give resins of unknown composition but with interesting properties. In the same category is the use of acrolein, usually as a cross-linking agent, to modify existing polymers (1239).

Finally, copolymerization offers a means of counteracting intractability. At low acrolein contents, the possibility of interaction between neighboring units is decreased, so that a larger fraction of the aldehyde groups survives in the copolymer than in the homopolymer and there is less cross-linking (1276).

As a polyaldehyde, polyacrolein is sensitive to oxidation and molecular-weight degradation may result. Amines are claimed to be effective stabilizers (1277).

53 has been studied less extensively but appears to be similar to 52 in polymerization characteristics. Free radicals (1248a,1249,1268,1278–1286), anions (1279,1287,1288), cations (1279,1289,1290), sodium (1245,1291), ionizing radiation (1252,1292), trialkylphosphines (1293), and nitrites (1294) act as initiators. The free-radical polymers have mainly a fused polypyran structure like that of polyacrolein (1295), and the high-molecular-weight products are insoluble. The preferred method for their preparation is again polymerization in water (1249,1252,1268,1281,1283, 1285), but alcohol is sometimes added to bring the monomer into solution (1280,1284). The heat of polymerization of 53 (bulk, free-radical) is 15.6 ± 0.3 kcal per mole (96).

Structure determinations on the products indicate that both cationic (1290) and anionic (1287,1288) polymerization of 53 proceed through the C=C and C=O double bonds with some 1,4 addition. The fused poly-pyran structure and, at higher reaction temperatures, lactonization by the Tischenko reaction were also observed in the anionic polymers. With sodium as initiator, 1,4 addition is the principal polymerization route (1245).

Polymerizable carbonyl derivatives of 53 are cyclic acetals (1053,1056, 1059–1062), the diacetate (1296), and the bisulfite adduct (1266). 53 has been graft copolymerized with vinyl polymers (1297) and cellulose (1298). Resinous compositions are formed by reaction with triazines (1299) and glycols, epoxides, and vinyl compounds or their mixtures (1300). Poly-methacroleins have been chemically modified by the Cannizzaro reaction (1301,1302), oxidation (1275,1279,1283), reduction (1303,1304), acylation (1305), acetalization (1284,1305), treatment with nitrogen bases (1268), bisulfite addition (1270), and conversion to the polyoxime (1279).

54 is a highly reactive monomer and polymerizes readily by free-radical (811,868,1278,1306–1320), anionic (1321-1330), and cationic (83,1331) mechanisms. The heat of polymerization is 17.7 ± 0.1 kcal per mole (1332). Polymers of 54 are also reactive and relatively unstable. This has limited their commercial development but stimulated interest, too. Much of the research on poly(methyl vinyl ketone) has had to do with its reactions.

Methyl vinyl ketone was among the monomers used by Marvel and coworkers to demonstrate the prevalence of the head-to-tail propagation mechanism in free-radical vinyl polymerization. Benzoyl peroxide initiation gave a polymer that lost water on pyrolysis at 270°. Cracking of the residue at 360° gave 3-methyl-2-cyclohexenone (62) and other cyclohexanone derivatives. These results indicated a head-to-tail orientation (61) —Eq. (124):

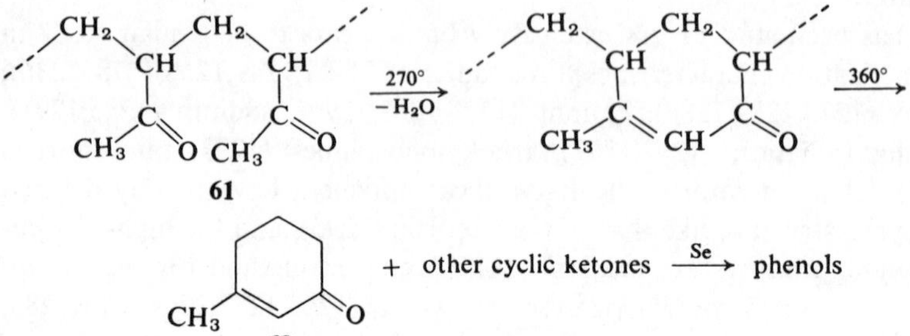

$$\xrightarrow{270°}_{-H_2O}$$

$$\xrightarrow{360°}$$

+ other cyclic ketones $\xrightarrow{Se}$ phenols   (124)

62

Furthermore, the polymer was not dehydrogenated by SeO_2 nor dehydrated by $ZnCl_2$ or P_2O_5 under mild conditions—strong evidence against the alternative head-to-head, tail-to-tail structure 63:

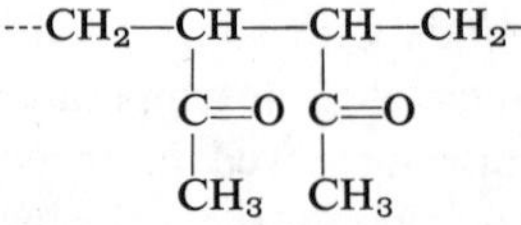

$$\cdots CH_2—CH—CH—CH_2—\cdots$$
$$\begin{array}{cc} C{=}O & C{=}O \\ | & | \\ CH_3 & CH_3 \end{array}$$
63

which would be expected to react readily (1308a). A polypyridine (64) was formed by long heating of the polymer's oxime with HCl in aqueous ethanol—Eq. (125)—confirming the head-to-tail structure (1308b):

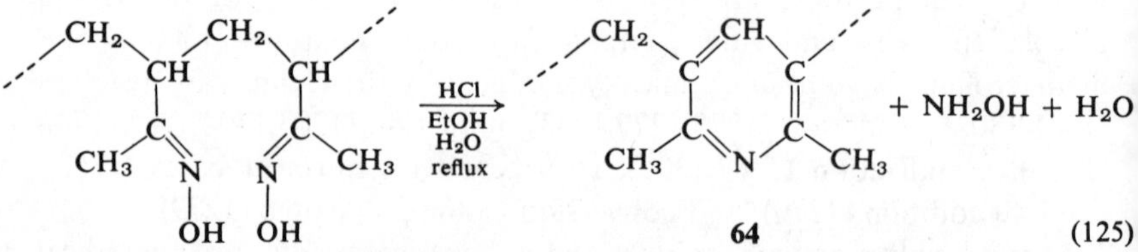

$$\xrightarrow[\substack{EtOH \\ H_2O \\ reflux}]{HCl}$$

+ NH_2OH + H_2O

64   (125)

The free-radical polymer develops color on heating. This has been explained by a continuation of the intramolecular dehydration process in Eq. (124) giving long runs of conjugated C=C double bonds (1319,1333). The process is accelerated by heating with $ZnCl_2$ or $BaCl_2$ (1320). Anionic polymers appear to have a head-to-tail structure (61) also and are still more sensitive to thermal color development, undoubtedly because the

basic conditions promote aldol condensation followed by dehydration—
Eq. (126) (1325,1329):

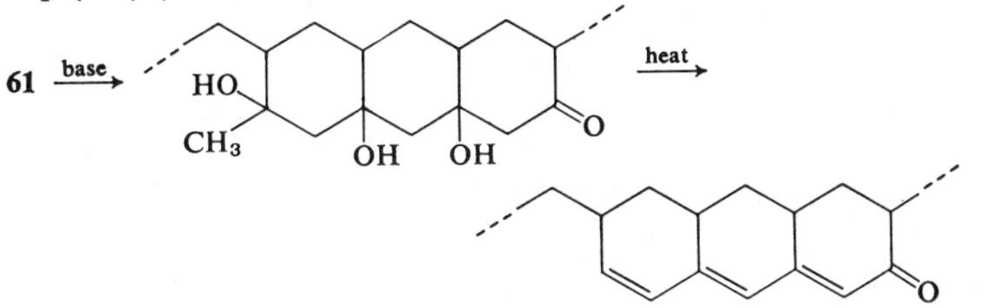

$$61 \xrightarrow{\text{base}}$$

(126)

Evidence of chain segments arising from repeated intermonomer Michael
additions—Eq. (127)—has been found in anionic polymers (1324):

$$n\text{CH}_2{=}\text{CHCOCH}_3 \xrightarrow{\text{base}} (-\text{CH}_2\text{CH}_2\text{COCH}_2{-})_n \qquad (127)$$

Crystalline, stereoregular homopolymers are obtained with organometallic
initiators (1322,1323,1325–1328,1330).

Many reactions of methyl vinyl ketone polymers and copolymers have
been studied. Treatment with hydroxylamine hydrochloride yields the
polyoxime (868,1304,1308,1334,1335a), which can be oxidized to poly-
acrylic acid (1308b) or reduced to the amine (1304,1335a). Reaction with
ammonia produces the polypyridine **64** (1336). Reduction of the carbonyl
group to the carbinol occurs with LiAlH$_4$ (1304,1318,1334,1335b,1337) or
alkali metal borohydrides (1304,1318). The addition of PCl$_3$ followed by hydrolysis produces a poly(hydroxyphosphonic acid)—Eq. (128) (1338, 1339):

$$(-\text{CH}_2\text{CH}{-})_n \xrightarrow{\text{PCl}_3} \xrightarrow{\text{hydrolysis}} (-\text{CH}_2\text{CH}{-})_n \qquad (128)$$

with COCH$_3$ group on the left and HOCPO(OH)$_2$ / CH$_3$ on the right.

The homopolymer undergoes cyanoethylation with acrylonitrile (1340).
Cationic polymerization of **54** in the presence of acetic anhydride gives a
poly(vinyl acetonyl ketone) that forms chelates with Cu(II) and UO$_2$(II)
(1331). Sodium methoxide effects copolymerization of **54** and ethyl
acrylate, which is followed by a Dieckmann condensation—Eq. (129)
(1341):

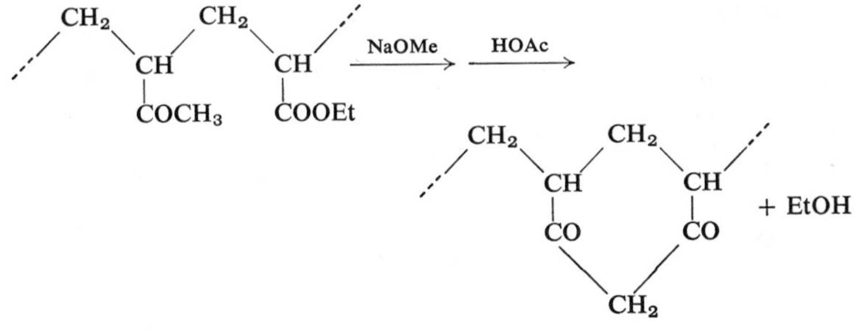

+ EtOH   (129)

TABLE 19
Literature References to Copolymerizations of Acrolein, Methacrolein, and Vinyl Ketones

Monomer	References
Acrolein	551,552,893,1239,1276,1350–1362
Methacrolein	551,632,1338,1361,1363–1373
Methyl vinyl ketone	632,896,1304,1334,1335,1337,1338,1341, 1345b,1365,1374–1408
Phenyl vinyl ketone	879,1408
Divinyl ketone	674,1409–1411

TABLE 20
Relative Reactivity Ratios for Free-radical Copolymerization of Acrolein (52), Methacrolein (53), and Methyl Vinyl Ketone (54) (M_1)

M_1	Comonomer (M_2)	r_1	r_2	Reference
52	Acrylamide[a]	2.0 ± 0.05	0.76 ± 0.02	1412
	Acrylamide[b]	1.7 ± 0.1	0.21 ± 0.02	1352
	Acrylamide	1.65 ± 0.1	0.19 ± 0.02	1413
	Acrylonitrile[a]	1.09 ± 0.05	0.8 ± 0.1	1412
	Acrylonitrile[b]	1.60 ± 0.04	0.52 ± 0.02	1352
	Methacrylonitrile[c]	0.72 ± 0.06	1.20 ± 0.08	1352
	Methyl acrylate[a]	~0	7.7 ± 0.2	1352
	Vinyl acetate[a]	3.3 ± 0.1	0.1 ± 0.05	1412
	2-Vinylpyridine[b]	~4	~0	1352
53	Acrylonitrile[a]	1.7 ± 0.3	0.15 ± 0.04	1369a
		2.0	0.06	1365
	Methacrylonitrile[b]	1.78 ± 0.06	0.40 ± 0.04	1352
54	Acrylonitrile	1.8 ± 0.2	0.61 ± 0.04	1383
	Butyl acrylate	1.6 ± 0.1	0.65 ± 0.07	1345b
	4,6-Diamino-2-vinyl-s-triazine	0.26 ± 0.04	1.2 ± 0.15	1395
	2,5-Dichlorostyrene	0.5	2.0	1414
	Phenyl vinyl ether	4.4	0.01	1407
	Styrene	0.35 ± 0.02	0.29 ± 0.04	1383
	Vinyl acetate	7.0	0.05	1365
	Vinyl chloride	8.3	0.10	1414
	Vinylidene chloride	1.8	0.55	1414

[a] In water.
[b] In dimethylformamide.
[c] In dioxane.

Reactions of the homopolymer with aldehydes (656,1342), nitrogen bases (1317,1343), alcohols (1317), mercaptans (1317), and ammonia (followed by hydrogenation) (1344) yield resins of ill-defined composition. Vulcanization occurs with strong bases or zinc oxide and sulfur (1345). Heating with metal halides, for example, a mixture of $AlBr_3$, $ZnCl_2$, and $CuCl_2$, produces a carbonaceous solid said to be useful as a semiconductor or heat-stable coating (1346). Other vinyl monomers have been graft-copolymerized with methyl vinyl ketone polymers (1347).

Phenyl vinyl ketone has been polymerized by free-radical (660,663,868, 879,1307) and anionic (1328,1330) initiation. Reaction of the homopolymer with hydroxylamine hydrochloride gives the polyoxime (868).

Divinyl ketone polymerizes very readily (671,672,880). Attempts to prepare the oxime and dimethyl ketal failed, because the monomer polymerized rapidly under the reaction conditions (672). An attempted synthesis of divinyl ketone from vinyllithium and carbon monoxide produced mostly polymer (1348). Divinyl ketone can be used as a cross-linking agent for linear polymers (1349).

Literature references to copolymerizations of the monomers in this section are listed in Table 19, copolymerization parameters in Tables 20 and 21. The relative rates of methyl radical addition to a number of vinyl monomers, including **54**, have been determined (1415,1416). Polymers of **52**, **53**, and **54** are discussed in a review article on reactive resins (1417).

TABLE 21

Q and e Values for Acrolein (**52**), Methacrolein (**53**), and Methyl Vinyl Ketone (**54**)

Monomer	Q	e	Reference
52	1.08	0.89	137
	0.85	0.73	146
	0.69	0.67	1352
	0.64	0.56	1352
	0.55	0.88	1412
	-2.86 kcal/mole (q)[a]	0.20×10^{-10}esu (ϵ)[a]	1352
	-2.8 kcal/mole (q)[a]	0.17×10^{-10}esu (ϵ)[a]	1352
53	1.75	-0.01	146
	1.70	-0.26	137
	1.59	0.36	1352
	1.3	0.2	1369a
	-3.6 kcal/mole (q)[a]	-0.10×10^{-10}esu (ϵ)[a]	147
	-3.4 kcal/mole (q)[a]	0.11×10^{-10}esu (ϵ)[a]	1352
54	1.0	0.7	321
	0.69	0.68	137,146
	-2.8 kcal/mole (q)[a]	0.19×10^{-10}esu (ϵ)[a]	147

[a] Schwan-Price resonance q and electrical ϵ factors (147).

G. Applications of Polymers

There is no evidence that polymers of **52**, **53**, or the vinyl ketones are being used commercially. The continuing high level of research activity, however, suggests that this may no longer be true 10 years from now. Some of the polymer applications that have been proposed in the literature are discussed below.

Acrolein polymers do not have outstanding physical properties and tend to be unstable. Either the reactive aldehyde group must be protected by chemical modification, or its high reactivity made use of in the application. Four approaches show promise: reaction of the carbonyl group with a polyol to form an acetal before or during polymerization, conversion of polyacrolein to the bisulfite adduct, condensation of acrolein monomer with amines, phenols, aldehydes, and related reagents, and use of acrolein polymers as cross-linking agents.

The reaction of **52** and pentaerythritol can be controlled so that either a soluble A-stage precondensate or the cyclic acetal **59** (R = H) is the product. The A-stage resin cross-links on heating, and the cyclic acetal does so on further reaction with a polyol. Systems of this type have been marketed in Europe under the trade name Ultralon, and similar products have been developed in the United States. The mechanical strength and adhesive properties of the cured resins are good. They have the disadvantages, however, of low softening points and the need for acidic catalysts and high curing temperatures (1239).

The bisulfite adduct of polyacrolein is water-soluble. It is said to be useful as a thickening agent, protective colloid, emulsifier, and in treatment of paper and textiles (551,552,1248,1254,1270).

Fabric-finishing agents from **52**, urea, and formaldehyde with the trade name Acrisin have been produced in Europe. Similar resins are claimed to be useful as adhesives, wood preservatives, and molding compositions (1239).

Acrolein polymers are capable of cross-linking a variety of polymers and of forming adducts with cellulose and polyamides. Therefore, they are of interest in the leather, textile, paper, coatings, and other industries that use reactive resins (552,1276). The cyanohydrin of polyacrolein releases HCN slowly and may be effective as a pesticide (1418). Other applications suggested for acrolein polymers are heavy metal complexing and absorption or exchange of materials that react with the aldehyde group (552).

Some of the same applications have been proposed for methacrolein polymers also. These are use of the bisulfite adduct in paper treatment (1248,1270), complex formation with heavy metals (1285,1419), and reaction of methacrolein with polyols to give tough, cross-linked resins

(1056,1059–1062). Vulcanized products are obtained by heating methacrolein polymers with hexamethylenetetramine (1420). Butadiene-methacrolein copolymers are useful as tire-cord dips (1370), rubbers (1361), and coatings with air-drying properties (1372). Polymeric hydroxyacids obtained from methacrolein polymers by the Cannizzaro reaction can be used as surface-active agents (1301), core binders (1302), components of oil-well cements (1371), and in textile treatment (1421). The reaction of polymethacrolein with ammonia followed by acidification yields a cationic emulsifier (1268). The homopolymer (1364), polyols obtained by its reduction (1303), and products from reaction with HCN (1366), amines (1368), and triazines (1299) are claimed to be useful in coatings.

The patent literature to 1953 on applications of methyl vinyl ketone polymers has been reviewed (896). The most frequently mentioned uses for both the unmodified polymer and products obtained by reactions of the carbonyl group are adhesives (896,1316,1335a), coatings (896,1342), and dye assistants (896,1393,1396,1397,1404,1422–1424). Ion-exchange resins have been made by the introduction of amine (896,1399,1425) or phosphonic acid (1339,1400,1425) groups. The homopolymer is an effective ion-excluding barrier, useful in the ultrafiltration of salt solutions (1426). The reaction of the polymer with alcohols, amines, or mercaptans yields pour-point depressants (1317). Copolymerization of **54** with styrene and reduction of the product to the alcohol imparts flexibility (1335b). The copolymer with butadiene is effective in bonding rubber to textiles (1427).

The use of phenyl vinyl ketone polymers in lacquers, adhesives, and molding compositions has been suggested (879). Methyl methacrylate–divinyl ketone copolymers are suitable for optical applications (1409). Divinyl ketone has also been used as the cross-linking agent in making ion-exchange resins (674,1410,1411).

V. VINYL SULFUR COMPOUNDS

$(CH_2=CH)_2S$ $(CH_2=CH)_2SO$ $(CH_2=CH)_2SO_2$
65   **66**   **67**

$CH_2=CHSR$ $CH_2=CHSOR$ $CH_2=CHSO_2R$
68   **69**   **70**

$CH_2=CHSO_2OH$ (R = alkyl or aryl)
71

A. Introduction

In 1887, Semmler (1428) claimed the isolation of vinyl sulfide (divinyl sulfide) (**65**) from the broad-leafed garlic plant, ramson (*Allium ursinum*), and observed its reactions, including polymerization. The boiling point of

Semmler's compound, however, is 15° higher than that of synthetic **65**, and the latter fails to undergo the reactions with silver oxide that Semmler described (1429). Kohler (1430) synthesized and studied the reactions of ethenesulfonic acid (vinylsulfonic acid) (**71**) from 1897 to 1899, and Strömholm (1431) prepared ethyl vinyl sulfide (**68**, R = C_2H_5) in 1900. There was no further activity in the field until the use of bis(2-chloroethyl) sulfide (mustard gas) as a chemical-warfare agent in the First World War stimulated research on this compound. Treatment with base was found (1432) to give **65**, but the reaction proved to be complex, and in only one instance (1432c) was the pure product isolated. At about the same time, phenyl vinyl sulfide (**68**, R = C_6H_5) (1433), butyl vinyl sulfide (**68**, R = C_4H_9) (1434), and ethenesulfonamide (1435) were synthesized. Vinyl sulfoxide (divinyl sulfoxide) (**66**) (1436) and vinyl sulfone (divinyl sulfone) (**67**) (1436,1437) were first described in 1930.

The earliest industrial interest in vinyl sulfur monomers was shown by I. G. Farben., with patents appearing in 1935 on the manufacture of alkyl and aryl vinyl sulfides (1438), their oxidation to the corresponding sulfoxides and sulfones (1439), and reactions of the oxidation products with amines (1440). The following year, patents were obtained on the polymerization of vinyl sulfides (1441).

As the polymer industry expanded rapidly after the Second World War, many new monomers were prepared and evaluated. Among the driving forces for this was the earlier success of compounds containing elements other than carbon, hydrogen, oxygen, and nitrogen, such as vinyl chloride, tetrafluoroethylene, and the chlorosilanes. It is understandable that many polymer chemists pictured another variation with unusual (and salable) properties based on a previously untried or incompletely evaluated constituent element. Sulfur and phosphorus were especially attractive because of their availability and low cost. In most instances, the results were disappointing from a commercial standpoint, but the work was useful in terms of the technical and theoretical knowledge gained. Furthermore, some of the products were found to be of value in applications other than polymerization.

The research on vinyl sulfur compounds followed such a pattern. Of the large number that have been studied, only ethenesulfonic acid (**71**) shows promise today as a commercial monomer. Divinyl sulfone (**67**) is used as a wash-and-wear treating agent for cellulosic textiles because of its ability to introduce cross-links through bifunctional reaction with hydroxyl groups (1442). Several of the compounds have shown activity as insecticides, nematocides, and fungicides. None is produced in large volume.

A large part of the work in this field has been done in the U.S.S.R. and Germany. The chemistry of vinyl sulfides was reviewed by Shostakovskii and coworkers (1443), and that of **71** and its esters by Distler (1444). The extensive research by Reppe and his associates at I. G. Farben. on vinyl sulfides, sulfoxides, and sulfones has also been summarized (1445). The only review in English is a discussion of **71** and its polymers by Kutner and Breslow (1446).

Farbwerke Hoechst produces sodium ethenesulfonate, which is sold as a weakly alkaline 25 percent aqueous solution. This company also produces poly(sodium ethenesulfonate) with a carefully controlled molecular-weight range (Pergalen) for use as a blood anticoagulant (1446).

B. Synthetic Methods

1. ETHENESULFONIC ACID (71) AND DERIVATIVES

Kohler (1430) first prepared **71** from 1,2-dibromoethane by the reactions in Eq. (130):

$$BrCH_2CH_2Br \xrightarrow{Na_2SO_3} NaO_3SCH_2CH_2SO_3Na \xrightarrow{PCl_5}$$

$$ClSO_2CH_2CH_2SO_2Cl + H_2O \longrightarrow CH_2{=}CHSO_3H + SO_2 + 2HCl \qquad (130)$$
$$\quad\;\;\mathbf{72} \qquad\qquad\qquad\qquad\qquad\qquad \mathbf{71}$$

The conversion of 1,2-bis(chlorosulfonyl)ethane (**72**) to **71** was also carried out with aqueous ethanol or sodium acetate in acetic acid. The method has been reinvestigated several times (10,1447–1449) and is still the most convenient laboratory procedure for making **71**.

Another good method, and one which is of more industrial interest, is the elimination of sulfuric acid from ethionic acid (**73**) by treatment with base—Eq. (131). This was originally described in a 1939 German patent (1450):

$$HOSO_2OCH_2CH_2SO_3H \xrightarrow{base} CH_2{=}CHSO_3H + H_2SO_4 \qquad (131)$$
$$\qquad\quad\mathbf{73} \qquad\qquad\qquad\qquad \mathbf{71}$$

A later detailed study of the process showed that **73** can be prepared from ethanol and sulfur trioxide in two steps—Eq. (132):

$$C_2H_5OH \xrightarrow[0°C]{SO_3} C_2H_5OSO_2OH \xrightarrow[50°C]{SO_3} \mathbf{73} \qquad (132)$$

On heating the product with aqueous NaOH and neutralizing excess base with H_2SO_4, the sodium salt of **71** is obtained in 80 to 84 percent yield, based on ethanol, as shown by quantitative hydrogenation of the resulting solution (1451). Saturation of the solution with HCl and distillation give **71** (1452). An alternative approach is the reaction of ethylene and SO_3 to form carbyl sulfate (74), which, on treatment with NaOH, yields sodium ethionate (75)—Eq. (133):

$$CH_2{=}CH_2 + 2SO_3 \longrightarrow \underset{\textbf{74}}{H_2C{<}\overset{\displaystyle CH_2}{\underset{\displaystyle O}{}}SO_2} \overset{NaOH}{\longrightarrow} \underset{\textbf{75}}{NaOSO_2OCH_2CH_2SO_3Na} \tag{133}$$

This reacts further with NaOH—Eq. (131)—to give the sodium salt of 71 in 80 percent yield, based on SO_3, and 94 percent yield, based on 74 (1453).

In his pioneering work, Kohler (1430) also synthesized 71 by pyrolysis of 2-bromoethanesulfonyl chloride and subsequent hydrolysis, and by pyrolysis of potassium 2-acetoxyethanesulfonate. There are no further literature references to either method. Simultaneous dehydrohalogenation and hydrolysis of 2-chloroethanesulfonyl chloride with aqueous base produces salts of 71 (1454), as does dehydrohalogenation of 2-chloro- or 2-bromoethanesulfonic acid salts (1455–1457). Dehydration of 2-hydroxyethanesulfonic acid or its salts by heating under vacuum in the presence of pyrophosphoric acid yields 71 (1458). The preparation of sodium ethenesulfonate from acetylene and $NaHSO_3$ with a persulfate catalyst has been patented (1459a). Esters of 71 are hydrolyzed to the acid by passing steam through their alcoholic solutions (1460).

The usual means of purifying 71 is distillation at reduced pressure (<1 mm), but treatment of the sodium salt, suspended in ether, with HCl has also been used (1461). Since poly(ethenesulfonic acid) is mainly of interest as a polar, water-soluble polyacid, polymerization of the unisolated monomer or its sodium salt in aqueous solution is often satisfactory.

Ethenesulfonyl chloride (76) was prepared by Kohler (1430) from the potassium salt of 71 and PCl_5—Eq. (134):

$$CH_2{=}CHSO_3K + PCl_5 \overset{CHCl_3}{\longrightarrow} \underset{\textbf{76}}{CH_2{=}CHSO_2Cl} + POCl_3 + KCl \tag{134}$$

The ammonium salt has also been used (1462). A novel synthesis of 76 is the reaction of divinyl disulfide and chlorine in acetic acid (1459b).

Esters of 71 are usually prepared from 2-chloroethanesulfonyl chloride, an alcohol or phenol, and a base that serves as dehydrochlorinating agent in the formation of both the C=C double bond and the ester moiety— Eq. (135):

$$ClCH_2CH_2SO_2Cl + ROH + 2B{:} \longrightarrow CH_2{=}CHSO_3R + 2BH^+Cl^- \tag{135}$$

Pyridine (1457,1463), the corresponding alkali metal alcoholate or phenolate (1444,1464a), NaOH (1444,1463,1464b,c), and K_2CO_3 (1463) have been used as bases. Esterification of **76** by a similar procedure is also effective (1447,1465). Treatment of carbyl sulfate (**74**) with an alcohol (1466) or phenol (1444,1467) and a strong base, such as NaOH or $Ca(OH)_2$, produces **71** esters—Eq. (136):

$$\text{H}_2\text{C}\overset{\displaystyle \text{CH}_2}{\underset{\displaystyle \text{O}}{\diagup}}\text{SO}_2 + \text{ROH} \xrightarrow{\text{base}} \text{HOSO}_2\text{OCH}_2\text{CH}_2\text{SO}_3\text{R} \xrightarrow{\text{base}}$$

$$\underset{\text{74}}{} \qquad \overset{\text{77}}{}$$

$$\text{CH}_2{=}\text{CHSO}_3\text{R} + \text{H}_2\text{SO}_4 \qquad (136)$$

The reaction apparently proceeds through the ethionate monoester (**77**) (this intermediate has been isolated in alkyl esterifications) from which H_2SO_4 is eliminated by further reaction with base.

The lower alkyl esters have been prepared from **71** by reaction with ketene diethylacetal (1468), ethyl orthoformate (1469), and diazomethane (1470)—Eqs. (137) to (139):

$$\text{CH}_2{=}\text{CHSO}_3\text{H} \xrightarrow{\begin{array}{c}\text{CH}_2{=}\text{C(OC}_2\text{H}_5)_2\\ \text{HC(OC}_2\text{H}_5)_3\\ \text{CH}_2\text{N}_2\end{array}}$$

$$\text{CH}_2{=}\text{CHSO}_3\text{C}_2\text{H}_5 + \text{CH}_3\text{COOC}_2\text{H}_5 \qquad (137)$$
$$\text{CH}_2{=}\text{CHSO}_3\text{C}_2\text{H}_5 + \text{C}_2\text{H}_5\text{OH} + \text{HCOOC}_2\text{H}_5 \qquad (138)$$
$$\text{CH}_2{=}\text{CHSO}_3\text{CH}_3 + \text{N}_2 \qquad (139)$$

71

Ethenesulfonamide (**78**) has been made by treatment of 1,2-bis(chlorosulfonyl)ethane (**72**) with ammonia (1435,1449), from 2-chloroethanesulfonyl chloride and ammonia (1471,1472), from **76** and an ammonium salt (1473), by dehydration of the ammonium salt of **71** with P_2O_5 (1459c), and by heating the isomeric anhydrotaurine (**79**) under vacuum (1459d):

$$\underset{\text{78}}{\text{CH}_2{=}\text{CHSO}_2\text{NH}_2} \qquad \underset{\underset{\text{79}}{\text{NH}-\text{SO}_2}}{\text{CH}_2-\text{CH}_2}$$

N-Arylamides have been prepared from 2-chloroethanesulfonyl chloride and the appropriate arylamine (1474). A wide variety of melting points is reported for **78**, ranging from 17° to 170°. By analogy with ethanesulfonamide and ethanesulfonanilide, both of which melt at 58°, the value of 87° originally assigned to **78** by Clutterbuck and Cohen (1435) is the most plausible (ethenesulfonanilide melts at 68° (1474)).

2. DIVINYL SULFIDE (65), SULFOXIDE (66), AND SULFONE (67)

The most common method of preparing **65** is dehydrohalogenation of a bis(2-haloethyl) sulfide. The first pure **65** was obtained in this manner from bis(2-chloroethyl) sulfide and KOH in ethanol (1432c), and others have used a similar procedure (1429,1475–1479). Substitution of bis(2-bromoethyl) sulfide for the chloro compound gives better yields (1429). Divinyl sulfide has also been made by dehydrochlorination of bis(1-chloroethyl) sulfide (from acetaldehyde, HCl, and H_2S) with N,N-diethylaniline (1480) and by dehydration of bis(2-hydroxyethyl) sulfide over KOH (1481).

Divinyl sulfoxide (**66**) and divinyl sulfone (**67**) were originally prepared by oxidation of **65** with perbenzoic acid (1436). As discussed above, **65** is obtained by an elimination reaction of a 2,2′-disubstituted diethyl sulfide. It is advantageous to reverse the order of the elimination and oxidation steps for two reasons. First, **65** is more susceptible to undesirable side reactions during oxidation than is a saturated sulfide, and, second, the electron-withdrawing —SO— or —SO$_2$— group in the substituted sulfoxide or sulfone favors the elimination reaction, so that it proceeds more readily than in the formation of **65** from the corresponding sulfide. All the other syntheses of **66** and **67** use this route—Eqs. (140) and (141):

$$(ACH_2CH_2)_2SO \longrightarrow (CH_2{=}CH)_2SO + 2HA \qquad (140)$$
$$\mathbf{66}$$

$$(ACH_2CH_2)_2SO_2 \longrightarrow (CH_2{=}CH)_2SO_2 + 2HA \qquad (141)$$
$$\mathbf{67}$$

Dehydrohalogenation (A = halogen) with triethylamine (1476), NaHCO$_3$ (1482,1483), zinc, zinc oxide (1437), or CaCO$_3$ (1484) as base has been used most often. The reaction of bis(2-chloroethyl) sulfoxide with NaHCO$_3$ and sodium hypochlorite gives a mixture of **66**, **67**, 2-chloroethyl vinyl sulfoxide, and 2-chloroethyl vinyl sulfone by a combination of dehydrochlorination and oxidation (1483). Dehydration (A = OH) with phosphoric acid (1485) or phosphoric acid and silica (1486) and acetate pyrolysis (A = OAc) (1485) are other methods that have been reported.

3. ALKYL AND ARYL VINYL SULFIDES (68), SULFOXIDES (69), AND SULFONES (70)

The first industrial synthesis of vinyl sulfides, by addition of mercaptans or thiophenols to acetylene, was developed at I. G. Farben. in the course of a broad investigation of acetylene reactions. The process is analogous to additions of alcohols, phenols, amines, and amides to acetylene, and, therefore, the reaction conditions are similar to those described in Sec. I for the commercial preparation of 9-vinylcarbazole (**1**) and the N-vinyl lactams (2). Thus, it was found (1438,1445) that mercaptans and thiophenols add to acetylene at 160°C and 15 atm in the presence of KOH, a basic oxide, or a basic salt to give the corresponding vinyl sulfide—Eq. (142). The same precautions to avoid explosions that are exercised in the syntheses of 1 and 2 are necessary here also. Excess acetylene must be used to avoid side reactions, specifically, the addition of the thiol to the desired product to form dithioethers and thioacetals—Eq. (143):

$$RSH + CH\!\equiv\!CH \xrightarrow{\text{base}} RSCH\!=\!CH_2 \tag{142}$$

$$RSCH\!=\!CH_2 + RSH\!-\!\begin{cases} \longrightarrow RSCH_2CH_2SR \\ \longrightarrow CH_3CH(SR)_2 \end{cases} \tag{143}$$

Formation of the former is promoted by zinc and cadmium salts, the latter by copper and mercury salts (1445). Others (1487–1490) have since investigated the process. Nedwick et al. (1490) found, as they have in other acetylene additions (45), that the use of a good solvent for acetylene, for example, methylal, as diluent is salutary.

Under similar conditions, H_2S and acetylene do not give 65, as might be expected, but mainly thioacetaldehyde trimer, undoubtedly formed from the primary addition product, vinyl mercaptan Eq. (144):

$$H_2S + CH\!\equiv\!CH \longrightarrow \left[CH_2\!=\!CHSH \rightleftharpoons CH_3C\underset{\displaystyle S}{\overset{\displaystyle H}{\big\backslash}} \right] \longrightarrow (CH_3CHS)_3 \tag{144}$$

When the reaction is run in a high-boiling polyethylene glycol, it is accompanied by reduction, and ethyl vinyl sulfide, ethyl mercaptan, or 1,2-bis(ethylthio)ethane is obtained, depending on the specific conditions, along with a resinous material (1445). The preparation of ethyl vinyl sulfide in this manner has been patented (1491).

A related but novel synthesis of alkyl vinyl sulfides uses an alkyl halide, thiourea, and acetylene as starting materials (1492). It is known that alkyl halides react with thiourea—Eq. (145)—to give S-alkyl isothiuronium salts (80). Treatment of these with acetylene and a strong aqueous base provides alkyl vinyl sulfides—Eq. (146)—in excellent yields (89 to 91 percent):

$$RX + S\!=\!C(NH_2)_2 \longrightarrow RSC\underset{\displaystyle NH_2}{\overset{\displaystyle NH_2{}^+}{\big\backslash}} \quad X^- \tag{145}$$
$$\textbf{80}$$

$$\textbf{80} + CH\!\equiv\!CH + H_2O \xrightarrow{\text{base}} RSCH\!=\!CH_2 + O\!=\!C(NH_2)_2 + HX \tag{146}$$
$$\textbf{68}$$

The mechanism is probably hydrolysis of **80** to urea and mercaptan, with subsequent addition of the thiol to acetylene, as in Eq. (142). Since **80** need not be isolated, however, the process is, in effect, a one-step synthesis of **68** from the alkyl halide.

2-Hydroxyethyl and 2-hydroxypropyl vinyl sulfide are obtained from acetylene, the appropriate alkylene oxide, and an alkali metal hydrosulfide (1493)—Eq. (147):

$$\text{RCH}\underset{\diagdown\diagup}{\text{——}}\text{CH}_2 + \text{CH}{\equiv}\text{CH} + \text{MSH} \xrightarrow[\text{H}_2\text{O}]{\overset{\text{heat}}{\text{pressure}}} \text{RCHOHCH}_2\text{SCH}{=}\text{CH}_2 \qquad (147)$$

$$(\text{R} = \text{H or CH}_3, \text{M} = \text{Na or K})$$

Another related synthesis, historically the earliest, is the reaction of sodium ethyl mercaptide and vinyl bromide to give **68** $(\text{R} = \text{C}_2\text{H}_5)$ (1431). Later, the same compound was obtained from 1,1-dibromoethane and sodium ethyl mercaptide (1494). These approaches were extended by I. G. Farben. (1445,1495) to the preparation of vinyl alkyl sulfides in general from the appropriate mercaptan and a vinyl halide in the presence of KOH or an alkali metal hydrosulfide—Eq. (148):

$$\text{RSH} + \text{CH}_2{=}\text{CHX} \xrightarrow{\text{base}} \text{RSCH}{=}\text{CH}_2 + \text{HX} \qquad (148)$$

and by using 1,1- or 1,2-dichloroethane in place of the vinyl halide—Eq. (149):

$$\text{RSH} + \text{ClCH}_2\text{CH}_2\text{Cl (or CH}_3\text{CHCl}_2) \xrightarrow{\text{base}} \underset{\mathbf{68}}{\text{RSCH}{=}\text{CH}_2} + 2\text{HCl} \qquad (149)$$

It is unlikely that the process shown in Eq. (148) involves direct displacement of X by RS (Williamson synthesis) because of the low reactivity of vinyl halides in such reactions. Probably, dehydrohalogenation is the first step, followed by addition of the mercaptan to the resulting acetylene— Eq. (142)—as proposed by Truce et al. (1496) in the preparation of **68** $(\text{R} = tert\text{-C}_4\text{H}_9)$ by this route. There are two possible mechanisms for the reaction in Eq. (149): displacement of Cl by RS to yield an alkyl chloroethyl sulfide with subsequent dehydrochlorination to **68**, or elimination of 2 moles of HCl from the dichloride to give acetylene followed by addition of the mercaptan, as in Eq. (142).

Dehydrohalogenation of a 2-haloethyl alkyl or aryl sulfide with base— sodium amyloxide (1497,1498) or KOH (1479,1499–1502)—has been employed most often in laboratory preparations of **68**. 1-Haloethyl alkyl or aryl sulfides have also been used as starting materials with a N,N-dialkylaniline as base (1503,1504).

Phenyl vinyl sulfide (1433,1505) and benzyl vinyl sulfide (1505) have been prepared by Hofmann elimination—Eq. (150):

$$RSCH_2CH_2\overset{+}{N}R_3' \ OH^- \xrightarrow{heat} RSCH=CH_2 + R_3'N + H_2O \qquad (150)$$

A novel synthesis of phenyl vinyl sulfide is the elimination of thiophenol from 1,2-bis(phenylthio)ethane by treatment with phenyllithium (1506).

2-Hydroxyethyl alkyl sulfides, which are conveniently prepared from the appropriate mercaptan and ethylene oxide (1507), have been dehydrated to alkyl vinyl sulfides (1507–1509). Phenyl (1506) and alkyl (1510) vinyl sulfides are formed on pyrolysis of 2-acetoxyethyl sulfides at 450 to 465°. Better yields (44 to 46 percent versus 24 to 30 percent) of **68** (R = alkyl) are obtained at 130 to 138° (1511).

Although thiols add readily to vinyl sulfides—Eq. (143)—the reaction is reversed at high temperatures (350 to 500°). **68** (R = C_6H_5) (1512) and **68** (R = CH_3) (1513) have been obtained in this way from 1,2-ethanedithiol ethers—Eq. (151):

$$RSCH_2CH_2SR' \xrightarrow{heat} RSCH=CH_2 + R'SH \qquad (151)$$

A clever modification is to decompose the dithioether—R = R', Eq. (151) —in the presence of acetylene and a base (1514). Under these conditions, the mercaptan that is produced adds to the acetylene to provide another mole of **68**.

Several interchange reactions give vinyl sulfides. Divinylmercury and thiols yield them by way of the intermediate alkyl- or arylthiovinylmercury —Eq. (152) (1515):

$$(CH_2=CH)_2Hg + RSH \longrightarrow CH_2=CHHgSR + CH_2=CH_2$$
$$\downarrow heat \qquad (152)$$
$$CH_2=CHSR + Hg$$

The familiar method of preparing vinyl esters from vinyl acetate and a carboxylic acid in the presence of mercury(II) acetate and H_2SO_4 (1516) has been applied to the synthesis of alkyl vinyl sulfides from mercaptans— Eq. (153) (1517):

$$RSH + CH_3COOCH=CH_2 \xrightarrow[H_2SO_4]{(CH_3COO)_2Hg} RSCH=CH_2 + CH_3COOH \qquad (153)$$

Interchange between tributyl thioborate and butyl vinyl ether is reported (1518) to give **68** (R = C_4H_9) in 71 percent yield—Eq. (154):

$$B(SC_4H_9)_3 + 3C_4H_9OCH=CH_2 \xrightarrow{heat} 3C_4H_9SCH=CH_2 + B(OC_4H_9)_3 \quad (154)$$

Vinyl sulfoxides (**69**) have been made by oxidation of the corresponding vinyl sulfides. A variety of oxidizing agents has been used, for example, peracids and their salts, peroxides, hypochlorites, and electrolysis, but care must be exercised to avoid further oxidation to the sulfone. Low temperatures, a stoichiometric (or lower) quantity of the oxidizer, and diluents are desirable. The oxidation should be carried out in a neutral or slightly alkaline medium to minimize side reactions (1439,1445,1508,1519,1520). Another method is dehydrohalogenation of a 2-haloethyl sulfoxide, which avoids the difficulties inherent in oxidizing a sensitive unsaturated compound. Treatment of 2-chloroethyl methyl sulfoxide with sodium *tert*-butoxide (1520) and 2-chloroethyl phenyl sulfoxide with aqueous KOH (1521) gives the corresponding **69**.

As implied above, vinyl sulfones (**70**) may also be prepared by the oxidation of vinyl sulfides. The same broad choice of oxidizing agents described for sulfoxide preparation is suitable for sulfones, too, but higher temperatures and excess reagent are used. Again, it is advisable to operate in a neutral or slightly alkaline medium to preclude extensive side reactions (condensation, polymerization, elimination of vinyl groups) (1439,1445,1501,1504,1508). Manganese dioxide and selenium dioxide have been used with hydrogen peroxide as auxiliary oxidizing agents (1519).

As in the preparation of **66** and **67**, the usual route to **70** is an elimination reaction—Eq. (155):

$$ACH_2CH_2SO_2R \longrightarrow CH_2{=}CHSO_2R + HA \qquad (155)$$

and for the same reasons (see Sec. V-B-2). The electron-withdrawing sulfone moiety favors the elimination, and the oxidation of a saturated thioether to the sulfone avoids side reactions that may occur when a vinyl sulfide is similarly oxidized. Among the syntheses of **70** by the general route shown in Eq. (155), dehydrohalogenation (A = halogen) with a base such as triethylamine (1484,1498,1522–1527), triethanolamine (1528), and aqueous solutions of alkali metal hydroxides (1529–1531) has been most widely used. A related process is elimination of H_2SO_4 from the sulfate ester (A = OSO_2OH) by means of aqueous NaOH (1532). Dehydration of 2-hydroxyethyl sulfones (A = OH) with Al_2O_3 (1533), H_3PO_4 (1534,1535), or calcium phosphate (1536) as catalyst and pyrolysis of 2-acetoxyethyl sulfones (A = CH_3COO) (1533,1537) are the other common synthetic methods. Hofmann elimination (A = $R_3N^+OH^-$) has been reported only once (1538).

Vinyl *p*-tolyl sulfone is obtained by reaction of *p*-toluenesulfinic acid with divinylmercury (1539).

C. Properties

1. Physical

Boiling points, melting points, refractive indices, and densities are listed in Table 22. The monomers are presented in the order in which they are discussed in the preceding section. Only representative examples of the many vinyl sulfides, vinyl sulfones, and esters of ethenesulfonic acid that have been reported are included.

With the exception of the ethenesulfonamides, phenyl ethenesulfonate, and aryl vinyl sulfones, the compounds in Table 22 are liquids at room temperature. Ethenesulfonic acid (71) is soluble in water, ethanol, and acetic acid but soluble only with difficulty in ether and chloroform (1430). Pure 71 is colorless but darkens in the presence of oxygen (1540). The esters of 71 and the vinyl sulfides are relatively nonpolar and, therefore, soluble in typical organic solvents but insoluble in water. The sulfoxides and sulfones are more polar and the lower homologs are water-soluble. For example, 65 is only slightly soluble, but 66 and 67 are readily soluble in water.

The dipole moment of divinyl sulfide was found to be $1.20D$ (1547). Values reported for methyl vinyl and phenyl vinyl sulfone are $4.82D$ and $5.29D$, respectively (1548).

2. PHYSIOLOGICAL

Divinyl sulfone is a powerful vesicant and highly toxic (1549). Intravenous injection into dogs and rabbits causes transient hypertension, followed by prolonged hypotension and sometimes death at a dosage of 80 mg per kg (1550). Alkyl vinyl sulfides (1445) and alkyl ethenesulfonates (1457) may cause severe dermatitis, and the latter are vesicants. Like the sulfoxide analog of mustard gas, divinyl sulfoxide is not a vesicant (1549).

3. ANALYTICAL METHODS AND SPECTRA

Two analytical procedures for divinyl sulfone (67) depend on its reactions with thiols. One is a gravimetric method based on the bicysteine derivative (1484); the other measures trace amounts of 67 in water by reaction with thiophenol in the presence of base and observation of the decrease in ultraviolet absorption at 262 mμ as the adduct precipitates (1551). The separation and determination of various sulfides, sulfoxides, and sulfones, including vinyl compounds, by thin-layer chromatography have been described (1552).

References to spectrometric studies are listed in Table 23. It is important to use peroxide-free solvents in determining the spectra of sulfides, since oxidation to the sulfoxide may otherwise occur (1553). Methyl vinyl sulfide, sulfoxide, and sulfone have been subjected to mass-spectral analysis (1564).

TABLE 22
Physical Properties of Vinyl Sulfur Compounds

Compound	bp, °C (mm)	Reference	n_D (°C)	Reference	d_4 (°C)	Reference
CH₂=CHSO₃H	93(0.07)	1458	1.4514(20)	1452	1.3921(25)	1447
	115(0.5)	1540	1.4496(25)	1447		
	130(0.9)	1527	1.4499(25)	1540		
	125(1)	1447	1.4505(25)	1458		
	127(1)	1452				
	132(1.4)	1452				
CH₂=CHSO₂Cl	56(1)	1462	1.4680(20)	1527	1.393(20)	1462
	31(1)	1527	1.4686(20)	1462,1541		
	44(5)	1541	1.4630(25)	1459b		
CH₂=CHSO₃R:						
R = CH₃	60(0.4)	1444	1.4136(20)	1457	1.248(20)	1457
	64(1)	1464a	1.4310(20)	1444		
	85(12)	1470				
	91(15)	1457				
R = C₂H₅	60(0.4)	1464a	1.4300(20)	1468	1.1831(25)	1447
	76(5)	1447	1.4325(20)	1466		
	89(10)	1466	1.4349(20)	1464a		
	96(14)	1468	1.4316(25)	1447		
R = n-C₃H₇	98(8)	1466	1.4361(20)	1466	1.156(20)	1457
	110(18)	1457	1.4368(20)	1457		
R = n-C₄H₉	82(0.4)	1464a	1.4392(20)	1466	1.122(20)	1457
	108.5(9)	1466	1.4416(20)	1457		
	117(15)	1457				
R = C₆H₅	91(0.1)	1464a,c	1.5171(40)	1444,1464a,c		
	100(0.5)	1467				
	mp 40°	1463				
	mp 42°	1444				
CH₂=CHSO₂HNR:						
R = H	114(0.1)	1471	1.488(20)	1472		
	125(0.5)	1472				

R	bp (mm) / mp		n_D		d	
		1474				
$R = p\text{-CH}_3\text{C}_6\text{H}_4$	mp 68°	1474				
$R = p\text{-C}_2\text{H}_5\text{OC}_6\text{H}_4$	mp 74°	1474				
$(\text{CH}_2{=}\text{CH})_2\text{S}$	mp 88°	1474				
	42.5(150)	1481	1.5076(20)	1481	0.9174(15)	1432c
	84	1429				
	85	1475				
	86	1432c,1480				
	93	1479				
$(\text{CH}_2{=}\text{CH})_2\text{SO}$	59(3.5)	1436	1.5100(20)	1436,1483	1.0867(17)	1436
	68(6)	1436			1.084(20)	1436
	81(16)	1483				
$(\text{CH}_2{=}\text{CH})_2\text{SO}_2$	91(0.8)	1542	1.4750(20)	1483	1.1821(17)	1436
	92(8)	1485	1.4780(20)	1542	1.1790(20)	1437
	102(8)	1436	1.4782(20)	1483	1.1794(20)	1436
	110(17)	1483	1.4799(20)	1436		
	125(18)	1484	1.4730(27)	1485		
	121(20)	1437				
	mp −16°	1484				
$\text{CH}_2{=}\text{CHSR}$:						
$R = \text{CH}_3$	67	1498a,1508	1.4835(20)	1498a	0.9026(20)	1508
	68	1492	1.4837(20)	1509	0.8986(25)	1492
	70	1509,1513	1.4845(20)	1508,1509		
	71	1504	1.4826(25)	1492		
$R = \text{C}_2\text{H}_5$	91	1438,1479, 1503,1507	1.4750(20)	1502	0.887(14)	1431
	91.5	1431	1.4756(20)	1488	0.8767(20)	1488
	92	1488,1492, 1504,1511	1.4763(20)	1503	0.876(20.5)	1503
	93	1502	1.4723(25)	1511	0.8734(25)	1492
			1.4735(25)	1492		
$R = n\text{-C}_3\text{H}_7$	43.5(50)	1488c	1.4734(20)	1488c	0.8723(20)	1488c
$R = n\text{-C}_4\text{H}_9$	46(20)	1518	1.4722(20)	1488c,1518	0.8698(20)	1488c
	48.5(21)	1488c	1.4738(25)	1492	0.8712(25)	1492
	70(60)	1492				

continued

TABLE 22 (continued)

Compound	bp, °C (mm)	Reference	n_D (°C)	Reference	d_4 (°C)	Reference
R = $C_6H_5CH_2$	90.5(7)	1488c	1.5794(20)	1488c	1.0347(20)	1488c
	98(10)	1492	1.5773(25)	1492	1.0360(25)	1492
	219	1505				
R = C_6H_5	56(4)	1543	1.5873(20)	1515	1.0386(20)	1498b
	64(5.5)	1506	1.5888(20)	1488c	1.0417(20)	1488c
	76(11)	1504	1.5890(20)	1543		
	94(25)	1527	1.5868(25)	1501		
	94.5(25)	1498b	1.5878(25)	1498b		
	102(45)	1498b	1.5847(27)	1506		
	195	1505				
	198	1498b				
	201	1433				
CH_2=CHSOR:						
R = CH_3	43.5(0.5)	1520	1.4912(20)	1508	1.0945(20)	1508
	52(3)	1520	1.4951(20)	1519	1.0876(20)	1519
	59(6)	1519	1.4925(25)	1520		
	87(15)	1508				
	82(21)	1544				
R = C_2H_5	54(2)	1519	1.4900(20)	1519	1.0422(20)	1519
R = n-C_3H_7	83(8)	1519	1.4870(20)	1519	1.0125(20)	1519
R = n-C_4H_9	90(6)	1519	1.4840(20)	1519	0.9891(20)	1519
R = C_6H_5	95(0.2)	1521	1.5869(20)	1519	1.1629(20)	1519
	133(7)	1519				
CH_2=CHSO$_2$R:						
R = CH_3	80(0.7)	1542	1.4636(20)	1508	1.2117(20)	1508
	97(8)	1545	1.4640(20)	1508	1.2146(20)	1508

R	bp (°C) (ref)	ref	n_D (ref)	ref	d (ref)	ref
	117(17)	1535,1544	1.4589(25)	1542	1.1407(20)	1519
	117(19)	1498a,1508, 1522			1.1457(20)	1534a
R = C₂H₅	120(19)	1527				
	85(1)	1531	1.4622(20)	1534b		
	88(2)	1519	1.4628(20)	1519		
	93(3)	1519	1.4635(20)	1519		
	95(4)	1534b	1.4640(20)	1534a		
	98(5)	1534a				
	106(8)	1523				
	108(11)	1504				
	117(17)	1538				
	119(22)	1528				
R = n-C₃H₇	107.5(5)	1519	1.4643(20)	1519	1.1060(20)	1519
R = n-C₄H₉	106(3)	1519	1.4620(20)	1519	1.0657(20)	1519
	117(5)	1519	1.4632(20)	1546		
	135(15)	1522	1.4640(20)	1519		
R = CH₂CH=CH₂	83.5(0.7)	1525	1.4815(25)	1525	1.1427(25)	1525
	120(9.5)	1525				
R = C₆H₅	mp 67.5°	1498b				
	mp 68°	1501,1519, 1537				
	mp 68.5°	1484				
	mp 69°	1504				
	mp 70°	1538,1542				
	mp 72°	1531,1532				
R = p-CH₃C₆H₄	mp 66°	1526,1539				
	mp 67°	1532				

[a] Other melting points that have been reported are 17° (1472), 24° (1471), and 120 to 170°, depending on rate of heating (1473). As discussed earlier (see Sec. V-B-1), these are less plausible values.

1479

TABLE 23
Literature References to Spectra of Vinyl Sulfur Compounds

Compound	UV	IR	Raman	NMR
$(CH_2{=}CH)_2S$	1477,1553,1554	1481,1555	1555	
$(CH_2{=}CH)_2SO$		1555	1555	
$(CH_2{=}CH)_2SO_2$		1555,1556	1555	73
$CH_2{=}CHSR$:				
R = CH_3		1508,1557		1558,1559
R = C_2H_5	1554,1560			1561
R = n-C_3H_7	1560		1560	
R = n-C_4H_9	1560		1560	
R = t-C_4H_9	1554,1560		1560	
R = C_6H_5	1498b,1499	1498b		73,1561
$CH_2{=}CHSOCH_3$	1520	1508,1520		
$CH_2{=}CHSO_2R$:				
R = CH_3		1508,1556,1557, 1562		
R = C_2H_5	1523			
R = C_6H_5	1498b,1563	1498b		
$CH_2{=}CHSO_3H$				73
$CH_2{=}CHSO_3C_6H_5$				73

D. Storage and Handling

Crystalline sodium ethenesulfonate polymerizes very readily at room temperature and an attempt to find a satisfactory stabilizer was unsuccessful. Sodium nitrite and, to a lesser degree, hydroquinone monobenzyl ether, however, inhibit its polymerization in aqueous solution (1451). The salt is marketed in Europe as a weakly alkaline aqueous solution containing no stabilizer (1446). Ethenesulfonamide has been stabilized with N-phenyl-2-naphthylamine (1471) and divinyl sulfone by adding triethylamine and storing in the dark (1484). Methyl vinyl sulfide (1513) and ethyl vinyl sulfone (1529) do not polymerize on standing at room temperature.

The toxicities of many of these monomers are not known, but, in light of the irritant and poisonous properties of some (see Sec. V-C-2), care should be taken to avoid skin contact and inhalation.

E. Chemical Reactions

Additions to the C=C double bond in vinyl sulfur compounds have been studied extensively. Nucleophilic reagents add to the $CH_2{=}CHSO$

moiety in the same sense as to α,β-unsaturated carbonyl compounds—Eq. (156):

$$CH_2{=}CH\overset{|}{S}O + HA \xrightarrow{\text{base}} ACH_2CH_2\overset{|}{S}O \qquad (156)$$

Treatment of **71** with aqueous base yields 2-hydroxyethanesulfonic acid— Eq. (157) (1428):

$$CH_2{=}CHSO_3H + 2OH^- \longrightarrow HOCH_2\bar{C}HSO_3^- \xrightarrow{2H^+} HOCH_2CH_2SO_3H \quad (157)$$

and the analogous reaction of ethyl vinyl sulfone gives ethyl 2-hydroxy-ethyl sulfone—Eq. (158) (1534a):

$$CH_2{=}CHSO_2C_2H_5 + OH^- \longrightarrow HOCH_2\bar{C}HSO_2C_2H_5 \xrightarrow{H^+}$$
$$HOCH_2CH_2SO_2C_2H_5 \quad (158)$$

For divinyl sulfoxide (1476) and divinyl sulfone (1437,1476,1524), the resulting bis-2-hydroxyethyl compounds are dehydrated under the reaction conditions to 1,4-thioxane derivatives—Eq. (159):

$$(CH_2{=}CH)_2SO_{(2)} + 2H_2O \xrightarrow{\text{base}} (HOCH_2CH_2)_2SO_{(2)} \longrightarrow$$

$$\begin{array}{c} CH_2CH_2 \\ O \diagup \diagdown SO_{(2)} + H_2O \quad (159) \\ \diagdown CH_2CH_2 \diagup \end{array}$$

Similar nucleophilic additions, in the manner of Eq. (156), are H_2S to **69** and **70** (1445), alcohols or phenols to **66** (1476), **67** (1437,1476), **70** (1534a, 1565), and esters of **71** (1566,1567), thiols to **66** (1549), **67** (1437,1549), **69** (1445,1521), and **70** (1445,1528,1534a), ammonia or amines to **67** (1476, 1524,1568), **69** (1440,1445), **70** (1440,1445,1565), and ethenesulfonamide (1471), sulfinic acids to **69** and **70** (1445), and $NaHSO_3$ to **67** (1437), **69** (1445), and **70** (1445). The thiosulfate anion adds to **67**—Eq. (160)—to give the *Bunte salt* **81** (1549):

$$(CH_2{=}CH)_2SO_2 + 2Na_2S_2O_3 + 2H_2O \longrightarrow$$
$$(NaO_3SSCH_2CH_2)_2SO_2 + 2NaOH \quad (160)$$
$$\mathbf{81}$$

Michael nucleophilic additions of carbanions, for example, the reaction of alkyl (1445,1522) and aryl (1526) vinyl sulfones with nitroalkanes—Eq. (161)—have been studied also:

$$2RSO_2CH{=}CH_2 + R'CH_2NO_2 \xrightarrow{\text{base}} (RSO_2CH_2CH_2)_2CR'NO_2 \quad (161)$$

Truce and Wellisch (1545) demonstrated that methyl vinyl sulfone under-goes Michael addition with a variety of active hydrogen compounds and called the process *sulfonethylation* because of its resemblance to cyano-ethylation with acrylonitrile (270). Diethyl malonate and nitromethane add in an analogous fashion to methyl ethenesulfonate (1470).

Nucleophilic addition to the bifunctional divinyl sulfone (**67**) by compounds with two active hydrogens, for example, diethyl malonate (1569a), a glycol or dithiol (1569b), or urea (1570), has been used to make polymers, as shown in Eq. (162):

$$n(CH_2{=}CH)_2SO_2 + nHYH \xrightarrow{\text{base}} (-YCH_2CH_2SO_2CH_2CH_2-)_n \quad (162)$$
$$(Y = C(COOC_2H_5)_2, \text{ O—R—O, S—R—S, or NHCONH})$$

The same type of process is involved when cotton textiles are treated with **67** to impart wash-and-wear qualities. A controlled degree of cross-linking is introduced by addition of two cellulosic hydroxyl groups to each **67** molecule.

The S—O bond in sulfoxides, sulfones, and sulfonic acid derivatives has some double-bond character because of participation of the sulfur $3d$ orbitals (expansion of the sulfur octet) (1506). It is, therefore, capable of conjugation with an adjacent C=C double bond, explaining the ease with which nucleophilic reagents add to vinyl sulfoxyl and sulfonyl compounds —Eq. (163):

$$CH_2{=}CH{-}\overset{O}{\underset{\underset{\text{nucleophile}}{|}}{S}}{-} + A^- \longrightarrow ACH_2CH{\cdots}\overset{O}{\underset{|}{\overset{||}{S}}}{-} \xrightarrow{H^+} ACH_2CH_2{-}\overset{O}{\underset{|}{S}}{-} \quad (163)$$

Divinyl sulfone undergoes addition of thiols at a higher rate than divinyl sulfoxide (1549), because two oxygen atoms on sulfur provide greater resonance stabilization of the resulting anion than one. The results of Parham et al. (1506,1571) indicate that there can be sulfur $3d$ orbital involvement in additions to vinyl sulfides also. The strong base, butyl-lithium, adds to phenyl vinyl sulfide—Eq. (164)—but not to phenyl vinyl ether. This is best explained by resonance stabilization of the anion by way of octet expansion—Eq. (165):

$$C_6H_5SCH{=}CH_2 + C_4H_9Li \longrightarrow C_6H_5SCHLiCH_2C_4H_9 \quad (164)$$

$$C_6H_5SCH{=}CH_2 + C_4H_9^- \longrightarrow$$
$$C_6H_5\overset{-}{S}CHCH_2C_4H_9 \longleftrightarrow C_6H_5\overset{-}{S}{=}CHCH_2C_4H_9 \quad (165)$$

The oxygen in phenyl vinyl ether is incapable of such an expansion.

Additions also occur in the absence of base but often at low rates. The sluggish hydration of **67** in pure water and its slow reaction with thiols when no base is present are examples of this behavior (1549). Phenol, thiophenol, and H_2S add to **67** without alkaline catalysis (1476). The H_2S addition gives mainly polymer, probably by the process in Eq. (162)

(Y = S), but also a small amount of 1,4-dithioxane 1,1-dioxide (**82**)—Eq. (166):

$$(CH_2\!\!=\!\!CH)_2SO_2 + H_2S \longrightarrow S\overset{\displaystyle CH_2CH_2}{\underset{\displaystyle CH_2CH_2}{\diagup\!\!\!\diagdown}}SO_2 + polymer \qquad (166)$$

82

Hydrogen sulfide, mercaptans, and sodium bisulfite add to alkyl vinyl sulfides without a catalyst, but the first two reactions are accelerated by addition of a base (1445). Shostakovskii et al. (1488b) report that ethyl vinyl sulfide and H_2S give 59 percent 2-ethylthioethyl mercaptan (**83**) and 40 percent bis(2-ethylthioethyl) sulfide (**84**) on long standing—Eq. (167):

$$C_2H_5SCH\!\!=\!\!CH_2 + H_2S \longrightarrow \underset{\textbf{83}}{C_2H_5SCH_2CH_2SH} + \underset{\textbf{84}}{(C_2H_5SCH_2CH_2)_2S} \quad (167)$$

The same authors found that ethyl mercaptan adds to ethyl vinyl sulfide at 0° to form 1,2-bis(ethylthio)ethane in 93 percent yield, but when hydroquinone is present, diethyl thioacetal is produced in 37 percent yield—Eq. (168):

$$C_2H_5SCH\!\!=\!\!CH_2 + C_2H_5SH \overset{0°}{\underset{\underset{\text{hydroquinone}}{0°}}{\Bigg[}} \quad \begin{array}{l} \rightarrow C_2H_5SCH_2CH_2SC_2H_5 \\[4pt] \rightarrow CH_3CH(SC_2H_5)_2 \end{array} \quad (168)$$

This reversal in mode of addition suggests that the high-yield reaction is a free-radical process. It will be recalled—Eq. (143)—that the same reversal can be effected by using different salts as catalysts (1445).

Halogenation of **65** and **66** produces the expected bis(1,2-dihaloethyl) compounds (1476). The addition of HCl to **65** yields a dichloride that, because it is not a vesicant (mustard gas), must be bis(1-chloroethyl) sulfide (1432c). Bromination of alkyl vinyl sulfides is not straightforward. The dibromide cannot be isolated (1434), and HBr is evolved (1494). The product of HBr addition to butyl vinyl sulfide appears to be a mixture of the 1- and 2-bromoethyl derivatives (1434). HCl does not add to **66**, and HBr and HI cause polymerization or decomposition (1476).

Iodine chloride adds to **71** in the expected manner, with the more electronegative Cl going to the β carbon atom (1572). Bromination of **67** (1524) and **71** (1430) gives products that indicate that the initially formed 1,2-dibromides are unstable—Eqs. (169) and (170):

$$(CH_2\!\!=\!\!CH)_2SO_2 + Br_2 \longrightarrow [(BrCH_2CHBr)_2SO_2] \overset{-2HBr}{\longrightarrow}$$

$$[(CH_2\!\!=\!\!CBr)_2SO_2] \overset{Br_2}{\longrightarrow} (BrCH_2CBr_2)_2SO_2 \quad (169)$$

$$CH_2\!\!=\!\!CHSO_3H + Br_2 \longrightarrow [BrCH_2CHBrSO_3H] \overset{-HBr}{\longrightarrow}$$

$$CH_2\!\!=\!\!CBrSO_3H \quad (170)$$

It is not unreasonable that the strongly electron-withdrawing sulfonyl group should facilitate dehydrobromination of the $BrCH_2CHSO_2$ structure. The reported failure of HCl to add to **67** is consistent with this, but the same authors (1476) found that HBr and HI add normally, and others have observed normal addition of hydrogen halides to both **67** (1437) and **71** (1430). There is also disagreement about the halogenation of **67**. One paper (1476) reports that bromine reacts but not chlorine, another (1437) that both react readily.

Alkyl and aryl vinyl sulfides are converted to 1,1-dichlorocyclopropane derivatives by dichlorocarbene—Eq. (171) (1573):

$$RSCH{=}CH_2 + :CCl_2 \xrightarrow[t\text{-}C_4H_9OK]{CHCl_3} RS{-}\overset{H \quad H}{\underset{Cl \quad Cl}{\triangle}}{-}H \qquad (171)$$

Phenyl vinyl sulfone and diazomethane give the pyrazoline **85**—Eq. (172) (1501):

$$C_6H_5SO_2CH{=}CH_2 + CH_2N_2 \xrightarrow{ether} C_6H_5SO_2{-}\underset{\underset{H}{N{-}N}}{\boxed{}} \qquad (172)$$

85

Aryl diazonium chlorides add to vinyl sulfones (1574,1575) and potassium ethenesulfonate (1575b) in the presence of $CuCl_2$ (Meerwein chloroarylation)—Eq. (173):

$$-SO_2CH{=}CH_2 + ArN_2{}^+Cl^- \xrightarrow[H_2O]{CuCl_2} -SO_2CHClCH_2Ar + N_2 \qquad (173)$$

Chloramine T and $HgCl_2$ form adducts with divinyl sulfide (1475). Ethyl vinyl sulfide also reacts with $HgCl_2$ to give a crystalline derivative, mp 34° (1488).

Ethenesulfonyl chloride (1527), methyl ethenesulfonate (1470), **68** ($R = C_2H_5$ and C_6H_5) (1576), **70** ($R = CH_3$) (1527), **70** ($R = C_2H_5$) (1534a), and **70** ($R = p\text{-}CH_3C_6H_4$) (1577) act as dienophiles in the Diels-Alder reaction. Ethenesulfonic acid does not react with anthracene and causes polymerization of other dienes (1527). Phenyl vinyl sulfide behaves as a dienophile toward acrolein—Eq. (174) (1578):

$$C_6H_5SCH{=}CH_2 + CH_2{=}CHCHO \longrightarrow C_6H_5S{-}\underset{O}{\boxed{}} \qquad (174)$$

Vinyl sulfides and sulfoxides are oxidized by a variety of reagents to the corresponding sulfones. Under milder conditions, oxidation of a vinyl sulfide can be controlled to yield the sulfoxide as the major product (see Secs. V-B-2 and -3). Treatment of 71 with $KMnO_4$ or $Ag(NH_3)_2^+$ produces CO_2, H_2O, and H_2SO_4 (1430).

The vinyl group is reduced to ethyl in alkyl vinyl sulfides by hydrogenation (1445), and in 71 by phosphorus and HI (1430). Lithium aluminum hydride reduces both the vinyl and sulfonyl groups in phenyl vinyl sulfone to give ethyl phenyl sulfide (1579). Divinyl sulfoxide and divinyl sulfone are reduced to divinyl sulfide by triphenyltin hydride (1580). Only the vinyl group of methyl vinyl sulfone undergoes polarographic reduction (1581).

Like other sulfonic acids, 71 has a very high dissociation constant. It catalyzes hydrolytic reactions, such as the inversion of sugar and hydrolysis of glycylglycine, more effectively than either HCl or H_2SO_4 (1582). A variety of 71 salts is known (1430,1454,1457), all of which have been prepared indirectly from 2-substituted ethanesulfonyl chlorides. Heating the ammonium or sodium salt and S-benzylthiuronium chloride in aqueous solution yields the S-benzylthiuronium salt, a convenient derivative (mp 146°) of 71 (1457).

The alkyl ethenesulfonates are excellent alkylating agents for the preparation of ethers from alcohols, phenols, and thiols, and tertiary amines from secondary amines. Phenol is converted to anisole in 90 percent yield in this manner (1583).

F. Polymerization

Ethenesulfonic acid (71) does not polymerize on storage at ambient temperatures. Its polymerization, however, is readily initiated by peroxides (1447,1449,1452,1582), ultraviolet light (1447,1452,1540,1582), and γ rays (1582). The salts polymerize much more easily. Ammonium ethenesulfonate melts at 156° and then resolidifies to a horny insoluble resin (1430). On heating at 70° for 4 hr, it becomes insoluble in ethanol (1457). The sodium salt polymerizes on standing at room temperature (1451) and in concentrated aqueous solution with peroxides or ultraviolet radiation as initiator (1584a,1585). Other derivatives of 71 that homopolymerize are the acid chloride (1447,1582), esters (1447,1465,1586), and amide (1459e,1471, 1587).

Copolymerizations of 71 have been carried out with vinyl acetate, methyl methacrylate, and methyl 2-chloroacrylate in bulk (1447), with methyl acrylate (1452) and acrylonitrile (1448,1452,1588) in bulk or solution, and with acrylic acid in water (1589). The sodium (1584, 1590,1591) and amine (1592) salts of 71 have been copolymerized with a variety of monomers in aqueous solution. For water-insoluble comonomers, dimethyl sulfoxide has been used as solvent (1585). Ethenesulfonyl chloride (1447,1593a), ethenesulfonyl fluoride (1594), and esters of **71** (1447,1465,1593b,1595–1597) have been copolymerized with other monomers by free-radical processes.

Divinyl sulfide homopolymerizes during distillation (1432b), on standing at room temperature (1432c), and on treatment with bases (1476). Benzoyl peroxide is a poor initiator because it oxidizes the sulfide linkage, but 2,2'-bisazoisobutyronitrile at 60° produces a CS_2-soluble solid polymer (1553). Emulsion copolymers with butadiene have been patented as rubber substitutes (1478). Copolymerizations with other monomers using free-radical initiators (1598) and with olefins using $AlCl_3$ or BF_3 (1599) are described in other patents.

Divinyl sulfoxide polymerizes on treatment with HBr (1476). No reference was found to homopolymerization of divinyl sulfone. It copolymerizes very reluctantly with methyl methacrylate and styrene (1600), and somewhat more readily with acrylonitrile (1601). In the latter system, the intrinsic viscosity of the copolymer decreases with increasing divinyl sulfone concentration.

Cationic polymerization of monovinyl sulfides is easily initiated by Lewis acids, but the products invariably have a low molecular weight. The first references to polymerization of these compounds are three I. G. Farben. patents (1441) claiming a large assortment of initiators, both cationic and free-radical, among which SO_2 seems to have been preferred. Reppe et al., in their summary of this and later work (1445), reported that SO_2, BF_3, and $ZnCl_2$ give viscous oils, waxes, or brittle solids, indicating short polymer chain length. Shostakovskii and coworkers also obtained low-molecular-weight (about 3,000) homopolymers with BF_3, $SnCl_4$, and $FeCl_3$ (1543,1602). Ease of polymerization decreases with increasing basicity of the thioether linkage in the monomer, presumably because of complexing with the acid catalyst. For example, ethyl vinyl sulfide polymerizes less readily than phenyl vinyl sulfide (1543).

Methyl vinyl sulfide does not polymerize on storage (1513), and exposure of ethyl vinyl sulfide to sunlight for 2.5 months gave only a low-molecular-weight polymer (1488b). Free-radical polymerization of monovinyl sulfides is initiated by 2,2'-bisazoisobutyronitrile (1488b), and the rates are higher than those for the corresponding vinyl ethers (1603). The molecular weights of the products are about 10,000. Pentachlorophenyl vinyl sulfide can be polymerized to a nonflammable resin (1489).

Cationic copolymerization of monovinyl sulfides with di- and polyolefins (1604), styrene, and vinyl ethers (1543) has been studied. As in homopolymerization, phenyl vinyl sulfide is more reactive than ethyl vinyl sulfide (1543). Free-radical copolymerizations with acrylonitrile (1500,

1605), methacrylonitrile, ethacrylonitrile (1605), styrene (1498,1543,1602, 1606,1607), methyl acrylate (1498,1602,1606), and methyl methacrylate (1500,1543,1602) have also been investigated. The sulfides exhibit greater reactivity than the corresponding vinyl ethers as in homopolymerization (1603). Copolymers of **68** (R = C_2H_5, C_4H_9, and C_6H_5) with styrene, methyl acrylate, and methyl methacrylate are brittle solids with mechanical strengths that decrease with increasing **68** content (1602).

There are no literature references to the homopolymerization of monovinyl sulfoxides. Methyl vinyl sulfoxide displays low reactivity and a retarding effect in peroxide-initiated copolymerizations with styrene and methyl methacrylate (1520).

Ethyl vinyl sulfone shows no tendency to polymerize at room temperature (1529), and with 2,2'-bisazoisobutyronitrile yields a low-molecular-weight (< 5,000) product (1534). Polymerization of alkyl vinyl sulfones with redox initiators in emulsion has been described (1608a), and the strong base, Triton B, converts ethyl vinyl sulfone into a brown, viscous oil (1534a). Treatment of methyl vinyl sulfone with potassium in bis(2-ethoxyethyl) ether or lithium in hexane or tetrahydrofuran gave no polymer. Only a dimeric product, $CH_3SO_2(CH_2)_4SO_2CH_3$, was obtained (1608b). Low reactivities are shown by the sulfones in free-radical copolymerizations with styrene (1498,1609), vinyl acetate (1498), acrylonitrile (1610), butadiene (1608a), methyl acrylate, and methacrylic acid (1534b). Butyl vinyl sulfone exhibits a much higher reactivity in anionic copolymerization with acrylonitrile, using sodium in liquid NH_3 as initiator (1546). Recalling that ethyl vinyl sulfone and divinyl sulfide are also polymerized by bases and that vinyl sulfides, sulfoxides, and sulfones undergo nucleophilic addition, it is surprising that anionic polymerization of these monomers, which show so little promise for the preparation of resins by free-radical or cationic processes, has received so little attention.

Copolymerization parameters are presented in Tables 24 and 25.

G. Applications of Polymers

Although considerable effort has been expended by the chemical industry on vinyl sulfur monomers, only **71** has achieved any measure of commercial success. The reasons for this are not difficult to find. The sulfides, sulfoxides, and sulfones do not polymerize easily by free-radical processes and yield low-molecular-weight products with poor mechanical properties by cationic polymerization. As suggested above, further research on anionic initiation would seem to be worthwhile because it could lead to high-molecular-weight resins and, therefore, determination of the optimum properties of which these systems are capable.

TABLE 24

Relative Reactivity Ratios for Copolymerization of Vinyl Sulfur Monomers (M_1)[a]

M_1	Comonomer (M_2)	r_1	r_2	Reference
$(CH_2{=}CH)_2S$	Methyl methacrylate	0.13 ± 0.05	0.85 ± 0.05	1553
	Styrene	0.47 ± 0.05	1.9 ± 0.1	1553
$(CH_2{=}CH)_2SO_2$	Acrylonitrile[b]	0.364	1.94	1601
	Methyl methacrylate	0.045	8.5	1600
	Styrene	0.01	1.3	1600
$CH_2{=}CHSR$:				
R = CH_3	Methyl acrylate	0.05	0.35	1498a
	Styrene	0.12	5.1	1498a
	Vinylene carbonate	11 ± 1	0.05 ± 0.04	1611
R = C_2H_5	Methyl acrylate	0.3 ± 0.1	2.7 ± 1.5	1543
	Styrene	0.25 ± 0.1	6.0 ± 1.5	1543
R = C_6H_5	Methyl acrylate	0.05	0.40	1498b
	Styrene	0.15	4.5	1498b
R = C_6Cl_5	Methyl acrylate	0.25	0.89	1606
	Styrene	0.24	3.9	1606
$CH_2{=}CHSOCH_3$	Methyl methacrylate	0	20 ± 10	315
	Styrene	0.01 ± 0.01	4.2 ± 0.2	1520
$CH_2{=}CHSO_2R$:				
R = CH_3	Methyl methacrylate	0	14 ± 2	1612
	Styrene	0.01	2.0	1498a
		0.01	1.40	1498a
		0.00 ± 0.02	2.4 ± 0.1	1612
	Vinyl acetate	0.4	0.3	1498a
		0.40 ± 0.08	0.00 ± 0.01	1612
R = C_4H_9	Acrylonitrile[c]	0.2 ± 0.1	1.1 ± 0.2	1546
R = C_6H_5	Styrene	0.01	3.3	1498b
	Vinyl acetate	0.35	0.28	1498b
$CH_2{=}CHSO_3H$	Acrylamide[d]	0	14.9	1584a
	Acrylate anion[d]	0	5.8	1584a
	Acrylonitrile	0.15	1.5	1448
			4.5[e]	1588
		0.12	4.5 ± 4.7	1613
	Methyl methacrylate	0.0	v. large	1365
$CH_2{=}CHSO_3\text{-}n\text{-}C_4H_9$	Methyl acrylate	0.11 ± 0.03	5.0 ± 1.5	1595
	Styrene	0.13 ± 0.03	2.5 ± 1	1595
	Vinyl acetate	0.20 ± 0.05	0.04 ± 0.01	1595
	Vinyl chloride	0.30 ± 0.05	0.35 ± 0.05	1595
	Vinylidene chloride	0.065 ± 0.007	7.5 ± 0.6	1595

[a] Copolymerizations are free-radical–initiated unless otherwise noted.
[b] Cyclization relative reactivity ratio (r_c) 0.067.
[c] Anionic initiation.
[d] Copolymerization in aqueous solution. M_1 is the ethenesulfonate anion.
[e] Copolymerization in aqueous zinc chloride solution.

TABLE 25
Q and e Values for Vinyl Sulfur Monomers

Monomer	Q	e	Reference
$(CH_2{=}CH)_2S$	0.68	-1.13	1553
	0.58	-1.11	137,146
	0.48	-1.08	1553
$(CH_2{=}CH)_2SO_2$	0.14	1.4	1600
	0.14	1.33	146
$CH_2{=}CHSR$:			
$R = CH_3$	0.35	-1.4	1498a
	0.34	-1.5	1498a
	0.32	-1.45	137,146
	-2.5 kcal/mole $(q)^a$	-0.43×10^{-10} esu $(\epsilon)^a$	147
$R = C_2H_5$	0.37	-0.12	137,146
	0.3	-1.3	1543
	0.08	-1.27	1543
$R = C_6H_5$	0.35	-1.4	1498b
	0.34	-1.40	137,146
	0.29	-1.4	1498b
	-2.4 kcal/mole $(q)^a$	-0.40×10^{-10} esu $(\epsilon)^a$	147
$R = C_6Cl_5$	0.23	-0.6	1606
	0.22	-0.58	137,146
$CH_2{=}CHSOCH_3$	0.10	0.9	1520
	0.057	0.98	137,146
	-1.6 kcal/mole $(q)^a$	0.28×10^{-10} esu $(\epsilon)^a$	147
$CH_2{=}CHSO_2R$:			
$R = CH_3$	0.15	1.3	1498a
	0.11	1.29	137,146
	0.11	1.2	1498a
	0.07	1.2	1498a
	-1.2 kcal/mole $(q)^a$	0.33×10^{-10} esu $(\epsilon)^a$	147
$R = C_6H_5$	0.07	1.2	1498b
	0.07	1.0	1498b
	0.069	1.18	137,146
	-1.2 kcal/mole $(q)^a$	0.33×10^{-10} esu $(\epsilon)^a$	147
$CH_2{=}CHSO_3H$	0.093	-0.02	146
	0.09	1.3	1448
	0.064^b	0.41^b	137
	0.048^b	0.51^b	1614
	0.036^b	0.06^b	1614
$CH_2{=}CHSO_3\text{-}n\text{-}C_4H_9$	0.13	1.19	137,146
	0.021	0.84	1595
	-0.8 kcal/mole $(q)^a$	0.37×10^{-10} esu $(\epsilon)^a$	147

[a] Schwan-Price resonance q and electrical ϵ factors (147).
[b] For copolymerization of the ethenesulfonate anion in aqueous solution.

1489

Ethenesulfonic acid (71) and its derivatives are the most promising monomers in the group. Commercial interest in them is based on the polarity or acidity they impart to polymers. Although there is lively competition in this field from such monomers as acrylic acid, *N*-vinylpyrrolidone, and vinylpyridines and from the introduction of sulfonic acid groups into inexpensive polystyrene by sulfonation, 71 and its derivatives may well find a market in special applications. They polymerize satisfactorily and are relatively cheap to produce. Poly(sodium ethenesulfonate) is sold by Farbwerke Hoechst under the trade name Pergalen as a blood anticoagulant (1446). Stable copolymer emulsions are obtained without the need of an auxiliary emulsifying agent by using small amounts of 71 or one of its salts in the formulation (1446,1615–1617). Dyeability of polyacrylonitrile is improved by incorporating low percentages of 71 as comonomer (1446,1448,1617,1618). Ethenesulfonyl chloride (1593a), ethyl ethenesulfonate (1593b), and vinyl sulfones (1610) are also useful for the same purpose. Salts of 71 polymers have been patented as plasticizers for concrete mortars that reduce the amount of water required (1619). Introduction of 71 into synthetic rubbers as a comonomer provides vulcanization sites (1446). Other uses that have been suggested for 71 polymers and copolymers are textile finishing (1446,1617), soil conditioning (1617), thickening (1617,1620), and cation exchange (1446,1597,1621).

VI. VINYL PHOSPHORUS COMPOUNDS

$$CH_2{=}CHPO(OR)_2 \qquad CH_2{=}CRPO(OR')_2 \qquad CH_2{=}CHCH{=}CHPO(OR)_2$$
$$\textbf{86} \qquad\qquad\qquad \textbf{87} \qquad\qquad\qquad\qquad \textbf{88}$$

$$CH_2{=}CHOPO(OR)_2 \qquad RPO(OR')OCH{=}CH_2$$
$$\textbf{89} \qquad\qquad\qquad \textbf{90}$$
$$(R \text{ and } R' = H, \text{ alkyl, or aryl})$$

A. Introduction

The first compound of types 86 to 90, 1-phenylvinylphosphonic acid (87, R = C_6H_5, R' = H), was reported by Conant et al. (1622a) in 1921. The 1-methyl analog (87, R = CH_3, R' = H) and some of its derivatives were described by Hamilton (1623) in 1944 and 1945. Kosolapoff (1624) discussed the preparation of diethyl 1,3-butadienylphosphonate (88, R = C_2H_5) in a 1945 patent. Two years later, Ford-Moore and Williams (1625) and Kabachnik (1626) reported the synthesis of diethyl vinylphosphonate (86, R = C_2H_5). Kabachnik also described di-2-chloroethyl vinylphosphonate in his paper. Upson (1627) claimed the preparation of divinyl phenylphosphonate (90, R = C_6H_5, R' = vinyl) and diethyl vinyl phosphate (89, R = C_2H_5) in a 1951 patent, but the latter product was subsequently shown to have been impure by Allen and Johnson (1628), who synthesized it by another route.

In the 25 years since the study of vinyl phosphorus monomers became active, many have been made and their properties examined. Most are vinylphosphonate esters and vinyl esters of phosphoric or phosphonic acids. The U.S.S.R., where organophosphorus chemistry has been stressed for many years, is the major source of the literature in this field, with the United States in second position. Like their sulfur analogs, vinyl phosphorus monomers have not achieved great commercial success. Di-2-chloroethyl vinylphosphonate (**86**, R = $ClCH_2CH_2$) is available from the Specialty Chemical Division of Stauffer Chemical Co. and was produced by Monsanto in the 1950s. There is no evidence that it is being used in quantity or that the other monomers described in this section are produced industrially today. Again, the problem is that these compounds are reluctant to undergo free-radical polymerization and yield low-molecular-weight products when they do polymerize, apparently because of chain transfer to monomer. The best results have been obtained by anionic polymerization of vinylphosphonate esters, so that any future commercial developments involving vinyl phosphorus monomers would be expected to occur in this direction.

B. Synthetic Methods

1. VINYLPHOSPHONIC ACID DERIVATIVES

Esters of vinylphosphonic acid are most frequently prepared by dehydrohalogenation of 2-haloethylphosphonate esters—Eq. (175):

$$XCH_2CH_2PO(OR)_2 \xrightarrow{\text{base}} CH_2{=}CHPO(OR)_2 + HX \qquad (175)$$

The starting materials are obtained by the Michaelis-Arbuzov reaction— Eq. (176) (1629):

$$RX + P(OR')_3 \longrightarrow [RP(OR')_3]^+X^- \xrightarrow{\text{heat}} RPO(OR')_2 + R'X \qquad (176)$$

in which an alkyl halide adds to a trialkyl phosphite to form the phosphonium halide intermediate, which eliminates alkyl halide on heating to give the phosphonate ester. An important special instance of the reaction is the rearrangement of tri-2-haloethyl phosphites, which have both the alkyl halide and trialkyl phosphite structures in the same molecule—Eq. (177) (1630a,b):

$$P(OCH_2CH_2X)_3 \xrightarrow{\text{heat}} XCH_2CH_2PO(OCH_2CH_2X)_2 \qquad (177)$$

These are readily available from PX_3 and ethylene oxide—Eq. (178) (1630b,c):

$$PX_3 + 3CH_2\overset{\diagdown}{\underset{O}{\diagup}}CH_2 \longrightarrow P(OCH_2CH_2X)_3 \qquad (178)$$

The resulting phosphonate esters (91) can be dehydrohalogenated directly to di-2-haloethyl vinylphosphonates. In addition, they are useful intermediates for the preparation of other esters of 2-haloethylphosphonic acids by conversion to the acid dichloride—Eq. (179) (1630a,b,d)—which is then allowed to react with the appropriate alcohol or phenol in the presence of a tertiary amine as acid acceptor—Eq. (180) (1631–1633):

$$XCH_2CH_2PO(OCH_2CH_2X)_2 + 2PCl_5 \longrightarrow$$
$$\underset{\textbf{91}}{}$$
$$XCH_2CH_2POCl_2 + 2POCl_3 + 2XCH_2CH_2Cl \quad (179)$$

$$XCH_2CH_2POCl_2 + 2ROH \xrightarrow{\text{base}} XCH_2CH_2PO(OR)_2 + 2HCl \quad (180)$$

The acid dichloride has also been made by hydrolysis of 91 to the acid and treatment of this with PCl_5 (1634). An acid acceptor is not essential in the preparation of aryl esters from the acid dichloride and phenols. Simply heating at 150 to 190° gives the desired reaction (1635).

Alternatively, alkyl 2-haloethylphosphonates may be prepared by the Michaelis-Arbuzov reaction, starting with a 1,2-dihaloethane—Eq. (181) (1625,1632,1636–1638):

$$XCH_2CH_2X + P(OR)_3 \longrightarrow XCH_2CH_2PO(OR)_2 + RX \qquad (181)$$

The alkyl halide is removed by distillation as rapidly as it is formed, and an excess of dihalide is used in order to minimize the formation of the Michaelis-Arbuzov by-products, $RPO(OR)_2$ and $(RO)_2POCH_2CH_2$-$PO(OR)_2$. Preparation of the trialkyl phosphite and its reaction with the dihalide can be combined into one step, using aluminum chloride as catalyst—Eq. (182) (1639):

$$PCl_3 + 3ROH + ClCH_2CH_2Cl \xrightarrow{AlCl_3}$$
$$ClCH_2CH_2PO(OR)_2 + 3HCl + RCl \quad (182)$$

2-Haloethylphosphonate esters have been dehydrohalogenated—Eq. (175)—with triethylamine (1625,1631,1633,1635,1638–1643), alcoholic potassium hydroxide (1626,1632,1636,1637,1644), and an alkali metal carbonate or carboxylate (1645–1647) as bases. Other syntheses of vinylphosphonate esters are alcoholysis of vinylphosphonyl dichloride (1648, 1649), dehalogenation of dialkyl 1,2-dihaloethylphosphonates with zinc dust (1650), and the reaction of vinylmagnesium bromide with a dialkyl chlorophosphite followed by oxidation—Eq. (183) (1651,1652):

$$CH_2{=}CHMgBr + (RO)_2PCl \longrightarrow CH_2{=}CHP(OR)_2 + MgBrCl$$
$$\underset{\xrightarrow{MnO_2}\; CH_2{=}CHPO(OR)_2}{} \quad (183)$$

Vinylphosphonyl dichloride is obtained by the reaction in Eq. 179 but starting with the vinylphosphonate ester (1653a,c) and by dehydrohalogenation—Eq. (184) (1648):

$$ClCH_2CH_2POCl_2 \xrightarrow[\substack{330-340° \\ N_2}]{BaCl_2} CH_2{=}CHPOCl_2 + HCl \qquad (184)$$

Vinylphosphonic acid has been prepared by hydrolysis of the acid dichloride (1648,1653b) or the trimethylsilyl ester—Eq. (185) (1654):

$$CH_2{=}CHPO(OR)_2 + 2Me_3SiCl \longrightarrow CH_2{=}CHPO(OSiMe_3)_2 + 2RCl$$
$$\downarrow H_2O \qquad (185)$$
$$CH_2{=}CHPO(OH)_2 + 2HOSiMe_3$$

The reaction of vinylphosphonyl dichloride and dimethylamine yields the corresponding diamide (1648). Di-2-chloroethyl vinylphosphonate and PCl_5 in 1,1,2-trichloroethane give 2-chloroethyl vinylphosphonochloridate —Eq. (186):

$$CH_2{=}CHPO(OCH_2CH_2Cl)_2 + PCl_5 \longrightarrow$$

$$CH_2{=}CHPO \underset{\diagdown OCH_2CH_2Cl}{\overset{\diagup Cl}{}} + POCl_3 + ClCH_2CH_2Cl \qquad (186)$$

from which monoamides and mixed esters can be prepared by treatment with amines and alcohols, respectively (1655).

Conant et al. (1622a) found that treatment of acetophenone with PCl_3 in acetic acid followed by hydrolysis produces a mixture of 1-hydroxy-1-phenylethylphosphonic and 1-phenylvinylphosphonic acids. Similar results were obtained with acetone—Eq. (187):

$$\underset{R}{\overset{H_3C}{\diagdown}}C{=}O + PCl_3 \xrightarrow{HOAc} \xrightarrow{H_2O}$$

$$\underset{R}{\overset{H_3C}{\diagdown}}\underset{PO(OH)_2}{\overset{OH}{C}} + CH_2{=}\underset{R}{\overset{}{C}}{-}PO(OH)_2 \qquad (187)$$

Hamilton (1623) later demonstrated that the hydroxy acid from acetone is dehydrated on heating, and so this is probably the source of the unsaturated acid in both instances. If the reaction of acetophenone and PCl_3 is carried out in the presence of HCl, the product is 1-chloro-1-phenylethylphosphonic acid, which undergoes dehydrochlorination at 180° to give 1-phenylvinylphosphonic acid in 90 percent yield (1622b).

The mechanisms of these processes have not been established, but it seems logical that the first step is nucleophilic addition of PCl_3 to the carbonyl group—Eq. (188):

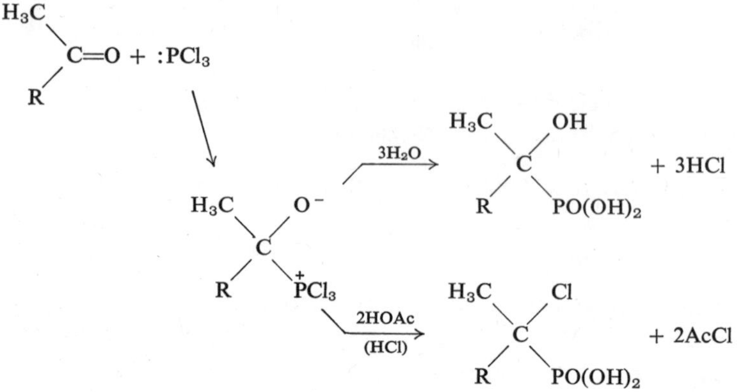

$$\underset{R}{\overset{H_3C}{\diagdown}}C=O + :PCl_3$$

$$\underset{R}{\overset{H_3C}{\diagdown}}\underset{\overset{+}{PCl_3}}{\overset{|}{\underset{|}{C}}} O^-$$

$$\xrightarrow{3H_2O} \quad \underset{R}{\overset{H_3C}{\diagdown}}\underset{PO(OH)_2}{\overset{OH}{\underset{|}{C}}} \quad + \ 3HCl$$

$$\xrightarrow[\text{(HCl)}]{2HOAc} \quad \underset{R}{\overset{H_3C}{\diagdown}}\underset{PO(OH)_2}{\overset{Cl}{\underset{|}{C}}} \quad + \ 2AcCl$$

(188)

Hamilton (1623) obtained the acid dichloride from crude 1-methyl-vinylphosphonic acid by treatment with PCl_5 or $SOCl_2$. From this he prepared the pure acid by hydrolysis, esters by alcoholysis in the presence of the corresponding sodium alcoholates, and amides by reaction with ammonia or amines. Diethyl 1-phenylvinylphosphonate has been prepared by a Grignard synthesis—Eq. (189) (1656):

$$C_6H_5CH_2PO(OC_2H_5)_2 + (CH_3)_2CHMgCl \longrightarrow$$

$$C_6H_5\underset{\overset{|}{MgCl}}{CHPO}(OC_2H_5)_2 + CH_3CH_2CH_3$$

$$C_6H_5\underset{\overset{|}{MgCl}}{CHPO}(OC_2H_5)_2 + CH_2O \longrightarrow$$

(189)

$$CH_2=\underset{\overset{|}{C_6H_5}}{C}-PO(OC_2H_5)_2 + HOMgCl$$

Alkyl 1-carboalkoxyvinylphosphonates are obtained by reaction of an acrylate ester with PCl_3 and oxygen to give the β-chlorophosphonyl dichloride, alcoholysis to the phosphonate ester, and dehydrochlorination —Eq. (190):

$$CH_2=CHCOOR + PCl_3 + \tfrac{1}{2}O_2 \longrightarrow ClCH_2\underset{\overset{|}{COOR}}{CHPO}Cl_2 \xrightarrow[-2HCl]{2R'OH}$$

$$ClCH_2\underset{\overset{|}{COOR}}{CHPO}(OR')_2 \xrightarrow{(C_2H_5)_3N} CH_2=\underset{\overset{|}{COOR}}{C}-PO(OR')_2 + HCl \quad (190)$$

Starting with acrylonitrile instead of an acrylate results in a mixture of 1- and 2-cyanovinylphosphonates (1644).

The addition of PCl_5 to 1-olefins—Eq. (191):

$$RCH{=}CH_2 + PCl_5 \xrightarrow{\text{benzene}} RCHClCH_2PCl_4 \qquad (191)$$

is a well-known reaction (1629). With butadiene, both 1,2 and 1,4 addition have been shown to occur—Eq. (192):

$$CH_2{=}CHCH{=}CH_2 + PCl_5 \underset{1,4}{\overset{1,2}{\lessgtr}} \begin{array}{l} CH_2{=}CHCHClCH_2PCl_4 \\[1em] ClCH_2CH{=}CHCH_2PCl_4 \end{array} \qquad (192)$$

Anisimov and Kolobova (1657) obtained formic acid by the oxidation of the adduct, after conversion to the phosphonyl dichloride with SO_2, clearly demonstrating the presence of a vinyl group and, therefore, 1,2 addition. The product obtained in the same manner by Pudovik and Konovalova (1658), however, gave chloroacetic acid on oxidation, and its infrared spectrum showed the presence of an internal (trans) $C{=}C$ bond, indicating 1,4 addition.

Kosolapoff (1624b) prepared 1,3-butadienylphosphonic acid by hydrolysis of the butadiene–PCl_5 adduct, dehydrochlorination occurring simultaneously without addition of base—Eq. (193):

$$\left\{ \begin{array}{l} CH_2{=}CHCHClCH_2PCl_4 \\ \quad\quad \text{or} \\ ClCH_2CH{=}CHCH_2PCl_4 \end{array} \right\} + 3H_2O \longrightarrow CH_2{=}CHCH{=}CHPO(OH)_2 + 5HCl \qquad (193)$$

He synthesized the diethyl ester by ethanolysis of the adduct, followed by dehydrochlorination with KOH—Eq. (194) (1624a):

$$\left\{ \begin{array}{l} CH_2{=}CHCHClCH_2PCl_4 \\ \quad\quad \text{or} \\ ClCH_2CH{=}CHCH_2PCl_4 \end{array} \right\} + 3C_2H_5OH \xrightarrow[-C_2H_5Cl]{-3HCl}$$

$$\left\{ \begin{array}{l} CH_2{=}CHCHClCH_2PO(OC_2H_5)_2 \\ \quad\quad \text{or} \\ ClCH_2CH{=}CHCH_2PO(OC_2H_5)_2 \end{array} \right\} \xrightarrow[C_2H_5OH]{KOH}$$

$$CH_2{=}CHCH{=}CHPO(OC_2H_5)_2 + HCl \qquad (194)$$

The corresponding phosphonyl dichlorides, which are obtained from the adduct by treatment with SO_2, also yield 1,3-butadienylphosphonate esters on alcoholysis and dehydrochlorination with base (1657,1658). Another method of preparing the intermediate chloroesters is the Michaelis-Arbuzov reaction of 1,4-dichloro-2-butene and a trialkyl phosphite—Eq. (195) (1624a,1658):

$$ClCH_2CH{=}CHCH_2Cl + (RO)_3P \longrightarrow$$

$$ClCH_2CH{=}CHCH_2PO(OR)_2 + RCl \qquad (195)$$

2. VINYL ESTERS OF PHOSPHORIC AND PHOSPHONIC ACIDS

The first reported synthesis of a vinyl phosphate and a vinyl phosphonate were by dehydrochlorination of the corresponding 2-chloroethyl esters—Eqs. (196) and (197) (1627):

$$(C_2H_5O)_2PO-OCH_2CH_2Cl \xrightarrow[C_2H_5OH]{KOH} (C_2H_5O)_2PO-OCH=CH_2 \quad (196)$$

$$C_6H_5PO(OCH_2CH_2Cl)_2 \xrightarrow[C_2H_5OH]{KOH} C_6H_5PO(OCH=CH_2)_2 \quad (197)$$

For the first reaction—Eq. (196)—however, it was later shown that the main product is triethyl phosphate, apparently because of ester interchange—Eq. (198) (1628,1659,1660):

$$(C_2H_5O)_2PO-OCH_2CH_2Cl + C_2H_5OH \xrightarrow{KOH}$$
$$(C_2H_5O)_3PO + HOCH_2CH_2Cl \quad (198)$$

Even with a better leaving group (Br), a stronger base (NaH), and ethanol-free ether as the diluent, the dehydrohalogenation of diethyl 2-bromoethyl phosphate gave only a 10 percent yield of diethyl vinyl phosphate (1661).

The Perkow reaction—Eq. (199) (1662,1663):

$$(RO)_3P + R'CO\overset{|}{\underset{|}{C}}-X \longrightarrow (RO)_2PO-OCR'=C\diagdown + RX \quad (199)$$

and related processes are much better synthetic methods. Allen and Johnson (1628) prepared pure diethyl vinyl phosphate and diethyl 1-methylvinyl phosphate for the first time by the Perkow route—Eqs. (200) and (201):

$$(C_2H_5O)_3P + ClCH_2CHO \longrightarrow (C_2H_5O)_2PO-OCH=CH_2 + C_2H_5Cl \quad (200)$$
$$(C_2H_5O)_3P + ClCH_2COCH_3 \longrightarrow$$
$$(C_2H_5O)_2PO-OC(CH_3)=CH_2 + C_2H_5Cl \quad (201)$$

It has been used by others to synthesize diethyl vinyl phosphate (1663, 1664,1665a), di-2-chloroethyl vinyl phosphate (1665a), and diethyl 1-carbethoxyvinyl phosphate (1663).

Related methods are the reactions of triethyl phosphite with $ClHgCH_2CHO$ or $BrHgCH_2COCH_3$ to give diethyl vinyl and diethyl 1-methylvinyl phosphate, respectively (1666), and with chloroethylene carbonate to give diethyl vinyl phosphate—Eq. (202) (1659):

$$(C_2H_5O)_3P + \underset{O}{\overset{Cl}{\big\lceil}} \underset{\underset{O}{\overset{\|}{C}}}{O} \xrightarrow[\text{reflux}]{\text{benzene}}$$
$$(C_2H_5O)_2PO-OCH=CH_2 + CO_2 + C_2H_5Cl \quad (202)$$

Since it has been shown (1667) that this carbonate ester decomposes to chloroacetaldehyde—Eq. (203):

$$\text{(cyclic chlorocarbonate)} \xrightarrow{\text{catalyst}} CO_2 + \underset{O}{CH_2{-}CHCl} \xrightarrow{\text{heat}} ClCH_2CHO \qquad (203)$$

the second process may be actually a Perkow reaction with a novel source of the haloaldehyde.

A general method of preparing vinyl esters of phosphorous, phosphonic, and phosphoric acids has been described by Gefter and Kabachnik —Eq. (204) (1668):

$$\diagdown\!P{-}Cl + CH_3CHO \xrightarrow{R_3N} \diagdown\!P{-}OCH{=}CH_2 + HCl$$

$$\diagdown\!PO{-}Cl + CH_3CHO \xrightarrow{R_3N} \diagdown\!PO{-}OCH{=}CH_2 + HCl \qquad (204)$$

The authors proposed the mechanism in Eq. (205), based on isolation of salts of type **92** from runs in which trimethylamine was the base used:

$$\diagdown\!PO{-}Cl + CH_3CHO \longrightarrow \diagdown\!PO{-}OCHClCH_3 \xrightarrow{R_3N}$$

$$\diagdown\!PO{-}\underset{\underset{\mathbf{92}}{R_3N^+Cl^-}}{OCHCH_3} \longrightarrow \diagdown\!PO{-}OCH{=}CH_2 + R_3NH^+Cl^- \qquad (205)$$

Diethyl vinyl phosphate has also been synthesized (90 percent yield) by an exchange reaction between divinylmercury and diethyl phosphate (1669).

C. Properties

1. PHYSICAL

Boiling points, melting points, refractive indices, and densities are presented in Table 26. For homologous series, only the lower members (through butyl) have been included. With the exceptions of 1-phenyl-vinylphosphonic acid and its anilide, the compounds are liquids at room temperature. This acid and amide and vinylphosphonic acid are water-soluble. The other monomers are generally soluble only in organic solvents. The pK_1 and pK_2 values of vinylphosphonic acid in aqueous ethanol were found (1652) to be 2.20 to 3.90 and 7.23 to 8.54, respectively, depending on the concentration of ethanol (7 to 80 percent).

TABLE 26
Physical Properties of Vinyl Phosphorus Compounds

Compound	bp, °C (mm)	Reference	n_D (°C)	Reference	d_4 (°C)	Reference
CH_2=$CHPO(OR)_2$:						
R = CH_3	72.5(10)	1648	1.4305(20)	1644	1.1405(20)	1648
	54(2)	1642	1.4330(20)	1648	1.1696(20)	1644
	33(0.2)	1638	1.4095(25)	1638		
R = C_2H_5	99(16)	1640	1.4320(15)	1625	1.0526(20)	1626
	92(12)	1641	1.4296(20)	1641	1.0529(20)	1641
	70(3)	1626,1647	1.4300(20)	1626,1642	1.0550(20)	1648
	63(2.5)	1625	1.4338(20)	1648	1.0569(20)	1642
	60.5(2)	1642	1.4260(25)	1632		
	60(2)	1641	1.4262(25)	1637		
	50(1)	1632	1.4268(25)	1640		
R = n-C_3H_7	83.5(3)	1639	1.4350(20)	1639	1.0057(20)	1639
R = i-C_3H_7	60(5)	1648	1.4263(20)	1639	0.9908(20)	1648
	59(2)	1639	1.4268(20)	1648	0.9948(20)	1639
R = n-C_4H_9	116(5)	1636	1.4365(20)	1651	0.9810(20)	1636
	89.5(2)	1651	1.4372(20)	1636	0.9812(20)	1651
	79(2)	1642	1.4390(20)	1642	0.9816(20)	1642
R = i-C_4H_9	78.5(3)	1642	1.4356(20)	1642	0.9730(20)	1642
R = CH=CH_2	50(2)	1668	1.4530(20)	1668	1.1020(20)	1668
R = CH_2CH=CH_2	88(1)	1633	1.4598(20)	1643	1.1234(20)	1643
	84(1)	1643	1.4555(20)	1633	1.1222(20)	1633
R = CH_2CH_2Cl	139(4)	1626	1.4787(20)	1631	1.3233(20)	1631
	137(4)	1631	1.4785(20)	1645	1.3205(20)	1645
	132(3)	1631	1.4772(20)	1626	1.3182(20)	1626

Compound	bp (°C/mm) or mp	ν (cm^{-1})	n_D	ν (cm^{-1})	d	ν (cm^{-1})
(continued)	122(2)	1645	1.475(25)	1670	1.314(25)	1670
(continued)	132(1)	1670	1.4759(25)	1671	1.3198(25)	1671
(continued)	130(1)	1671	1.4742(20)	1672	1.2593(20)	1672
(continued)	112(0.4)	1646	1.5519(20)	1642	1.1647(20)	1642
(continued)	116(0.008)	1672	1.5555(20)	1635	1.1930(20)	1648
$R = CH_2CH{-}CH_2$ (O, epoxide)	$62(<10^{-3})$	1642	1.5571(20)	1648	1.1947(20)	1635
$R = CH_2C_6H_5$	142(2–3)	1635	1.5681(20)	1648	1.3422(20)	1648
$R = C_6H_5$	$110(10^{-4})$	1648	1.5512(20)	1648	1.1439(20)	1653b
$R = p\text{-}ClC_6H_4$	$189(10^{-4})$	1648	1.4737(20)	1652	1.398(20)	1652
$R = m,p\text{-}Me_2C_6H_3$	$192(10^{-4})$	1648	1.4735(20)	1653b	1.3892(20)	1648
$CH_2{=}CHPO(OH)_2$	69(21)	1648	1.4710(20)	1648	1.3888(20)	1648
$CH_2{=}CHPOCl_2$	61(11)	1653	1.4808(20)	1653	1.4092(20)	1648
$CH_2{=}CHPO(NMe_2)_2$	82(2)	1648	1.4795(20)	1648	1.0257(20)	1648
$CH_2{=}CMePO(OR)_2$: R = CH$_3$	46(2)	1623a	1.4732(20)	1623a	1.1268(20)	1644
R = $n\text{-}C_4H_9$	87(0.25)	1623a	1.4340(20)	1623a	1.1621(20)	1644
$CH_2{=}C(COOMe)PO(OR)_2$: R = CH$_3$	106(2)	1644	1.4376(20)	1644		
R = C_2H_5	119(3)	1644	1.4412(20)	1644		
$CH_2{=}CPhPO(OH)_2$	mp 113°	1622				
$CH_2{=}CPhPO(OH)_2$	mp 115°	1673				
$CH_2{=}CMePOCl_2$	86(32)	1623a	1.4348(20)	1623a		
$CH_2{=}CPhPO(NMe_2)_2$	80(3)	1623a	1.4735(20)			
$CH_2{=}CPhPO(NHPh)_2$	mp 181°	1622a				

continued

TABLE 26 (continued)

Compound	bp, °C (mm)	Reference	n_D (°C)	Reference	d_4 (°C)	Reference
$CH_2=CHCH=CHPO(OR)_2$:						
R = CH_3	78.5(3)	1657	1.4835(20)	1657	1.1021(20)	1657
R = C_2H_5	123(13)	1658	1.4728(20)	1657	1.0451(20)	1658
	87(2)	1657	1.4749(20)	1658	1.0465(20)	1657
R = $n\text{-}C_3H_7$	104(1)	1657	1.4700(20)	1657	1.0163(20)	1657
R = $i\text{-}C_3H_7$	84.5(2)	1657	1.4632(20)	1657	1.0030(20)	1657
R = $n\text{-}C_4H_9$	129(2)	1657	1.4680(20)	1657	0.9922(20)	1657
R = $i\text{-}C_4H_9$	113(2)	1657	1.4642(20)	1657	0.9819(20)	1657
R = $CH_2CH=CH_2$	100(2)	1657	1.4872(20)	1657	1.0541(20)	1657
$CH_2=CHOPO(OR)_2$:						
R = C_2H_5	105(25)	1666	1.4109(20)	1669	1.0926(20)	1669
	99(20)	1666	1.4134(20)	1663	1.0724(35)	1628,1665
	89(10)	1664	1.4109(25)	1659		
	79(6)	1628,1665	1.4100(35)	1628,1665		
	59(1)	1669				
	56(1)	1659				
	50(0.2)	1663				
R = CH_2CH_2Cl	148(1)	1665				
$RPO(OCH=CH_2)_2$:						
R = CH_3	65(8)	1668	1.4394(20)	1668	1.1097(20)	1668
R = C_2H_5	61(5.5)	1668	1.4409(20)	1668	1.0707(20)	1668
R = CH_2Cl	67(1)	1668	1.4636(20)	1668	1.2458(20)	1668
R = C_6H_5	111(2)	1668	1.5144(20)	1668	1.1589(20)	1668
$(CH_2=CHO)_3PO$	85(10)	1668	1.4314(20)	1668	1.1240(20)	1668

2. PHYSIOLOGICAL

The physiological properties of only one of the monomers discussed in this section have been reported. Di-2-chloroethyl vinylphosphonate has an oral LD_{50} for rats of 1.40 g per kg of body weight. The minimum lethal dose for rabbits, when applied undiluted to the unabraded skin, is 4.2 to 6.6 g per kg. Undiluted monomer causes mild blistering of the skin and severe irritation and swelling of the eyes in rabbits. The eye irritation persists for several days but there is no apparent permanent damage. Rats survived a 6-hr exposure to an atmosphere saturated with the monomer, the only observed effects being excessive salivation, labored breathing, and lethargy (1670).

3. ANALYTICAL METHODS AND SPECTRA

No specific analytical procedures for these monomers were found in the literature. Kosolapoff (1674) has briefly discussed analytical methods for organophosphorus compounds in general, including atomic refraction and parachor values. References to spectrometric studies are listed in Table 27. The mass spectrum of diethyl vinylphosphonate has been determined (1678).

TABLE 27
Literature References to Spectra of Vinyl Phosphorus Compounds

Compound	Type of spectrum			
	UV	IR	Raman	NMR
$CH_2=CHPO(OR)_2$	1675,1676	1676–1678	1675	1676
$CH_2=CHPOCl_2$	1675	1677	1675	
$CH_2=CHPO(NMe_2)_2$	1675		1675	
$CH_2=CPhPO(OH)_2$		1673		1673
$CH_2=CHCH=CHPO(OR)_2$		1677		
$CH_2=CHOPO(OR)_2$				73

D. Storage and Handling

Because some nerve gases and insecticides are organophosphorus compounds, there is a widely held belief that they are highly toxic in general. Although this is a misconception, it is perhaps the best philosophy to adopt in handling any organophosphorus compound of unknown toxicity (1674). All but one of the monomers in this section are in this category.

The exception is di-2-chloroethyl vinylphosphonate. Tests on animals (see Sec. VI-C-2) indicate that this compound should be classified as slightly toxic with regard to both oral ingestion and absorption through the skin. Its irritant effects on skin and eyes make avoidance of repeated or prolonged skin contact and of any eye contact advisable. Voluntary exposure to high vapor concentrations is unlikely because of their irritating nature. A polymerization inhibitor is not required for storage (1670).

E. Chemical Reactions

Bromine and chlorine add to the $C{=}C$ double bond in dialkyl vinylphosphonates (1626,1648), 1-phenylvinylphosphonic acid (1622b), and diethyl vinyl phosphate (1661,1665b). The main product from bromination of diethyl vinylphosphonate (1648) is diethyl 1-bromovinylphosphonate—Eq. (206):

$$CH_2{=}CHPO(OC_2H_5)_2 + Br_2 \xrightarrow{CHCl_3} BrCH_2CHBrPO(OC_2H_5)_2$$

$$\downarrow {-HBr} \qquad (206)$$

$$CH_2{=}CBrPO(OC_2H_5)_2$$
$$\text{Major product}$$

undoubtedly because of dehydrobromination of the 1,2-dibromo compound. Diethyl vinyl phosphate and a mixture of Cl_2 and Br_2 yield diethyl 1-chloro-2-bromoethyl phosphate (1661,1665b). Hydrogen bromide adds to 1-phenylvinylphosphonic acid to give a mixture of 1-bromo- and 2-bromo-1-phenylethylphosphonic acid (1622b).

Hydrogenation of diethyl vinylphosphonate over PtO_2 (1637) and of diethyl vinyl phosphate over Pd (1661) produce the corresponding saturated compounds. The former undergoes electrolytic reductive coupling in aqueous tetraethylammonium p-toluenesulfonate—Eq. (207) (1679):

$$2CH_2{=}CHPO(OC_2H_5)_2 + 2H \xrightarrow[\text{reduction}]{\text{electrolytic}}$$
$$(C_2H_5O)_2POCH_2CH_2CH_2CH_2PO(OC_2H_5)_2 \quad (207)$$

Nucleophilic additions to dialkyl vinylphosphonates occur in the same manner as to α,β-unsaturated carbonyl compounds—Eq. (208):

$$CH_2{=}CHPO(OR)_2 + HA \xrightarrow{\text{base}} ACH_2CH_2PO(OR)_2 \quad (208)$$

Reagents that have been shown to add in this way are alcohols (1680a, 1681), ammonia and amines (1680b,1682), hydrogen sulfide and thiols (1636,1680a), malonic esters (1680a,1683), cyanoacetic and acetoacetic esters (1683), dialkyl phosphites (1682,1684a), and thioacetic acid (1681). The relative ease with which dialkyl vinylphosphonates undergo anionic

polymerization is another example of nucleophilic addition. In consideration of the poor results obtained with free-radical initiation of these
monomers, it is surprising that the demonstrably better anionic route has
not been investigated more thoroughly.

Dialkyl vinylphosphonates also undergo the Meerwein reaction with
benzenediazonium halides—Eq. (209) (1685):

$$CH_2=CHPO(OC_2H_5)_2 + C_6H_5N_2{}^+X^- \longrightarrow$$
$$C_6H_5CH_2CHXPO(OC_2H_5)_2 + N_2 \quad (209)$$

and the Diels-Alder reaction with dienes (1636,1684b,1686,1687), as
exemplified by Eq. (210) for butadiene (1686):

$$CH_2=CHPO(OC_2H_5)_2 + \underset{5 \text{ hr}}{\overset{150°}{\longrightarrow}} \quad (210)$$

Amine salts of vinylphosphonic acid have been made from the acid
(1654) and from its sodium salt (1655). Salts of calcium, barium, magnesium, zinc, and other divalent metals can be prepared from the metal
chloride and an alkyl (1688a) or 2-chloroethyl (1688b) vinylphosphonate—
Eq. (211):

$$2CH_2=CHPO(OR)_2 + MCl_2 \xrightarrow{\text{heat}} M^{++}[CH_2=CHPO(OR)O]_2{}^- + 2RCl \quad (211)$$

An anhydroester has been synthesized from diethyl vinylphosphonate
and $SOCl_2$—Eq. (212) (1689):

$$2CH_2=CHPO(OC_2H_5)_2 + SOCl_2 \longrightarrow$$
$$(CH_2=CHP(OC_2H_5)O)_2O + SO_2 + 2C_2H_5Cl \quad (212)$$

Hydrolysis of diethyl vinyl phosphate with aqueous ethanolic HCl at
80° attacks only the vinyloxy group—Eq. (213). Sodium iodide in ethyl
acetate attacks an ethoxy group, and acidification of the product in ether
forms a solution of ethyl vinyl phosphate. Evaporation of the solvent,
however, leaves only monoethyl phosphate as a residue—Eq. (214)
(1690):

$$(C_2H_5O)_2PO(OCH=CH_2) \xrightarrow[\text{HCl}]{\text{EtOH—H}_2\text{O}} (C_2H_5O)_2PO(OH) + CH_3CHO \quad (213)$$

$$(C_2H_5O)_2PO(OCH=CH_2) \xrightarrow[\text{EtOAc}]{\text{NaI}} C_2H_5OPO(OCH=CH_2)O^-Na^+$$

$$\xrightarrow[\text{ether}]{\text{HCl}} C_2H_5OPO(OCH=CH_2)OH \xrightarrow[\text{ether}]{\text{evaporate}} C_2H_5OPO(OH)_2 \quad (214)$$
$$\text{In solution}$$

F. Polymerization

In 1945, Kosolapoff (1624a) reported that diethyl 1,3-butadienyl-phosphonate polymerizes with benzoyl peroxide as initiator to a sulfur-vulcanizable gum. Later, he described the polymerization of the corresponding acid to a water-soluble elastomer (1624b). In 1947, Kabachnik (1626) stated that diethyl vinylphosphonate forms a viscous polymer when heated with benzoyl peroxide.

Since these earliest publications, most of the research on free-radical polymerization of vinyl phosphorus monomers has been on dialkyl vinylphosphonates (1632,1637–1639,1642,1644,1649,1691–1699). The products invariably have a low molecular weight, and the monomers are unreactive in copolymerizations. (Kosolapoff (1632) referred to diethyl vinylphosphonate as having "mildly expressed polymerizability.") Pike and Cohen (1693) were the first to explore the reaction in some depth, establishing by infrared spectrometry that polymerization does occur through the vinyl double bond. They attributed the low molecular weights to facile chain transfer by way of the alkoxy groups attached to phosphorus. Later kinetic studies (1649,1696b,1698) and the observation (1700) that diphenyl vinylphosphonate–styrene copolymers have a high molecular weight confirmed this explanation. The activation energy for polymerization of dimethyl vinylphosphonate is reported to be 22 kcal per mole (1649) and that of di-2-chloroethyl vinylphosphonate 30.4 kcal per mole (1696a). Not surprisingly, the trifunctional divinyl vinylphosphonate gives a cross-linked product on polymerization with benzoyl peroxide (1668).

A novel combination of vinyl and condensation polymerization of di-2-chloroethyl vinylphosphonate has been described (1701). Heating the monomer yields a polyester by elimination of 1,2-dichloroethane—Eq. (215):

$$n CH_2{=}CHPO(OCH_2CH_2Cl)_2 \xrightarrow[220-230°]{N_2}$$
$$ClCH_2CH_2{\{}O{-}PO{-}OCH_2CH_2{\}}_n Cl + (n-1)ClCH_2CH_2Cl \quad (215)$$
$$\overset{|}{CH{=}CH_2}$$

Free-radical polymerization of this intermediate through the vinyl groups produces a cross-linked infusible resin. The order of these operations can be reversed with the radical polymerization of the monomer first, followed by cross-linking by way of pyrolytic elimination of $ClCH_2CH_2Cl$.

The most promising results have been obtained with ionic initiators. Di-2-chloroethyl vinylphosphonate gives scarcely any polymer with $SnCl_4$ or $TiCl_4$ (1699), but a crystalline polymer, soluble in DMF and infusible at the charring temperature (620°), is obtained with a Ziegler-Natta catalyst (1702). Diethyl vinylphosphonate polymerizes to a tough

TABLE 28
Literature References to the Polymerization of Vinyl Phosphorus Monomers

Monomer	Type of study[a]	Reference
CH_2=CHPO(OH)$_2$	P, C	1704
	C	1705
CH_2=CHPOCl$_2$	P, C	1704
CH_2=CHPO(OR)OH and salts	P, C	1706
CH_2=CMePO(OMe)$_2$	C	1707,1708
CH_2=CPhPO(OH)$_2$	P	1709
	C	1707,1710–1713
CH_2=CPhPO(OR)OH	C	1714
CH_2=CPhPO(OR)$_2$	C	1715
CH_2=C(COOMe)PO(OR)$_2$	P, C	1644
RPO(OCH=CH$_2$)$_2$ and R$_2$PO—OCH=CH$_2$[b]	P	1668
(EtO)$_2$PO—OCH=CH$_2$	P	1664
(EtO)$_2$PO—OC(COOEt)=CH$_2$	P	1663

[a] P denotes polymerization and C copolymerization.
[b] R = alkyl, ClCH$_2$, ClCH$_2$CH$_2$, alkoxy, or aryloxy.

elastomer in the presence of triethylaluminum (1703). Both of these monomers yield high polymers with anionic initiators such as Grignard reagents and sodium naphthalene (1697,1699).

Polymerization studies of other vinyl phosphorus monomers are summarized in Table 28. Unsaturated esters of phosphorus acids have been polymerized with a TiCl$_4$–Et$_3$Al catalyst (1716). Copolyesters of vinylphosphonic acid and diethylene glycol are cross-linkable through the vinyl groups (1717). Copolymerization parameters are listed in Table 29.

G. Applications of Polymers

There is no hard evidence that any vinyl phosphorus monomer is being used industrially. The only one that is commercially available, di-2-chloroethyl vinylphosphonate, has been suggested for use in flame-resistant polyester formulations (1671), in extreme pressure lubricants (1671), as a comonomer in polyacrylonitrile fibers to impart dyeability and self-extinguishing characteristics (1692), and in polypropylene, by grafting, to improve adhesion to metals (1718). Vinylphosphonic acid copolymers have been suggested as flame-resistant treating agents in textiles and coatings (1705) and as cation-exchange resins (1713,1719). Monoesters of vinylphosphonic acids and their salts are said to be useful in plastics, textiles, coatings, and as emulsifiers (1706).

TABLE 29
Relative Reactivity Ratios for Copolymerization of Vinyl Phosphorus Monomers (M_1)

M_1	Comonomer (M_2)	r_1	r_2	Reference
CH_2=$CHPO(OR)_2$:				
R = CH_3	Styrene	0.4	4.61	1698
R = C_2H_5[a]	Styrene	0.06 ± 0.02	8.87 ± 0.14	1697
		0	4.1	1698
		0	3.25	1637
R = n-C_3H_7	Styrene	0.9	4.24	1698
R = i-C_3H_7	Styrene	0	2.39	1698
R = n-C_4H_9	Styrene	0.3	5.4	1698
R = i-C_4H_9	Styrene	0.5	4.4	1698
R = CH_2CH_2Cl	Methacrylic acid	0.1 ± 0.1	1.7 ± 0.5	1696b
	Styrene	0.2 ± 0.2	2.2 ± 0.4	1696b
		0.03	2.43	1699
R = C_6H_5	Styrene	0	2.03	1698
CH_2=$CPhPO(OH)_2$	Acrylic acid[b]	0.44 ± 0.03	0.98 ± 0.08	1711
	Acrylonitrile[b]	0.32 ± 0.07	0.69 ± 0.18	1711
	Methacrylic acid	0.36 ± 0.12	3.5 ± 0.2	1710
	Methyl methacrylate	0.06 ± 0.04	3.3 ± 0.2	1710

[a] Q values of 0.13 (137) and 0.09 (146) and e values of 0.26 (137) and 0.25 (146) have been calculated for this monomer.

[b] Q and e values of 0.80 ± 0.02 and 0.76 ± 0.04, respectively, were calculated from these data for M_1.

VII. VINYLFURAN, VINYLTHIOPHENE, AND NONCOMMERCIAL STYRENES

93   94

A. Introduction

Styrenes, 2-vinylfuran (vinylfuran) (93), and 2-vinylthiophene (vinyl-thiophene) (94) are the vinyl aromatic monomers that have received the most attention, aside from vinylpyridines. Commercially available styrenes have already been discussed in Part 2, Chap. 1, and will not be dealt with further here. Also because of this earlier discussion, the treatment of noncommercial styrenes will be concerned mainly with monomer preparation and physical properties. Vinylthiophene is included here rather than in Sec. V, because it is more closely related to vinylfuran and styrenes in polymerization behavior than to other sulfur-containing monomers.

Vinylfuran (93) was first prepared by Liebermann (1720) in 1894 by decarboxylation of 3-(2-furyl)acrylic acid—Eq. (216):

$$\text{(furan)}-CH{=}CHCOOH \xrightarrow{\text{heat}} \text{(furan)}-CH{=}CH_2 + CO_2 \qquad (216)$$

Its styrene-like odor and polymerization in the presence of light and air were noted by Moureu, Dufraisse, and Johnson (1721) in 1927. The preparation and properties of vinylthiophene (94) were first described by Kuhn and Dann (1722) in 1941. Both compounds were among the co-monomers evaluated in synthetic diene rubbers during and just after the Second World War (1723). This application and other wartime require-ments also provided the stimulus for the preparation and evaluation of many substituted styrenes. Vinylfuran, vinylthiophene, and most of the substituted styrenes discussed in this section polymerize and copolymerize readily. They are not used in industry, however, because their polymers are similar to those obtained from cheaper, commercially available styrenes.

The reactions of styrenes (1724a) and the preparation of substituted styrenes by methods not involving hydrocarbon cracking (1724b) were reviewed by Emerson in 1949.

B. Synthetic Methods

1. VINYLFURAN (93)

The most common synthesis of 93 is the Perkin reaction (1725) of furfural with acetic anhydride—Eq. (217):

$$\text{(furan)}-CHO + (CH_3CO)_2O \xrightarrow{\text{NaOAc}}$$

$$\text{(furan)}-CH{=}CHCOOH + CH_3COOH \qquad (217)$$

followed by decarboxylation of the product, 3-(2-furyl)acrylic acid—Eq. (216) (1721,1726–1729). A modification in which malonic acid and quino-line are the acid component and basic catalyst in the Perkin step has also been used (1730,1731)—Eq. (218):

$$\text{(furan)}-CHO + CH_2(COOH)_2 \xrightarrow{\text{quinoline}} \text{(furan)}-CH{=}C(COOH)_2 \xrightarrow{\text{heat}}$$

$$\text{(furan)}-CH{=}CH_2 + 2\,CO_2 \qquad (218)$$

The decarboxylation—Eq. (216)—is run at 250 to 320° (1720,1721,1726). An inert diluent, such as methane or nitrogen, may be used (1732). Less polymerization occurs if the pyrolysis is done in quinoline with copper sulfate added as an inhibitor (1727,1728). Overall yields of 64 to 71 percent of **93** from furfural have been reported (1727,1730).

A related synthesis is the passage of a furfural–acetic anhydride mixture over H_2SO_4-activated bentonite that produces both **93** and 2-ethylfuran (1733). There is no evidence of the intermediate formation of 3-(2-furyl)-acrylic acid as in the Perkin reaction.

Vinylfuran has also been prepared from furfural and ketene by way of 3-(2-furyl)-3-propiolactone (1734)—Eq. (219):

$$\text{furan-CHO} + CH_2{=}C{=}O \xrightarrow{\text{catalyst}} \text{furan-CHCH}_2\text{CO (O)} \longrightarrow$$

$$\text{furan-CH}{=}CH_2 + CO_2 \qquad (219)$$

The furfural-ketene condensation is catalyzed by boric acid, its esters and acetyl derivatives, and salts of carboxylic acids. Dehydration of 1-(2-furyl)- and 2-(2-furyl)ethanol over alumina at 390 to 400° yields a mixture of **93** and 2-ethylfuran (1735).

2. VINYLTHIOPHENE (94)

A synthetic method that could most readily be scaled up to an industrial process is alkylation of thiophene with ethylene to 2-ethylthiophene and dehydrogenation of the latter to **94** (1736,1737). This is analogous to commercial methods of preparing styrene and 2-methyl-5-vinylpyridine.

A good laboratory procedure is the chloroethylation of thiophene with paraldehyde and HCl, followed by dehydrochlorination (1738)—Eq. (220):

$$\text{thiophene} + (CH_3CHO)_n + HCl \longrightarrow \text{thiophene-CHClCH}_3 \xrightarrow{\text{pyridine}}$$

$$\text{thiophene-CH}{=}CH_2 \qquad (220)$$

Another is dehydration of a 2-(hydroxyethyl)thiophene. The starting materials for this approach have been obtained in a variety of ways. 2-(2-Hydroxyethyl)thiophene is prepared from a 2-halothiophene by reaction with ethylene oxide in the presence of sodium amalgam (1739) or by way of the Grignard reagent (1740–1743). It is also produced, along with **94**, from 2-methylthiophene and formalin by passage over H_3PO_4

on silica-alumina at 200 to 230° (1744). Potassium hydroxide is the preferred reagent for dehydrating 2-(2-hydroxyethyl)thiophene to **94**. 2-(1-Hydroxyethyl)thiophene is obtained by Meerwein-Ponndorf-Verley reduction of 2-acetylthiophene (1722,1745,1746) or from 2-thienylmagnesium bromide and acetaldehyde (1747). Dehydration occurs on heating the alcohol at 130 to 145° with hydroquinone as inhibitor (1722, 1745) or at 250 to 310° and 30 to 100 mm over activated alumina (1746, 1747).

Pyrolysis of a 1,1-diarylethane, a good method for the preparation of styrenes (see Sec. VII-B-3-f), has not been successful when applied to the synthesis of **94**. Although analysis of the crude product from 1,1-di(2-thienyl)ethane indicated a 90 percent yield, only a very small amount of pure **94** was isolated on work-up of the mixture (1748).

Vinylthiophene has also been prepared from 2-thienylmagnesium bromide and vinyl chloride in the presence of cobalt(II) chloride (1749), from 2-thiophenecarboxaldehyde and acetic anhydride passed in the vapor phase over H_2SO_4-activated bentonite (1733), and from 2-acetylthiophene and diethyl phosphite (1750)—Eq. (221):

$$\text{thiophene-COCH}_3 + HPO(OC_2H_5)_2 \xrightarrow{NaNH_2} \text{thiophene-}\underset{\underset{OH}{|}}{\overset{\overset{CH_3}{|}}{C}}PO(OC_2H_5)_2 \xrightarrow{heat}$$

$$\text{thiophene-}CH{=}CH_2 + HOPO(OC_2H_5)_2 \qquad (221)$$

3. SUBSTITUTED STYRENES

a. Dehydration of an Aryl Methyl Carbinol—Eq. (222)

$$ArCHOHCH_3 \longrightarrow ArCH{=}CH_2 + H_2O \qquad (222)$$

This is the most common method of preparing substituted styrenes. The reaction is usually accomplished by heating the alcohol at reduced pressure in the presence of a dehydrating agent, such as $KHSO_4$, P_2O_5, or Al_2O_3, and a polymerization inhibitor, and distilling the styrene as it is formed.

b. Dehydration of a 2-Arylethanol—Eq. (223)

$$ArCH_2CH_2OH \longrightarrow ArCH{=}CH_2 + H_2O \qquad (223)$$

The procedure is similar to 3a, but the dehydrating agent is generally KOH or Al_2O_3.

c. Dehydrohalogenation of an Arylhaloethane—Eq. (224)

$$ArC_2H_4X \longrightarrow ArCH{=}CH_2 + HX \qquad (X = Br\ or\ Cl) \qquad (224)$$

Both 1- and 2-arylhaloethanes are used. The dehydrohalogenation can sometimes be effected by heat alone, but a base, such as an amine or KOH, is usually employed.

d. Elimination of Acetic Acid from an Arylethyl Acetate—Eq. (225)

$$ArCH(OAc)CH_3 \longrightarrow ArCH{=}CH_2 + HOAc \qquad (225)$$

The reaction normally proceeds smoothly on pyrolysis of the pure ester at 250° or higher. However, a base, for example, methanolic KOH, has also been used to encourage the elimination.

e. Decarboxylation of a Cinnamic Acid—Eq. (226)

$$ArCH{=}CHCOOH \xrightarrow{heat} ArCH{=}CH_2 + CO_2 \qquad (226)$$

Better yields are obtained by conducting the reaction in a high-boiling tertiary amine in the presence of copper or $CuSO_4$ as polymerization inhibitor. A variation is the dehydrobromination and decarboxylation of a 3-bromo-3-arylpropionic acid without isolation of the intermediate cinnamic acid—Eq. (227):

$$ArCHBrCH_2COOH \xrightarrow[heat]{base} ArCH{=}CH_2 + HBr + CO_2 \qquad (227)$$

f. Pyrolysis of a 1,1-Diarylethane—Eq. (228)

$$Ar_2CHCH_3 \longrightarrow ArCH{=}CH_2 + ArH \qquad (228)$$

The usual catalyst is a hydrated aluminum silicate (clay).

g. Dehydrogenation of an Arylethane—Eq. (229)

$$ArCH_2CH_3 \longrightarrow ArCH{=}CH_2 + H_2 \qquad (229)$$

Typical reaction conditions are 600 to 700° in steam over a metallic oxide or activated carbon catalyst.

h. Dehalogenation of an Aryl-1,2-dihaloethane—Eqs. (230) and (231)

$$ArCHXCH_2X + Zn \longrightarrow ArCH{=}CH_2 + ZnX_2 \qquad (230)$$

$$ArCHXCH_2X + 2KI \longrightarrow ArCH{=}CH_2 + 2KX + I_2 \qquad (231)$$
$$(X = Br\ or\ Cl)$$

i. Pyrolysis of a 1-Aryl-(1-hydroxy)ethylphosphonate Ester—Eq. (232)

$$ArC(OH)(CH_3)PO(OEt)_2 \longrightarrow ArCH{=}CH_2 + HOPO(OEt)_2 \qquad (232)$$

The reaction probably occurs by way of rearrangement of the phosphonate to the phosphate, $ArCH(CH_3)OPO(OEt)_2$.

j. Pyrolysis of an Arylethyl Sulfite—Eq. (233)

$$(ArCH_2CH_2O)_2SO \xrightarrow{290-295°} 2ArCH{=}CH_2 + H_2O + SO_2 \qquad (233)$$

k. Condensation of a Grignard Reagent and an Organic Halide—Eq. (234)

$$ArMgX + CH_2{=}CHX \longrightarrow ArCH{=}CH_2 + MgX_2 \qquad (234)$$
$$(\text{or } CH_2{=}CHMgX + ArX)$$
$$(X = \text{Br or Cl})$$

l. Reaction of an Aryl Methyl Ketone and a Primary Alcohol—Eq. (235)

$$ArCOCH_3 + C_2H_5OH \xrightarrow[\text{oxide catalyst}]{200-500°} ArCH{=}CH_2 + CH_3CHO + H_2O \quad (235)$$

The ketone is reduced by the alcohol to the aryl methyl carbinol, which is dehydrated under the reaction conditions.

m. Condensation of a Benzaldehyde and Acetic Anhydride—Eq. (236)

$$ArCHO + Ac_2O \longrightarrow ArCH{=}CH_2 + HOAc + CO_2 \qquad (236)$$

n. Pyrolysis of a 3-Aryl-3-propiolactone—Eq. (237)

$$ArCHCH_2CO \longrightarrow ArCH{=}CH_2 + CO_2 \qquad (237)$$
$$\underset{\displaystyle O}{\underline{\qquad\qquad}}$$

o. Pyrolysis of a Polystyrene—Eq. (238)

$$(-CHArCH_2-)_n \longrightarrow nArCH{=}CH_2 \qquad (238)$$

The starting material for this method is polystyrene. Groups are introduced into the aromatic rings of the polymer by conventional substitution reactions, and pyrolysis then yields the substituted styrene.

p. Addition of a Substituted Benzene to Acetylene—Eq. (239)

$$ArH + CH{\equiv}CH \xrightarrow[\text{H}_2\text{O,5°}]{\text{HF, HgO}} ArCH{=}CH_2 \qquad (239)$$

The following procedures are not general, being useful only for specific monomers.

q. Demethanation of a Substituted Cumene—Eq. (240)

$$ArCH(CH_3)_2 \longrightarrow ArCH{=}CH_2 + CH_4 \qquad (240)$$

This reaction is limited to cumenes with a substituent, for example, CH_3, ortho to the isopropyl group.

r. *Elimination of Water and Isobutene from a* tert-*Butylaryl Methyl Carbinol—Eq.* (241)

$$+ H_2O + (CH_3)_2C{=}CH_2 \qquad (241)$$

The starting material is obtained by Friedel-Crafts acetylation and the *tert*-butyl group serves to locate the acyl function between the methyl substituents.

s. *Pyrolysis of Chlorinated 4-Vinylcyclohexene (Butadiene Dimer)—Eq.* (242)

$$+ H_2 + HCl \qquad (242)$$

t. *Reduction of Divinylbenzene—Eq.* (243)

u. *Chlorination of Methylstyrene—Eq.* (244)

$$CH_3C_6H_4CH{=}CH_2 + Cl_2 \longrightarrow ClCH_2C_6H_4CH{=}CH_2 + HCl \qquad (244)$$

v. *Chloromethylation of Styrene—Eq.* (245)

$$C_6H_5CH{=}CH_2 + CH_2O + HCl \longrightarrow ClCH_2C_6H_4CH{=}CH_2 + H_2O \qquad (245)$$

In Table 31 the designations above (3a–3v) indicate the methods by which the individual monomers have been made.

4. SUBSTITUTED α-METHYLSTYRENES

a. Dehydration of an Aryl Dimethyl Carbinol—Eq. (246)

$$ArC(CH_3)_2OH \longrightarrow ArC(CH_3)=CH_2 + H_2O \qquad (246)$$

The usual reaction conditions are similar to those for method 3a above.

b. Dehydrogenation of a Substituted Cumene—Eq. (247)

$$ArCH(CH_3)_2 \longrightarrow ArC(CH_3)=CH_2 + H_2 \qquad (247)$$

This reaction is run under approximately the same conditions as in method 3g for substituted styrenes. An exception is the dehydrogenation of *p*-cymene with chloranil, which requires a temperature of only 140° (1751).

c. Dehydrochlorination of a 1-Chloro-2-arylpropane—Eq. (248)

$$ArCH(CH_3)CH_2Cl \xrightarrow{KOH} ArC(CH_3)=CH_2 + HCl \qquad (248)$$

d. Dehydrogenation of a Terpene Hydrocarbon

This method yields only *p*-methyl-α-methylstyrene. The hydrocarbons that have been used as starting materials are *p*-menthane (**95**), dipentene (**96**), α-pinene (**97**), and terpinolene (**98**). A related procedure is dehydrogenation and dehydration of α-terpineol (**99**) (method 4e):

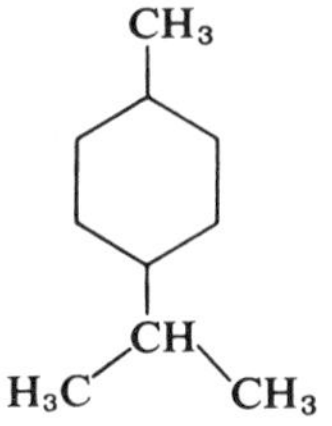

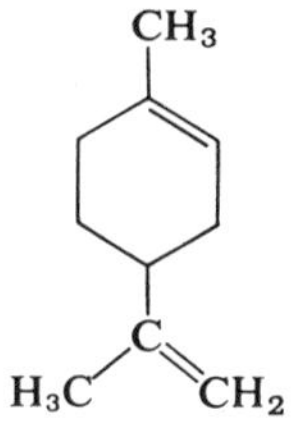

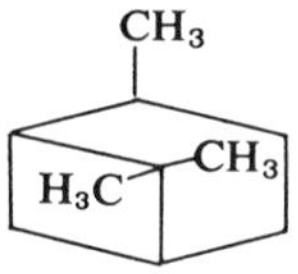

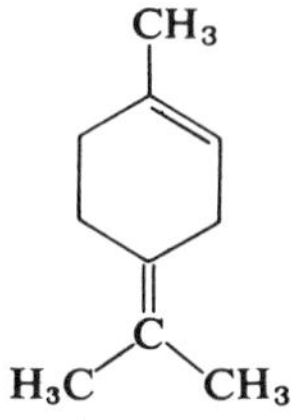

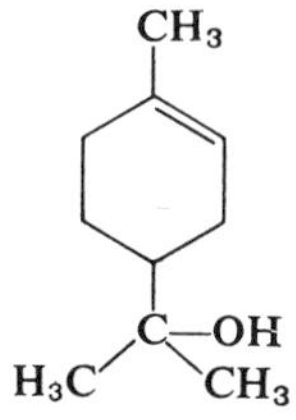

Other procedures, each of which has been reported only once, are pyrolysis of poly(3,4-dichloro-α-methylstyrene) (method 4f), acetate ester pyrolysis (method 4g), decarboxylation of 3-(*p*-tolyl)crotonic acid (method

4h), decarboxylation and dehydration of 3-hydroxy-3-(*p*-tolyl)butyric acid—Eq. (249) (method 4i):

$$CH_3C(OH)CH_2COOH \qquad CH_3C{=}CH_2$$

$$\xrightarrow{\text{heat}} \qquad\qquad + H_2O + CO_2 \qquad (249)$$

and dimerization of isopropenylacetylene—Eq. (250) (method 4j):

$$CH_3C{=}CH_2$$

$$CH_2{=}C(CH_3)C{\equiv}CH \xrightarrow{120°} \qquad\qquad (250)$$

$$CH_3$$

In Table 32 these designations (4a to 4j) are used to show the methods by which the various monomers have been prepared.

5. α-ETHYLSTYRENE (100)

This compound was first synthesized (1752) by decarboxylation of 3-phenyl-2-pentenoic acid—Eq. (251):

$$CH_3CH_2CPh{=}CHCOOH \xrightarrow[\text{heat}]{50\% \ H_2SO_4} CH_3CH_2CPh{=}CH_2 + CO_2 \qquad (251)$$

100

and later by dehydration of 2-phenylbutanol over KOH (1753,1754). It is also obtained by way of a rearrangement in the elimination of arylsulfonic acid from 1-methyl-2-phenylpropyl *p*-tosylate or *p*-brosylate—Eq. (252) (1754):

$$CH_3CHPhCH(CH_3)Y \xrightarrow{\text{HOAc or } CH_3CN} CH_3CH_2CPh{=}CH_2 + HY \qquad (252)$$

$$(Y = p\text{-}CH_3C_6H_4SO_2O\text{—} \ \text{or} \ p\text{-}BrC_6H_4SO_2O\text{—})$$

100 is obtained in good yield by pyrolysis of 2-phenylbutyl acetate (79 percent) (1755) and dehydrochlorination of 2-phenylbutyl chloride with ethanolic KOH (77 percent) (1756). It is the main product of the catalytic dehydrogenation of 2-phenylbutane (1757,1758). Two novel and related reactions that produce **100** are the decompositions of 5-ethyl-5-phenyl-1,3-dioxan-2-one—Eq. (253) (1759)—and 2-ethyl-2-phenyl-1,3-propylene bromohydrin—Eq. (254) (1760):

$$\xrightarrow[200-210°]{KCN} EtCPh{=}CH_2 + CH_2O + CO_2 \qquad (253)$$

$$\underset{Ph}{\overset{Et}{\diagdown}}\underset{CH_2OH}{\overset{CH_2Br}{\diagup}}C \xrightarrow[H_2O]{KOH} EtCPh{=}CH_2 + CH_2O + HBr \qquad (254)$$

C. Properties

1. PHYSICAL

Boiling points, melting points, refractive indices, and densities of vinylfuran, vinylthiophene, and α-ethylstyrene are listed in Table 30, those of substituted styrenes in Table 31, and those of substituted α-methylstyrenes in Table 32. The last two tables also indicate methods of preparation.

TABLE 30

Physical Properties of 2-Vinylfuran (93), 2-Vinylthiophene (94), and α-Ethylstyrene (100)

Compound	bp, °C (mm)	Reference	n_D (°C)	Reference	d_4 (°C)	Reference
93	19(17)	1721	1.4992(18.5)	1721,1761	0.950(15)	1762
			1.5000(20)	1733		
	25(27)	1721	1.5007(20)	1763	0.9445(18.5)	1721,1761
	33(57)	1721	1.4985(26)	1762	0.9436(20)	1763
	50(130)	1721	1.4994(28)	1763	0.945(20)	1733,1762
	98(735)	1734b			0.941(25)	1762
	99(760)	1720,1730, 1731			0.9392(28)	1763
	99.8(760)	1733				
	100(760)	1721,1761			0.937(30)	1762
	97(764)	1742,1763				
	mp −94°	1721				
94	51(28)	1747	1.5618(20)	1722	1.0410(20)	1742
			1.5679(20)	1733		
	59(46)	1764	1.5720(20)	1739	1.0429(20)	1745
	66.5(48)	1739,1742	1.5722(20)	1742	1.043(20)	1733
	63(50)	1722,1736, 1746	1.5612(23.4)	1745	1.044(20)	1722
	63.5(50)	1733				
	67(50)	1738	1.5697(25)	1749		
	73(69)	1749	1.5698(25)	1746		
	78(70)	1741	1.5701(25)	1738		
	86(109)	1748	1.5707(25)	1764		
	141.0(760)	1733	1.5731(25)	1747		
100	82(20)	1752,1756	1.5270(20)	1759	0.8868(25)	1755
	180(742)	1753	1.5262(22)	1753	0.8984(25)	1754
	180	1759,1760	1.5262(25)	1754		
	181	1755	1.5264(25)	1755		
	182	1754	1.5265(25)	1765		
	183	1765				

TABLE 31

Physical Properties of and Synthetic Methods for Substituted Styrenes,

R-C₆H₄-CH=CH₂ (R at ring position)

R	Synthetic method[a]	Reference	bp, °C (mm)	Reference	n_D (°C)	Reference	d_4 (°C)	Reference
o-C_2H_5	3t	1766	68.5(12)	1766	1.53805(20)	1767	0.90576(20)	1767
			100(45)	1768	1.5422(22)	1766	0.9119(22)	1766
			104(57)	1768	1.53565(25)	1767	0.90168(25)	1767
			109(66)	1768				
			112(76)	1768				
			128(132)	1768				
			mp −75.60°	1767				
m-C_2H_5	3a	1746	74(14)	1746	1.53512(20)	1767	0.89449(20)	1767
			107(57)	1768	1.5315(25)	1746	0.89045(25)	1767
			115(76)	1768	1.53250(25)	1767		
			128(124)	1768				
			158(316)	1768				
			mp −101.3°	1767				
p-C_2H_5	3a	1746,1769–1775	40(1)	1775	1.5377(15)	1776	0.9074(13)	1777
	3b	1778	45(1)	1779	1.5340(20)	1780	0.8953(15)	1776
	3c	1776,1777	61(5)	1781	1.5352(20)	1774	0.89249(20)	1767
	3f	1748,1781	67(9)	1774	1.5365(20)	1779	0.8927(20)	1774
	3g	1782	68(9)	1780	1.5371(20)	1781	0.8935(20)	1781
	3i	1780	73(10)	1772	1.53763(20)	1767	0.8997(20)	1780
			68(11)	1776	1.53484(25)	1767	0.88845(25)	1767
			71(14)	1778	1.5349(25)	1772	0.8950(25)	1778
			68(16)	1746,1775	1.5350(25)	1746,1775		
			87(17)	1773	1.5366(25)	1778		
			86(20)	1770,1771, 1777	1.5349(25.5)	1775		
			117(79)	1768				
			177	1778				
			mp 49.72°	1767				

1516

$m\text{-}i\text{-}C_3H_7$	3d	1787	79(13)	1.5253(25)	1787	0.8497(20)	1767
$p\text{-}i\text{-}C_3H_7$	3a	1773,1774,1788–1791	44(0.95)	1.5281(20)	1792		1791
	3b	1793	45(1.3)	1.5283(20)	1774	0.8860(20)	1791
	3f	1748	57(2)	1.5288(20)	1791	0.8861(22)	1792
			63(4)	1.52891(20)	1767	0.88101(25)	1767
			68(5)	1.5290(20)	1779,1790	0.8835(25)	1793
			80(10)	1.5265(24)	1773	0.8867(25)	1788
			86(12)	1.5245(25)	1788		
			88(14)	1.5246(25)	1793		
			120(57)	1.5253(25)	1787		
			128(78)	1.52650(25)	1767		
			142(127)				
			173(315)				
			195(770)				
$m\text{-}tert\text{-}C_4H_9$	3a	1794	mp −44.66°				
			75(5)	1.5234(20)	1794	0.897(20)	1794
			100(17)	1.5237(20)	1779		
$p\text{-}tert\text{-}C_4H_9$	3a	1746,1795–1797	54(0.7)	1.5260(20)	1779	0.8866(20)	1797
	3f	1748	59(1.2)	1.5267(20)	1797	0.8883(20)	1795
	3g	1796	86(7.5)	1.5270(20)	1795	0.897(20)	1794
			91.5(9)	1.5245(25)	1746	0.8972(20)	1798
			97(13)		1779		
			100(14)		1746		
$p\text{-}CH_2Cl$	3a	1799	mp −26.9°		1795		
CH_2Cl^b	3c	1800,1801	79(2)	1.5725(25)	1799		1799
	3u	1802					
	3v	1803					
$o\text{-}F$	3a	1774,1804	29(3)	1.5197(20)	1804	1.0253(20)	1805a
	3e	1805a,1806	34(3)	1.5200(20)	1774	1.0282(20)	1774
			30(4)	1.5201(20)	1779,1806	1.030(20)	1804
			46(32)				

continued

TABLE 31 (*continued*)

R	Synthetic method[a]	Reference	bp, °C (mm)	Reference	n_D (°C)	Reference	d_4 (°C)	Reference
m-F	3a	1774,1804,1807	32(2)	1807,1808	1.5170(20)	1774,1808	1.0177(20)	1774
			31(4)	1804	1.5173(20)	1804	1.025(20)	1804
			41(13)	1774				
p-F	3a	1774,1804,1805a, 1809–1813	29(4)	1811	1.5140(20)	1774	1.0220(20)	1812
	3b	1810	30(4)	1804	1.5150(20)	1812	1.0225(20)	1774
	3c	1814	36(7)	1805a	1.5156(20)	1808	1.024(20)	1804
			45(15)	1810	1.5129(25)	1813	1.029(20)	1805a
			59(25)	1814	1.5130(25)	1809	1.0178(25)	1809
			64(40)	1813	1.5131(25)	1810	1.019(25)	1810
			67.4(50)	1812	1.5158(25)	1804		
			69(52)	1809				
			mp −34.5°	1774				
				1809				
o-Br	3a	1807,1815,1816	65(3)	1807,1808	1.592(15)	1815	1.409(15)	1815
	3e	1805a	65(4)	1816	1.5914(20)	1808	1.3721(20)	1805a
			98(20)	1805a	1.59268(20)	1767	1.41601(20)	1767
			104(22)	1805a,1815	1.5893(25)	1816	1.41024(25)	1767
			110(38)	1768	1.59014(25)	1767		
			117(47)	1768				
			121(57)	1768				
			126(66)	1768				
			129(76)	1768				
			mp −52.75°	1767				
m-Br	3a	1807,1817	48.5(0.5)	1818	1.5855(20)	1817	1.4059(20)	1817
	3e	1818	75(3)	1807,1808	1.5900(20)	1818		
			94(20)	1817	1.5903(20)	1808		
p-Br	3a	1774,1805a,1807, 1815,1819–1824	57(2)	1822	1.599(18)	1820	1.401(18)	1820

1518

Substituent	Compound	Ref	bp °(mm)/mp	Ref	n_D	Ref	d	Ref
	3i	1750	84.5(11)	1819	1.5933(20)	1826	1.39838(20)	1767
	3k	1825	88(12)	1807,1808, 1820	1.59472(20)	1767	1.400(20)	1826
			88(13)	1826	1.5961(20)	1819	1.401(20)	1819
			90(15)	1824	1.5962(20)	1774	1.4025(20)	1774
			102(20)	1805a	1.59212(25)	1767	1.3976(23)	1821
			104(20)	1815			1.39263(25)	1767
			120(47)	1768				
			125(57)	1768				
			129(66)	1768				
			133(76)	1768				
			147(124)	1768				
			mp 4.5°	1819				
			mp 5.5°	1774				
			mp 6°	1821				
			mp 7.67°	1767				
Br^b	3b	1827						
	3o	1828						
	3e	1829–1832	mp 13.5°	1829				
o-NO$_2$			mp 14°	1830,1832				
m-NO$_2$	3a	1833–1835	81(1.2)	1835	1.5810(20)	1733	1.171(20)	1733
					1.5818(20)	1779,1833		
					1.5830(20)	1837	1.1552(32.5)	1834
	3e	1831,1836–1838	100(3)	1779,1833	1.5836(20)	1838		
			101.0(3)	1733	1.5810(25)	1835		
	3m	1733	96(3.5)	1837,1838	1.5801(27)	1838		
			107(8)	1834	1.5806(32.5)	1834		
			mp −10°	1834				
			mp −5°	1836				
p-NO$_2$	3c	1749,1839–1842	87(2)	1843				
	3e	1830,1831,1844	109(5)	1845				
	3h	1843,1845,1846	mp 20°	1839				
	3n	1844	mp 21°	1830				

continued

TABLE 31 (continued)

R	Synthetic method[a]	Reference	bp, °C (mm)	Reference	n_D (°C)	Reference	d_4 (°C)	Reference
2,4-(CH₃)₂			mp 21.4°	1749,1840				
			mp 23°	1841				
			mp 24°	1843,1845,1846				
			mp 29°	1842,1844				
	3a	1774,1847,1848	57(4)	1848	1.539(20)	1849	0.9049(20)	1848
	3b	1849	74(10)	1774	1.5302–	1850	0.905(20)	1849
					1.5400(20)			
	3c	1776	76(12)	1851	1.5405(20)	1851	0.9062(20)	1774
	3f	1748,1850,1852,1853	80(12)	1776	1.5416(20)	1774	0.9022(21.5)	1776
	3j	1849	85(14)	1849	1.5423(20)	1779,1847	0.8999(25)	1852
			90(25)	1779,1847	1.5426(20)	1848	0.904(25)	1847
			95.6(30)	1852	1.5398(25)	1852		
			mp − 64.32°	1848				
2,5-(CH₃)₂	3a	1774,1847,1848, 1854	69(10)	1776,1848	1.5395(20)	1779,1847, 1848	0.9072(17.5)	1776
	3c	1776	72(10)	1774	1.5397(20)	1774	0.9032(20)	1848
	3f	1748,1852	83(23)	1847	1.5370(25)	1852	0.9043(20)	1774
			94(30)	1852			0.899(25)	1847
			mp − 35.41°	1848			0.8990(25)	1852
2,6-(CH₃)₂	3a	1848	66(10)	1848,1855	1.5314(20)	1848,1855	0.9078(20)	1848
	3r	1855	mp − 38.65°	1848,1855				
3,4-(CH₃)₂	3a	1774,1847	75(10)	1774	1.5459(20)	1774	0.9062(20)	1780
	3f	1748,1780,1852	88(23)	1779	1.5463(20)	1847	0.9080(20)	1774
			96(26)	1847	1.5465(20)	1779,1780	0.9024(25)	1852

Substituent	Isomer	Ref.	bp/mp		n_D		d	
3,5-(CH₃)₂			100.7(30)	1852	1.5466(25)	1852	0.906(25)	1847
			106(36)	1780				
	3a	1847	56(4)	1848	1.5374(20)	1774	0.8934(20)	1848
	3b	1774,1848,1856	58(4)	1779,1847	1.5382(20)	1779,1847, 1848	0.8957(20)	1774
(CH₃)₂[b]	3a	1857	62.3(5)	1774,1848	1.5328(23)	1856	0.894(25)	1847
	3c	1858	mp −60.19°	1848				
	3f	1859–1861						
	3g	1862,1863a						
	3p	1864						
	3q	1865						
2,4,5-(CH₃)₃	3a	1866	97(13)	1776	1.5379(17)	1776	0.916(16.5)	1866
	3c	1776,1867	97(22)	1867	1.5462(20)	1866	0.9137(17)	1776
	3d	1867	214	1867				
			216.8	1866				
				1866				
2,4,6-(CH₃)₃	3a	1774,1866–1869	mp 2.5°	1869	1.5296(17.5)	1776,1777	0.9073(17.5)	1776,1777
	3c	1776,1777,1867	75(5)	1868	1.5291(20)	1774	0.9057(20)	1776,1777
			88(12)	1776,1777	1.5293(20)	1869	0.9062(20)	1869
			92(14)	1774	1.5305(20)	1851	0.906(21)	1866
			94(17)	1851	1.5323(20)	1866,1868		
			104(25)	1776,1777				
			207(755)	1869				
			208(754)	1866				
			209	1866				
2,3-Cl₂	3a	1807,1870,1871	mp −37.0°	1870	1.5834(20)	1808	1.2826(20)	1871
			61(1)	1807,1871	1.5848(20)	1871	1.264(25)	1870
			94(5)	1808	1.5780(25)	1870		
			95(5)					

continued

TABLE 31 (continued)

R	Synthetic method[a]	Reference	bp, °C (mm)	Reference	n_D (°C)	Reference	d_4 (°C)	Reference
2,4-Cl₂	3a	1807,1870–1874	69(2.5)	1870	1.5828(20)	1808,1818, 1871	1.24(20)	1873
	3e	1818	81(6)	1807,1808, 1871–1874	1.5812(25)	1870	1.249(20)	1872
							1.243(25)	1871
							1.246(25)	1870
							1.246(20)	1872
2,5-Cl₂	3a	1804,1870,1872, 1875–1877	60(1.5)	1878	1.5798(20)	1804,1879	1.245(25)	1870
	3c	1875,1877,1879, 1880	73(2)	1804,1872	1.5785(25)	1875	1.246(25)	1875,1879
	3f	1748	74(3)	1870	1.5786(25)	1877		
	3I	1878	77.7(3.2)	1877	1.5788(25)	1870		
	3s	1881	93(5)	1879	1.5737(30)	1878		
			94(5)	1872				
			mp 8°	1882				
			mp 8.3°	1875				
			mp 8.5°	1877				
2,6-Cl₂	3a	1807,1870,1871	59(2)	1870	1.5752(20)	1871	1.2631(20)	1871
	3c	1883	65(3)	1807,1871	1.5754(20)	1808	1.280(25)	1870
			72(5)	1808	1.5724(25)	1870		
					1.5729(25)	1883		
3,4-Cl₂	3a	1746,1804,1807, 1870–1872	64(1.5)	1746	1.5847(20)	1818	1.256(20)	1804,1872
	3d	1871	76(3)	1870	1.5851(20)	1871	1.243(25)	1870
	3e	1818	84(4)	1808	1.5857(20)	1804		

Compound	Designation	Refs	bp/mp	Refs	n_D^{25}	Refs	d^{25}	Refs
	3f	1748	83.5(5)	1828	1.5839(25)	1746		
	3k	1884	95(5)	1807,1871	1.5840(25)	1870		
	3o	1828	92(6)	1872				
3,5-Cl$_2$	3a	1807,1870,1871	53.5(1)	1870	1.5745(25)	1808,1870, 1871	1.225(25)	1871
			59(1)	1807,1808, 1871			1.237(25)	1870
Cl$_2$[b]	3a	1885						
	3c	1886,1887						
	3f	1852						
	3q	1888						
2,3,4,5,6-Cl$_5$	3a	1889a,1890,1891	142(0.5)	1808				
	3c	1889,1890,1892, 1893	312	1892				
			mp 112°	1894				
			mp 114°	1891				
			mp 114.5°	1889,1890				
			mp 123°	1893				

[a] The designation refers to the subdivision of Sec. VII-B-3 in which the method is described.
[b] Mixed isomers or unspecified isomer.

1523

TABLE 32

Physical Properties of and Synthetic Methods for Substituted α-Methylstyrenes,  R—⬡—C(CH$_3$)=CH$_2$

R	Synthetic method[a]	Reference	bp, °C (mm)	Reference	n_D (°C)	Reference	d_4 (°C)	Reference
o-CH$_3$	4a	1895–1900	54(11)	1899	1.521(15)	1897	0.9076(0)	1896
			83(18.5)	1901	1.5149(20)	1898	0.9180(15)	1897
			173(756)	1895				
			173(760)	1901				
			169	1896,1898				
			175	1897				
m-CH$_3$	4a	1896–1898,1902, 1903	79(11)	1903	1.5332(20)	1903	0.9112(0)	1904
	4b	1863	74(14)	1904	1.5335(20)	1898	0.9115(0)	1896
	4j	1904	89(25)	1898	1.530(22)	1897	0.9076(14)	1904
			185	1896			0.9054(20)	1903
			186	1897,1902			0.9034(22)	1897
				1899			0.9120(0)	1897
p-CH$_3$	4a	1897,1899,1900, 1905–1911	56.5(4)	1899	1.528(16)	1897	0.9122(0)	1896
	4b	1751,1757,1912– 1917	73(11.5)	1918			0.9073(16)	1897
			74(13)	1909	1.5345(18.7)	1906		
	4c	1919,1920	83(18.5)	1901	1.5356(19.5)	1921	0.9024(18.5)	1906
	4d	1922–1924	78(19)	1920	1.5268(20)	1924	0.8969(20)	1924
	4e	1925	101.5(29)	1906	1.5283(20)	1926	0.8936(23)	1927
					1.5334(20)	1918		
	4h	1928	98(30)	1910	1.5329(20.5)	1899	0.895(25)	1920

1524

	Ref	bp(mp)	Year	n_D	Year	d	Year
4i	1927	189(748)	1927	1.5340(22)	1910	1.054(25)	1931
		187(760)	1924	1.5283(23)	1927		
		187(780)	1905	1.5290(25)	1920		
		185	1896				
		186	1897				
		191	1929				
		mp −28°	1918				
		mp −20°	1905				
o-Cl 4a	1899,1930,1931	54.5(4)	1899	1.5344(20)	1899		
4g	1932	67(10)	1899	1.5329(25)	1931		
		73(14)	1930	1.5324(25)	1930		
		75(14)	1931				
		184	1932				
m-Cl 4a	1930,1933	62(4)	1933	1.5536(20)	1933		
		100(24)	1930	1.5506(25)	1930		
		108(26)	1934				
p-Cl 4a	1899,1930,1931, 1935,1936	80(8)	1935	1.5537(20)	1797	1.0723(20)	1935
4b	1937	84(9)	1899	1.5550(20)	1899	1.070(25)	1931
4c	1920	83(10)	1920	1.5559(20)	1935	1.076(25)	1920
		86(10)	1931	1.5528(25)	1899		
		85(15)	1909	1.5529(25)	1920		
		89(15)	1930	1.5540(25)	1931		
		95(20)	1797	1.5543(25)	1930		
		mp 3.5°	1935	1.5558(25)	1936		
Cl^b 4a	1938						
2,4-(CH$_3$)$_2$ 4a	1909,1939	83(17)	1909	1.515(20)	1939	0.888(20)	1939
		91(24)	1939				

continued

TABLE 32 (continued)

R	Synthetic method[a]	Reference	bp, °C (mm)	Reference	n_D (°C)	Reference	d_4 (°C)	Reference
2,5-$(CH_3)_2$	4a	1900,1909,1939	68(7)	1900	1.517(20)	1939	0.886(20)	1939
			104(30)	1939	1.5210(20)	1900	0.9067(20)	1900
			105(30)	1909				
3,4-$(CH_3)_2$	4a	1900,1940	86(8)	1940	1.5320(20)	1900	0.9330(20)	1900
			97(15)	1900	1.5376(25)	1940		
			mp −21°	1940				
2,4-Cl_2	4a	1931	104(15)	1931	1.5460(25)	1931	1.179(25)	1931
2,5-Cl_2	4a	1899,1930	64(3)	1930	1.5490(20)	1941	1.1908(20)	1941
	4b	1941	76.5(5)	1899	1.5492(25)	1899,1930		
			100(10)	1941				
3,4-Cl_2	4a	1828,1930,1931, 1940	94(3)	1828,1931	1.5732(25)	1828,1931	1.215(24)	1940
	4b	1940,1942	93(4)	1930	1.5733(25)	1930	1.2206(25)	1828
	4f	1828	100(9)	1828	1.5740(25)	1940	1.221(25)	1931
			141(43)	1940	1.5746(25)	1828		
			242	1940				
3,5-Cl_2	4a	1940	111(12)	1940	1.5660(25)	1940	1.196(25)	1940
	4b	1940						
Cl_2[b]	4a	1938						

[a] The designation refers to the subdivision of Sec. VII-B-4 in which the method is described.
[b] Mixed isomers or unspecified isomer.

With the exceptions of *p*-nitro- and pentachlorostyrene, the monomers discussed in this section are liquids at room temperature. They are soluble in typical organic solvents but insoluble in water. The heat of combustion of 2-vinylfuran is 767.9 kcal per mole at constant volume and 768.5 kcal per mole at constant pressure (1761). Its surface tension is 29.8 dynes per cm and viscosity 5.35 mP at 26° (1762). The following dipole moments have been determined:

Compound	μ, D	Reference
p-Ethylstyrene	0.61	1943
p-Bromostyrene	1.35	1943
	1.52	1840
p-Nitrostyrene	4.23	1840
o-Methyl-α-methylstyrene	0.8	1934
m-Chloro-α-methylstyrene	1.89	1934

2. PHYSIOLOGICAL

Dichlorostyrene causes pathological changes in the lung, liver, spleen, and thyroid and symptoms such as loss of reflexes, retardation of respiration, and partial paralysis when administered to laboratory animals. The minimum concentration in air that produces an observable change in respiration rate is 0.31 to 0.62 mg per liter for rabbits and 0.5 mg per liter for humans. A maximum permissible concentration of 0.05 to 0.1 mg per liter has been suggested (1944). The single oral dose LD_{50} for rats has been determined to be 4.0 g per kg (1945).

Nitrostyrene is toxic to rabbits (1946) and causes skin and eye irritation. The LD_{50} for rats is 185 mg per kg, and for mice 44 mg per kg (1947).

p-Methyl-α-methylstyrene has been isolated from the essential oil of Egyptian hashish (1948).

3. ANALYTICAL METHODS AND SPECTRA

A general method for the determination of vinyl aromatic compounds is reaction with mercury(II) acetate (67) (see Sec. I-C-3). The mixed products from dehydrogenation of diethylbenzene, which include ethylstyrenes, have been analyzed by gas-liquid chromatography (1786,1949,1950). An analytical method for *p*-methyl-α-methylstyrene in the presence of styrenes and *p*-cymene is based on addition of HCl or HBr and estimation of the resulting tertiary halide by its preferential hydrolysis (1921). Polyalkylated styrene monomers can be determined in their polymers by polarographic reduction (1951).

The mass spectrum of 2-vinylfuran has been investigated (1952). Literature references to other spectrometric studies of the monomers in this section are summarized in Table 33.

TABLE 33
Literature References to Spectra of 2-Vinylfuran, 2-Vinylthiophene, and Substituted Styrenes

Compound	Type of spectrum			
	UV	IR	Raman	NMR
2-Vinylfuran	1953–1955	1956	1957,1958	
2-Vinylthiophene	1722	1959		
α-Ethylstyrene	1755	1754		
Substituted styrenes:				
p-C_2H_5		1773		
p-C_3H_7		1773		
m-F	1954			
o-, m-, and p-Br	1954			
p-NO_2	1960			
2,4-$(CH_3)_2$	1848			73
2,5-$(CH_3)_2$	1848			
2,6-$(CH_3)_2$	1848			73
3,5-$(CH_3)_2$	1848,1954			
2,4,5-$(CH_3)_3$	1961	1961		
2,4,6-$(CH_3)_3$	1961,1962	1961,1963		
2,3-Cl_2	1954			
2,4-Cl_2	1954			73
2,5-Cl_2	1954			
2,6-Cl_2	1954			
3,4-Cl_2	1954			
Substituted α-methylstyrenes:				
o-CH_3	1901,1964	1901		
p-CH_3	1901,1921	1901		

D. Storage and Handling

There are no published data on the subject, but it would seem prudent to handle the unhalogenated monomers in this section in the same manner as styrene, which they resemble in a general way. Similarly, handling practices for chlorostyrene are reasonable guides for the bromo- and dichlorostyrenes (see Part 2, Chap. 1). Pyrogallol and hydroquinone are effective polymerization inhibitors for vinylfuran (1965) and 1-nitroso-2-naphthol is excellent for vinylthiophene (1738).

E. Chemical Reactions

2-Vinylfuran can be hydrogenated catalytically to 2-ethylfuran (1966–1970). Under more severe conditions, the furan ring is also reduced (1968, 1971). Halogens add to the vinyl double bond in vinylfuran (1721) and vinylthiophene (1738), and thiocyanogen adds to the latter, giving the 1,2-dithiocyanate (1738). Vinylfuran (1972) and vinylthiophene (1740, 1741) undergo the Diels-Alder reaction with maleic anhydride. Both monomers form adducts with metallic sodium that are effective anionic polymerization initiators (1973,1974).

Vinylfuran condenses with tetrafluoroethylene to yield a cyclobutane derivative—Eq. (255) (1975):

$$\text{furan-CH}{=}\text{CH}_2 + \text{CF}_2{=}\text{CF}_2 \xrightarrow{\text{heat}} \text{furan-CH}{-}\text{CH}_2 \text{ (with } \text{CF}_2{-}\text{CF}_2 \text{ ring)} \qquad (255)$$

Secondary amines in the presence of sodium—Eq. (256) (1976)—and silanes with chloroplatinic acid as catalyst—Eq. (257) (1977)—add to vinylfuran:

$$\text{furan-CH}{=}\text{CH}_2 + \text{R}_2\text{NH} \xrightarrow{\text{Na}} \text{furan-CH}_2\text{CH}_2\text{NR}_2 \qquad (256)$$

$$\text{furan-CH}{=}\text{CH}_2 + \text{HSiR}_3 \xrightarrow{\text{H}_2\text{PtCl}_6} \text{furan-CH}_2\text{CH}_2\text{SiR}_3 \qquad (257)$$

Acid-catalyzed addition of vinylfuran to α,β-unsaturated ketones has been reported—Eq. (258) (1731). The monomer dimerizes on long heating

$$\text{furan-CH}{=}\text{CH}_2 + \text{CH}_2{=}\text{CHCOR} \xrightarrow{\text{H}_2\text{SO}_4}$$

$$\text{CH}_2{=}\text{CH-furan-CH}_2\text{CH}_2\text{COR} \qquad (258)$$

at 130° in toluene or bromobenzene (1978). Oxidation of vinylfuran with potassium ferricyanide gives a low yield of furoic acid (1979).

The reactions of styrenes have been reviewed by Emerson (1724a) (also see Part 2, Chap. 1).

F. Polymerization

Vinylfuran does not polymerize appreciably in the absence of catalysts at 100 to 150° but resinifies with decomposition (gas evolution) at 200°

(1726). It polymerizes readily in the presence of light and air because of formation of peroxides that act as initiators (1721,1965). Conventional peroxidic or azo catalysts are effective in bulk (1980–1982) and emulsion (1728,1983,1984a) polymerizations. The energy of activation for peroxide-initiated homopolymerization of vinylfuran is 17 kcal per mole (1742a, 1743,1985).

Freshly prepared free-radical–initiated homopolymers are thermoplastic and soluble in most solvents except water, methanol, and petroleum ether. On exposure to air, however, they develop color and become insoluble with absorption of oxygen (1982,1983a). This behavior, which can be counteracted by addition of antioxidants (1983a), is probably due to the susceptibility of the furan ring to oxidation (1979,1986). It has been proposed (1986) that the following reactions of poly(vinylfuran) occur in the presence of oxygen: (1) formation of poly(vinylfuran peroxide), (2) decomposition of the peroxide, (3) oxidative cleavage of the furan ring, and (4) condensations of the cleavage products to form cross-links. The fact that the highest molecular weight (100,000) that has been reported for poly(vinylfuran) was obtained by redox emulsion polymerization (1987) may be related to this. It is reasonable that the reducing agent in the initiating system could inhibit disruptive side reactions (oxidative cross-linking, premature consumption of oxidizer) and thus allow chain propagation to proceed in a normal manner.

Photopolymerization with ultraviolet light in the absence of oxygen starts immediately on irradiation and gives a product up to 90 percent soluble in dioxane. The soluble portion has a molecular weight of 25,000. In the presence of oxygen, there is an induction period before polymerization begins (1988). γ-Irradiation of vinylfuran gives a polymer that is 32 percent cross-linked gel and high in oxygen content, as shown by its infrared spectrum (1989).

Cationic initiators, such as diethylaluminum chloride, give polymers in which the C=C bonds in the furan ring have undergone reaction, as indicated by the infrared spectra (1982). Anionic initiation with sodium naphthalene (1987) and with sodium in tetrahydrofuran (1990) has been reported. The former gave a homopolymer having a molecular weight of 10,000, and a block copolymer with methyl methacrylate was also prepared using this initiator.

Copolymerizations of vinylfuran with acrylonitrile, butadiene, methacrylonitrile, styrene, and vinyl ketones are described in the 1938 patent literature (1983). Vinylfuran was evaluated in the United States as a comonomer in synthetic rubber during the Second World War and found to copolymerize with butadiene in emulsion less readily than styrene. The product differed from SBR rubber made by the same process in being benzene-insoluble and giving vulcanizates of higher modulus, lower extensibility, and poorer flex-cracking resistance (1723). These properties probably reflect the oxidative cross-linking reaction of vinylfuran polymers mentioned above. Reikhsfel'd (1991) obtained similar results using a conventional emulsion procedure. By suitable modifications in the latex formulation, however, he produced rubber of very good quality. Among these variations were the addition of a softener (isopropylxanthogen disulfide) and use of excess base and an oil-soluble iron naphthenate to increase the copolymerization rate.

In copolymerizations with styrene, vinylfuran was found to decrease the reaction rate and the molecular weight of the product (1992). Other monomers with which vinylfuran has been copolymerized are vinylidene chloride (1993a,1994), chloroprene (1993b), α-alkylacrylonitriles (1993c), and drying oils (1984b). Two patents (1993d,1995) describe a variety of copolymerizations.

2-Vinylthiophene polymerizes on standing or, more rapidly, on heating (1722,1745). Its activation energy for peroxide-initiated polymerization is 16 kcal per mole (1742,1743). An attempt to prepare an isotactic homopolymer by Ziegler catalysis was unsuccessful. X-ray diffraction showed the product to be completely amorphous (1764).

Like vinylfuran, vinylthiophene copolymerizes with butadiene more slowly in emulsion than does styrene. The product is very similar to SBR rubber in properties (1723,1996).

Many substituted styrenes polymerize as readily as styrene. Koton (1997) has compared the relative bulk polymerization rates of 40 substituted styrenes at 100°. Those discussed in this section, listed in order of decreasing rate, which polymerize more rapidly than styrene are 2,5-dichloro-, 2,4-dichloro-, 3,4-dichloro-, o-bromo-, o-fluoro-, 2,5-dimethyl-, 2,4-dimethyl-, and p-bromostyrene. 3,4-Dimethyl- and p-fluorostyrene polymerize only a little more slowly than styrene. In general, an alkyl substituent increases the polymerization rate (1998). m-Nitro- (1833) and 2,4,6-trimethylstyrene (1851,1999) polymerize with great reluctance. Substituted α-methylstyrenes do not homopolymerize by free-radical initiation but dimerize in the presence of acid catalysts (1929,2000).

The energies of activation for free-radical polymerization of p-fluoro- (2001), p-bromo- (1805b,2001), 2,4-dichloro-, 3,4-dichloro-, 2,5-dichloro- (2002), 2,4-dimethyl- (1851), and 2,4,5-trimethylstyrene (2003) are in the range 14 to 18 kcal per mole, that is, similar to that of styrene. The same is true of the heats of polymerization (16.3 to 16.7 kcal per mole) of ethyl-, 2,5-dichloro- (1882), and 2,4,6-trimethylstyrene (2004).

The polymerization of many substituted styrenes with Ziegler-Natta catalysts has been studied (2005–2007).

For polymers of substituted styrenes, ortho and branched substituents give the highest softening points, and meta substituents the lowest (1997, 1998). Homopolymers of p-$tert$-butylstyrene (1795,1796) and 2,5-dichlorostyrene (1880) have heat-distortion temperatures of 115° or higher.

Over 50 substituted styrenes were evaluated in the United States as monomers during the Second World War, particularly as comonomers with dienes in the government's synthetic-rubber program (1779,1808, 2008).

Relative reactivity ratios for vinylfuran, vinylthiophene, and substituted styrenes are listed in Table 34; Q and e values in Table 35. Based on α-methylstyrene = 1.00, the relative reactivities of p-methyl- and p-chloro-α-methylstyrene toward maleic anhydride were found to be 1.7 ± 0.1 and 0.79 ± 0.02, respectively (1918).

TABLE 34

Relative Reactivity Ratios for Copolymerization of 2-Vinylfuran, 2-Vinylthiophene, and Substituted Styrenes (M_1)[a]

M_1	Comonomer	r_1	r_2	Reference
2-Vinylfuran	Acrylonitrile	0.82 ± 0.06	0.037 ± 0.004	1982
	Butadiene	4.52	0.11	1991
	Styrene	1.9 ± 0.1	0.25 ± 0.05	1982
	Vinylidene Chloride	11.7 ± 0.1	0.15 ± 0.01	1994
2-Vinylthiophene	Styrene	3.1 ± 0.45	0.35 ± 0.025	310
p-Ethylstyrene	p-Chlorostyrene[b]	4.1 ± 0.5	0.29 ± 0.04	1775
	Styrene[c]	1.05 ± 0.1	0.95 ± 1	2009
		1.0 ± 0.2	1.0 ± 0.2	2005b
p-Isopropylstyrene	Styrene	0.54 ± 0.01	1.11 ± 0.01	2010
p-Fluorostyrene	Styrene	0.9	0.7	2011
	Styrene[c]	0.7 ± 0.1	1.5 ± 0.2	2005b
		0.6 ± 0.1	1.5 ± 0.1	2009
m-Bromostyrene	Methyl methacrylate	1.17 ± 0.25	0.48 ± 0.02	2012
	Styrene	1.05 ± 0.2	0.55 ± 0.03	2012
p-Bromostyrene	p-Chlorostyrene[b]	1.0 ± 0.1	1.0 ± 0.1	2013
	p-Methoxystyrene	1.1	0.43	2014
	Methyl methacrylate	1.10 ± 0.25	0.395 ± 0.02	2012
	Styrene	1.1	0.60	2014
		1.05 ± 0.05	0.71 ± 0.02	2015
		0.99 ± 0.07	0.695 ± 0.02	2012
	Styrene[c]	0.55 ± 0.2	1.75 ± 0.01	2009

M_1	Comonomer	r_1	r_2	Reference
		0.5 ± 0.2	1.7 ± 0.3	2005b
-Nitrostyrene	p-Chlorostyrene	1.3 ± 0.1	0.25 ± 0.05	2016
	Methyl methacrylate	0.85 ± 0.2	0.35 ± 0.05	2016
	Styrene	0.85 ± 0.1	0.45 ± 0.05	2016
	Styrene[b]	0.03 ± 0.03	20 ± 4	2013
-Nitrostyrene	p-Chlorostyrene	0.9 ± 0.4	0.70 ± 0.08	2012
	Styrene	1.15 ± 0.2	0.19 ± 0.02	2012
5-Dimethylstyrene	2,5-Dichlorostyrene	0.27	1.55	1365
4,6-Trimethylstyrene	Acrylonitrile	0.16 ± 0.02	0.98 ± 0.02	2017
	p-Chlorostyrene	0.06	10	2017
	Methyl methacrylate	0.08	1.4	2017
		0.05 ± 0.01	1.6 ± 0.3	2017
	Styrene[c]	0.004	1.060	2018
5-Dichlorostyrene	Acrylonitrile	0.09 ± 0.02	0.26 ± 0.02	2019
		0.07 ± 0.06	0.23 ± 0.15	2019
		0.07 ± 0.05	0.22 ± 0.05	2019
	Butadiene	0.20 ± 0.04	0.65 ± 0.1	2020
	Butadiene[d]	0.46 ± 0.01	0.46 ± 0.01	2021
	2,5-Dimethylstyrene	1.55	0.27	1365
	Methyl acrylate	4	0.15	2022
	Methyl methacrylate	2.25	0.44	2023
	α-Methylstyrene	3	0.14	1414
	Methyl vinyl ketone	2.0	0.5	1414
	Styrene	2.2	0.29	2024
		2.2	0.23	2024
		1.9	0.31	2024
		1.8	0.30	2024
		1.77	0.37–0.38	2025
		0.8	0.2	2026
		0.23 ± 0.11	0.19 ± 0.08	2019
		0.07 ± 0.06	0.32 ± 0.06	2019
		0.05 ± 0.03	0.40 ± 0.03	2019
	Styrene[b]	0.3 ± 0.2	15 ± 2	2027a
	Vinyl acetate	...	<0.04	2028
	9-Vinylcarbazole	8 ± 0.5	0.016 ± 0.002	131
	Vinylidene cyanide	0.031	0.0092	2029
	2-Vinylpyridine	0.9	1.1	305
4-Dichlorostyrene	Styrene[b]	0.0–0.48	2.8–7.2	2027b
Dichlorostyrene[e]	Methyl acrylate	4.3 ± 0.3	0.25 ± 0.04	308
entachlorostyrene	Methyl methacrylate	0.35 ± 0.05	4.0 ± 0.4	1894
	Styrene	0.10 ± 0.02	1.3 ± 0.2	1894
	Vinyl chloride	5.3	0.43	2030

[a] Copolymerizations are free-radical–initiated unless otherwise indicated.
[b] Cationic initiation.
[c] Anionic or Ziegler-Natta initiation.
[d] Emulsion polymerization.
[e] Mixed isomers.

TABLE 35
Q and e Values for 2-Vinylfuran, 2-Vinylthiophene, and Substituted Styrenes

Monomer	Q	e	Reference
2-Vinylfuran	2.0 ± 0.2	0.0 ± 0.2	1982
2-Vinylthiophene	3.0	-0.8	321
	2.86	-0.80	137,146
	-3.8 kcal/mole $(q)^a$	-0.24×10^{-10}esu $(\epsilon)^a$	147
m-Bromostyrene	1.20	-0.4	321
	1.07	-0.21	137,146
	0.98	-0.1	321
	-3.1 kcal/mole $(q)^a$	-0.04×10^{-10}esu $(\epsilon)^a$	147
p-Bromostyrene	1.27	-0.5	321
	1.10	-0.35	137
	1.04	-0.32	146
	0.88	-0.2	321
	-3.1 kcal/mole $(q)^a$	-0.13×10^{-10}esu $(\epsilon)^a$	147
m-Nitrostyrene	3.47	0.81	137
	2.47	0.81	146
	-3.4 kcal/mole $(q)^a$	0.10×10^{-10}esu $(\epsilon)^a$	147
p-Nitrostyrene	1.86	0.4	321
	1.63	0.39	137,146
	1.06	0.4	321
	-3.5 kcal/mole $(q)^a$	0.13×10^{-10}esu $(\epsilon)^a$	147
2,5-Dimethylstyrene	0.95	-0.84	137
	-2.8 kcal/mole $(q)^a$	-0.33×10^{-10}esu $(\epsilon)^a$	147
2,5-Dichlorostyrene	1.67	0.4	321
	1.60	0.09	137,146
	-3.0 kcal/mole $(q)^a$	-0.04×10^{-10}esu $(\epsilon)^a$	147
Pentachlorostyrene	0.22	0.52	137,146
	0.2	0.25	1894
	-2.2 kcal/mole $(q)^a$	0.20×10^{-10}esu $(\epsilon)^a$	147

^{a} Schwan-Price resonance q and electrical ϵ factors (147).

References

1. G. R. Clemo and W. H. Perkin, Jr., *J. Chem. Soc.*, **125**:1804 (1924).
2. (a) W. Reppe and E. Keyssner (to I. G. Farbenindustrie A.-G.), German Patent 618,120, Sept. 2, 1935; *Chem. Abstr.*, **30**:110 (1936). (b) J. W. Reppe, "Acetylene Chemistry" (PB Report 18852–S), Charles A. Meyer and Co., Inc., New York, 1949.
3. (a) French Patent 865,354 (to I. G. Farbenindustrie A.-G.), May 21, 1941. (b) German Patent 744,414 (to I. G. Farbenindustrie A.-G.), Nov. 18, 1943. (c) W. Reppe, H. Krzikalla, O. Dornheim, and R. Sauerbier (vested in the Alien Property Custodian), U.S. Patent 2,317,804, Apr. 27, 1943; *Chem. Abstr.*, **37**: 6057 (1943).
4. French Patent 792,820 (to I. G. Farbenindustrie A.-G.), Jan. 11, 1936; *Chem. Abstr.*, **30**:4178 (1936).

5. E. Jenckel, *Z. Phys. Chem.*, **A190**:24 (1941).
6. Patents assigned to I. G. Farbenindustrie A.-G.: (a) W. Reppe, C. Schuster, and A. Hartmann, U.S. Patent 2,265,450, Dec. 9, 1941; *Chem. Abstr.*, **36**:2052 (1942). (b) W. Reppe, C. Schuster, and A. Hartmann, German Patent 737,663, Jun. 10, 1943; *Chem Abstr.*, **38**:3757 (1944).
7. H. Beck, *Kunststoffe*, **27**:90 (1937).
8. H. Weese, G. Hecht, and W. Reppe (to I. G. Farbenindustrie A.-G.), German Patent 738,994, Aug. 5, 1943; *Chem. Abstr.*, **39**:5408 (1945).
9. G. Hecht and H. Weese, *Münch. Med. Wochschr.*, **90**:11 (1943); *Chem. Abstr.*, **38**:1607 (1944).
10. C. E. Schildknecht, "Vinyl and Related Polymers," Wiley, New York, 1952.
11. W. M. Shine, *Mod. Plast.*, **25**(1):130 (1947).
12. W. F. Busse, J. M. Lambert, C. McKinley, and H. R. Davidson, *Ind. Eng. Chem.*, **40**:2271 (1948).
13. "Vinylpyrrolidone," Technical Bulletin 7543-037, General Aniline & Film Corp.
14. J. L. Jezl, H. M. Khelghatian, and L. D. Hague (to Sun Oil Co.), U.S. Patent 3,100,764, Aug. 13, 1963; *Chem. Abstr.*, **59**:11687 (1963).
15. Y. P. Kozlov and A. I. Gorin, *Nauchn. Dokl. Vysshei Shkoly, Biol. Nauki*, **1964**(2):91 and earlier papers; through *Chem. Abstr.*, **63**:3292 (1965). B. P. Ivannik, N. A. Klipson, T. G. Mamedova, N. I. Ryabachenko, M. V. Sklobovskaya, and A. G. Yaskevich, *Vestn. Akad. Med. Nauk SSSR*, **20**(9):18 (1965); through *Chem. Abstr.*, **64**:11521 (1966).
16. H. F. Cross and F. M. Snyder, *Soap Sanit. Chemicals*, **25**(2):135 (1949). H. G. Guy (to Koppers Co., Inc.), U.S. Patent 2,606,139, Aug. 26, 1952; *Chem. Abstr.*, **46**:11565 (1952).
17. W. E. Hanford and D. L. Fuller, *Ind. Eng. Chem.*, **40**:1171 (1948).
18. E. Keyssner (to I. G. Farbenindustrie A.-G.), German Patent 642,939, Mar. 19, 1937; *Chem. Abstr.*, **31**:5816 (1937).
19. H. Beller, R. E. Christ, and F. Wuerth (to General Aniline & Film Corp.), U.S. Patent 2,472,085, Jun. 7, 1949; *Chem. Abstr.*, **43**:7036 (1949). H. Beller, R. E. Christ, and F. Wuerth (to General Aniline & Film Corp.), British Patent 641,437, Aug. 9, 1950; *Chem. Abstr.*, **45**:8044 (1951).
20. G. M. Kline, *Mod. Plast.*, **23**(2):152A (1945); *ibid.*, **24**(3):157 (1946); *Plast. and Resins*, **4**(12):13 (1946).
21. O. Solomon, C. Ionescu, and I. Ciutá, *Chem. Tech. (Berlin)*, **9**:202 (1957); *Chem. Abstr.*, **51**:15493 (1960).
22. K. Yamamoto et al. (to Mitsui Chemical Industries Co.), Japanese Patent 1,714, Mar. 30, 1951; *Chem. Abstr.*, **47**:4917 (1953).
23. H. Davidge, *J. Appl. Chem.*, **9**:241 (1959).
24. W. Reppe et al., *Ann. Chem.*, **601**:128 (1956).
25. C. Schuster and F. Hanusch (to Badische Anilin- & Soda-Fabrik A.-G.), German Patent 940,981, Mar. 29, 1956; *Chem. Abstr.*, **52**:14696 (1958).
26. T. Takizawa and K. Yonetani, *Mem. Inst. Sci. Ind. Research Osaka Univ.*, **5**:110 (1947); through *Chem. Abstr.*, **47**:2748 (1953).
27. M. Amagasa, I. Yamaguchi, and R. Shioya, Japanese Patent 954, Apr. 26, 1962; *Chem. Abstr.*, **58**:3399 (1963).
28. W. Reppe and E. Keyssner (to I. G. Farbenindustrie A.-G.), U.S. Patent 2,066,160, Dec. 29, 1936; *Chem. Abstr.*, **31**:1040 (1937).
29. E. Keyssner and W. Wolff (to I. G. Farbenindustrie A.-G.), German Patent

642,424, Mar. 9, 1937; *Chem. Abstr.*, **31**:3504 (1937). E. Keyssner and W. Wolff (to I. G. Farbenindustrie A.-G.), U.S. Patent 2,123,734, Jul. 12, 1938; *Chem. Abstr.*, **32**:7055 (1938).

30. E. Keyssner and W. Wolff (to I. G. Farbenindustrie A.-G.), U.S. Patent 2,123,733, Jul. 12, 1938; *Chem. Abstr.*, **32**:7055 (1938).

31. British Patent 1,017,604 (to Pullman, Inc.), Jan. 19, 1966; *Chem. Abstr.*, **64**:14098 (1966).

32. S. A. Miller and H. Davidge (to British Oxygen Co. Ltd.), U.S. Patent 2,830,059, Apr. 8, 1958; *Chem. Abstr.*, **52**:14696 (1958).

33. H. Otsuki, I. Okano, and T. Takeda, *J. Soc. Chem. Ind. Japan*, **49**:169 (1946); through *Chem. Abstr.*, **42**:6354 (1948). H. Otsuki et al., Japanese Patent 174,356, Dec. 16, 1946; *Chem. Abstr.*, **44**:1544 (1950).

34. V. P. Lopatinskii, E. E. Sirotkina, I. P. Zherebtsov, and M. A. Leiman, *Tr. Tomskogo Gos. Univ. Ser. Khim.*, **170**:29 (1964); *Metody Polucheniya Khim. Reaktivov i Preparatov*, **11**:37 (1964); through *Chem. Abstr.*, **63**:565 (1965) and **65**:2203 (1966).

35. (a) R. G. Flowers, H. F. Miller, and L. W. Flowers, *J. Am. Chem. Soc.*, **70**:3019 (1948). (b) H. F. Miller and R. G. Flowers (to General Electric Co.), U.S. Patent 2,426,465, Aug. 26, 1947; *Chem. Abstr.*, **42**:224 (1948).

36. British Patents 620,733 and 620,734 (to British Thomson-Houston Co. Ltd.), Mar. 29, 1949; *Chem. Abstr.*, **43**:6669 (1949).

37. J. W. Copenhaver and M. H. Bigelow, "Acetylene and Carbon Monoxide Chemistry," Reinhold, New York, 1949.

38. *Chem. Eng.*, **58**(6):176 (1951).

39. R. A. Labine, *ibid.*, **67**(4):112 (1960).

40. E. Späth and J. Lintner, *Chem. Ber.*, **69B**:2727 (1936).

41. I. Hirao and Y. Miyazu, *J. Chem. Soc. Japan, Ind. Chem. Sect.*, **57**:450 (1954); through *Chem. Abstr.*, **49**:15860 (1955).

42. J. J. Nedwick (to Rohm and Haas Co.), U.S. Patent 2,806,847, Sept. 17, 1957; *Chem. Abstr.*, **52**:2931 (1958).

43. (a) M. F. Shostakovskii, F. P. Sidel'kovskaya, and M. G. Zelenskaya, *Izv. Akad. Nauk SSSR, Otd. Khim. Nauk*, **1957**:1457; through *Chem. Abstr.*, **52**:7270 (1958). (b) M. F. Shostakovskii, F. P. Sidel'kovskaya, and M. G. Zelenskaya, *ibid.*, **1952**:690; through *Chem. Abstr.*, **47**:9917 (1953). (c) M. F. Shostakovskii, F. P. Sidel'kovskaya, and M. G. Zelenskaya, *ibid.*, **1959** : 516, 892; through *Chem. Abstr.*, **53**:18937 (1959) and **54**:1286 (1960). (d) M. F. Shostakovskii, F. P. Sidel'kovskaya, and M. G. Zelenskaya, *ibid.*, **1954** : 689; through *Chem. Abstr.*, **49**:10853 (1955). (e) M. F. Shostakovskii, F. P. Sidel'kovskaya, and M. G. Zelenskaya, *ibid.*, **1956**: 615; through *Chem. Abstr.*, **50**:16568 (1956).

44. S. A. Miller and W. O. Jones (to British Oxygen Co. Ltd.), British Patent 799,924, Aug. 13, 1958; *Chem. Abstr.*, **53**:5286 (1959).

45. J. J. Nedwick, *Ind. Eng. Chem., Proc. Des. Develop.*, **1**:137 (1962).

46. French Patent 1,340,350 (to Farbwerke Hoechst A.-G.), Oct. 18, 1963; *Chem. Abstr.*, **61**:1837 (1964).

47. W. O. Jones (to British Oxygen Co. Ltd.), British Patent 846,575, Aug. 31, 1960; *Chem. Abstr.*, **55**:8431 (1961).

48. N. F. Kononov, V. V. Zarutskii, A. A. El'bert, and F. P. Sidel'kovskaya, U.S.S.R. Patent 173,776, Aug. 6, 1965; *Chem. Abstr.*, **64**:2063 (1966).

49. Spanish Patent 203,923 (to Laboratorios Primex, SL), Jun. 17, 1952; *Chem. Abstr.*, **49**:379 (1955).

50. *Chem. Eng. News,* **45**(36):82 (1967).
51. A. W. Schnizer (to Celanese Corp. of America), U.S. Patent 2,669,570, Feb. 16, 1954; *Chem. Abstr.,* **49**:2515 (1955). British Patent 717,799 (to Celanese Corp. of America), Nov. 3, 1954; *Chem. Abstr.,* **49**:5532 (1955).
52. (a) B. Puetzer, L. Katz, and L. Horowitz, *J. Am. Chem. Soc.,* **74**:4959 (1952). (b) B. Puetzer, L. Katz, and L. Horowitz (to Schenley Industries, Inc.), U.S. Patent 2,775,599, Dec. 25, 1956; *Chem. Abstr.,* **51**:9703 (1957). (c) B. Puetzer, L. Katz, and L. Horowitz (to Farbenfabriken Bayer A.-G.), German Patent 941,846, Apr. 19, 1956; *Chem. Abstr.,* **51**:12147 (1957).
53. (a) S. N. Ushakov, V. V. Davidenkova, and V. B. Lushchik, U.S.S.R. Patent 125,567, Jan. 15, 1960; *Chem. Abstr.,* **54**:15403 (1960). (b) S. N. Ushakov, V. V. Davidenkova, and V. B. Lushchik, *Izv. Akad. Nauk SSSR, Otd. Khim. Nauk,* **1961**: 901; through *Chem. Abstr.,* **55**:22287 (1961).
54. French Patent 1,421,336 (to Farbwerke Hoechst A.-G.), Nov. 8, 1965; *Chem. Abstr.,* **65**:15239 (1966).
55. W. E. Walles, W. F. Tousignant, and T. Houtman, Jr. (to The Dow Chemical Co.), U.S. Patent 2,891,058, Jun. 16, 1959; *Chem. Abstr.,* **54**:2359 (1960). British Patent 881,814 (to The Dow Chemical Co.), appl. May 15, 1959; *Chem. Abstr.,* **56**:14304 (1962).
56. M. F. Shostakovskii, N. A. Medzykhovskaya, and M. G. Zelenskaya, *Izv. Akad. Nauk SSSR, Otd. Khim. Nauk,* **1952**:682; through *Chem. Abstr.,* **47**:10479 (1953).
57. M. F. Shostakovskii, N. A. Medzykhovskaya, and M. G. Zelenskaya, *Akad. Nauk SSSR, Inst. Org. Khim., Sintezy Org. Soedinenii, Sbornik,* **2**:44 (1952); through *Chem. Abstr.,* **48**:584 (1954).
58. R. I. Longley, Jr., W. S. Emerson, and T. C. Shafer, *J. Am. Chem. Soc.,* **74**:2012 (1952).
59. O. Solomon, M. Dimonie, and C. Ambrus, *Rev. Chim. (Bucharest),* **11**:520 (1960); through *Chem. Abstr.,* **57**:12707 (1962).
60. O. F. Solomon, M. Dimonie, and M. Tomescu, *Makromol. Chem.,* **56**:1 (1962).
61. A. Chapiro and G. Hardy, *J. Chim. Phys.,* **59**:993 (1962).
62. H. P. Franck, *J. Polymer Sci.,* **13**:187 (1954).
63. I. R. Tabershaw and J. B. Skinner, *J. Ind. Hyg. Toxicol.,* **26**:313 (1944).
64. M. L. Rylova, *Gigiena i Sanit.,* **1953**(10):27 through *Chem. Abstr.,* **48**:2915 (1954).
65. H. Zeller, *Arch. Exptl. Pathol. Pharmakol.,* **232**:239 (1957); through *Chem. Abstr.,* **52**:4930 (1958).
66. Y. P. Ponomarev, V. N. Dimitrieva, and V. D. Bezuglyi, *Zh. Anal. Khim.,* **18**:654 (1963); through *Chem. Abstr.,* **59**:5265 (1963).
67. (a) R. P. Marquardt and E. N. Luce, *Anal. Chem.,* **20**:751 (1948); *ibid.,* **21**:1194 (1949). (b) R. W. Martin, *ibid.,* **21**:921 (1959). (c) E. N. Luce, Analysis of Styrene Monomer, in R. H. Boundy and R. F. Boyer (eds.), "Styrene," p. 142, Reinhold, New York, 1952.
68. F. P. Sidel'kovskaya, T. Y. Ogibina, and V. G. Arakelyan, *Zh. Prikl. Khim.,* **37**:182 (1964); through *Chem. Abstr.,* **60**:11385 (1964).
69. V. D. Bezuglyi and Y. P. Ponomarev, *Zh. Anal. Khim.,* **20**:1231 (1965); through *Chem. Abstr.,* **64**:8928 (1966).
70. C. E. R. Jones, in *Gas Chromatog., Proc. Symposium, 3rd, Edinburgh,* 401 (1960).

71. W. J. Potts, Jr. and R. A. Nyquist, *Spectrochim. Acta*, **1959**:679.
72. G. Oster and E. H. Immergut, *J. Am. Chem. Soc.*, **76**:1393 (1954).
73. W. Brügel, T. Ankel, and F. Krückeberg, Z. *Elektrochem.*, **64**:1121 (1960).
74. C. N. Banwell, *Mol. Phys.*, **3**:511 (1960).
75. G. E. Maciel, *J. Phys. Chem.*, **69**:1947 (1965).
76. P. P. Shorygin, T. N. Shkurina, M. F. Shostakovskii, F. P. Sidel'kovskaya, and M. G. Zelenskaya, *Izv. Akad. Nauk SSSR, Otd. Khim. Nauk*, **1959**:2208; through *Chem. Abstr.*, **54**:10516 (1960).
77. W. Freudenberg (to General Aniline & Film Corp.), U.S. Patent 2,414,407, Jan. 14, 1947; *Chem. Abstr.*, **41**:2276 (1947). W. Freudenberg (to General Aniline & Film Corp.), British Patent 604,011, Jun. 28, 1948; *Chem. Abstr.*, **43**:907 (1949).
78. W. O. Ney (to General Aniline & Film Corp.), U.S. Patent 2,449,951, Sept. 21, 1948; *Chem. Abstr.*, **43**:2037 (1949).
79. C. E. Barnes (to General Aniline & Film Corp.), U.S. Patent 2,483,962, Oct. 4, 1949; *Chem. Abstr.*, **44**:2798 (1950).
80. E. V. Hort and D. E. Graham (to General Aniline & Film Corp.), U.S. Patent 2,883,393, Apr. 21, 1959; *Chem. Abstr.*, **53**:14585 (1959).
81. British Patent 825,614 (to General Aniline & Film Corp.), Dec. 16, 1959; *Chem. Abstr.*, **54**:20333 (1960). C. P. Albus and G. G. Stoner (to General Aniline & Film Corp.), U.S. Patent 3,028,396, Apr. 3, 1962; *Chem. Abstr.*, **57**:2424 (1962).
82. H. Craubner, A. Hrubesch, H. Burger, and R. Krzikalla (to Badische Anilin- & Soda-Fabrik A.-G.), German Patent 1,151,506, Jul. 18, 1963; *Chem. Abstr.*, **60**:1898 (1964).
83. C. E. Schildknecht, A. O. Zoss, and F. Grosser, *Ind. Eng. Chem.*, **41**:2891 (1949).
84. C. E. H. Bawn, A. Ledwith, and Y. Shih-Lin, *Chem. Ind. (London)*, **1965**:769.
85. L. P. Ellinger, *Polymer*, **5**:559 (1964).
86. M. F. Shostakovskii and F. P. Sidel'kovskaya, *Zh. Obshch. Khim.*, **24**:1576 (1954); through *Chem. Abstr.*, **49**:13095 (1955).
87. M. F. Shostakovskii, F. P. Sidel'kovskaya, M. G. Zelenskaya, T. N. Shkurina, and T. Y. Ogibina, *Izv. Akad. Nauk SSSR, Otd. Khim. Nauk*, **1961**:482; through *Chem. Abstr.*, **55**:27267 (1961).
88. V. S. Abramov and L. A. Shapshinskaya, *Zh. Obshch. Khim.*, **22**:1450 (1952); through *Chem. Abstr.*, **47**:10488 (1953).
89. F. P. Sidel'kovskaya, M. G. Zelenskaya, and M. F. Shostakovskii, *Izv. Akad. Nauk SSSR, Otd. Khim. Nauk*, **1961**:128; through *Chem. Abstr.*, **55**:18702 (1961).
90. J. W. Breitenbach, O. F. Olaj, and F. Wehrmann, *Monatsh. Chem.*, **95**:1007 (1964).
91. S. McKinley, J. V. Crawford, and C. H. Wang, *J. Org. Chem.*, **31**:1963 (1966).
92. L. P. Ellinger, J. Fenney, and A. Ledwith, *Monatsh. Chem.*, **96**:131 (1965).
93. J. W. Breitenbach, F. Galinovsky, H. Nesvadba, and E. Wolf, *Naturwissenschaften*, **42**:155, 440 (1955); *Monatsh. Chem.*, **87**:580 (1956).
94. H. Fikentscher and K. Herrle (to Badische Anilin- & Soda-Fabrik A.-G.), German Patent 1,040,031, Oct. 2, 1958; *Chem. Abstr.*, **55**:6497 (1961).
95. G. C. Clark, *Proc. S. Dakota Acad. Sci.*, **40**:226 (1961); *Chem. Abstr.*, **57**:7211 (1962).
96. R. M. Joshi, *Makromol. Chem.*, **55**:35 (1962).

97. D. L. Nicol, M. Kaufman, and S. A. Miller (to British Oxygen Co. Ltd.), British Patent 718,912, Nov. 24, 1954; *Chem. Abstr.*, **49**:6654 (1955).

98. S. A. Miller and H. Davidge (to British Oxygen Co. Ltd.), U.S. Patent 2,830,059, Apr. 8, 1958; *Chem. Abstr.*, **52**:14696 (1958).

99. H. Fikentscher and R. Fricker (to Badische Anilin- & Soda-Fabrik A.-G.), German Patents 931,731, Aug. 16, 1955, and 936,421, Dec. 15, 1955; British Patent 739,438 (to Badische Anilin- & Soda-Fabrik A.-G.), Oct. 26, 1955; *Chem. Abstr.*, **52**:12458, 1684 (1958) and **50**:17532 (1956).

100. H. Davidge, *J. Appl. Chem. (London)*, **9**:553 (1959); British Patent 831,913 (to British Oxygen Co., Ltd.), Apr. 6, 1960; *Chem. Abstr.*, **54**:16925 (1960).

101. L. P. Ellinger, *J. Appl. Polymer Sci.*, **10**(4):551, 575 (1966).

102. (a) E. H. Cornish, *Plastics (London)*, **27**(301):132 (1962). (b) *ibid.*, **28**(305):61 (1963).

103. (a) A. Chapiro and G. Hardy, *J. Chim. Phys.*, **59**:993 (1962). (b) A. Chapiro, *U.S. Atomic Energy Comm.*, TID–7643, 136 (1962).

104. J. Kroh and W. Pekala, *Bull. Acad. Pol. Sci., Ser. Sci. Chim.*, **14**(1):55 (1966).

105. D. E. Sargent (to General Aniline & Film Corp.), Canadian Patent 460,270, Oct. 11, 1949; *Chem. Abstr.*, **44**:5645 (1950).

106. O. F. Solomon and M. Dimonie, *J. Polymer Sci., Pt. C*, **4**:969 (1964); O. F. Solomon, I. Z. Ciutá, and N. Cobianu, *ibid., Pt. B*, **2**(3):311 (1964). O. F. Solomon, I. Z. Ciutá, N. Cobianu, and M. Georgescu, *Bul. Inst. Politeh. Bucuresti*, **27**(1):59 (1965); O. F. Solomon, N. Cobianu, and I. Z. Ciutá, *ibid.*, **27**(2):65 (1965); through *Chem. Abstr.*, **64**:2169, 3693 (1966).

107. A. Gandini and P. H. Plesch, *J. Chem. Soc., Phys. Org.*, **1966**(1):7.

108. O. Solomon, M. Dimonie, C. Ambrus, and M. Tomescu, *Mezhdunarod. Simpozium po Makromol. Khim., Dokl.*, **1960**, *Moscow*, **1960**, *Sektsiya*, 1, 131; O. Solomon, M. Dimonie, and C. Ambrus, *Rev. Chim. (Bucharest)*, **11**:520 (1960); through *Chem. Abstr.*, **55**:6015 (1961) and **57**:12707 (1962).

109. A. S. Teot (to The Dow Chemical Co.), U.S. Patent 3,024,225, Mar. 6, 1962; *Chem. Abstr.*, **56**:14481 (1962).

110. J. Heller, D. O. Tieszen, and D. B. Parkinson, *J. Polymer Sci., Pt. A*, **1**:125 (1963).

111. O. F. Solomon, M. Dimonie, and M. Tomescu, *Makromol. Chem.*, **56**:1 (1962).

112. E. H. Cornish (to Standard Telephones and Cables Ltd.), British Patent 1,007,040, Oct. 13, 1965; *Chem. Abstr.*, **64**:836 (1966).

113. Spanish Patent 290,150 (to Standard Electrica, SA), Nov. 14, 1963; E. H. Cornish (to Standard Telephones and Cables Ltd.), British Patent 1,003,910, Sept. 8, 1965; *Chem. Abstr.*, **61**:13497 (1964) and **63**:16491 (1965).

114. J. W. Breitenbach and C. Srna, *J. Polymer Sci., Pt. B*, **1**(5):263 (1963); J. W. Breitenbach and O. F. Olaj, *ibid.*, **2**(7):685 (1964).

115. H. Scott and M. M. Labes, *ibid.*, **1**(8):413 (1963); H. Scott, T. P. Konen, and M. M. Labes, *ibid.*, **2**(7):689 (1964).

116. L. P. Ellinger, *Chem. Ind. (London)*, **1963**:1982; British Patent 1,005,116 (to British Oxygen Co. Ltd.), Sept. 22, 1965; *Chem. Abstr.*, **63**:18294 (1965).

117. H. Scott, G. A. Miller, and M. M. Labes, *Tetrahedron Letters*, **1963**(17):1073.

118. K. Takakura, K. Hayashi, and S. Okamura, *J. Polymer Sci., Pt. B*, **3**(7):568 (1965). M. Nishii, K. Tsuyi, K. Takakura, K. Hayashi, and S. Okamura, *Nippon Hoshasen Kobunshi Kenkyu Kyokai Nenpo*, **6**:181, 205 (1964–1965);

through *Chem. Abstr.*, **64**:14281 (1966). *J. Polymer Sci., Pt. A-1*, **4**(8):2028 (1966).

119. C. H. Wang, *Chem. Ind. (London)*, **1964**:751.
120. E. H. Cornish, E. L. Bush, and M. Kumar, *Plastics (London)*, **31**(340):157 (1966).
121. H. Fikentscher and K. Herrle, *Mod. Plast.*, **23**(3):157 (1945).
122. W. O. Ney, Jr., W. R. Nummy, and C. E. Barnes (to Arnold Hoffman and Co., Inc.), U.S. Patent 2,634,259, Apr. 7, 1953; *Chem. Abstr.*, **47**:6096 (1953).
123. H. Beller (to General Aniline & Film Corp.), U.S. Patent 2,665,271, Jan. 5, 1954; *Chem. Abstr.*, **48**:5555 (1954).
124. British Patent 725,674 (to General Aniline & Film Corp.), Mar. 9, 1955; *Chem. Abstr.*, **49**:12037 (1955).
125. British Patent 1,021,121 (to General Aniline & Film Corp.), Feb. 23, 1966; *Chem. Abstr.*, **64**:19821 (1966).
126. J. F. Voeks and T. G. Traylor (to The Dow Chemical Co.), U.S. Patent 2,982,762, May 2, 1961; *Chem. Abstr.*, **55**:17099 (1961).
127. B. J. Luberoff and W. Gersumky (to American Cyanamid Co.), U.S. Patent 3,162,625, Dec. 22, 1964; *Chem. Abstr.*, **62**:6593 (1965).
128. A. R. Mukherjee, P. Ghosh, S. C. Chadha, and S. R. Palit, *Makromol. Chem.*, **80**:208 (1964).
129. R. Resz and H. Bartl (to Farbenfabriken Bayer A.-G.), French Patent 1,392,354, Mar. 12, 1965; *Chem. Abstr.*, **63**:13503 (1965).
130. Belgian Patents 655,504 and 655,506 (to Farbwerke Hoechst A.-G.), May 10, 1965; *Chem. Abstr.*, **64**:19898, 19899 (1966).
131. T. Alfrey, Jr., J. J. Bohrer, and H. Mark, "Copolymerization" ("High Polymers," vol. VIII), p. 39, Interscience, New York, 1952.
132. R. Hart, *Makromol. Chem.*, **47**:143 (1961).
133. T. Alfrey, Jr. and S. L. Kapur, *J. Polymer Sci.*, **4**:215 (1949).
134. S. N. Ushakov and A. F. Nikolaev, *Izv. Akad. Nauk SSSR, Otd. Khim. Nauk,* **1956**:83,226; *Bull. Acad. Sci. USSR, Div. Chem. Sci.*, **1956**:79,217.
135. G. van Paesschen and G. Smets, *Bull. Soc. Chim. Belges*, **64**:173 (1955).
136. D. J. Kahn and H. H. Horowitz, *J. Polymer Sci.*, **54**:363 (1961).
137. L. J. Young, *ibid.*, **54**:411 (1961).
138. J. F. Bork and L. E. Coleman, *ibid.*, **43**:413 (1960).
139. R. M. Pike and D. L. Bailey, *ibid.*, **27**:55 (1956).
140. K. Hayashi and G. Smets, *ibid.*, **27**:275 (1958).
141. M. Vrancken and G. Smets, *Makromol. Chem.*, **30**:197 (1959).
142. J. W. Breitenbach and H. Edelhauser, *Ric. Sci.*, **25A**:242 (1955).
143. F. P. Sidel'kovskaya, M. A. Askarov, and F. Ibrazimov, *Vysokomol. Soedin.*, **6**(10):1810 (1964); through *Chem. Abstr.*, **62**:6563 (1965).
144. F. P. Sidel'kovskaya, M. F. Shostakovskii, F. Ibrazimov, and M. A. Askarov, *ibid.*, **6**(9):1585 (1964); through *Chem. Abstr.*, **61**:16164 (1964).
145. L. Ghosez and G. Smets, *J. Polymer Sci.*, **35**:215 (1959).
146. G. E. Ham (ed.), "Copolymerization" ("High Polymers," vol. XVIII), appendix B, Interscience, New York, 1964.
147. T. C. Schwan and C. C. Price, *J. Polymer Sci.*, **40**:457 (1959).
148. W. Reppe, "Polyvinylpyrrolidon," Verlag Chem., Weinheim, 1954; *Angew. Chem.*, **65**:577 (1953).
149. I. Greenfield, *Ind. Chemist*, **32**:11 (1956).
150. J. Remond, *Rev. Prod. Chim.*, **59**:127, 260 (1956).

151. E. Ferraris, *Materie Plastiche*, **25**:208 (1959).

152. W. W. Myddleton, *Mfg. Chemist*, **34**(7):316 (1963).

153. "Polyvinylpyrrolidone," Tech. Bull. 7543–113, General Aniline and Film Corp.

154. "Polyvinylpyrrolidone," Tech. Bull., Badische Anilin- & Soda-Fabrik A.-G. (available from BASF Colors and Chemicals, Inc., New York).

155. H. R. Jacobi, *Kunststoffe*, **43**:381 (1953).

156. H. Hoegl, *J. Phys. Chem.*, **69**(3):755 (1965).

157. British Patent 988,363 (to Gevaert Photo-Producten NV), Apr. 7, 1965; *Chem. Abstr.*, **65**:1662 (1966).

158. Netherlands Appl. 6,515,152 (to Rank-Xerox Ltd.), May 24, 1966; *Chem. Abstr.*, **65**:10006 (1966).

159. Netherlands Applns. 6,515,554–5, Jun. 13, 1966, and 6,516,580, Jul. 1, 1966 (to Kalle A.-G.); *Chem. Abstr.*, **65**:13064, 16303 (1966).

160. G. E. Ham (to Chemstrand Corp.), U.S. Patents 2,769,793, Nov. 6, 1956, and 2,850,477, Sept. 2, 1958; *Chem. Abstr.*, **51**:6178 (1957) and **53**:4763 (1959).

161. (a) A. Ladenburg, *Chem. Ber.*, **20**:1643 (1887). (b) A. Ladenburg, *ibid.*, **22**:2583 (1889). (c) A. Ladenburg, *Ann. Chem.*, **301**:117 (1898).

162. A. Einhorn and P. Lehnkering, *ibid.*, **246**:160 (1888).

163. T. Methner, *Chem. Ber.*, **27**:2689 (1894).

164. G. Prausnitz, *ibid.*, **23**:2725 (1890); *ibid.*, **25**:2394 (1892).

165. J. Meisenheimer, J. Neresheimer, P. Finn, and W. Schneider, *Ann. Chem.*, **420**:190 (1920).

166. H. A. Iddles, E. H. Lang, and D. C. Gregg, *J. Am. Chem. Soc.*, **59**:1945 (1937).

167. French Patent 849,126 (to I. G. Farbenindustrie A.-G.), Nov. 14, 1939; *Chem. Abstr.*, **35**:6358 (1941). W. Gumlich (to I. G. Farbenindustrie A.-G.), German Patent 695,098, Jul. 18, 1940; *Chem. Abstr.*, **35**:5220 (1941).

168. J. M. Folz, J. E. Mahan, and D. H. White, *Petroleum Processing*, **7**:1802 (1952).

169. J. R. Haws, *Rubber Chemistry and Technology*, **30**:1387 (1957).

170. E. R. Wallsgrove, *Mfg. Chemist*, **30**(5):206 (1959).

171. A. E. Chichibabin, *J. Russ. Phys.-Chem. Soc.*, **37**:1229 (1905); through *Chem. Zentr.*, **77**:1438 (1906).

172. R. L. Frank, J. R. Blegen, R. J. Dearborn, R. L. Myers, and F. E. Woodward, *J. Am. Chem. Soc.*, **68**:1368 (1946).

173. R. Graf, W. Langer, and K. Haumeder, *J. Prakt. Chem.*, **150**:153 (1938).

174. J. E. Mahan, S. D. Turk, A. M. Schnitzer, R. P. Williams, and G. D. Sammons, *Ind. Eng. Chem., Chem. Eng. Data Ser.*, **2**:76 (1957).

175. G. B. Gechele, A. Nenz, C. Garbuglio, and S. Pietra, *Chim. Ind. (Milan)*, **42**:959 (1960); through *Chem. Abstr.*, **55**:10431 (1961).

176. J. E. Mahan (to Phillips Petroleum Co.), U.S. Patent 2,769,811, Nov. 6, 1956; *Chem. Abstr.*, **51**:7433 (1957).

177. British Patent 873,998 (to Phillips Petroleum Co.), appl. May 6, 1959; *Chem. Abstr.*, **56**:7285 (1962).

178. G. Cevidalli, J. Herzenberg, and A. Nenz (to Sicedison Società per Azioni), Italian Patent 596,924, Aug. 11, 1959; *Chem. Abstr.*, **57**:12443 (1962).

179. Patents assigned to Reilly Tar & Chemical Corp.: (a) F. E. Cislak, U.S. Patent 2,716,119, Aug. 23, 1955; *Chem. Abstr.*, **50**:5770 (1956). (b) F. E. Cislak, U.S. Patent 2,716,118, Aug. 23, 1955; *Chem. Abstr.*, **50**:5770 (1956). (c) F. E. Cislak, U.S. Patent 2,749,349, Jun. 5, 1956; *Chem. Abstr.*, **51**:4442 (1957). (d) F. E. Cislak, U.S. Patent 2,854,455, Sept. 30, 1958; *Chem. Abstr.*, **53**:5292 (1959).

180. C. R. Wagner (to Phillips Petroleum Co.), U.S. Patent 2,732,376, Jan. 24, 1956; *Chem. Abstr.*, **50**:9450 (1956).
181. J. T. Hays (to Hercules Powder Co.), U.S. Patent 2,611,769, Sept. 23, 1952; *Chem. Abstr.*, **47**:9367 (1953).
182. Patents assigned to Union Carbide Corp.: (a) J. T. Dunn and D. T. Manning, U.S. Patent 2,980,684, Apr. 18, 1961; *Chem. Abstr.*, **55**:19956 (1961). (b) J. T. Dunn and D. T. Manning, U.S. Patent 3,158,615, Nov. 24, 1964; *Chem. Abstr.*, **62**:11972 (1965).
183. Patents assigned to Yoshitomi Pharmaceutical Industries Ltd.: (a) K. Saruto and H. Maekawa, Japanese Patent 10,696, Aug. 9, 1962; *Chem. Abstr.*, **59**:3898 (1963). (b) K. Saruto and H. Maekawa, Japanese Patent 11,035, Aug. 14, 1962; *Chem. Abstr.*, **59**:10006 (1963).
184. L. A. Burrows and G. H. Kalb (to E. I. du Pont de Nemours & Co.), U.S. Patent 2,677,688, May 4, 1954; *Chem. Abstr.*, **51**:15603 (1957).
185. W. E. Burns (to Phillips Petroleum Co.), U.S. Patents 2,757,130, Jul. 31, 1956, and 2,769,773, Nov. 6, 1956; *Chem. Abstr.*, **51**:5843, 8147 (1957).
186. H. A. Larson (to Phillips Petroleum Co.), U.S. Patent 3,151,046, Sept. 29, 1964; *Chem. Abstr.*, **61**:14447 (1964). J. J. Moon and H. A. Larson (to Phillips Petroleum Co.), U.S. Patent 3,151,047, Sept. 29, 1964; *Chem. Abstr.*, **61**:13287 (1964).
187. B. Skinner (to Phillips Petroleum Co.), U.S. Patent 2,868,696, Jan. 13, 1959; *Chem. Abstr.*, **53**:18569 (1959).
188. O. Ajika and T. Kawano (to Takeda Yakuhin Kogyo Co.), Japanese Patent 7,720, Jul. 10, 1962; *Chem. Abstr.*, **59**:13956 (1963).
189. J. J. Costolow (to Phillips Petroleum Co.), U.S. Patent 3,239,433, Mar. 8, 1966; *Chem. Abstr.*, **64**:15850 (1966).
190. D. M. Haskell and D. L. McKay (to Phillips Petroleum Co.), U.S. Patent 2,768,169, Oct. 23, 1956; *Chem. Abstr.*, **51**:11393 (1957).
191. C. A. Ray, Jr. (to Phillips Petroleum Co.), U.S. Patent 2,922,754, Jan. 26, 1960; *Chem. Abstr.*, **54**:12164 (1960).
192. S. D. Turk and B. D. Simpson (to Phillips Petroleum Co.), U.S. Patent 2,996,509, appl. Jul. 7, 1959; *Chem. Abstr.*, **56**:4740 (1962).
193. D. L. McKay and H. W. Goard, *Chem. Eng. Progr.*, **61**:99 (1965).
194. D. M. Haskell (to Phillips Petroleum Co.), U.S. Patent 2,716,120, Aug. 23, 1955; *Chem. Abstr.*, **50**:5770 (1956). E. N. Pennington (to Phillips Petroleum Co.), U.S. Patent 2,879,272, Mar. 24, 1959; *Chem. Abstr.*, **53**:17152 (1959).
195. K. H. Hachmuth (to Phillips Petroleum Co.), U.S. Patent 2,755,282, Jul. 17, 1956; *Chem. Abstr.*, **51**:2056 (1957). N. L. Stalder and G. H. Dale (to Phillips Petroleum Co.), U.S. Patent 2,772,269, Nov. 27, 1956; *Chem. Abstr.*, **51**:8147 (1957). J. J. Moon (to Phillips Petroleum Co.), U.S. Patent 2,853,489, Sept. 23, 1958; *Chem. Abstr.*, **53**:10257 (1959).
196. R. N. Lacey and B. Yeomans (to Distillers Co. Ltd.), British Patent 852,129, Oct. 26, 1960; B. Yeomans (to Distillers Co. Ltd.), British Patent 852,130, Oct. 26, 1960; *Chem. Abstr.*, **55**:13452 (1961).
197. R. A. Findlay (to Phillips Petroleum Co.), U.S. Patents 2,731,468, Jan. 17, 1956, and 2,799,677, Jul. 16, 1957; *Chem. Abstr.*, **50**:14001 (1956) and **52**:450 (1958).
198. Y. Matsuda, Y. Nakahara, R. Kato, T. Yasuda, Y. Moriyama, and Y. Ito (to Dainippon Celluloid Co. Ltd.), Japanese Patent 7,728, Sept. 3, 1959; *Chem. Abstr.*, **54**:15406 (1960).

199. K. Omae and H. Yamamoto (to Japan Synthetic Chem. Ind. Co. Ltd.), Japanese Patent 11,141, Aug. 15, 1962; *Chem. Abstr.*, **59**:10005 (1963).

200. D. H. White and J. M. Folz (to Phillips Petroleum Co.), U.S. Patent 2,962,498, Nov. 29, 1960; *Chem. Abstr.*, **55**:10478 (1961).

201. C. R. Wagner (to Phillips Petroleum Co.), U.S. Patent 2,592,625, Apr. 15, 1952; *Chem. Abstr.*, **47**:6446 (1953).

202. D. F. Runge, G. Naumann, and M. Morgner (to VEB Farbenfabrik Wolfen), British Patent 828,205, Feb. 17, 1960; *Chem. Abstr.*, **54**:13149 (1960).

203. C. W. Tullock and S. M. McElvain, *J. Am. Chem. Soc.*, **61**:961 (1939).

204. R. N. Lacey (to Distillers Co. Ltd.), British Patent 850,114, Sept. 28, 1960; *Chem. Abstr.*, **55**:8434 (1961).

205. J. E. Mahan (to Phillips Petroleum Co.), U.S. Patents 2,512,660, Jun. 27, 1950 and 2,698,848, Jan. 4, 1955; *Chem. Abstr.*, **44**:9987 (1950) and **50**:1088 (1956).

206. E. Profft, *Chem. Tech.* (*Berlin*), **7**:511 (1955).

207. S. Chrzczonowicz, J. Michalski, K. Studniarski, and H. Zajac, Polish Patent 42,386, Oct. 15, 1959; *Chem. Abstr.*, **55**:6501 (1961).

208. K. Winterfeld and C. Heinen, *Ann. Chem.*, **573**:85 (1951).

209. L. F. Salisbury (to E. I. du Pont de Nemours & Co.), British Patent 632,661, Nov. 28, 1949; *Chem. Abstr.*, **44**:4513 (1950).

210. H. L. Dimond, L. J. Fleckenstein, and M. O. Shrader (to Pittsburgh Coke & Chemical Co.), U.S. Patent 2,848,456, Aug. 19, 1958; *Chem. Abstr.*, **53**:1384 (1959).

211. British Patent 956,398 (to Ruetgerswerke A.-G.), Apr. 29, 1964; *Chem. Abstr.*, **61**:5618 (1964).

212. J. E. Mahan (to Phillips Petroleum Co.), U.S. Patent 2,534,285, Dec. 19, 1950; *Chem. Abstr.*, **45**:3425 (1951).

213. F. E. Cislak and W. R. Wheeler (to Reilly Tar & Chemical Corp.), U.S. Patent 2,786,846, Mar. 26, 1957; *Chem. Abstr.*, **51**:13941 (1957).

214. A. F. MacLean and A. W. Schnizer (to Celanese Corp. of America), U.S. Patent 2,754,300, Jul. 10, 1956; *Chem. Abstr.*, **51**:2878 (1957).

215. (a) R. L. Frank, C. E. Adams, J. R. Blegen, P. V. Smith, A. E. Juve, C. H. Schroeder, and M. M. Goff, *Ind. Eng. Chem.*, **40**:879 (1948). (b) G. B. Bachman and D. D. Micucci, *J. Am. Chem. Soc.*, **70**:2381 (1948). (c) H. C. Brown and N. R. Eldrid, *ibid.*, **70**:2878 (1948). (d) G. Buchmann and O. Wolniak, *J. Prakt. Chem.*, **25**:101 (1964).

216. Y. I. Chumakov and Y. P. Shapovalova, *Zh. Org. Khim.*, **1**:940 (1965); through *Chem. Abstr.*, **63**:6960 (1965).

217. G. J. Janz and N. E. Duncan, *J. Am. Chem. Soc.*, **75**:5389 (1953).

218. V. Boekelheide and W. Feely, *ibid.*, **80**:2217 (1958).

219. R. H. Hall and I. D. Fleming (to Distillers Co. Ltd.), British Patent 866,380, Apr. 26, 1961; *Chem. Abstr.*, **55**:24795 (1961).

220. C. F. Woodward, A. Eisner, and P. G. Haines, *J. Am. Chem. Soc.*, **66**:911 (1944).

221. (a) W. von E. Doering and R. A. N. Weil, *ibid.*, **69**:2461 (1947). (b) W. von E. Doering and S. J. Rhoads, *ibid.*,**75**:4738 (1953).

222. M. Yoshida and H. Kumagae, *Kogyo Kagaku Zasshi*, **59**:196 (1956); through *Chem. Abstr.*, **51**:10516 (1957).

223. V. G. Ostroverkhov, I. S. Vakarchuk, and V. G. Sinyavskii, *Vysokomol. Soedin.*, **3**:1197 (1961); through *Chem. Abstr.*, **56**:8921 (1962).

1544 VINYL AND DIENE MONOMERS

224. R. P. Mariella, L. F. A. Peterson, and R. C. Ferris, *J. Am. Chem. Soc.*, **70**:1494 (1948).
225. E. Profft, *Chem.-Ztg.*, **81**:427 (1957).
226. (a) M. M. Koton and O. K. Surnina, *Dokl. Akad. Nauk SSSR*, **113**:1063 (1957); *Proc. Acad. Sci. USSR*, **113**:347 (1957). (b) M. M. Koton, *J. Polymer Sci.*, **30**:331 (1958). (c) M. M. Koton, *Khim. Tekhnol. i Primenenie Proizvodnykh Piridina i Khinolina, Materialy Soveshchaniya, Inst. Khim. Akad. Nauk Latv. SSR, Riga,* **1957**:119; through *Chem. Abstr.*, **55**:16546 (1961).
227. (a) W.-H. Yang and K.-K. Wu, *Chung-Kuo K'o Hsueh Yuan Ying Yung Hua Hsueh Yen Chiu So Chi K'an,* **1963**(7):10; *Communist Chinese Sci. Abstr.*, **83**:50 (April, 1965); through *Chem. Abstr.*, **64**:2798 (1966). (b) P.-T. Li, Y.-T. Liu, H.-F. Kao, and C.-Y. Kung, *Chung-Kuo K'o Hsueh Yuan Ying Yung Hua Hsueh Yen Chiu So Chi K'an,* **1963**(7):7; *Communist Chinese Sci. Abstr.*, **83**:49 (April, 1965); through *Chem. Abstr.*, **64**:1356 (1966). (c) P.-T. Li, H.-F. Kao, and C.-Y. Kung, *Chung-Kuo K'o Hsueh Yuan Ying Yung Hua Hsueh Yen Chiu So Chi K'an,* **1963**(7):17; *Communist Chinese Sci. Abstr.*, **88**:65 (June, 1965); through *Chem. Abstr.*, **63**:14984 (1965).
228. Phillips Petroleum Company, Property Data and Shipping Information Sheet on 2-Methyl-5-vinylpyridine.
229. Reilly Tar & Chemical Corp., Product Information Sheet on 2-Vinylpyridine, Nov. 1, 1965.
230. A. Einhorn, *Ann. Chem.*, **265**:208 (1891).
231. A. F. Frolov, M. A. Laginova, and M. M. Kiseleva, *Zh. Fiz. Khim.*, **35**:1784 (1961); through *Chem. Abstr.*, **56**:63 (1962).
232. (a) A. A. Petrov and W. Ludwig, *Zh. Obshch. Khim.*, **25**:739 (1955); *J. Gen. Chem. USSR*, **25**:703 (1955). (b) A. A. Petrov and W. Ludwig, *Zh. Obshch. Khim.*, **26**:51 (1956); *J. Gen. Chem. USSR*, **26**:49 (1956).
233. P. F. Onyon, *Trans. Faraday Soc.*, **51**:400 (1955).
234. G. Favini, *Gazz. Chim. Ital.*, **93**:635 (1963); through *Chem. Abstr.*, **59**:13456 (1963).
235. R. H. Linnell, *J. Org. Chem.*, **25**:290 (1960).
236. A. A. Balandin, E. I. Klabunovskii, A. P. Oberemok-Yakubova, and I. I. Brusov, *Izv. Akad. Nauk SSSR, Otd. Khim. Nauk,* **1960**:784; through *Chem. Abstr.*, **56**:10998 (1962).
237. A. I. Dukhovnaya, *Gigiena Truda i Prof. Zabolevaniya,* **10**(3):9 (1966); through *Chem. Abstr.*, **65**:4523 (1966).
238. H. F. Smyth, Jr., C. P. Carpenter, C. S. Weil, and U. C. Pozzani, *Arch. Ind. Hyg. Occupational Med.*, **10**:61 (1954); *Chem. Abstr.*, **48**:13951 (1954).
239. V. D. Bezuglyi, V. N. Dmitrieva, T. A. Alekseeva, and G. G. Belous, *Zh. Anal. Khim.*, **16**:477 (1961); through *Chem. Abstr.*, **56**:2002 (1962).
240. M. Yoshida, *Kogyo Kagaku Zasshi,* **63**:893 (1960); through *Chem. Abstr.*, **57**:10533 (1962).
241. H. Kamio, M. Nishikawa, and T. Kanzawa, *Bunseki Kagaku,* **10**:851 (1961); through *Chem. Abstr.*, **56**:6669 (1962).
242. K. Löffler, *Chem. Ber.*, **37**:161 (1904).
243. K. W. Wilson, F. E. Anderson, and R. W. Donohoe, *Anal. Chem.*, **23**:1032 (1951).
244. M. L. Swain, A. Eisner, C. F. Woodward, and B. A. Brice, *J. Am. Chem. Soc.*, **71**:1341 (1949).
245. J. P. Wibaut and H. C. Beyerman, *Rec. Trav. Chim. Pays-Bas,* **70**:977 (1951).

246. H. A. Laitinen, F. A. Miller, and T. D. Parks, *J. Am. Chem. Soc.*, **69**:2707 (1947).

247. W. Brügel, *Z. Elektrochem.*, **66**:159 (1962).

248. Patents assigned to Phillips Petroleum Co.: (a) J. E. Mahan and M. F. Potts, U.S. Patent 2,732,377, Jan. 24, 1956; (b) C. W. Mertz, U.S. Patents 2,745,834, May 15, 1956, 2,842,551, Jul. 8, 1958, 2,860,140, Nov. 11, 1958, 2,866,789, Dec. 30, 1958, 2,900,387, Aug. 18, 1959, and 2,907,770, Oct. 6, 1959; (c) W. L. Smith, M. F. Potts, and P. S. Hudson, U.S. Patents 2,748,131, May 29, 1956, 2,773,874, Dec. 11, 1956, 2,775,594, Dec. 25, 1956, 2,776,975, Jan. 8, 1957, 2,824,105, Feb. 18, 1958, and 2,874,159, Feb. 17, 1959; (d) R. E. Reusser and A. M. Schnitzer, U.S. Patents 2,761,864, Sept. 4, 1956, 2,812,329, Nov. 5, 1957, and 2,857,389, Oct. 21, 1958; (e) P. F. Warner, U.S. Patent 2,861,997, Nov. 25, 1958; (f) W. B. Reynolds and R. E. Reusser, U.S. Patent 2,861,998, Nov. 25, 1958; (g) H. R. Snyder, U.S. Patent 2,862,927, Dec. 2, 1958; (h) P. F. Warner and J. E. Duke, U.S. Patent 2,920,078, Jan. 5, 1960; (i) G. Kraus, U.S. Patent 2,993,903, Jul. 25, 1961. *Chem. Abstr.*, **50**:14268 (1956); **51**:1295 (1957); **52**:19253 (1958); **53**:6260, 6687 (1959); **54**:1562, 12163 (1960); **50**:14268 (1956); **52**:791 (1958); **51**:4759, 4759 (1957); **52**:10178 (1958); **53**:12310 (1959); **51**:766 (1957); **52**:3402 (1958); **53**:5296, 11414, 11308, 8168 (1959); **54**:12164 (1960); **55**:27981 (1961).

249. British Patent 809,320 (to Chemische Werke Hüls A.-G.), Feb. 18, 1959; *Chem. Abstr.*, **53**:13985 (1959).

250. Patents assigned to Sicedison Società per Azioni: (a) G. B. Gechele, A. Nenz, and G. Barberis, Italian Patent 601,912, Feb. 16, 1960, British Patent 883,808, Dec. 6, 1961, and U.S. Patent 3,020,284, Feb. 6, 1962; *Chem. Abstr.*, **55**:12934 (1961) and **56**:11789 (1962). (b) G. B. Gechele, A. Nenz, and G. Barberis, Italian Patent 611,047, Oct. 20, 1960; *Chem. Abstr.*, **55**:21673 (1961). (c) A. Nenz and G. B. Gechele, Italian Patent 601,911, Feb. 16, 1960, and British Patent 883,809, Dec. 6, 1961; *Chem. Abstr.*, **55**:12934 (1961).

251. M. Naito and M. Katayama (to Nitto Boseki Co. Ltd.), Japanese Patent 7,268 (1960), appl. Dec. 25, 1957; *Chem. Abstr.*, **56**:2585 (1962).

252. British Patent 758,954 (to Chemstrand Corp.), Oct. 10, 1956; *Chem. Abstr.*, **51**:11761 (1957).

253. K. Saruwatari and M. Matsushima (to Yoshitomi Drug Manufg. Co.), Japanese Patent 5,872, Jul. 4, 1959; *Chem. Abstr.*, **54**:14274 (1960).

254. M. I. Farberov, B. F. Ustavshchikov, A. M. Kut'in, T. P. Vernova, and E. V. Yarosh, *Izv. Vyssh. Ucheb. Zaved., Khim. Khim. Tekhnol.*, **1958**:92; through *Chem. Abstr.*, **53**:11364 (1959).

255. A. Nenz and G. B. Gechele, *Chim. Ind. (Milan)*, **43**:142 (1961); through *Chem. Abstr.*, **55**:19921 (1961).

256. K. H. Nelson and M. D. Grimes, *Anal. Chem.*, **30**:1928 (1958).

257. J. E. Mahan, S. D. Turk, and R. P. Williams (to Phillips Petroleum Co.), U.S. Patent 2,826,581, Mar. 11, 1958; *Chem. Abstr.*, **52**:9665 (1958).

258. British Patent 967,156 (to U.S. Rubber Co.), Aug. 19, 1964; *Chem. Abstr.*, **61**:13487 (1964). M. Farber and R. Miller (to U.S. Rubber Co.), U.S. Patent 3,248,363, Apr. 26, 1966; *Chem. Abstr.*, **65**:877 (1966).

259. S. Ohki and Y. Noika, *J. Pharm. Soc. Japan*, **72**:490 (1952); through *Chem. Abstr.*, **47**:6418 (1953).

260. S. Tazuke, N. Sato, and S. Okamura, *J. Polymer Sci., Pt. A-1*, **4**(10):2461 (1966).

261. J. F. Pudvin and J. A. Mattern, *J. Am. Chem. Soc.*, **78**:2104 (1956).

262. A. A. Artamonov, A. A. Balandin, G. M. Marukyan, and M. I. Kotalenets, *Dokl. Akad. Nauk SSSR*, 163:359 (1965); through *Chem. Abstr.*, 63:11484 (1965).

263. J. T. Dunn and D. L. Heywood (to Union Carbide Corp.), U.S. Patent 3,062,824, Nov. 6, 1962; *Chem. Abstr.*, 58:9030 (1963).

264. V. Boekelheide and R. Scharrer, *J. Org. Chem.*, 26:3802 (1961).

265. T. Tamikado, T. Sakai, and K. Sagisaka, *Makromol. Chem.*, 50:244 (1961).

266. H. H. Sisler, W. C. L. Ming, E. Metter, and F. R. Hurley, *J. Am. Chem. Soc.*, 75:446 (1953).

267. L. N. Ferguson and A. J. Levant, *Nature*, 167:817 (1951).

268. M. M. Baizer (to Monsanto Co.), U.S. Patent 3,218,245, Nov. 16, 1965; *Chem. Abstr.*, 64:17554 (1966).

269. P. W. Solomon (to Phillips Petroleum Co.), U.S. Patent 3,192,218, Jun. 29, 1965; *Chem. Abstr.*, 63:11424 (1965).

270. H. A. Bruson, Cyanoethylation, in R. Adams (ed.), "Organic Reactions," vol. V., p. 79, Wiley, New York, 1949.

271. (a) R. Levine and M. H. Wilt, *J. Am. Chem. Soc.*, 74:342 (1952). (b) M. H. Wilt and R. Levine, *ibid.*, 75:1368 (1953). (c) H. E. Reich and R. Levine, *ibid.*, 77:4913 (1955). (d) H. E. Reich and R. Levine, *ibid.*, 77:5434 (1955). (e) G. Magnus and R. Levine, *ibid.*, 78:4127 (1956). (f) G. Magnus and R. Levine, *J. Org. Chem.*, 22:270 (1957).

272. (a) V. Boekelheide and S. Rothchild, *J. Am. Chem. Soc.*, 69:3149 (1947). (b) V. Boekelheide and S. Rothchild, *ibid.*, 71:879 (1949). (c) V. Boekelheide and E. J. Agnello, *ibid.*, 72:5005 (1950). (d) V. Boekelheide and J. H. Mason, *ibid.*, 73:2356 (1951).

273. (a) E. Profft, *J. Prakt. Chem.* (4), 4:19 (1956). (b) E. Profft and G. Busse, *Z. Chem.*, 1:19 (1961). (c) E. Profft and S. Lojack, *Rev. Chim., Acad. Rep. Populaire Roumaine*, 7:405 (1962); through *Chem. Abstr.*, 59:8696 (1963).

274. A. M. Clifford (to Wingfoot Corp.), U.S. Patents 2,579,419, Dec. 18, 1951, and 2,615,892, Oct. 28, 1952; *Chem. Abstr.*, 46:7593 (1952) and 47:10011 (1953).

275. A. P. Phillips, *J. Am. Chem. Soc.*, 78:4441 (1956).

276. W. H. Vinton (to E. I. du Pont de Nemours & Co.), U.S. Patents 2,607,775 and 2,607,776, Aug. 19, 1952; *Chem. Abstr.*, 47:6989 (1953).

277. J. H. Boyer, *J. Am. Chem. Soc.*, 73:5248 (1951).

278. E. Maruszewska-Wieczorkowska and J. Michalski, *Bull. Acad. Pol. Sci., Ser. Sci. Chim. Géol. et Géograph.*, 6:19 (1958); *J. Org. Chem.*, 23:1886 (1958).

279. L. Bauer and L. A. Gardella, Jr., *J. Org. Chem.*, 26:82 (1961).

280. J. S. Meek, R. T. Merrow, D. E. Ramey, and S. J. Cristol, *J. Am. Chem. Soc.*, 73:5563 (1951); J. S. Meek, R. T. Merrow, and S. J. Cristol, *ibid.*, 74:2667 (1952).

281. D. B. Pattison and M. Carmack, *ibid.*, 68:2033 (1946). M. Carmack and D. F. DeTar (to Rohm and Haas Co.), U.S. Patent 2,495,567, Jan. 24, 1950; *Chem. Abstr.*, 44:7868 (1950).

282. F. E. Cislak and W. H. Rieger (to Reilly Tar & Chemical Corp.), U.S. Patent 2,402,020, Jun. 11, 1946; *Chem. Abstr.*, 40:4907 (1946).

283. C. C. Price and C. E. Greene, *J. Polymer Sci.*, 6:111 (1951).

284. J. Harmon (to E. I. du Pont de Nemours & Co.), U.S. Patent 2,491,472, Dec. 20, 1949; *Chem. Abstr.*, 44:2797 (1950).

285. M. F. Potts and P. S. Hudson (to Phillips Petroleum Co.), U.S. Patent 2,767,159, Oct. 16, 1956; *Chem. Abstr.*, **51**:4054 (1957).
286. S. Tazuke and S. Okamura, *J. Polymer Sci.*, *Pt. B*, **3**(2):135 (1965); *ibid.*, *Pt. A-1*, **4**(1):141 (1966).
287. G. Champetier, M. Fontanille, A. C. Korn, and P. Sigwalt, *ibid.*, **58**:911 (1962).
288. J. Smid and M. Szwarc, *ibid.*, **61**:31 (1962).
289. C. L. Lee, J. Smid, and M. Szwarc, *Trans. Faraday Soc.*, **59**:1192 (1963).
290. P. P. Spiegelman and G. Parravano, *J. Polymer Sci.*, *Pt. A*, **2**(5):2245 (1964).
291. G. Greber and G. Egle, *Makromol. Chem.*, **54**:136 (1962).
292. G. Natta, G. Mazzanti, G. Dall'Asta, and P. Longi, *ibid.*, **37**:160 (1960); *J. Polymer Sci.*, **51**:487 (1961). G. Natta, G. Mazzanti, P. Longi, G. Dall'Asta, and F. Bernardini (to " Montecatini " Società Generale per l'Industria Mineraria e Chimica), German Patent 1,114,638, Oct. 5, 1961; *Chem. Abstr.*, **57**:1076 (1962).
293. G. Geuskens, J. C. Lubikulu, and C. David, *Polymer*, **7**(1):63 (1966).
294. R. C. Schulz and J. Schwaab, *Makromol. Chem.*, **85**:297 (1965).
295. R. M. Fuoss and U. P. Strauss, *J. Polymer Sci.*, **3**:246 (1948); R. M. Fuoss, *Science*, **108**:545 (1948).
296. E. B. Fitzgerald and R. M. Fuoss, *Ind. Eng. Chem.*, **42**:1603 (1950).
297. W. N. Maclay and R. M. Fuoss, *J. Polymer Sci.*, **6**:511 (1951).
298. U. P. Strauss and E. G. Jackson, *ibid.*, **6**:649 (1951).
299. R. H. Sprague and L. G. S. Brooker (to Eastman Kodak Co.), U.S. Patent 2,484,430, Oct. 11, 1949; *Chem. Abstr.*, **44**:9729 (1950).
300. J. T. Dunn and D. T. Manning (to Union Carbide Corp.), U.S. Patent 3,159,611, Dec. 1, 1964; *Chem. Abstr.*, **62**:7892 (1965).
301. R. M. Joshi, *J. Polymer Sci.*, **56**:313 (1962).
302. M. Matsuoka, M. Otsuka, K. Takemoto, and M. Imoto, *Kogyo Kagaku Zasshi*, **69**(1):137 (1966); through *Chem. Abstr.*, **65**:15515 (1966).
303. Y. Iwakura, T. Tamikado, M. Yamaguchi, and K. Takei, *J. Polymer Sci.*, **39**:203 (1959).
304. B. L. Funt and E. A. Ogryzlo, *ibid.*, **25**:279 (1957).
305. Ref. 131, p. 40.
306. C. C. Price and T. F. McKeon, *J. Polymer Sci.*, **41**:445 (1959).
307. T. Alfrey, Jr. and H. Morawetz, *J. Am. Chem. Soc.*, **74**:436 (1952).
308. S. L. Aggarwal and F. A. Long, *J. Polymer Sci.*, **11**:127 (1953).
309. T. Tamikado, *ibid.*, **43**:489 (1960).
310. C. Walling, E. R. Briggs, and K. B. Wolfstirn, *J. Am. Chem. Soc.*, **70**:1543 (1948).
311. T. Alfrey, Jr., J. Bohrer, H. Haas, and C. Lewis, *J. Polymer Sci.*, **5**:719 (1950).
312. A. V. Ryabov, Y. D. Semchikov, N. N. Slavnitskaya, and V. N. Vakhrusheva, *Dokl. Akad. Nauk SSSR*, **154**(5):1135 (1964); *Proc. Acad. Sci. USSR*, **154**:163 (1964).
313. S. Okamura and K. Uno, *Chem. High Polymers* (*Japan*), **8**:467 (1951); through *Chem. Abstr.*, **47**:9663 (1953).
314. R. M. Fuoss and G. I. Cathers, *J. Polymer Sci.*, **4**:97 (1949).
315. Ref. 146, appendix A.
316. T. Yamamoto, *Kogyo Kagaku Zasshi*, **62**:476 (1959); through *Chem. Zentr.*, **131**:6516 (1960).
317. T. Tamikado, *Makromol. Chem.*, **38**:85 (1960).

318. V. L. Tsailingol'd, M. I. Farberov, and G. A. Bugrova, *Vysokomol. Soedin.*, 1(3):415 (1959); through *Chem. Abstr.*, 54:5157 (1960).
319. C. Aso and M. Sogabe, *Kogyo Kagaku Zasshi*, 68(10):1970 (1965); through *Chem. Abstr.*, 64:11324 (1966).
320. I. Sakurada, *Chem. High Polymers (Tokyo)*, 18:496 (1961); through *Makromol. Chem.*, 49:251 (1961).
321. C. C. Price, *J. Polymer Sci.*, 3:772 (1948).
322. F. Leavitt, V. Stannett, and M. Szwarc, *ibid.*, 31:193 (1958); F. Carrack and M. Szwarc, *J. Am. Chem. Soc.*, 81:4138 (1959).
323. G. B. Gechele and G. Convalle, *J. Appl. Polymer Sci.*, 5:203 (1961).
324. S. Chrzczonowicz and Z. Michalska, *Polimery*, 7(5):162 (1962); through *Chem. Abstr.*, 58:2507 (1963).
325. H. W. Schlipkoeter and A. Brockhaus, *Klin. Wochschr.*, 39:1182 (1961); *Chem. Abstr.*, 56:7955 (1962). H. W. Schlipkoeter, R. Dolgner, and A. Brockhaus, *Deut. Med. Wochschr.*, 88(39):1895 (1963); *Chem. Abstr.*, 60:2248 (1964).
326. A. C. Allison, J. S. Harington, and M. Birbeck, *J. Exptl. Med.*, 124(2):141 (1966).
327. P. F. Holt and E. T. Nasrallah, *Nature*, 211:878 (1966).
328. G. Natta, E. C. Vigliani, F. Danusso, B. Pernis, P. Ferruti, and M. A. Marchisio, *Atti Accad. Nazl. Lincei, Classe Sci. Fis., Mat. Nat.*, 40:11 (1966); through *Chem. Abstr.*, 65:11222 (1966).
329. S. N. Ushakov and A. M. Itenberg, *J. Gen. Chem. USSR*, 7:2495 (1937); through *Chem. Abstr.*, 32:2083 (1937).
330. D. T. Hurd, *J. Am. Chem. Soc.*, 67:1813 (1945).
331. E. G. Rochow, *ibid.*, 67:963 (1945).
332. D. T. Hurd and E. G. Rochow, *ibid.*, 67:1057 (1945).
333. L. H. Sommer and F. C. Whitmore, Abstracts, National Meeting of the American Chemical Society, Cleveland, Ohio, April 1944, p. 34 M; *J. Am. Chem. Soc.*, 68:485 (1946).
334. D. T. Hurd (to General Electric Co.), U.S. Patent 2,420,912, May 20, 1947; *Chem. Abstr.*, 41:5145 (1947).
335. British Patent 752,700 (to Midland Silicones Ltd.), Jul. 11, 1956; *Chem. Abstr.*, 51:7402 (1957).
336. G. H. Wagner (to Union Carbide & Carbon Corp.), U.S. Patents 2,632,013, Mar. 17, 1953, and 2,637,738, May 5, 1953; *Chem. Abstr.*, 48:2760, 8254 (1954).
337. G. H. Wagner and W. G. Whitehead, Jr. (to Union Carbide Corp.), U.S. Patent 2,851,473, Sept. 9, 1958; *Chem. Abstr.*, 53:3060 (1959).
338. H. Normant, *Compt. Rend.*, 239:1510 (1954).
339. H. E. Ramsden, J. R. Leebrick, S. D. Rosenberg, E. H. Miller, J. J. Walburn, A. E. Balint, and R. Cserr, *J. Org. Chem.*, 22:1602 (1957).
340. S. D. Rosenberg, J. J. Walburn, T. D. Stankovich, A. E. Balint, and H. E. Ramsden, *ibid.*, 22:1200 (1957).
341. H. E. Ramsden (to Metal and Thermit Corp.), U.S. Patent 2,838,508, Jun. 10, 1958; *Chem. Abstr.*, 53:6054 (1959).
342. "Selection Guide to Dow Corning Reactive Organosilicon Chemicals," Form No. 03–017, Dow Corning Corporation, Chemical Products Division, Midland, Mich.
343. E. P. Plueddemann, H. A. Clark, L. E. Nelson, and K. R. Hoffman, *Mod. Plast.*, 39:135 (Aug., 1962).

344. R. Steinman (to Libbey-Owens-Ford Glass Co.), U.S. Patent 2,688,006, Aug. 31, 1954; *Chem. Abstr.*, **49**:4253 (1955).

345. M. C. Brooks (to U.S. Rubber Co.), U.S. Patent 2,754,237, Jul. 10, 1956; *Chem. Abstr.*, **50**:16183 (1956).

346. L. P. Biefeld (to Owens-Corning Fiberglas Corp.), U.S. Patent 2,763,573, Sept. 18, 1956; *Chem. Abstr.*, **51**:3104 (1957).

347. A. Gottfurcht (to Libbey-Owens-Ford Glass Fibers Co.), U.S. Patent 2,763,629, Sept. 18, 1956; *Chem. Abstr.*, **51**:3105 (1957).

348. H. A. Clark (to Dow Corning Corp.), U.S. Patent 2,762,717, Sept. 11, 1956; *Chem. Abstr.*, **51**:3955 (1957).

349. K. L. Sayre (to Bjorksten Research Laboratories, Inc.), U.S. Patent 2,785,085, Mar. 12, 1957; *Chem. Abstr.*, **51**:7059 (1957).

350. E. H. Balz, L. F. Ornella, and J. D. Villwock (to Libbey-Owens-Ford Glass Fibers Co.), U.S. Patent 2,798,020, Jul. 2, 1957; *Chem. Abstr.*, **51**:15180 (1957).

351. W. Hinz and G. Solow, *Silikat Tech.*, **8**:178 (1957); through *Chem. Abstr.*, **51**:17225 (1957).

352. M. H. Jellinek (to Union Carbide Corp.), U.S. Patent 2,834,693, May 13, 1958; *Chem. Abstr.*, **52**:15958 (1958).

353. K. Imamura and T. Sakayori, *Asahi Garasu Kenkyu Hokoku*, **8**:13 (1958); through *Chem. Abstr.*, **53**:1819 (1959).

354. M. H. Jellinek (to Union Carbide Corp.), British Patent 796,574, Jun. 11, 1958; *Chem. Abstr.*, **53**:5739 (1959).

355. L. L. Yeager (to Bjorksten Research Laboratories, Inc.), German Patent 1,034,075, Jul. 10, 1958; *Chem. Abstr.*, **54**:26023 (1960).

356. W. L. Morgan (to Owens-Corning Fiberglas Corp.), U.S. Patent 3,013,915, Dec. 19, 1961; *Chem. Abstr.*, **56**:11807 (1962).

357. W. Moebes and A. Wende, *Plaste Kautschuk*, **9**:232 (1962); through *Chem. Abstr.*, **63**:4468 (1965).

358. J. Gaehde and A. Wende, *ibid.*, **13**(1):23 (1966); through *Chem. Abstr.*, **64**:11387 (1966).

359. H. A. Clark and E. P. Plueddemann, *Mod. Plast.*, **40**:133 (June, 1963).

360. British Patent 850,419 (to Lord Manufacturing Co.), Oct. 5, 1960; *Chem. Abstr.*, **55**:9946 (1961).

361. V. G. Simpson (to Compagnie Française Thomson-Houston), French Patent 1,419,909, Dec. 3, 1965; *Chem. Abstr.*, **65**:10761 (1966).

362. Belgian Patent 611,687 (to Société des Usines Chimiques Rhône-Poulenc), Jun. 18, 1962; *Chem. Abstr.*, **57**:12688 (1962).

363. T. A. Te Grotenhuis (1/5 to General Tire & Rubber Co.), U.S. Patent 2,742,378, Apr. 17, 1956; *Chem. Abstr.*, **50**:11682 (1956).

364. H. F. Jordan and W. V. Smith (to U.S. Rubber Co.), U.S. Patent 2,952,595, Sept. 13, 1960; *Chem. Abstr.*, **55**:5032 (1961).

365. S. Sterman and J. G. Marsden, *Polymer Eng. Sci.*, **6**(2):97 (1966); through *Chem. Abstr.*, **64**:17794 (1966).

366. A. A. Chuiko, G. E. Pavlik, V. A. Artemov, and I. E. Neimark (to L. V. Pisarzhevskii Institute of Physical Chemistry, Academy of Sciences, Ukrainian S.S.R.), U.S.S.R. Patent 176,414, Nov. 2, 1965; *Chem. Abstr.*, **64**:11406 (1966).

367. A. M. Gessler, H. K. Wiese, and J. Rehner, Jr. (to Esso Research and Engineering Co.), U.S. Patent 2,906,722, Sept. 29, 1959; *Chem. Abstr.*, **54**:10377 (1960).

368. T. A. Te Grotenhuis (1/5 to General Tire & Rubber Co.), U.S. Patent 2,751,369, Jun. 19, 1956; *Chem. Abstr.*, **50**:13471 (1956).

369. British Patent 1,012,558 (to General Electric Co.), Dec. 8, 1965; *Chem. Abstr.*, **64**:8237 (1966).

370. T. H. Ferrigno (to Mineral & Chemicals Philipp Corp.), U.S. Patent 3,231,404, Jan. 25, 1966; *Chem. Abstr.*, **64**:9954 (1966).

371. M. M. Olson and R. M. Christenson (to Pittsburgh Plate Glass Co.), U.S. Patent 2,894,922, Jul. 14, 1959; *Chem. Abstr.*, **53**:20838 (1959).

372. J. Wynstra and B. E. Godard (to Union Carbide Corp.), U.S. Patent 3,075,941, Jan. 29, 1963; *Chem. Abstr.*, **58**:10405 (1963).

373. R. N. Hazelwood (to McGraw Electric Co.), U.S. Patent 2,727,880, Dec. 20, 1955; *Chem. Abstr.*, **50**:6091 (1956).

374. B. Baum and J. Harding (to Union Carbide Corp.), U.S. Patent 2,921,870, Jan. 19, 1960; *Chem. Abstr.*, **54**:9356 (1960).

375. British Patent 618,608 (to British Thomson-Houston Co. Ltd.), Feb. 24, 1949; *Chem. Abstr.*, **43**:6220 (1949).

376. E. K. Drechsel (to American Cyanamid Co.), U.S. Patent 2,804,439, Aug. 27, 1957; *Chem. Abstr.*, **52**:1679 (1958).

377. D. D. Gagliardi (to Union Carbide Corp.), British Patent 1,020,052, Feb. 16, 1966; *Chem. Abstr.*, **64**:16055 (1966).

378. H. S. Lehman, W. A. Pliskin, and R. L. Ruggles, Jr. (to International Business Machines Corp.), British Patent 997,164, Jul. 7, 1965; *Chem. Abstr.*, **63**:12847 (1965).

379. N. P. Smetankina, N. S. Chernaya, V. Y. Oprya, V. P. Kuznetsova, and L. E. Karborskaya (to Institute of the Chemistry of Polymers and Monomers, Academy of Sciences, Ukrainian S.S.R.), U.S.S.R. Patent 172,997, Jul. 7, 1965; *Chem. Abstr.*, **64**:2242 (1966).

380. German Patent 1,057,746 (to Libbey-Owens-Ford Glass Co.), May 21, 1959; *Chem. Abstr.*, **55**:10833 (1961).

381. Patents assigned to Harris-Intertype Corp.: (a) M. J. Astle, U.S. Patent 2,991,204, Jul. 4, 1961; *Chem. Abstr.*, **55**:27918 (1961). (b) British Patent 894,186 Apr. 18, 1962; *Chem. Abstr.*, **57**:9985 (1962). (c) D. N. Adams and J. L. Sorkin, U.S. Patent 3,163,534, Dec. 29, 1964; *Chem. Abstr.*, **62**:4823 (1965).

382. A. T. Troshchenko and A. Bairgozhin, U.S.S.R. Patent 116,493, Jan. 19, 1959; *Chem. Abstr.*, **53**:19340 (1959).

383. H. H. Ender (to Union Carbide Corp.), U.S. Patent 3,190,775, Jun. 22, 1965; *Chem. Abstr.*, **63**:8114 (1965).

384. S. Witz and E. Mishuck (to Aerojet-General Corp.), U.S. Patent 3,260,631, Jul. 12, 1966; *Chem. Abstr.*, **65**:10418 (1966).

385. British Patent 824,107 (to Pyrene Co. Ltd.), Nov. 25, 1959; *Chem. Abstr.*, **54**:14690 (1960).

386. R. P. Cox and K. R. Guenther (to Bjorksten Research Laboratories, Inc.), U.S. Patent 2,983,719, May 9, 1961; *Chem. Abstr.*, **55**:17066 (1961). R. P. Cox (to Bjorksten Research Laboratories, Inc.), U.S. Patent 3,029,225, Apr. 10, 1962; *Chem. Abstr.*, **57**:2386 (1962).

387. British Patent 896,656 (to National Distillers and Chemical Corp.), May 16, 1962; *Chem. Abstr.*, **57**:4886 (1962).

388. L. W. Tyran (to E. I. du Pont de Nemours & Co.), U.S. Patent 2,532,583, Dec. 5, 1950; *Chem. Abstr.*, **45**:2264 (1951).

389. Belgian Patent 611,687 (to Société des Usines Chimiques Rhône-Poulenc), Jun. 18, 1962; *Chem. Abstr.*, **57**:12688 (1962).

390. D. T. Hurd and G. E. Roedel, *Ind. Eng. Chem.*, **40**:2078 (1948).

391. B. C. Carlson and R. C. Hartlein (to Dow Corning Corp.), Belgian Patent 622,451, Mar. 14, 1963; *Chem. Abstr.*, **59**:1370 (1963).

392. G. T. Gmitter and E. V. Braidich (to General Tire & Rubber Co.), U.S. Patent 3,050,477, Aug. 21, 1962; *Chem. Abstr.*, **57**:16897 (1962).

393. D. L. Bailey and R. Y. Mixer (to Union Carbide Corp.), U.S. Patent 2,777,869, Jan. 15, 1957; *Chem. Abstr.*, **52**:794 (1958).

394. R. S. Nelson and C. W. West (to Union Carbide Corp.), U.S. Patent 2,862,885, Dec. 2, 1958; *Chem. Abstr.*, **53**:9645 (1959).

395. W. F. Tousignant and W. E. Walles (to The Dow Chemical Co.), U.S. Patent 3,004,950, Oct. 17, 1961; *Chem. Abstr.*, **56**:14484 (1962). British Patent 868,188 (to Midland Silicones Ltd.), May 17, 1961; *Chem. Abstr.*, **55**:22905 (1961).

396. D. T. Hurd (to General Electric Co.), U.S. Patent 2,878,195, Mar 17, 1959; *Chem. Abstr.*, **53**:12739 (1959).

397. "Borden's Catalogue and Price List," G-652, The Borden Chemical Co., Monomer-Polymer Labs., 5000 Langdon St., Philadelphia, Pa.

398. A. L. Klebanskii and V. S. Fikhtengol'ts, *Zh. Obshch. Khim.*, **27**:2648 (1957); through *Chem. Abstr.*, **52**:7131 (1958).

399. M. F. Shostakovskii and D. A. Kochkin, *Izv. Akad. Nauk SSSR, Otd. Khim. Nauk*, **1954**:174; through *Chem. Abstr.*, **49**:6090 (1955).

400. G. S. Popeleva, I. V. Trofimova, K. A. Andrianov, and S. A. Golubtsov, *Khim. i Prakt. Primenenie Kremneorgan. Soedin. tr. Konf. Leningrad*, **1958**(6): 90 (Publ. 1961); through *Chem. Abstr.*, **57**:8604 (1962).

401. K. A. Andrianov, I. V. Trofimova, G. S. Popeleva, and S. A. Golubtsov, *Plast. Massy*, **1959**(3):25; through *Chem. Abstr.*, **59**:1669 (1963).

402. C. O. Strother and G. H. Wagner (to Linde Air Products Co.), U.S. Patent 2,532,430, Dec. 5, 1950; *Chem. Abstr.*, **45**:2968 (1951).

403. M. F. Shostakovskii, D. A. Kochkin, and L. V. Mustova, *Izv. Akad. Nauk SSSR, Otd. Khim. Nauk*, **1957**:1493; through *Chem. Abstr.*, **52**:7132 (1958).

404. S. A. Golubtsov, G. S. Popeleva, K. A. Andrianov, and N. I. Zaslavskaya, *Plast. Massy*, **1962**(10):21; through *Chem. Abstr.*, **58**:11393 (1963).

405. S. Yamada, E. Yasunaga, and R. Sakamoto (to Institute of Industrial Research), Japanese Patent 7,720, Oct. 24, 1955; *Chem. Abstr.*, **51**:17980 (1957).

406. E. Yasunaga and R. Sakamoto (to Bureau of Industrial Technics), Japanese Patent 8,761, Oct. 2, 1958; *Chem. Abstr.*, **54**:5466 (1960).

407. V. F. Mironov, *Collect. Czech. Chem. Commun.*, **25**:2167 (1960); *Chem. Abstr.*, **55**:1413 (1961).

408. V. F. Mironov, U.S.S.R. Patent 130,884, Aug. 20, 1960; *Chem. Abstr.*, **55**:6376 (1961).

409. V. F. Mironov, U.S.S.R. Patent 128,882, Mar. 10, 1960; *Chem. Abstr.*, **54**:19485 (1960).

410. E. A. Chernyshev, V. F. Mironov, and A. D. Petrov, *Izv. Akad. Nauk SSSR, Otd. Khim. Nauk*, **1960**:2147; through *Chem. Abstr.*, **55**:14345 (1961).

411. V. F. Mironov and A. D. Petrov, *ibid.*, **1958**:787; through *Chem. Abstr.*, **52**:19909 (1958).

412. S. A. Golubtsov, K. A. Andrianov, and G. S. Popeleva, *Intern. Symp. Organosilicon Chem., Sci. Common, Prague*, **1965**:115; through *Chem. Abstr.*, **65**:8946 (1966).

413. C. L. Agre and W. Hilling, *J. Am. Chem. Soc.*, **74**:3895 (1952).
414. C. L. Agre, *ibid.*, **71**:300 (1949).
415. L. H. Sommer, D. L. Bailey, and F. C. Whitmore, *ibid.*, **70**:2869 (1948).
416. E. G. Rochow, in H. S. Booth (ed.), "Inorganic Syntheses," vol. I, p. 42, McGraw-Hill, New York, 1939.
417. I. I. Shtetter, U.S.S.R. Patent 44,934, Nov. 30, 1935; through *Chem. Abstr.*, **32**:2958 (1938).
418. C. A. Burkhard and R. H. Krieble, *J. Am. Chem. Soc.*, **69**:2687 (1947).
419. M. F. Shostakovskii and D. A. Kochkin, *Izv. Akad. Nauk SSSR, Otd. Khim. Nauk*, **1956**:1150; through *Chem. Abstr.*, **51**:4935 (1957).
420. M. F. Shostakovskii, I. A. Shikhiev, and N. M. Komarov, *Dokl. Akad. Nauk Azerb. SSR*, **13**:277 (1957); through *Chem. Abstr.*, **52**:245 (1958).
421. M. Gaignon and M. Lefort (to Rhône-Poulenc SH), French Patent 1,390,999, Mar. 5, 1965; *Chem. Abstr.*, **62**:16298 (1965).
422. R. A. Pike (to Union Carbide Corp.), U.S. Patent 3,057,902, Oct. 9, 1962; *Chem. Abstr.*, **58**:2472 (1963).
423. D. T. Hurd, in L. F. Audrieth (ed.), "Inorganic Syntheses," vol. III, p. 58, McGraw-Hill, New York, 1950.
424. V. F. Mironov, V. V. Nepomnina, and L. A. Leites, *Izv. Akad. Nauk SSSR, Otd. Khim. Nauk*, **1960**:461; through *Chem. Abstr.*, **54**:22328 (1960).
425. G. H. Wagner (to Union Carbide & Carbon Corp.), U.S. Patent 2,737,520, Mar. 6, 1956; *Chem. Abstr.*, **51**:2020 (1957).
426. A. N. Pines and E. R. York (to Union Carbide & Carbon Corp.), U.S. Patent 2,735,860, Feb. 21, 1956; *Chem. Abstr.*, **50**:13986 (1956).
427. V. F. Mironov and V. V. Nepomnina, *Izv. Akad. Nauk SSSR, Otd. Khim. Nauk*, **1961**:920; through *Chem. Abstr.*, **55**:22100 (1961).
428. G. H. Wagner (to Union Carbide & Carbon Corp.), U.S. Patent 2,737,521, Mar. 6, 1956; *Chem. Abstr.*, **51**:2020 (1957).
429. G. H. Wagner (to Union Carbide & Carbon Corp.), U.S. Patent 2,735,859, Feb. 21, 1956; *Chem. Abstr.*, **50**:13986 (1956).
430. G. H. Wagner and A. N. Pines (to Union Carbide & Carbon Corp.), U.S. Patent 2,752,380, Jun. 26, 1956; *Chem. Abstr.*, **51**:2020 (1957).
431. G. H. Wagner, D. L. Bailey, A. N. Pines, M. L. Dunham, and D. B. McIntire, *Ind. Eng. Chem.*, **45**:367 (1953).
432. D. B. Hatcher (to Libbey-Owens-Ford Glass Co.), U.S. Patent 2,574,390, Nov. 6, 1951; *Chem. Abstr.*, **46**:1814 (1952).
433. R. Müller and K. Schnurrbusch, *Chem. Ber.*, **91**:1805 (1958).
434. A. D. Petrov, S. I. Sadykh-Zade, E. A. Chernyshev, and V. F. Mironov, *Zh. Obshch. Khim.*, **26**:1248 (1956); through *Chem. Abstr.*, **50**:14516 (1956).
435. R. Müller, M. Weist, and K. Schnurrbusch, German (East) Patent 13,480, Jul. 20, 1957; *Chem. Abstr.*, **53**:9059 (1959).
436. R. Müller, M. Weist, and K. Schnurrbusch (to Institut für Silikon- und Fluorkarbon-Chemie), German Patent 1,029,377, May 8, 1958; *Chem. Abstr.*, **54**:17269 (1960).
437. S. Munkelt and R. Müller, German (East) Patent 17,052, Jun. 5, 1959; *Chem. Abstr.*, **54**:18357 (1960).
438. S. Munkelt and R. Müller (to Institut für Silikon- und Fluorkarbon-Chemie), German Patent 1,056,609, May 6, 1959; *Chem. Abstr.*, **55**:6376 (1961).
439. S. Munkelt and R. Müller, *Chem. Ber.*, **92**:1012 (1959).
440. I. M. T. Davidson, C. Eaborn, and M. N. Lilly, *J. Chem. Soc.*, **1964**:2624.

441. F. S. Kipping, *Proc. Chem. Soc.*, 20:15 (1904).

442. M. F. Shostakovskii and D. A. Kochkin, *Dokl. Akad. Nauk SSSR*, 109:113 (1956); through *Chem. Abstr.*, 51:1826 (1957).

443. V. A. Ponomarenko, V. G. Cherkaev, A. D. Petrov, and N. A. Zadorozhnyi, *Izv. Akad. Nauk SSSR, Otd. Khim. Nauk*, 1958:247; through *Chem. Abstr.*, 52:12751 (1958).

444. V. F. Mironov, V. A. Ponomarenko, S. I. Sadykh-Zade, and E. A. Chernyshev, *Izv. Akad. Nauk SSSR, Otd. Khim. Nauk*, 1958:954; through *Chem. Abstr.*, 53:1120 (1959).

445. A. D. Petrov, V. A. Ponomarenko, V. G. Cherkaev, and N. A. Zadorozhnii, U.S.S.R. Patent 114,156, Jul. 30, 1958; *Chem. Abstr.*, 53:14003 (1959).

446. V. F. Mironov and V. V. Nepomnina, *Izv. Akad. Nauk SSSR, Otd. Khim. Nauk*, 1959:1231; through *Chem. Abstr.*, 54:1268 (1960).

447. V. Bazant, V. Chvalovsky, and J. Rathousky, "Organosilicon Compounds," Publishing House of the Czechoslovak Acad. of Sciences, Prague, Academic, New York, 1965.

448. G. M. Guzman and B. F. Suarez, *An. Real Soc. Espan. Fis. Quim.*, 56B:665 (1960); through *Chem. Abstr.*, 55:9261 (1961).

449. R. Nagel, C. Tamborski, and H. W. Post, *J. Org. Chem.*, 16:1768 (1951).

450. C. E. Scott and C. C. Price, *J. Am. Chem. Soc.*, 81:2670 (1959).

451. C. A. Burkhard, E. G. Rochow, H. S. Booth, and J. Hart, *Chem. Rev.*, 41:97 (1947).

452. H. A. Schuyten, J. W. Weaver, and J. D. Reid, *J. Am. Chem. Soc.*, 69:2110 (1947).

453. K. C. Frisch, P. A. Goodwin, and R. E. Scott, *ibid.*, 74:4584 (1952).

454. R. Nagel and H. W. Post, *J. Org. Chem.*, 17:1382 (1952).

455. F. S. Kipping, *J. Chem. Soc.*, 101:2108 (1912).

456. V. F. Mironov, *Izv. Akad. Nauk SSSR, Otd. Khim. Nauk*, 1962:1884; *Bull. Acad. Sci. USSR, Div. Chem. Sci.*, 1962:1797.

457. "Silanes," SF-1160B, Union Carbide Corp., Silicones Div., 270 Park Ave., New York, August, 1966.

458. "Organosilicon Chemicals," Technical Data Book S-15A, General Electric Co., Silicone Products Dept., Waterford, N.Y.

459. A. C. Jenkins and G. F. Chambers, *Ind. Eng. Chem.*, 46:2367 (1954).

460. A. D. Petrov, V. A. Ponomarenko, B. A. Sokolov, and Y. P. Egorov, *Izv. Akad. Nauk SSSR, Otd. Khim. Nauk*, 1957:310; through *Chem. Abstr.*, 51:14588 (1957).

461. (a) Preliminary Data Sheet, No. 7738, Borden Chemical Co., Monomer-Polymer Labs., 5000 Langdon St., Philadelphia, Aug. 31, 1967. (b) *ibid.*, No. 7739.

462. French Patent 1,142,551 (to Société des Usines Chimiques Rhône-Poulenc), Sept. 19, 1957; *Chem. Abstr.*, 54:1303 (1960).

463. R. Y. Mixer and D. L. Bailey, *J. Polymer Sci.*, 18:573 (1955).

464. D. O. Gumbotov, V. N. Kostryukov, and Y. K. Shaulov, *Izv. Akad. Nauk Azerb. SSR, Ser. Fiz.-Tekh. Mat. Nauk*, 1965(1):53; through *Chem. Abstr.*, 63:6385 (1965).

465. H. Reuther and G. Reichel, *Chem. Tech. (Berlin)*, 17(12):752 (1965); through *Chem. Abstr.*, 64:9754 (1966).

466. A. C. Jenkins and A. J. Reid, *Ind. Eng. Chem.*, 46:2566 (1954).

467. M. Wurst, *Collect. Czech. Chem. Commun.*, 30(6):2038 (1965); *Chem. Abstr.*, 63:10692 (1965).

468. A. D. Petrov, Y. P. Egorov, V. F. Mironov, G. I. Nikishin, and A. A. Bugorkova, *Izv. Akad. Nauk SSSR, Otd. Khim. Nauk*, **1956**:50; through *Chem. Abstr.*, **50**:8327 (1956).

469. Y. P. Egorov, *ibid.*, **1960**:1553; through *Chem. Abstr.*, **55**:9037 (1961).

470. V. A. Ponomarenko, V. G. Cherkaev, and N. A. Zadorozhnyi, *ibid.*, **1960**: 1610; through *Chem. Abstr.*, **55**:9262 (1961).

471. N. A. Chumaevskii, *Opt. Spektrosk.*, **10**(1):69 (1961); through *Chem. Abstr.*, **55**:10062 (1961).

472. V. F. Mironov and N. A. Chumaevskii, *Dokl. Akad. Nauk SSSR*, **146**:1117 (1962); *Proc. Acad. Sci. USSR*, **146**:720 (1962).

473. Y. Y. Mikhailenko, L. P. Senetskaya, and E. G. Kutyrina, *Tr. Komis. po Analit. Khim. Nauk SSSR, Inst. Geokhim. i Analit. Khim.*, **13**:383 (1963); through *Chem. Abstr.*, **59**:8778 (1963).

474. A. P. Kreshkov, Y. Y. Mikhailenko, L. P. Senetskaya, and D. A. Uklonskii, *Plast. Massy*, **1963**(9):44; through *Chem. Abstr.*, **60**:665 (1964).

475. I. V. Obreimov and N. A. Chumaevskii, *Zh. Strukt. Khim.*, **5**(1):137 (1964); through *Chem. Abstr.*, **60**:12784 (1964).

476. Y. P. Egorov and G. G. Kirei, *Zh. Obshch. Khim.*, **34**(11):3615 (1964); *J. Gen. Chem. USSR*, **34**:3665 (1964).

477. A. P. Kreshkov, Y. Y. Mikhailenko, and L. P. Senetskaya, *Plast. Massy*, **1965**(8):48; through *Chem. Abstr.*, **63**:15559 (1965).

478. L. A. Leites, I. D. Pavlova, and Y. P. Egorov, *Teor. Eksp. Khim.*, **1**(3):311 (1965); through *Chem. Abstr.*, **63**:13024 (1965).

479. M. C. Henry and J. G. Noltes, *J. Am. Chem. Soc.*, **82**:555 (1960).

480. V. A. Petukhov, V. F. Mironov, and A. L. Kravchenko, *Izv. Akad. Nauk SSSR, Ser. Khim.*, **1966**(1):156; through *Chem. Abstr.*, **64**:15192 (1966).

481. L. V. Vilkov, V. S. Mastryukov, and P. A. Akishin, *Zh. Strukt. Khim.*, **5**(2):183 (1964); through *Chem. Abstr.*, **61**:2563 (1964).

482. R. T. Hobgood, Jr., J. H. Goldstein, and G. S. Reddy, *J. Chem. Phys.*, **35**:2038 (1961).

483. R. T. Hobgood, Jr. and J. H. Goldstein, *Spectrochim. Acta*, **19**:321 (1963); *Chem. Abstr.*, **58**:4069 (1963).

484. R. Summitt, J. J. Eisch, J. T. Trainor, and M. T. Rogers, *J. Phys. Chem.*, **67**(11):2362 (1963).

485. G. A. Razuvaev, A. N. Egorochkin, M. L. Khidelkel, and V. F. Mironov, *Izv. Akad. Nauk SSSR, Ser. Khim.*, **1964**(5):928; *Bull. Acad. Sci. USSR, Div. Chem. Sci.*, **1964**:869.

486. R. A. Benkeser and H. R. Krysiak, *J. Am. Chem. Soc.*, **75**:2421 (1953).

487. L. H. Sommer, D. L. Bailey, G. M. Goldberg, C. E. Buck, T. S. Bye, F. J. Evans, and F. C. Whitmore, *ibid.*, **76**:1613 (1954).

488. M. Kanazashi, *Bull. Chem. Soc. Japan*, **28**:44 (1955); *Chem. Abstr.*, **52**:4556 (1958).

489. H. Soffer and T. De Vries, *J. Am. Chem. Soc.*, **73**:5817 (1951).

490. J. J. Eisch and J. T. Trainor, *J. Org. Chem.*, **28**:487 (1963).

491. "Information about Organosilicon Compounds," Bulletin 03–021, Chemical Products Div., Dow Corning Corp., Midland, Mich., February, 1967.

492. J. F. Hyde, *Science*, **147**:829 (1965).

493. M. A. Glaser, *Ind. Eng. Chem.*, **46**:2334 (1954).

494. M. Momonoi and N. Suzuki, *Nippon Kagaku Zasshi*, **78**:1324 (1957); *Chem. Abstr.*, **54**:5434 (1960).

495. R. M. Ismail, *Helv. Chim. Acta*, **47**(8):2405 (1964); *Chem. Abstr.*, **62**:7789 (1965).

496. A. D. Petrov, V. F. Mironov, and V. G. Glukhovtsev, *Zh. Obshch. Khim.*, **27**:1535 (1957); through *Chem. Abstr.*, **52**:3669 (1958).

497. Z. V. Belyakova, M. G. Pomerantseva, K. A. Andrianov, S. A. Golubtsov, and G. S. Popeleva, *Izv. Akad. Nauk SSSR, Ser. Khim.*, **1964**(11):2068; through *Chem. Abstr.*, **62**:7790 (1965).

498. R. E. Scott and K. C. Frisch, *J. Am. Chem. Soc.*, **73**:2599 (1951).

499. H. Minami and S. Nishizaki, *Kogyo Kagaku Zasshi*, **63**:366 (1960); through *Chem. Abstr.*, **56**:2467 (1962).

500. L. Goodman, R. M. Silverstein, and C. Williard, *J. Org. Chem.*, **22**:576 (1957).

501. V. M. Vdovin, N. S. Nametkin, E. S. Finkel'shtein, and V. D. Oppengeim, *Izv. Akad. Nauk SSSR, Ser. Khim.*, **1964**(3):458; through *Chem. Abstr.*, **60**:15902 (1964).

502. (a) C. Eaborn, "Organosilicon Compounds," p. 11, Butterworth, London, 1960. (b) C. Eaborn, *ibid.*, p. 12.

503. R. Nagel and H. W. Post, *J. Org. Chem.*, **17**:1379 (1952).

504. M. Cohen and J. R. Ladd, *J. Am. Chem. Soc.*, **75**:988 (1953).

505. M. Cohen and J. R. Ladd (to General Electric Co.), U.S. Patent 2,716,638, Aug. 30, 1955; *Chem. Abstr.*, **50**:7502 (1956).

506. M. Momonoi and N. Suzuki, *Nippon Kagaku Zasshi*, **78**:581 (1957); *Chem. Abstr.*, **53**:5169 (1959).

507. S. L. Davydova, Y. A. Purinson, B. D. Lavrukhin, and N. A. Plate, *Izv. Akad. Nauk SSSR, Ser. Khim.*, **1965**(2):387; through *Chem. Abstr.*, **62**:14717 (1965).

508. S. Tannenbaum, S. Kaye, and G. F. Lewenz, *J. Am. Chem. Soc.*, **75**:3753 (1953).

509. J. W. Curry, *ibid.*, **78**:1686 (1956).

510. British Patent 851,013 (to General Electric Co.), Oct. 12, 1960; *Chem. Abstr.*, **55**:9281 (1961).

511. H. Jenkner (to Kali-Chemie A.-G.), German Patent 1,091,566, Oct. 27, 1960; *Chem. Abstr.*, **55**:20996 (1961).

512. R. Mueller, H. Witte, and C. Dathe, *Z. Chem.*, **3**(10):391 (1963); through *Chem. Abstr.*, **61**:680 (1964).

513. E. Schnell, *Monatsh. Chem.*, **88**:1004 (1957); *Chem. Abstr.*, **52**:9950 (1958).

514. D. Y. Zhinkin, E. A. Semenova, and N. V. Markova, *Zh. Obshch. Khim.*, **33**(11):3736 (1963); through *Chem. Abstr.*, **60**:8056 (1964).

515. A. P. Kreshkov, D. A. Karateev, and V. Fyurst, *Zh. Prikl. Khim.*, **34**:2711 (1961); through *Chem. Abstr.*, **56**:12924 (1962).

516. A. P. Kreshkov, D. A. Karateev, and V. Fyurst, U.S.S.R. Patent 140,799, Sept. 9, 1961; *Chem. Abstr.*, **56**:12946 (1962).

517. G. R. Glowacki and H. W. Post, *J. Org. Chem.*, **27**:634 (1962).

518. K. A. Andrianov and V. V. Severnyi, *Izv. Akad. Nauk SSSR, Otd. Khim. Nauk*, **1961**:1788; through *Chem. Abstr.*, **56**:10183 (1962). *Izv. Akad. Nauk SSSR, Ser. Khim.*, **1964**(7):1268; through *Chem. Abstr.*, **61**:12027 (1964).

519. L. K. Freidlin, I. F. Zhukova, and V. F. Mironov, *Izv. Akad. Nauk SSSR, Otd. Khim. Nauk*, **1961**:1269; *Bull. Acad. Sci. USSR, Div. Chem. Sci.*, **1961**:1180.

520. P. D. George, M. Prober, and J. R. Elliot, *Chem. Rev.*, **56**:1065 (1956).

521. C. L. Agre and W. Hilling, *J. Am. Chem. Soc.*, **74**:3899 (1952).
522. V. F. Mironov, A. D. Petrov, and N. G. Maksimova, *Izv. Akad. Nauk SSSR, Otd. Khim. Nauk*, **1959**:1954; *Chem. Abstr.*, **54**:9731 (1960).
523. C. Tamborski and H. W. Post, *J. Org. Chem.*, **17**:1397 (1952).
524. K. A. Andrianov, A. A. Zhdanov, and V. A. Odinets, *Zh. Obshch. Khim.*, **32**:1126 (1962); through *Chem. Abstr.*, **58**:1484 (1963).
525. K. A. Andrianov, A. A. Zhdanov, and V. A. Odinets, *ibid.*, **31**:4033 (1961); through *Chem. Abstr.*, **57**:9874 (1962).
526. A. F. Gordon (to Dow Corning Corp.), U.S. Patent 2,715,113, Aug. 9, 1955; *Chem. Abstr.*, **50**:7131 (1956).
527. J. D. Park, J. D. Groves, and J. R. Lacher, *J. Org. Chem.*, **25**:1628 (1960).
528. G. W. Holbrook, A. F. Gordon, and O. R. Pierce, *J. Am. Chem. Soc.*, **82**:825 (1960).
529. M. Kleiman (to Arvey Corp.), U.S. Patent 2,697,089, Dec. 14, 1954; *Chem. Abstr.*, **49**:14027 (1955).
530. V. F. Mironov and L. L. Shchukovskaya, *Izv. Akad. Nauk SSSR, Otd. Khim. Nauk*, **1960**:760; through *Chem. Abstr.*, **54**:22327 (1960).
531. G. D. Odabashyan, T. A. Zhuravleva, and A. D. Petrov, *Dokl. Akad. Nauk SSSR*, **142**:604 (1962); *Proc. Acad. Sci. USSR*, **142**:70 (1962).
532. M. C. Henry and J. G. Noltes, *J. Am. Chem. Soc.*, **82**:558 (1960).
533. M. Prober (to General Electric Co.), U.S. Patent 2,835,690, May 20, 1958; *Chem. Abstr.*, **52**:18216 (1958).
534. H. Niebergall (to Koppers Co., Inc.), German Patent 1,113,827, Sept. 14, 1961; *Chem. Abstr.*, **56**:14475 (1962).
535. C. J. Albisetti, Jr. and M. J. Hogsed (to E. I. du Pont de Nemours & Co.) U.S. Patent 2,728,785, Dec. 27, 1955; *Chem. Abstr.*, **50**:12098 (1956).
536. R. Mueller and W. Mueller, *Chem. Ber.*, **98**(9):2916 (1965); *Chem. Abstr.*, **63**:14895 (1965).
537. A. D. Petrov, A. M. Polyakova, A. A. Sakharova, V. V. Korshak, V. F. Mironov, and G. I. Nikishin, *Dokl. Akad. Nauk SSSR*, **99**:785 (1954); through *Chem. Abstr.*, **49**:15727 (1955).
538. R. Y. Mixer and D. L. Bailey (to Union Carbide Corp.), U.S. Patent 2,777,868, Jan. 15, 1957; *Chem. Abstr.*, **52**:794 (1958).
539. Y. Tabata, H. Kimura, and H. Sobue, *Kogyo Kagaku Zasshi*, **67**(4):620 (1964); through *Chem. Abstr.*, **61**:10786 (1964).
540. M. Yuseum (to U.S. Dept. of the Navy), U.S. Patent 2,870,120, Jan. 20, 1959; *Chem. Abstr.*, **53**:7661 (1959).
541. D. T. Hurd (to Canadian General Electric Co. Ltd.), Canadian Patent 590,859, Jan. 19, 1960; *Chem. Abstr.*, **54**:14777 (1960).
542. N. Zutty (to Union Carbide Corp.), U.S. Patent 3,225,018, Dec. 21, 1965; *Chem. Abstr.*, **64**:8403 (1966).
543. B. R. Thompson, *J. Polymer Sci.*, **19**:373 (1956).
544. T. R. Santelli (to Owens-Illinois Glass Co.), U.S. Patent 3,075,948, Jan. 29, 1963; *Chem. Abstr.*, **58**:10326 (1963).
545. French Patent Addn. 73,034 to French Patent 1,181,893 (to Société des Usines Chimiques Rhône-Poulenc), Sept. 22, 1960; *Chem. Abstr.*, **56**:15684 (1962).
546. H. Zeiss, "Organometallic Chemistry," A.C.S. Monograph Series, Reinhold, New York, 1965.
547. E. G. Rochow, "An Introduction to the Chemistry of Silicones," Wiley, New York, 1951.

548. A. D. Petrov, V. F. Mironov, V. A. Ponomarenko, E. A. Chernyshev, and N. D. Zelinskii, "Synthesis of Organosilicon Monomers," Plenum, New York, 1964.

549. C. W. Smith (ed.), "Acrolein," Wiley, New York, 1962.

550. H. Schulz and H. Wagner, Angew. Chem., 62:105 (1950).

551. H. R. Guest, B. W. Kiff, and H. A. Stansbury, Jr., Acrolein and Derivatives, in A. Standen (ed.), "Encyclopedia of Chemical Technology," 2d ed., p. 225, Interscience, New York, 1962.

552. R. C. Schulz, Acrolein Polymers, in N. M. Bikales (ed.), "Encyclopedia of Polymer Science and Technology," vol. I, p. 160, Interscience, New York, 1964.

553. R. C. Schulz, Angew. Chem., 76(9):357 (1964).

554. N. G. Koral'nik, Sb. Nauchn.-Issled. Rabot, Khim. i Khim. Tekhnol. Vysokomolekul. Soedin., Tashkentsk. Tekstil'n. Inst., 1:5 (1964); through Chem. Abstr., 63:17882 (1965).

555. C. Moureu et al., Ann. Chim (Paris), 15:158 (1921).

556. J. Redtenbacher, Ann. Chem., 47:113 (1843).

557. (a) E. Lennemann, ibid. 125:310 (1863). (b) Simpson, Z. Chem., 1867:376 through "Beilsteins Handbuch Der Organischen Chemie," vol. 1, p. 725, Springer, Berlin, 1918. (c) E. v. Meyer, J. Prakt. Chem., 10(2):113 (1874).

558. G. Deniges, Compt. Rend., 126:1147 (1898).

559. (a) C. Moureu and C. Dufraisse, Compt. Rend., 169:621 (1919). (b) C. Moureu and A. Lepape, ibid., 169:705 (1919). (c) C. Moureu and A. Lepape, ibid., 169:885 (1919). (d) C. Moureu, C. Dufraisse, and P. Robin, ibid., 169:1068 (1919). (e) C. Moureu, C. Dufraisse, P. Robin, and J. Pougnet, ibid., 170:26 (1920). (f) C. Moureu, C. Dufraisse, and J. Pougnet, British Patent 141,361, Apr. 8, 1920; Chem. Abstr., 14:2495 (1920). (g) C. Moureu, U.S. Patent 1,436,047, Nov. 21, 1923; Chem. Abstr., 17:564 (1923). (h) C. Moureu and C. Dufraisse, Compt. Rend., 175:127 (1932). (i) C. Moureu and C. Dufraisse, ibid., 176:797 (1923). (j) C. Moureu, C. Dufraisse, and M. Badoche, ibid., 187:1092 (1928).

560. H. P. A. Groll and H. W. de Jong (to Shell Development Co.), U.S. Patent 2,042,220, May 26, 1936; Chem. Abstr., 30:4870 (1936).

561. H. P. A. Groll and G. Hearne (to Shell Development Co.), U.S. Patent 2,106,347, Jan. 25, 1938; Chem. Abstr., 32:2542 (1938).

562. H. Schulz and H. Wagner (to Chemical Marketing Co., Inc.), U.S. Patent 2,277,887, Mar. 31, 1942; Chem. Abstr., 36:4832 (1942).

563. H. Schulz and H. Wagner (to Deutsche Gold- und Silber-Scheideanstalt vorm. Roessler), German Patent 707,021, May 8, 1941; Chem. Abstr., 36:1955 (1942).

564. H. Wagner (to Chemical Marketing Co., Inc.), U.S. Patent 2,288,306, Jun. 30, 1942; Chem. Abstr., 37:201 (1943).

565. M. Gallagher and R. L. Hasche (to Eastman Kodak Co.), U.S. Patents 2,245,582 and 2,246,037, Jun. 17, 1941; Chem. Abstr., 35:5907 (1941).

566. A. D. Macullum (to E. I. du Pont de Nemours & Co.), U.S. Patent 2,197,258, Apr. 16, 1940; Chem. Abstr., 34:5468 (1940).

567. K. M. Herstein (to Acrolein Corp.), British Patent 531,001, Dec. 27, 1940; K. M. Herstein (to Acrolein Corp.), U.S. Patent 2,270,705, Jan. 20, 1942. Chem. Abstr., 35:7982 (1941) and 36:3191 (1942).

568. J. G. M. Bremner, D. G. Jones, and S. Beaumont, J. Chem. Soc., 1946:1018. J. G. M. Bremner and D. G. Jones (to Imperial Chemical Industries Ltd.),

U.S. Patent 2,451,712, Oct. 19, 1948, and British Patent 573,507, Nov. 23, 1945. J. G. M. Bremner (to Imperial Chemical Industries Ltd.), British Patent 608,538, Sept. 16, 1948; *Chem. Abstr.*, **43**:1795, 4702 (1948).

569. W. B. Converse (to Shell Development Co.), U.S. Patent 2,309,576, Jan. 26, 1943; *Chem. Abstr.*, **37**:3767 (1943).

570. A. Clark and R. S. Shutt (to Battelle Memorial Inst.), U.S. Patent 2,383,711, Aug. 28, 1945; *Chem. Abstr.*, **40**:349 (1946).

571. G. W. Hearne and M. L. Adams (to Shell Development Co.), U.S. Patent 2,451,485, Oct. 19, 1948; *Chem. Abstr.*, **43**:2222 (1949).

572. *Chem. Eng.*, **67**(2):48 (1960).

573. *Chem. Eng. News*, **39**(41):56 (1961).

574. "Acrolein Dimer," Technical Bulletin PD-123R, Shell Chemical Co., New York, 1963.

575. W. F. Gresham and C. E. Schweitzer (to E. I. du Pont de Nemours & Co.), U.S. Patent 2,485,236, Oct. 18, 1949; *Chem. Abstr.*, **44**:2017 (1950).

576. J. B. Rust (to Montclair Research Corp.), U.S. Patent 2,415,039, Jan. 28, 1947; *Chem. Abstr.*, **41**:2253 (1947).

577. S. Hayworth and J. R. Holker (to Cotton Silk and Man-Made Fibres Research Assoc.), British Patent 1,025,274, Apr. 6, 1966; *Chem. Abstr.*, **64**:17780 (1966).

578. (a) K. U. Usmanov, Y. P. Putiev, T. Gafurov, and Y. Tashpulatov, *Dokl. Akad. Nauk Uz. SSR*, **19**(7):40(1962); through *Chem. Abstr.*, **58**:5167 (1963). (b) K. Dustmukhamedov, M. M. Tulyaganov, T. G. Gafurov, and K. U. Usmanov, *Uzbeksk. Khim. Zh.*, **8**(5):74 (1964); through *Chem. Abstr.*, **62**:6611 (1965).

579. A. N. Stergiu, Y. L. Pogosov, and B. I. Aikhodzhaev, *Uzbeksk. Khim. Zh.*, **9**(2):43 (1965); through *Chem. Abstr.*, **63**:3154 (1965).

580. Y. K. Kirilenko, L. A. Vol'f, A. I. Meos, and V. V. Girdyuk, *Zh. Prikl. Khim.*, **38**(7):1638 (1965); *J. Appl. Chem. USSR*, **38**:1604 (1965).

581. J. R. McPhee and M. Lipson, *Austr. J. Chem.*, **7**:387 (1954).

582. G. Nishikawa, H. Kakiki, S. Takahashi, H. Irie, K. Kamimura, and Y. Mori (to Dainippon Printing Ink Manufg. Co. Ltd. and Japan Reichhold Chemicals, Inc.), Japanese Patent 6,999, Jul. 3, 1962; *Chem. Abstr.*, **59**:802 (1963).

583. H. Irie, Y. Uemura, Y. Mori, G. Nishikawa, H. Kakimoto, and S. Takahashi (to Dainippon Printing Ink Manufg. Co. Ltd., and Japan Reichhold Chemicals, Inc.), Japanese Patent 5,299, Jun. 20, 1962; *Chem. Abstr.*, **59**:11722 (1963).

584. H. Irie, Y. Mori, Y. Uemura, G. Nishikawa, and S. Takahashi (to Dainippon Printing Ink Manufg. Co. Ltd., and Japan Reichhold Chemicals, Inc.), Japanese Patent 4,594, Jun. 14, 1962; *Chem. Abstr.*, **59**:11722 (1963).

585. S. Tajima, E. Kawai, and T. Matsumoto (to Sumitomo Chemical Co. Ltd.), Japanese Patent 25,199, Nov. 27, 1963; *Chem. Abstr.*, **60**:5694 (1964).

586. N. P. Rowell and J. Wharton (to Courtaulds Ltd.), U.S. Patent 3,173,752, Mar. 16, 1965; *Chem. Abstr.*, **62**:13310 (1965).

587. F. A. Patty and R. R. Sayers, *Bur. of Mines, Rept. of Investigations*, **2027** (1930); H. H. Schrenck, F. A. Patty, and W. P. Yant, *ibid.*, **3031** (1930); through *Chem. Abstr.*, **24**:4561 (1930).

588. W. A. Kuenzli (to Servel Inc.), U.S. Patent 1,808,604, Jun. 2, 1931; *Chem. Abstr.*, **25**:4329 (1931).

589. V. F. Bruns, R. R. Yeo, and H. F. Arle, *U.S. Dept. Agr., Tech. Bull.*, no. 1299 (1964); through *Chem. Abstr.*, **61**:4889 (1964).

590. D, F. Langridge, *J. Agr. (Victoria, Australia)*, **63**:349 (1965); through *Chem. Abstr.*, **63**:13956 (1965).

591. R. D. Blackburn, D. E. Seaman, and L. W. Weldon, *Proc. Southern Weed Conf.*, 14th, 302 (1961); through *Chem. Abstr.*, **56**:1796 (1962).

592. L. S. Jordan, B. E. Day, and R. T. Hendrixon, *Hilgardia*, **32**:433 (1962); through *Chem. Abstr.*, **57**:6355 (1962).

593. R. D. Blackburn, *Weeds*, **11**(1):21 (1963); through *Chem. Abstr.*, **58**:9574 (1963).

594. W. G. L. Austin, *Outlook Agr.*, **4**(1):35 (1963); through *Chem. Abstr.*, **61**:6296 (1965).

595. R. D. Blackburn and L. W. Weldon, *Weeds*, **12**(4):295 (1964); through *Chem. Abstr.*, **62**:2190 (1965).

596. V. T. Stack, Jr., *Ind. Eng. Chem.*, **49**:913 (1957).

597. M. Legator (to Shell Internationale Research Maatschappij NV), French Patent 1,399,162, May 14, 1965; *Chem. Abstr.*, **64**:1884 (1966).

598. W. A. Kreutzer (to Shell Oil Co.), U.S. Patent 3,028,304, Apr. 3, 1962; *Chem. Abstr.*, **56**:14671 (1962).

599. C. L. Judson, Y. Hokama, and A. D. Bray, *J. Econ. Entomol.*, **55**:805 (1962); through *Chem. Abstr.*, **59**:3273 (1963).

600. J. P. de Villiers and J. G. MacKenzie, *Bull. World Health Organ.*, **29**(3):424 (1963); through *Chem. Abstr.*, **60**:9842 (1964).

601. J. A. Amant, W. C. Johnson, and M. J. Whalls, *Progr. Fish Culturist*, **26**(2):84 (1964); through *Chem. Abstr.*, **61**:4903 (1964).

602. British Patent 873,800 (to Shell Internationale Research Maatschappij NV), Appl. Apr. 4, 1960; *Chem. Abstr.*, **56**:1302 (1962).

603. Anon., *Federal Register*, **27**:46 (Jan. 4, 1962); through *Chem. Abstr.*, **56**:7755 (1962).

604. M. Legator (to Shell Oil Co.), U.S. Patent 3,240,605, Mar. 15, 1966; *Chem. Abstr.*, **64**:18109 (1966).

605. G. Bonjour (to Société Lumière, SA), French Patent Addn. 79,890 to French Patent 1,258,356, Feb. 8, 1963; *Chem. Abstr.*, **59**:1218 (1963).

606. L. D. Taylor (to International Polaroid Corp.), German Patent 1,182,526, Nov. 26, 1964; *Chem. Abstr.*, **62**:4820 (1965).

607. W. Himmelmann and A. Riebel (to Agfa AG), German Patent 1,203,604, Oct. 21, 1965; *Chem. Abstr.*, **63**:17372 (1965).

608. H. D. Mayor and L. E. Jordan, *J. Cell. Biol.*, **18**(1):207 (1963); through *Chem. Abstr.*, **60**:844 (1964).

609. D. T. Janigan, *Lab. Invest.*, **13**(9):1038 (1964); through *Chem. Abstr.*, **62**:4259 (1965).

610. D. T. Janigan, *J. Histochem. Cytochem.*, **13**(6):476 (1965); through *Chem. Abstr.*, **63**:16665 (1965).

611. N. Feder and M. K. Wolf, *J. Cell. Biol.*, **27**(2):327 (1965).

612. K. Motycka and L. Lacko, *Z. Krebs-forsch.*, **66**(5):491 (1965); through *Chem. Abstr.*, **63**:4830 (1965).

613. C. W. Cater, *J. Soc. Leather Trade's Chemists*, **47**:259 (1963); through *Chem. Abstr.*, **59**:8987 (1963).

614. J. H. Bowes and C. W. Cater, *J. Appl. Chem. (London)*, **15**:296 (1965).

615. G. I. Baeras, G. N. Bondarev, L. F. Chelpanova, and I. S. Okhrimenko, *Vysokomol. Soedin.*, **6**(10):1821 (1964); through *Chem. Abstr.*, **62**:2875 (1965).

616. W. Kern, H. Deibig, H. Cherdron, and V. Jaacks (to Deutsche Gold- und Silber-Scheideanstalt vorm. Roessler), German Patent 1,180,520, Oct. 29, 1964; *Chem. Abstr.*, **62**:4171 (1965).

617. K. H. Rink (to Deutsche Gold- und Silber-Scheideanstalt vorm. Roessler), German Patent 1,087,806, Appl. Feb. 10, 1958; *Chem. Abstr.*, **56**:8875 (1962).

618. G. V. Lund (to Courtaulds North America, Inc.), U.S. Patent 3,039,167, Jun. 19, 1962; *Chem. Abstr.*, **57**:8761 (1962).

619. R. F. Fischer and C. W. Smith (to Shell Oil Co.), U.S. Patent 3,183,054, May 11, 1965; *Chem. Abstr.*, **63**:4456 (1965).

620. B. D. Ostrow and F. I. Nobel, U.S. Patent 2,870,069, Jan. 20, 1959; *Chem. Abstr.*, **53**:5925 (1959).

621. A. B. Stamler and B. E. Geller, *Izv. Vyssh. Ucheb. Zaved., Khim. Khim. Tekhnol.*, **6**(5):879 (1963); through *Chem. Abstr.*, **60**:13329 (1964).

622. T. Hunt (to Distillers Co. Ltd.), British Patent 924,459, Apr. 24, 1963; *Chem. Abstr.*, **59**:4135 (1963).

623. S. Zeisel and M. Daniek, *Monatsh. Chem.*, **30**:727 (1909).

624. H. P. A. Groll (to Shell Development Co.), U.S. Patent 2,011,317, Aug. 13, 1935; *Chem. Abstr.*, **29**:6606 (1935).

625. French Patent 788,921 (to NV de Bataafsche Petroleum Maatschappij), Oct. 19, 1935; *Chem. Abstr.*, **30**:1806 (1936).

626. H. P. A. Groll and H. W. de Jong, British Patent 436,840, Oct. 18, 1935; *Chem. Abstr.*, **30**:2200 (1936).

627. C. T. Kautter (to Röhm & Haas A.-G.), German Patent 634,501, Aug. 28, 1936; *Chem. Abstr.*, **31**:420 (1937).

628. H. P. A. Groll and G. Hearne (to Shell Development Co.), U.S. Patent 2,042,224, May 26, 1936; *Chem. Abstr.*, **30**:4870 (1935).

629. G. Hearne and H. W. de Jong, *Ind. Eng. Chem.*, **33**:940 (1941).

630. M. I. Farberov and G. S. Mironov, *Dokl. Akad. Nauk SSSR*, **148**:1095 (1963); *Proc. Acad. Sci. USSR*, **148**:156 (1963).

631. British Patent 747,093 (to Farbenfabriken Bayer A.-G.), Mar. 28, 1956; *Chem. Abstr.*, **50**:12552 (1956).

632. C. N. Wolf (to Ethyl Corp.), French Patent 1,355,103, Mar. 13, 1964; *Chem. Abstr.*, **62**:11939 (1965).

633. E. E. Blaise and M. Maire, *Bull. Soc. Chim. France*, **3**(4):265 (1908).

634. S. Krapiwin, *Bull. Soc. Imp. Nat. Moscow*, **1908**:1; through *Chem. Zentr.*, **I**, 1335 (1910).

635. Patents assigned to Farbenfabriken vorm. F. Bayer & Co.: (a) German Patent 222,551, Jan. 8, 1909; *Chem. Abstr.*, **4**:2865 (1910). (b) German Patent 242,612, Jun. 3, 1910; *Chem. Abstr.*, **6**:2291 (1912).

636. A. Wohl and A. Prill, *Ann. Chem.*, **440**:139 (1924).

637. French Patent 719,309 (to E. I. du Pont de Nemours & Co.), Jun. 30, 1931; *Chem. Abstr.*, **26**:3265 (1932).

638. A. S. Carter (to E. I. du Pont de Nemours & Co.), U.S. Patent 1,896,161, Feb. 27, 1933; *Chem. Abstr.*, **27**:2458 (1933).

639. R. F. Conaway (to E. I. du Pont de Nemours & Co.), U.S. Patent 1,967,225, Jul. 24, 1934; *Chem. Abstr.*, **28**:5834 (1934).

640. N. S. Kozlov and N. P. Krechkov, U.S.S.R. Patent 42,073, Mar. 31, 1935; *Chem. Abstr.*, **30**:8248 (1936).

641. A. N. Churbakov and V. N. Ryazantsev, *J. Appl. Chem. USSR*, **13**:1464 (in French 1469) (1940); through *Chem. Abstr.*, **35**:3961 (1941).

642. A. N. Churbakov and V. N. Ryazantsev, *Org. Chem. Ind. USSR*, **7**:663 (1940); through *Chem. Abstr.*, **35**:5094 (1941).

643. French Patent 82,777 (to I. G. Farbenindustrie A.-G.), Feb. 16, 1938; *Chem. Abstr.*, **32**:6259 (1938).

644. British Patent 508,080 (to I. G. Farbenindustrie A.-G.), Jun. 23, 1939; *Chem. Abstr.*, **34**:453 (1940).

645. French Patent 836,149 (to Consortium für Elektrochemische Ind. GmbH), Jan. 11, 1939; *Chem. Abstr.*, **33**:5007 (1939).

646. C. C. Allen and V. E. Haury (to Shell Development Co.), U.S. Patent 2,225,542, Dec. 17, 1940; *Chem. Abstr.*, **35**:2160 (1941).

647. J. R. Long (to Wingfoot Corp.), U.S. Patent 2,256,149, Sept. 16, 1941; *Chem. Abstr.*, **36**:100 (1942).

648. A. C. Neish, *Can. Chem. Process Inds.*, **28**:862, 864, 866 (1944); through *Chem. Abstr.*, **39**:777 (1945).

649. J. J. Kolfenbach, E. F. Tuller, L. A. Underkofler, and E. I. Fulmer, *Ind. Eng. Chem.* **37**:1178 (1945).

650. N. A. Milas, E. Sakal, J. T. Plati, J. T. Rivers, J. K. Gladding, F. X. Grossi, Z. Weiss, M. A. Campbell, and H. F. Wright, *J. Am. Chem. Soc.*, **70**:1597 (1948).

651. R. G. Sims, *Mfg. Chemist*, **28**:128 (1957); through *Chem. Abstr.*, **51**:9083 (1957).

652. H. Arai, *Koryo*, **1961**(64):51; through *Chem. Abstr.*, **57**:6041 (1962).

653. W. B. Geiger, *Arch. Biochem.*, **16**:423 (1948).

654. L. J. Meuli (to The Dow Chemical Co.), U.S. Patent 2,775,066, Dec. 25, 1956; *Chem. Abstr.*, **51**:4638 (1957).

655. J. B. Rust (to Montclair Research Corp.), U.S. Patent 2,415,040, Jan. 28, 1947; *Chem. Abstr.*, **41**:2253 (1947).

656. H. Hopff (to I. G. Farbenindustrie A.-G.), German Patent 554,668, Oct. 5, 1930; *Chem. Abstr.*, **26**:6165 (1932).

657. Patents assigned to I. G. Farbenindustrie A.-G.: (a) British Patent 478,899, Jan. 27, 1938; *Chem. Abstr.*, **32**:5111 (1938). (b) French Patent 825,417, Mar. 3, 1938; *Chem. Abstr.*, **32**:6365 (1938).

658. H. Hopff (to I. G. Farbenindustrie A.-G.), U.S. Patent 2,196,452, Apr. 9, 1940; *Chem. Abstr.*, **34**:5569 (1940).

659. K. Vierling and H. Hopff (to I. G. Farbenindustrie A.-G.), U.S. Patent 2,201,750, May 21, 1940; *Chem. Abstr.*, **34**:6386 (1940).

660. H. Schafer and B. Tollens, *Chem. Ber.*, **39**:2181 (1906).

661. E. P. Kohler, *J. Am. Chem. Soc.*, **42**:375 (1909).

662. H. Beaufour, *Bull. Soc. Chim. France*, **13**(4):349 (1913).

663. F. Straus and A. Berkow, *Ann. Chem.*, **401**:121 (1913).

664. F. F. Blicke and J. H. Burckhalter, *J. Am. Chem. Soc.*, **64**:451 (1942).

665. Y. S. Kao, P. C. Pan, S. H. Loh, C. H. Chen, and H. Y. Hsu, *Hua Hsueh Hsueh Pao*, **24**:162 (1958); through *Chem. Abstr.*, **53**:7088 (1959).

666. W. B. Geiger and J. E. Conn, *J. Am. Chem. Soc.*, **67**:112 (1945).

667. M. Kamoda and N. Ito, *J. Agr. Chem. Soc. Japan*, **28**:799 (1954); through *Chem. Abstr.*, **49**:5573 (1955).

668. M. Kamoda, T. Chiba, K. Mori, and N. Ito, *Botyu-Kagaku*, **18**:117 (1953); through *Chem. Abstr.*, **48**:778 (1954).

669. F. Dickens and H. E. N. Jones, *Brit. J. Cancer*, **19**(2):392 (1965); through *Chem. Abstr.*, **63**:10441 (1965).

670. H. Bohme and P. Heller (to Farbwerke Hoechst A.-G.), German Patent 952,802, Nov. 22, 1956; *Chem. Abstr.*, **53**:1759 (1959).

671. I. N. Nazarov and I. V. Torgov, *Bull. Acad. Sci. USSR, Classe Sci. Chim.*, 1946:495; through *Chem. Abstr.*, 42:7735 (1948).

672. S. F. Reed, *J. Org. Chem.*, 27:4116 (1962).

673. G. S. Mironov, M. I. Farberov, and I. M. Orlova, *Zh. Obshch. Khim.*, 33(5):1512 (1963); *J. Gen. Chem. USSR*, 33:1476 (1963).

674. J. R. Millar and E. M. Wilkinson (to Permutit Co. Ltd.), British Patent 789,128, Jan. 15, 1958; *Chem. Abstr.*, 52:9669 (1958).

675. Y. A. Arbuzov, E. M. Klimov, and A. M. Korolev, *Zh. Obshch. Khim.*, 32:3681 (1962); *J. Gen. Chem. USSR*, 32:3610 (1962).

676. C. Moureu and A. Lepape, British Patent 141,057, Mar. 31, 1920; *Chem. Abstr.*, 14:2203 (1920).

677. H. Adkins and W. H. Hartung, in H. Gilman and A. H. Blatt (eds.), "Organic Syntheses," 2d ed., coll. vol. I, p. 15, Wiley, New York, 1951.

678. G. F. Bergh, *J. Prakt. Chem.*, 79(2):351 (1909).

679. J. B. Senderens, *Compt. Rend.*, 151:530 (1910).

680. A. Wohl and B. Mylo, *Chem. Ber.*, 45:2046 (1912).

681. E. J. Witzemann, *J. Am. Chem. Soc.*, 36:1766 (1914).

682. W. L. Evans and H. B. Haas, *ibid.*, 48:2703 (1926).

683. E. Freund (to Chemische Fabrik auf Aktien vorm. E. Schering), U.S. Patent 1,672,378, Jun. 5, 1928; *Chem. Abstr.*, 22:2571 (1928).

684. French Patent 695,931 (to Schering-Kahlbaum A.-G.), May 21, 1930; E. Schwenk, M. Gehrke, and F. Archer (to Schering-Kahlbaum A.-G.), U.S. Patent 1,916,743, Jul. 4, 1933; *Chem. Abstr.*, 25:2740 (1931) and 27:4547 (1933).

685. T. Kuwata and F. Matsubara, Japanese Patent 8,575, appl. Sept. 18, 1958; *Chem. Abstr.*, 56:4621 (1962).

686. M. Ciha, M. Cihova, and V. Macho, Czechoslovakian Patent 110,143, Mar. 15, 1964; *Chem. Abstr.*, 61:2975 (1964).

687. J. B. Williamson (to Distillers Co. Ltd.), British Patent 966,794, Aug. 19, 1964; *Chem. Abstr.*, 61:11894 (1964).

688. Netherlands Patent Appl. 6,402,495 (to Imperial Chemical Industries Ltd.), Sept. 14, 1964; *Chem. Abstr.*, 62:10341 (1965).

689. J. B. Williamson (to Distillers Co. Ltd.), British Patent 918,186, Feb. 13, 1963; *Chem. Abstr.*, 58:13796 (1963).

690. G. W. Godin and J. B. Williamson (to Distillers Co. Ltd.), British Patent 920,278, Mar. 6, 1963; *Chem. Abstr.*, 59:1491 (1963).

691. Belgian Patent 613,740 (to Nippon Sekiyu Kabushiki Kaisha), Feb. 28, 1962; *Chem. Abstr.*, 59:11261 (1963).

692. Y. Fujiwara and S. Masaki (to Japan Oil Co. Ltd.), U.S. Patent 3,172,914, Mar. 9, 1965; *Chem. Abstr.*, 62:12955 (1965).

693. Belgian Patent 660,006 (to Imperial Chemical Industries Ltd.), Aug. 19, 1965; *Chem. Abstr.*, 64:598 (1966).

694. T. Ishikawa and T. Kamio, *Yuki Gosei Kagaku Kyokaishi*, 20:56 (1962); through *Chem. Abstr.*, 56:11433 (1962).

695. T. Ishikawa and T. Kamio, *Tokyo Kogyo Shikensho Hokoku*, 58:40 (1963); through *Chem. Abstr.*, 61:4202 (1964).

696. S. Malinowski, H. Jedrzejewska, S. Basinski, and Z. Lipski, *Rocz. Chem.*, 31:71 (1957); through *Chem. Abstr.*, 51:14557 (1957).

697. S. Malinowski, H. Jablczynska-Jedrzejewska, S. Basinski, and S. Benbenek, *Chim. Ind. (Paris)*, 85:885 (1961); through *Chem. Abstr.*, 56:2321 (1962).

698. S. Malinowski and S. Basinski (to Politechnika Warszawska Zaklad Tech-

nologii Nieorganicznej I), Polish Patents 45,675, Apr. 16, 1962, and 45,582, May 2, 1962; *Chem. Abstr.*, **59**:9801 (1963).

699. S. Basinski and S. Malinowski, *Rocz. Chem.*, **38**(4):635 (1964); through *Chem. Abstr.*, **61**:9374 (1964).

700. British Patent 640,383 (to NV de Bataafsche Petroleum Maatschappij), Jul. 19, 1950; *Chem. Abstr.*, **45**:1619 (1951).

701. D. J. Hadley, R. Heap, and R. J. Nichol (to Distillers Co. Ltd.), British Patent 655,210, Jul. 11, 1951; *Chem. Abstr.*, **46**:6667 (1952).

702. British Patent 674,860 (to Standard Oil Development Co.), Jul. 2, 1952; *Chem. Abstr.*, **47**:4899 (1953).

703. G. W. Hearne and M. L. Adams (to Shell Development Co.), U.S. Patent 2,486,842, Nov. 1, 1949; *Chem. Abstr.*, **44**:2012 (1950).

704. British Patent 805,776 (to Continental Oil Co.), Dec. 10, 1958; *Chem. Abstr.*, **53**:17905 (1959).

705. J. T. Hackmann (to Shell Development Co.), U.S. Patent 2,690,457, Sept. 28, 1954; *Chem. Abstr.*, **49**:14798 (1955).

706. D. J. Hadley (to Distillers Co. Ltd.), British Patent 694,354, Jul. 22, 1953; *Chem. Abstr.*, **48**:10057 (1954).

707. G. Marullo, M. Agamennone, and L. Corsi (to Montecatini Società Generale per l'Industria Mineraria e Chimica), Italian Patent 580,541, Aug. 6, 1958; *Chem. Abstr.*, **54**:1305 (1960).

708. M. Agamennone, *Chim. Ind. (Milan)*, **43**:875 (1961); through *Chem. Abstr.*, **56**:1331 (1962).

709. J. L. Callahan, R. W. Foreman, and F. Veatch (to Standard Oil Co.), U.S. Patent 2,941,007, Jun. 14, 1960; *Chem. Abstr.*, **54**:19487 (1960).

710. J. L. Barclay, D. J. Hadley, and D. G. Stewart (to Distillers Co. Ltd.), British Patent 873,712, appl. Feb. 13 and Jun. 10, 1959; *Chem. Abstr.*, **56**:1349 (1962).

711. J. L. Barclay, J. R. Bethell, J. B. Bream, D. J. Hadley, R. H. Jenkins, D. G. Stewart, and B. Wood (to Distillers Co. Ltd.), British Patent 864,666, Apr. 6, 1961; *Chem. Abstr.*, **56**:1349 (1962).

712. A. Hausweiler, K. Schwarzer, and R. Stroh (to Farbenfabriken Bayer A.-G.), British Patent 839,808, Jun. 29, 1960; *Chem. Abstr.*, **55**:8293 (1961).

713. Patents assigned to Distillers Co. Ltd.: (a) D. J. Hadley, Belgian Patent 616,979, Oct. 29, 1962; (b) E. J. Gasson and R. H. Jenkins, Belgian Patent 630,159, Oct. 21, 1963; (c) F. C. Newman and C. J. Brown, Belgian Patent 633,024, Oct. 21, 1963; (d) J. L. Barclay, R. H. Jenkins, and E. J. Gasson, Belgian Patent 632,411, Nov. 18, 1963; (e) E. J. Gasson and J. Bahemen, British Patent 963,610, Jul. 15, 1964; (f) Netherlands Patent Appl. 6,408,649, Feb. 1, 1965; (g) E. J. Gasson and R. H. Jenkins, French Patent 1,395,072, Apr. 9, 1965; (h) W. I. Van der Meer and B. Wood, British Patent 991,085, May 5, 1965. *Chem. Abstr.*, **58**:12423 (1963); **60**:12705, 15737 (1964); **61**:4221, 9404 (1964); **63**:4116, 8202, 14707 (1965).

714. Patents assigned to Stamicarbon NV: (a) Belgian Patent 611,002, May 30, 1962; (b) Belgian Patent 614,584, Sept. 3, 1962; (c) French Patent 1,316,876, Feb. 1, 1963; (d) B. Phielix and J. J. T. M. Geerards, German Patent 1,160,838, Jan. 9, 1964; (e) Netherlands Patent 106,994, Dec. 16, 1963; (f) Netherlands Patent 110,366, Dec. 15, 1964. *Chem. Abstr.*, **57**:13616 (1962); **59**:449, 2651 (1963); **61**:10592 (1964); **62**:7641, 11691 (1965).

715. Patents assigned to Imperial Chemical Industries Ltd.: (a) French Patent 1,372,357, Sept. 11, 1964; (b) French Patent 1,364,810, Jun. 26, 1964; (c)

French Patent 1,380,884, Dec. 14, 1964; (d) B. P. Whim and G. C. Fettis, British Patent 989,401, Apr. 14, 1965. *Chem. Abstr.*, **62**:5195, 5196, 13048 (1965); **63**:4165 (1965).

716. Patents assigned to Shell Internationale Research Maatschappij NV: (a) British Patent 912,686, Dec. 12, 1962. (b) H. H. Voge and W. E. Armstrong, French Patent 1,342,963, Nov. 15, 1963. (c) C. R. Adams, T. J. Jennings, L. B. Ryland, and H. H. Voge, German Patent 1,161,552, Jan. 23, 1964. *Chem. Abstr.*, **60**:2773 (1964); **61**:6922, 5519 (1964).

717. Patents assigned to Standard Oil Co. of Ohio: (a) British Patent 963,611, Jul. 15, 1964. (b) J. L. Callahan, French Patent 1,375,538, Oct. 16, 1964. (c) J. L. Callahan, R. K. Grasselli, and W. R. Knipple, Belgian Patent 641,142, Apr. 1, 1964. *Chem. Abstr.*, **61**:8194 (1964); **62**:6396 (1965); **63**:4077 (1965).

718. Patents assigned to Chemical Investors SA: (a) D. M. Coyne and R. P. Cahoy, French Patent 1,362,890, Jun. 5, 1964; *Chem. Abstr.*, **61**:14533 (1964). (b) D. M. Coyne and R. P. Cahoy, French Patent 1,366,300, Jul. 10, 1964; *Chem. Abstr.*, **62**:447 (1965). (c) D. M. Coyne and R. P. Cahoy, French Patent 1,367,801, Jul. 24, 1964; *Chem. Abstr.*, **62**:1570 (1965).

719. Patents assigned to Eastman Kodak Co.: (a) H. J. Hagemeyer, Jr. and M. Statman, French Patent 1,351,913, Feb. 7, 1964; (b) E. L. McDaniel and H. S. Young, French Patent 1,409,061, Aug. 20, 1965. *Chem. Abstr.*, **60**:15736 (1964); **64**:1963 (1966).

720. Patents assigned to Knapsack-Griesheim A.-G.: (a) K. Sennewald, K. Gehrmann, W. Vogt, and S. Schaefer, German Patent 1,125,901, Mar. 22, 1962; *Chem. Abstr.*, **57**:12325 (1962). (b) K. Sennewald, K. Gehrmann, W. Vogt, and S. Schaefer, German Patent 1,137,427, Oct. 4, 1962; *Chem. Abstr.*, **58**:10082 (1963).

721. A. C. Shotts and A. H. Stephan (to Cities Service Research & Development Co.), U.S. Patent 3,009,960, appl. Oct. 3, 1958; *Chem. Abstr.*, **56**:8567 (1962).

722. E. Munekata, S. Minekawa, and K. Shibata (to Asahi Chemical Industry Co. Ltd.), Japanese Patent 7,511, 1961; *Chem. Abstr.*, **56**:2334 (1962).

723. A. Hendrickx (to Union Chimique Belge SA), British Patent 881,335, Nov. 1, 1961; *Chem. Abstr.*, **57**:2512 (1962).

724. L. Y. Margolis and O. V. Isaev, U.S.S.R. Patent 143,385, Jan. 24, 1962; *Chem. Abstr.*, **57**:9667 (1962).

725. R. W. Etherington, Jr. (to Petro-Tex Chemical Corp.), U.S. Patent 3,029,288, Apr. 10, 1962; *Chem. Abstr.*, **58**:1351 (1963).

726. Belgian Patent 613,157 (to Farbwerke Hoechst A.-G.), Jul. 26, 1962; *Chem. Abstr.*, **58**:1352 (1963).

727. O. Roelen and W. Rottig (to Ruhrchemie A.-G.), German Patent 1,146,046, Mar. 28, 1963; *Chem. Abstr.*, **59**:8597 (1963).

728. T. A. Kock (to E. I. du Pont de Nemours & Co.), French Patent 1,369,893, Aug. 21, 1964; *Chem. Abstr.*, **62**:6396 (1965).

729. French Patent 1,393,374 (to Sumitomo Chemical Co. Ltd.), Mar. 26, 1965; *Chem. Abstr.*, **63**:6863 (1965).

730. M. Sudo and S. Kusunoki (to Mitsubishi Rayon Co. Ltd.), Japanese Patent 23,646, Oct. 18, 1965; *Chem. Abstr.*, **64**:3359 (1966).

731. Netherlands Patent Appl. 6,500,418 (to B. F. Goodrich Co.), Jul. 21, 1965; *Chem. Abstr.*, **64**:3361 (1966).

732. French Patent 1,410,564 (to Mitsubishi Rayon Co. Ltd.), Sept. 10, 1965; *Chem. Abstr.*, **64**:4946 (1966).

733. W. Gruber (to Röhm & Haas GmbH), German Patent 1,207,367, Dec. 23, 1965; *Chem. Abstr.*, **64**:8039 (1966).

734. Netherlands Patent Appl. 6,506,246 (to Allied Chemical Corp.), Nov. 19, 1965; *Chem. Abstr.*, **64**:11088 (1966).

735. C. R. Adams, H. H. Voge, C. Z. Morgan, and W. E. Armstrong, *J. Catalysis*, 3(4):379 (1964).

736. Patents assigned to Shell Internationale Research Maatschappij NV: (a) K. D. Detling and H. H. Voge, Belgian Patent 618,222, Nov. 28, 1962; (b) A. T. Kister, Belgian Patent 618,223, Nov. 28, 1962. *Chem. Abstr.*, **59**:1490, 5024 (1963).

737. J. L. Barclay, D. J. Hadley, and D. G. Stewart (to Distillers Co. Ltd.), British Patent 1,011,599, Dec. 1, 1965; *Chem. Abstr.*, **64**:4944 (1966).

738. Belgian Patent 614,364 (to Distillers Co. Ltd.), Aug. 27, 1962; *Chem. Abstr.*, **58**:9666 (1963).

739. Y. B. Gorokhovatskii, *Kataliz i Katalizatory Akad. Nauk Ukr. SSR, Resp. Mezhved. Sb.*, **1965**:71; through *Chem. Abstr.*, **64**:4894 (1966).

740. E. N. Popova and B. Gorokhovatskii, *Dokl. Akad. Nauk SSSR*, **145**:570 (1962); *Proc. Acad. Sci. USSR*, **145**:626 (1962).

741. L. M. Kaliberdo, V. A. Pilosyan, and N. I. Popova, *Kinetika i Kataliz*, 3:237 (1962); *Kinetics and Catalysis*, 3:202 (1962).

742. A. M. Garnish, L. M. Shafranskii, N. P. Skvortsov, E. A. Zvezdina, and V. F. Stepanovskaya, *ibid.*, 3:257 (1962); 3:220 (1962).

743. A. G. Polkovnikova, B. D. Kruzhalov, A. N. Shatalova, and L. L. Tseitina, *ibid.*, 3:252 (1962); 3:216 (1962).

744. N. I. Popova and V. P. Latyshev, *Dokl. Akad. Nauk SSSR*, **147**:1382 (1962); *Proc. Acad. Sci. USSR*, **147**:1116 (1962).

745. O. V. Isaev and L. Y. Margolis, *Kinetika i Kataliz*, 1:237 (1960); *Kinetics and Catalysis*, 1:214 (1960).

746. Y. B. Gorokhovatskii and E. N. Popova, *ibid.*, 5:134 (1964); 5:111 (1964).

747. Y. B. Gorokhovatskii, E. N. Popova, and M. A. Rubanik, *ibid.*, 3:230 (1962); 3:196 (1962).

748. Y. B. Gorokhovatskii, M. A. Rubanik, and E. N. Popova, *ibid.*, 3:133 (1962); 3:111 (1962).

749. Y. B. Gorokhovatskii, E. N. Popova, and M. A. Rubanik, *Zh. Prikl. Khim.*, 36(12):2725 (1963); *J. Appl. Chem. USSR*, 36(12):2639 (1963).

750. A. G. Polkovnikova, A. N. Shatalova, and L. L. Tseitina, *Neftekhimiya*, 3(2):246 (1963); through *Chem. Abstr.*, **59**:6243 (1963).

751. N. Kominami, A. Shibata, and S. Minekawa, *Kogyo Kagaku Zasshi*, **65**:1510 (1962); *ibid.*, **65**:1514, 1517, 1520, 1522, 1525, 1528 (1962); through *Chem. Abstr.*, **59**:12630, 12631 (1963).

752. B. D. Kruzhalov, E. S. Shestukhin, and A. M. Garnish, *Kinetika i Kataliz*, 3:247 (1962); *Kinetics and Catalysis*, 3:211 (1962).

753. C. C. McCain, G. Gough, and G. W. Godin, *Nature*, 198(4884):989 (1963).

754. F. L. J. Sixma, E. F. J. Duynstee, and J. L. J. P. Hennekens, *Rec. Trav. Chim. Pays-Bas*, 82(9–10):901 (1963).

755. C. R. Adams and T. J. Jennings, *J. Catalysis*, 2(1):63 (1963).

756. W. M. H. Sachtler and N. H. de Boer, *Proc. Intern. Congr. Catalysis, 3rd, Amsterdam*, 1:252 (1964) (Pub. 1965); through *Chem. Abstr.*, **63**:13007 (1965).

757. N. I. Popova, R. N. Stepanova, and R. N. Stukova, *Kinetika i Kataliz*, 2:916 (1961); *Kinetics and Catalysis*, 2:836 (1961).

758. L. N. Kutseva and L. Y. Margolis, *Zh. Obshch. Khim.*, **32**:102 (1962); *J. Gen. Chem. USSR*, **32**:100 (1962).

759. M. Y. Rubanik, K. M. Kholyavenko, A. V. Gershingorina, and V. I. Lazukin, *Kinetika i Kataliz*, **5**(4):666(1964); *Kinetics and Catalysis*, **5**(4):588 (1964).

760. O. V. Isaev and L. Y. Margolis, *Nauchn. Osnovy Podbera i Proizv. Katalizatorov, Akad. Nauk SSSR, Sibirsk Otd.*, **1964**:210; through *Chem. Abstr.*, **63**:6847 (1965).

761. T. Ishikawa et al., *Tokyo Kogyo Shikensho Hokoku*, **59**(9):401, 407, 414, 422, 432 (1964); through *Chem. Abstr.*, **61**:16040, 14041 (1965). *Kogyo Kagaku Zasshi*, **67**(7):1015, 1018, 1021 (1964); through *Chem. Abstr.*, **61**:11884, 11885 (1965).

762. Patents assigned to Shell Internationale Research Maatschappij NV: (a) British Patent 911,035, Nov. 21, 1962; *Chem. Abstr.*, **58**:5440 (1963). (b) M. L. Courter and D. S. Thayer, German Patent 1,147,932, May 2, 1963; *Chem. Abstr.*, **60**:2772 (1964).

763. Netherlands Patent 80,852 (to NV de Bataafsche Petroleum Maatschappij), Mar. 15, 1956; *Chem. Abstr.*, **51**:2850 (1957).

764. Patents assigned to Ruhrchemie A.-G.: (a) Belgian Patent 612,201, Jul. 2, 1962; *Chem. Abstr.*, **57**:15357 (1962). (b) O. Roelen and W. Rottig, German Patents 1,134,979 and 1,134,980, Aug. 23, 1962; *Chem. Abstr.*, **58**:1352 (1963). (c) O. Roelen and W. Rottig, German Patent 1,155,109, Oct. 3, 1963; *Chem. Abstr.*, **60**:2774 (1964). (d) O. Roelen and W. Rottig, German Patent 1,160,837, Jan. 9, 1964; *Chem. Abstr.*, **60**:10550 (1964).

765. V. N. Vostrikova, R. E. Gurovich, M. E. Aerov, G. L. Motina, and R. G. Zahyaletdinova, *Neftekhimiya*, **3**(2):254 (1963); through *Chem. Abstr.*, **59**:6243 (1963).

766. H. Feichtinger and H. Linden (to Ruhrchemie A.-G.), German Patent 1,155,110, Oct. 3, 1963; *Chem. Abstr.*, **60**:1597 (1964).

767. H. De V. Finch (to Shell Oil Co.), U.S. Patent 2,991,233, Jul. 4, 1961; *Chem. Abstr.*, **56**:3358 (1962).

768. J. Howlett and C. A. Lamburd (to Distillers Co. Ltd.), British Patent 693,167, Jun. 24, 1953, and U.S. Patent 2,766,192, Oct. 9, 1956; *Chem. Abstr.*, **48**:10058 (1954) and **51**:7404 (1957).

769. M. I. Farberov, G. S. Mironov, and M. A. Korshunov, *Zh. Prikl. Khim.*, **35**:2483 (1962); *J. Appl. Chem. USSR*, **35**:2382 (1962).

770. H. E. Ramsden (to Metal and Thermit Corp.), British Patent 806,710, Dec. 31, 1958; *Chem. Abstr.*, **54**:2264 (1960).

771. H. S. Bloch (to Universal Oil Products Co.), U.S. Patent 3,028,419, Apr. 3, 1962; *Chem. Abstr.*, **57**:13620 (1962).

772. I. M. Sulima and D. K. Tolopko, *Khim. Prom., Inform. Nauk-Tekhn. Zb.*, **1965**(2):16; through *Chem. Abstr.*, **63**:8184 (1965).

773. F. E. Mertz and O. C. Dermer, *Proc. Okla. Acad. Sci.*, **30**:134 (1949); through *Chem. Abstr.*, **46**:7041 (1952).

774. British Patent 885,587 (to Cities Service Research & Development Co.), Dec. 28, 1961; *Chem. Abstr.*, **57**:7108 (1962).

775. M. B. Green and W. J. Hickinbottom, *J. Chem. Soc.*, **1957**:3262.

776. A. F. MacLean and B. G. Frenz (to Celanese Corp. of America), U.S. Patent 2,848,499, Aug. 12, 1958; *Chem. Abstr.*, **53**:3062 (1959).

777. H. J. Hagemeyer, Jr. (to Eastman Kodak Co.), U.S. Patent 2,639,295, May 19, 1953; *Chem. Abstr.*, **48**:3386 (1954).

778. M. I. Farberov, G. S. Mironov, M. A. Korshunov, E. P. Tepenitsyna, E. N. Shkarnikova, and V. D. Zav'yalov, U.S.S.R. Patent 144,164, Feb. 6, 1962; *Chem. Abstr.*, **57**:11024 (1962).

779. G. S. Mironov, M. I. Farberov, and M. A. Korshunov, *Uch. Zap. Yaroslavsk. Tekhnol. Inst.*, **1962**(1):33; through *Chem. Abstr.*, **61**:568 (1964).

780. Patents assigned to Eastman Kodak Co.: (a) C. W. Hargis, H. S. Young, and J. E. Williams, French Patent 1,340,385, Oct. 18, 1963. (b) C. W. Hargis and H. S. Young, German Patent 1,154,082, Sept. 12, 1963. *Chem. Abstr.*, **60**:6751 (1964); **61**:1759 (1964).

781. C. W. Hargis and H. S. Young, *Ind. Eng. Chem., Prod. Res. Develop.*, **5**(1):72 (1966).

782. W. F. Gresham (to E. I. du Pont de Nemours & Co.), U.S. Patent 2,549,457, Apr. 17, 1951; *Chem. Abstr.*, **45**:8549 (1951).

783. N. I. Popova, F. A. Mil'man, and L. E. Latysheva, *Izv. Sib. Otd. Akad. Nauk SSSR*, **1961**(7):77; through *Chem. Abstr.*, **56**:3338 (1962).

784. N. I. Popova, L. E. Latysheva, E. E. Vermel, and F. A. Mil'man, *Sintez i Svoistva Monomerov, Akad. Nauk SSSR, Inst. Neftekhim. Sinteza, Sb. Rabot 12-oi (Dvenadtsatoi) Konf. po Vysokomolekul. Soedin.*, **1962**:178 (Pub. 1964); through *Chem. Abstr.*, **62**:6387 (1965).

785. J. Aschenbrenner (to Badische Anilin- & Soda-Fabrik A.-G.), German Patent 857,951, Dec. 1, 1952; *Chem. Abstr.*, **52**:5453 (1958).

786. T. Mitsui, M. Kitahara, and Y. Miyatake, *Rika Gaku Kenkyusho Hokoku*, **38**:205 (1962); through *Chem. Abstr.*, **59**:3762 (1963).

787. T. Mitsui, M. Kitahara, and Y. Miyatake (to Physical and Chemical Research Institute), Japanese Patent 23,159, Oct. 31, 1963; *Chem. Abstr.*, **60**:2775 (1964).

788. Patents assigned to Distillers Co. Ltd.: (a) D. J. Hadley, British Patent 648,386, Jan. 3, 1951. (b) D. J. Hadley and R. J. Nichol, British Patent 694,362, Jul. 22, 1953. (c) D. J. Hadley and R. Heap, British Patent 704,388, Feb. 24, 1954. (d) D. J. Hadley and C. A. Woodcock, British Patent 695,028, Aug. 5, 1953. *Chem. Abstr.*, **45**:7134 (1951); **48**:10058 (1954); **49**:9028, 11685 (1955).

789. French Patent 1,345,265 (to Kanegafuchi Spinning Co. Ltd.), Dec. 6, 1963; *Chem. Abstr.*, **60**:10551 (1964).

790. British Patent 847,564 (to Montecatini Società Generale per l'Industria Mineraria e Chimica), Sept. 7, 1960; *Chem. Abstr.*, **55**:7291 (1961).

791. M. Agamennone, *Petrol. Mater. Prima Ind. Chim. Mod., Comun. Giornata (Milan)*, **1961**:209; through *Chem. Abstr.*, **56**:11892 (1962).

792. G. Marullo and M. Agamennone, *Chim. Ind. (Milan)*, **46**(4):376 (1964); through *Chem. Abstr.*, **61**:4202 (1964).

793. D. A. Dowden and A. M. U. Caldwell (to Imperial Chemical Industries, Ltd.), British Patent 828,812, Feb. 24, 1960; *Chem. Abstr.*, **54**:19488 (1960).

794. (a) M. Kitahara (to Physical and Chemical Research Institute), Japanese Patent 10,962, 1961; *Chem. Abstr.*, **56**:4621 (1962). (b) M. Kitahara and F. Moriya, *Rika Gaku Kenkyusho Hokoku*, **38**:45, 51, 57 (1962); through *Chem. Abstr.*, **59**:428, 429 (1963).

795. Patents assigned to Petro-Tex Chemical Corp.: (a) R. O. Kerr and R. W. Etherington, Jr., U.S. Patent 3,038,942, Jun. 12, 1962. (b) R. W. Etherington, Jr., and C. N. Wolf, U.S. Patent 3,031,508, Apr. 24, 1962. *Chem. Abstr.*, **57**:13616, 14941 (1962).

796. Patents assigned to Stamicarbon NV: (a) Belgian Patent 614,350, Aug. 23, 1962;

Chem. Abstr., **58**:1352 (1963). (b) Netherlands Patent 104,305, Mar. 15, 1963; *Chem. Abstr.*, **60**:2772 (1964).

797. D. M. Coyne and R. P. Cahoy (to Chemical Investors SA), French Patent 1,366,301, Jul. 10, 1964; *Chem. Abstr.*, **62**, 1570 (1965).

798. Netherlands Patent Appl. 6,408,142 (to Halcon International, Inc.), Jan. 26, 1965; *Chem. Abstr.*, **63**:11364 (1965).

799. (a) L. P. Shapovalova, Y. B. Gorokhovatski, and M. Y. Rubanik, *Dokl. Akad. Nauk SSSR*, **152**(3): 640 (1963); *Proc. Acad. Sci. USSR, Sect. Chem.*, **152**(3):763 (1963). (b) *Kinetika i Kataliz*, **5**(2):330 (1964); *Kinetics and Catalysis*, **5**(2):295 (1964).

800. R. S. Mann and D. J. Rouleau, in "Advances in Chemistry Series," no. 51, American Chemical Society, Washington, 1965, p. 40; through *Chem. Abstr.*, **64**:12489 (1966).

801. R. M. Cole, C. L. Dunn, and G. J. Pierotti (to Shell Development Co.), U.S. Patent 2,606,932, Aug. 12, 1952; *Chem. Abstr.*, **47**:4899 (1953).

802. B. M. Marks (to E. I. du Pont de Nemours & Co.), U.S. Patent 3,098,798, Jul. 23, 1963; *Chem. Abstr.*, **59**:13825 (1963).

803. G. Hearne, M. Tamele, and W. Converse, *Ind. Eng. Chem.*, **33**:805 (1941).

804. D. H. Hey, R. J. Nicholls, and C. W. Pritchett, *J. Chem. Soc.*, **1944**: 97.

805. (a) J. Ficini and H. Normant, *Compt. Rend.*, **247**:1627 (1958). (b) *Bull. Soc. Chim. France*, **1964**(6):1294.

806. A. Krattiger, *Bull. Soc. Chim. France*, **1953** 222.

807. R. Kitaoka (to Japan Bureau of Industrial Technics), Japanese Patent 19,959, Sept. 30, 1963; *Chem. Abstr.*, **60**:2774 (1964).

808. M. Hinder, H. Schinz, and C. F. Seidel, *Helv. Chim. Acta*, **30**:1495 (1947).

809. L. W. Kissinger, T. M. Benzinger, and R. K. Rohwer, *Nitro Compds. Proc. Intern. Symp. Warsaw*, **1963**:317 (Pub. 1964); through *Chem. Abstr.*, **64**:731 (1966).

810. J. Smidt, R. Sieber, W. Hafner, and R. Jira (to Consortium für Elektrochemische Industrie GmbH), German Patent 1,176,141, Aug. 20, 1964; *Chem. Abstr.*, **62**:447 (1965).

811. T. White and R. N. Haward, *J. Chem. Soc.*, **1943**:25.

812. J. T. Hays, G. F. Hager, H. M. Engelmann, and H. M. Spurlin, *J. Am. Chem. Soc.*, **73**:5369 (1951).

813. N. Murata, H. Arai, and M. Tanaka, *Kogyo Kagaku Zasshi*, **60**:1206 (1957); through *Chem. Abstr.*, **53**:14926 (1959).

814. Patents assigned to Celanese Corp. of America: (a) British Patent 783,458, Sept. 25, 1957; *Chem. Abstr.*, **52**:5458 (1958). (b) British Patent 833,666, Apr. 27, 1960; *Chem. Abstr.*, **54**:22369 (1960).

815. E. D. Bergmann, R. Ikan, and H. Weiler-Feilchenfeld, *Bull. Soc. Chim. France*, **1957**:290.

816. A. Y. Drinberg and A. A. Bulygina, *J. Appl. Chem. USSR*, **13**:1680 (1940); through *Chem. Abstr.*, **35**:4228 (1941).

817. H. Ezaki et al. (to Marumiya Co.), Japanese Patent 4,469, Sept. 9, 1953; *Chem. Abstr.*, **48**:9402 (1954).

818. S. Malinowski, S. Benbenek, J. Pasynkiewicz, and E. Wojciechowska, *Rocz. Chem.*, **32**:1089 (1958); through *Chem. Abstr.*, **53**:7974 (1959).

819. (a) M. I. Kogan, R. V. Fedorova, and O. D. Belova, U.S.S.R. Patent 137,510, Aug. 2, 1960; *Chem. Abstr.*, **56**:328 (1962). (b) M. I. Kogan, R. V. Fedorova, and

O. D. Belova, *Tr. Vses. Nauchn.-Issled. Vitamin Inst.*, 7:54 (1961); through *Chem. Abstr.*, **59**:1472 (1963).

820. G. S. Mironov, M. I. Farberov, and I. M. Orlova, *Zh. Prikl. Khim.*, **36**:654 (1963); *J. Appl. Chem. USSR*, **36**:622 (1963).

821. M. I. Farberov and G. S. Mironov, U.S.S.R. Patent 154,256, Jul. 24, 1963; *Chem. Abstr.*, **60**:6751 (1964).

822. H. J. Hagemeyer, Jr., *J. Am. Chem. Soc.*, **71**:119 (1949).

823. W. Wiesemann and H. W. Schwecten (to Farbenfabriken Bayer A.-G.), German Patent 877,606, May 26, 1953; *Chem. Abstr.*, **52**:8184 (1958).

824. A. M. Shur, *Neftekhimiya*, 2:600 (1962); through *Chem. Abstr.*, **58**:6735 (1963).

825. H. L. Yale and G. W. Hearne (to Shell Development Co.), U.S. Patent 2,398,685, Apr. 16, 1946; *Chem. Abstr.*, **40**:3768 (1946).

826. W. G. Toland, Jr. (to California Research Corp.), U.S. Patent 2,623,073, Dec. 23, 1952; *Chem. Abstr.*, **47**:9348 (1953).

827. T. Tsuchihara and H. Shingu, Japanese Patent 7,269, 1960; *Chem. Abstr.*, **56**:7139 (1962).

828. B. I. Chernyak, R. V. Kucher, and A. N. Nikolaevskii, *Neftekhimiya*, 4(3):452 (1964); through *Chem. Abstr.*, **61**:6887 (1964).

829. J. A. Mieuwland and F. J. Sowa (to E. I. du Pont de Nemours & Co.), British Patent 463,545, Mar. 30, 1937; *Chem. Abstr.*, **31**:6677 (1937).

830. N. G. Karapetyan, A. S. Tarkhanyan, and A. N. Lyubimova, *Izv. Akad. Nauk Arm. SSR, Khim. Nauki*, **17**(4):398 (1964); through *Chem. Abstr.*, **61**:15950 (1964).

831. E. Eberhardt, R. Stadler, A. Schweizer, and A. Wegerich (to Badische Anilin- & Soda-Fabrik A.-G.), German Patent 926,007, Apr. 4, 1955; *Chem. Abstr.*, **52**:13778 (1958).

832. H. Lange and O. Horn (to I. G. Farbenindustrie A.-G.), U.S. Patent 2,265,177, Dec. 9, 1941; *Chem. Abstr.*, **36**:1620 (1942).

833. French Patent 843,891 (to Consortium für Elektrochemische Industrie GmbH), Jul. 12, 1939; *Chem. Abstr.*, **34**:6948 (1940).

834. K. A. Grigoryan, A. A. Panfilov, and V. I. Isagulyants, *Dokl. Akad. Nauk Arm. SSR*, **35**(1):33 (1962); through *Chem. Abstr.*, **58**:2362 (1963).

835. Y. A. Gorin and L. P. Bogdanova, *Zh. Obshch. Khim.*, **28**:657 (1958); through *Chem. Abstr.*, **52**:17095 (1958).

836. Patents assigned to I. G. Farbenindustrie A.-G.: (a) British Patent 505,559, May 8, 1939. (b) H. Lange and O. Horn, U.S. Patent 2,267,829, Dec. 30, 1941, and German Patent 711,430, Aug. 28, 1941. (c) British Patent 499,034, Jan. 17, 1939. (d) H. Lange and O. Horn, U.S. Patent 2,208,296, Jul. 16, 1940. (e) French Patent 827,141, Apr. 20, 1938. (f) H. Lange and O. Horn, U.S. Patent 2,158,290, May 16, 1939, and German Patent 706,168, Apr. 10, 1941. (g) H. Lange and O. Horn, U.S. Patent 2,210,838, Aug. 6, 1940, and German Patent 709,725, Jul. 17, 1941. *Chem. Abstr.*, **33**:7822 (1939); **36**:2569 (1942) and **37**:4078 (1943); **33**:4604 (1939); **35**:137 (1941); **33**:993 (1939); **33**:6350 (1939) and **36**:1947 (1942); **35**:137 (1941) and **37**:3454 (1943).

837. W. J. Hale and L. A. Underkofler (to Natl. Agrol Co.), U.S. Patent 2,371,577, Mar. 13, 1945; *Chem. Abstr.*, **39**:3309 (1945).

838. J. J. Kolfenbach, *Iowa State Coll. J. Sci.*, **19**:35 (1944); through *Chem. Abstr.*, **39**:833 (1945).

839. British Patent 498,973 (to NV Bataafsche Petroleum Maatschappij), Jan. 17, 1939; *Chem. Abstr.*, **33**:4600 (1939).

840. W. J. Hale and H. Miller (to Natl. Agrol Co.), U.S. Patent 2,400,409, May 14, 1946; *Chem. Abstr.*, **40**:4744 (1946).
841. E. Arundale and H. O. Mottern (to Standard Oil Development Co.), U.S. Patent 2,620,357, Dec. 2, 1952; *Chem. Abstr.*, **47**:8089 (1953).
842. J. Lichtenberger and R. Lichtenberger, *Bull. Soc. Chim. France*, **1948**:1002.
843. J. Decombe, *Compt. Rend.*, **202**:1685 (1936).
844. O. C. Dermer and J. Newcombe, *J. Am. Chem. Soc.*, **74**:3417 (1952).
845. Y. A. Arbuzov and Y. P. Volkov, *Zh. Obshch. Khim.*, **29**:3279 (1959); *J. Gen. Chem. USSR*, **29**:3242 (1959).
846. British Patent 551,521 (to Wingfoot Corp.), Feb. 26, 1943; *Chem. Abstr.*, **38**:2349 (1944).
847. T. N. Nazarov, S. I. Zav'yalov, M. S. Burmistrova, I. A. Gurvich, and L. I. Shmonina, *Zh. Obshch. Khim.*, **26**:441 (1956); through *Chem. Abstr.*, **50**:13847 (1956).
848. S. G. Matsoyan, G. A. Chukhadzhyan, and S. A. Vartanyan, *Zh. Obshch. Khim.*, **29**:451 (1959); *J. Gen. Chem. USSR*, **29**:453 (1959).
849. S. G. Matsoyan and S. A. Vartanyan, *Izv. Akad. Nauk Arm. SSR, Ser. Fiz-Nat., Estestven. i Tekh. Nauk*, **8**(2):31 (1955); through *Chem. Abstr.*, **50**:4917 (1956).
850. H. Weber (to Chemische Werke Hüls A.-G.), German Patent 1,003,719, Mar. 7, 1957; *Chem. Abstr.*, **53**:21668 (1959).
851. I. E. Levine and W. G. Toland, Jr. (to California Research Corp.), U.S. Patent 2,464,244, Mar. 15, 1949; *Chem. Abstr.*, **43**:4287 (1949).
852. A. S. Carter and F. W. Johnson (to E. I. du Pont de Nemours & Co.), U.S. Patent 2,263,379, Nov. 18, 1941; *Chem. Abstr.*, **36**:1336 (1942).
853. K. Kalina, Czechoslovakian Patent 102,024, Dec. 15, 1961; *Chem. Abstr.*, **60**:1599 (1964).
854. S. O. Lawesson, E. H. Larsen, G. Sundstrom, and H. J. Jakobsen, *Acta Chem. Scand.*, **17**(8):2216 (1963); through *Chem. Abstr.*, **60**:5325 (1964).
855. W. Reppe and E. Joost (to Badische Anilin- & Soda-Fabrik A.-G.), German Patent 859,888, Dec. 18, 1952; *Chem. Abstr.*, **47**:11226 (1953).
856. W. Reppe et al., *Ann. Chem.*, **596**:38 (1955).
857. W. M. Schubert, T. H. Liddicoet, and W. A. Lanka, *J. Am. Chem. Soc.*, **76**:1929 (1954).
858. J. G. M. Bremner and D. G. Jones (to Imperial Chemical Industries Ltd.), British Patent 601,922, May 14, 1948; *Chem. Abstr.*, **42**:7319 (1948).
859. W. B. Guenther, *J. Am. Chem. Soc.*, **80**:1071 (1958).
860. F. G. Ponomarev and E. A. Vodop'yanova, *Nauch. Dokl. Vysshei Shkoly, Khim. i Khim. Tekhnol.*, **1959**(2):316; through *Chem. Abstr.*, **54**:267 (1960).
861. C. Mannich and G. Heilner, *Chem. Ber.*, **55B**:356 (1922).
862. H. Jager and M. Arenz, *ibid.*, **83**:182 (1950).
863. A. C. B. Smith and W. Wilson, *J. Chem. Soc.*, **1955**:1342.
864. A. Y. Yakubovich, V. V. Razumovskii, and I. N. Belyaeva, *Zh. Obshch. Khim.*, **28**:680 (1958); through *Chem. Abstr.*, **52**:17159 (1958).
865. W. G. Young and J. D. Roberts, *J. Am. Chem. Soc.*, **68**:649 (1946).
866. A. S. Angeloni and M. Tramontini, *Ann. Chim. (Rome)*, **54**(8-9):745 (1964).
867. G. A. Levy and H. B. Nisbet, *J. Chem. Soc.*, **1938**:1053.
868. C. S. Marvel and D. J. Casey, *J. Org. Chem.*, **24**:957 (1959).
869. (a) C. F. H. Allen and M. P. Bridgess, *J. Am. Chem. Soc.*, **51**:2151 (1929). (b) C. F. H. Allen and W. E. Barker, *ibid.*, **54**:736 (1932). (c) C. F. H. Allen and

H. W. J. Cressman, *ibid.*, **55**:2953 (1933). (d) C. F. H. Allen and A. C. Bell, *Can. J. Research*, **11**:40 (1934); through *Chem. Abstr.*, **29**:150 (1935).

870. H. Feuer and R. Harmetz, *J. Org. Chem.*, **26**:1061 (1961).
871. H. B. Couch and J. F. Norris, *J. Am. Chem. Soc.*, **42**:2329 (1920).
872. T. Matsumoto and K. Hata, *ibid.*, **79**:5506 (1957).
873. T. Matsumoto, K. Hata, and T. Nishida, *J. Org. Chem.*, **23**:106 (1958).
874. A. V. Dombrovskii and M. I. Shevchuk, *Zh. Obshch. Khim.*, **34**(1):192 (1964); *J. Gen. Chem. USSR*, **34**(1):190 (1964).
875. T. L. Gresham, J. E. Jansen, F. W. Shaver, and R. A. Bankert, *J. Am. Chem. Soc.*, **71**:2807 (1949).
876. R. A. Bankert (to B. F. Goodrich Co.), U.S. Patent 2,510,364, Jun. 6, 1950; *Chem. Abstr.*, **44**:8373 (1950).
877. K. Nagakubo, Y. Iwakura, K. Takei, and T. Okada, *Nippon Kagaku Zasshi*, **78**:1209 (1957); through *Chem. Abstr.*, **54**:5456 (1960).
878. G. Wittig, R. Mangold, and G. Felletschin, *Ann. Chem.*, **560**:116 (1948).
879. French Patent 802,499 (to Deutsche Celluloid Fabrik), Sept. 5, 1936; British Patent 461,495, Feb. 11, 1937. *Chem. Abstr.*, **31**:1912, 5065 (1937).
880. N. Jones and H. T. Taylor, *J. Chem. Soc.*, **1961**:1345.
881. H. A. P. de Jongh and H. Wynberg, *Tetrahedron*, **20**(11):2553 (1964).
882. J. R. Catch, D. F. Elliot, D. H. Hey, and E. R. H. Jones, *J. Chem. Soc.*, **1948**:278.
883. G. Baddley, H. T. Taylor, and W. Pickles, *ibid.*, **1953**:124.
884. I. N. Nazarov and A. A. Akhren, *Izv. Akad. Nauk SSSR, Otd. Khim. Nauk*, **1950**:621; through *Chem. Abstr.*, **45**:8516 (1951).
885. I. N. Nazarov and S. I. Zav'yalov, *Zh. Obshch. Khim.*, **23**:1703 (1953); through *Chem. Abstr.*, **48**:13669 (1954).
886. "Aldehydes," Tech. Bull., Union Carbide Corp., New York, 1965.
887. M. T. Rogers, *J. Am. Chem. Soc.*, **69**: (a) 1243. (b) 2544 (1947).
888. C. Moureu, A. Boutaric, and C. Dufraisse, *J. Chim. Phys.*, **18**:333 (1921).
889. C. Y. Chen and R. J. W. Le Fevre, *J. Chem. Soc.*, **1964**:234.
890. J. B. Bentley, K. B. Everard, R. J. B. Marsden, and L. E. Sutton, *ibid.*, **1949**:2957.
891. D. I. Coomber and J. R. Partington, *ibid.*, **1938**:1444.
892. E. A. Anderson and G. C. Hood, Physical Properties, in ref. 549, p. 7.
893. "Acrolein," Tech. Bull. SC: 59–66, Shell Chemical Co., New York.
894. W. H. Hartung and H. Adkins, *J. Am. Chem. Soc.*, **49**:2517 (1927).
895. W. F. Forbes and W. A. Mueller, *Can. J. Chem.*, **34**:1542 (1956).
896. "Methyl Vinyl Ketone," Data Sheet No. 512, Chas. Pfizer and Co., Inc., Brooklyn, N.Y., 1956.
897. A. Terada, *Nippon Kagaku Zasshi*, **81**:612 (1960); through *Chem. Abstr.*, **56**:1446 (1962).
898. G. Hart and K. T. Potts, *J. Org. Chem.*, **27**:2940 (1962).
899. W. Cooper, T. B. Bird, and E. Catterall, *Proc. 3rd Rubber Technol. Conf.*, *London*, **1953**:150 (Pub. 1956).
900. R. P. Mariella and R. R. Raube, *J. Am. Chem. Soc.*, **74**:518 (1952).
901. H. A. P. de Jongh and H. Wynberg, *Rec. Trav. Chim. Pays-Bas*, **82**:202 (1963).
902. E. Dyer, S. C. Brown, and R. W. Medeiros, *J. Am. Chem. Soc.*, **81**:4243 (1959).
903. J. Jakubicek, *Collect. Czech. Chem. Commun.*, (a) **28**:3180 (1963). (b) **26**:300 (1961).

904. "Azeotropic Data, Advances in Chemistry Series," no. 6, L. H. Horsley (ed.), American Chemical Society, Washington, 1952, p. 16.
905. V. N. Vostrikova, M. E. Aerov, R. E. Gurovich, and R. M. Solomantina, *Zh. Prikl. Khim.*, **37**(10):2210 (1964); through *Chem. Abstr.*, **62**:11201 (1965).
906. N. B. Hannay and C. P. Smyth, *J. Am. Chem. Soc.*, **68**:1357 (1946).
907. R. Wagner, J. Fine, J. W. Simmons, and J. H. Goldstein, *J. Chem. Phys.*, **26**:634 (1957).
908. P. D. Foster, V. M. Rao, and R. F. Curl, Jr., *ibid.*, **43**(3):1064 (1965).
909. H. Mackle and L. E. Sutton, *Trans. Faraday Soc.*, **47**:691 (1951).
910. J. Kossanyi, (a) *Compt. Rend.*, **256**(25):5308 (1963). (b) *Bull. Soc. Chim. France*, **1965**(3):704.
911. T. B. Albin, Handling and Toxicology, in ref. 549, p. 234.
912. S. D. Murphy, D. A. Klingshirn, and C. E. Ulrich, *J. Pharmacol. Exptl. Therap.*, **141**:79 (1963); *Chem. Abstr.*, **59**:14477 (1963).
913. "Acrolein Toxicity Data Sheet," Industrial Hygiene Bulletin SC: 57–76, Shell Chemical Corporation, New York, 1958.
914. "Acrolein Handling and Storage," Tech. Bull. SC: 59–105, Shell Chemical Corporation, New York, 1959.
915. E. D. Peters, Analytical Methods, in ref. 549, p. 240.
916. L. Nebbia and B. Pagani, *Chim. Ind.* (*Milan*), **46**(8):957 (1964); through *Chem. Abstr.*, **61**:15354 (1964).
917. S-C. Tung and E-K. Wang, *Hua Hsueh Hsueh Pao*, **29**(1):(1963); through *Chem. Abstr.*, **59**:2164 (1963).
918. V. P. Latyshev and N. I. Popova, *Teoriya i Praktike Polyarograf. Analiza, Akad. Nauk Moldavsk. SSR, Materialy Pervogo Vses. Soveshch.*, **1962**:406; through *Chem. Abstr.*, **59**:5771 (1963).
919. E. M. Bevilacqua, E. S. English, and J. S. Gall, *Anal. Chem.*, **34**:861 (1962).
920. M. Kitahara and T. Konishi, *Rika Gaku Kenkyusho Hokoku*, **38**:904 (1962); through *Chem. Abstr.*, **58**:5021 (1963).
921. G. Mizuno, E. McMeans, and J. R. Chipault, *Anal. Chem.*, **37**(1):151 (1965).
922. M. P. Stevens, *ibid.*, **37**(1):167 (1965).
923. R. W. McKinney and R. L. Jordan, *J. Gas Chromatog.*, **3**(9):317 (1965).
924. R. K. Sharma, D. R. McLean, and J. Bardwell, *Indian J. Technol.*, **3**(7):206 (1965).
925. C. F. H. Allen, *J. Am. Chem. Soc.*, **52**:2955 (1930).
926. V. P. Gryaznov and N. G. Polozhentseva, *Tr. Tsentr. Nauchn.-Issled. Inst. Spirt. i Likero-Vodochn. Prom.*, **1960**(9):84; through *Chem. Abstr.*, **57**:10356 (1962).
927. D. A. Forss and E. H. Ramshaw, *J. Chromatog.*, **10**(3):267 (1963).
928. A. Zamojski and F. Zamojska, *Chem. Anal.* (*Warsaw*), **9**(3):589 (1964); through *Chem. Abstr.*, **62**:8392 (1965).
929. V. M. Kisarov, *Zavod. Lab.*, **29**(2):163 (1963); through *Chem. Abstr.*, **59**:4472 (1963).
930. R. B. Wearn, W. M. Murray, Jr., M. P. Ramsey, and N. Chandler, *Anal. Chem.*, **20**:922 (1948).
931. S. J. Circle, L. Stone, and C. S. Boruff, *Ind. Eng. Chem.*, *Anal. Ed.*, **17**:259 (1945).
932. R. F. Graner and J. M. Garcia-Marquina, *Galenica Acta* (*Madrid*), **15**(5):343 (1962); through *Chem. Abstr.*, **59**:9068 (1963).

933. N. A. Yurko and Z. A. Volkova, *Khim. Prom., Inform. Nauk-Tekhn. Zb.*, **1964**(2):77; through *Chem. Abstr.*, **62**:2611 (1965).

934. W. Kaszper, *Med. Pracy*, **16**(6):453 (1965); through *Chem. Abstr.*, **64**:14852 (1966).

935. E. Sprague, *Univ. Kan. J. Home Econ.*, **11**:480 (1919); through *Chem. Abstr.*, **14**:302 (1920).

936. C. Moureu and E. Boismenu, *J. Pharm. Chim.*, **27**:49, 89 (1923).

937. J. Tavernier and P. Jacquin, *Inds. Agr. et Aliment. (Paris)*, **66**:357 (1950); *Chem. Abstr.*, **44**:6081 (1950).

938. L. Tsalapatanis, *An. Soc. Quim. Argentina*, **5**:244 (1917); through *Chem. Abstr.*, **12**:1034 (1918).

939. L. Rosenthaler and G. Vegezzi, *Z. Lebensm.-Unters. Forsch.*, **99**:352 (1954); *Chem. Abstr.*, **49**:2020 (1955).

940. I. R. Cohen and A. P. Altshuller, *Anal. Chem.*, **33**:726 (1961).

941. A. P. Altshuller and S. P. McPherson, *J. Air Pollution Control Assoc.*, **13**(3):109 (1963); *Chem. Abstr.*, **59**:3253 (1963).

942. W. C. Powick, *J. Agr. Research*, **26**:323 (1923); *Chem. Abstr.*, **18**:1580 (1924).

943. J. Pritzker, *Helv. Chim. Acta*, **11**:445 (1928).

944. I. Uzdina, *Chim. Ind. (Paris)*, **40**:260 (1938).

945. V. K. Pavolva, *Zh. Anal. Khim.*, **17**:368 (1962); through *Chem. Abstr.*, **57**:9214 (1962).

946. T. Mitsui and Y. Miyatake, *Rika Gaku Kenkyusho Hokoku*, **38**:189, 434, 446, 456 (1962); through *Chem. Abstr.*, **58**:13138 (1963), **60**:6216 (1964).

947. E. Sawicki, T. W. Stanley, and J. Pfaff, *Anal. Chim. Acta*, **28**:156 (1963).

948. T.-W. Kwon and B. M. Watta, *Anal. Chem.*, **35**:733 (1963).

949. L. Rosenthaler, *Pharm. Acta Helv.*, **26**:343 (1951); *Chem. Abstr.*, **46**:2962 (1952).

950. M. K. Berezova, *Hig. i Sanit. (USSR)*, **1940**(10):31 through *Chem. Abstr.*, **36**:6949 (1942).

951. I. Antener, *Mitt. Geb. Lebensmittelunters. Hyg.*, **28**:305 (1937); *Chem. Abstr.*, **32**:3178 (1938).

952. K. J. Bombaugh, *J. Chromatog.*, **11**:27 (1963).

953. J. Brodsky, M. Macka, and O. Mikl, *Chem. Prum.*, **10**:460 (1960); through *Chem. Abstr.*, **55**:4253 (1961).

954. K. J. Hughes, R. W. Hurn, and F. G. Edwards, *Gas Chromatog., Intern. Symposium, 2nd, E. Lansing, Mich.*, **1959**:171 (Pub. 1961).

955. H. Eustache, C. L. Guillemin, and F. Auricourt, *Bull. Soc. Chim. France*, **1965**(5):1386.

956. R. Zavodny, M. Konupeik, and M. Liska, Czechoslovakian Patent 113,210, Jan. 15, 1965; *Chem. Abstr.*, **63**:12983 (1965).

957. E. I. Fulmer, J. J. Kolfenbach, and L. A. Underkofler, *Ind. Eng. Chem., Anal. Ed.*, **16**:469 (1944).

958. B. Budesinsky, K. Mnoucek, F. Jancik, and E. Kraus, *Chem. Listy*, **51**:1819 (1957); through *Chem. Abstr.*, **52**:1860 (1958).

959. V. A. Devyatnin and I. A. Solunina, *Tr. Vses. Nauchn.-Issled. Vitamin Inst.*, **7**:104 (1961); through *Chem. Abstr.*, **59**:29 (1963).

960. A. P. Zozulya and E. V. Novikova, *Zavod. Lab.*, **29**(5):543 (1963); through *Chem. Abstr.*, **59**:4548 (1963).

961. T. Yoshida and T. Hirono, *Yaki Gosei Kagaku Kyokaishi*, **14**:508 (1956); through *Chem. Abstr.*, **51**:8001 (1957).

962. B. D. Modi and J. L. Bose, *Indian J. Chem.*, **3**(5):236 (1965).
963. H. Hata and K. Okada, *Bunseki Kagaku*, **10**:165 (1961); through *Chem. Abstr.*, **55**:23185 (1961).
964. J. Mitchell, Jr. and D. M. Smith, *Anal. Chem.*, **22**:746 (1950).
965. E. C. Dunlop, *Ann. N. Y. Acad. Sci.*, **53**:1087 (1951).
966. F. Ramirez and A. F. Kirby, *J. Am. Chem. Soc.*, (a) **75**:6026 (1953); (b) **76**:1037 (1954).
967. C. J. Timmons, *J. Chem. Soc.*, **1957**:2613.
968. A. Luthy, *Z. Phys. Chem.*, **107**:285 (1923).
969. V. Henri, *Compt. Rend.*, **178**:844 (1924).
970. H. W. Thompson and J. W. Linnett, *Nature*, **134**:937 (1934); *J. Chem. Soc.*, **1935**:1452.
971. F. E. Blacet, W. G. Young, and J. G. Roof, *J. Am. Chem. Soc.*, **59**:608 (1937).
972. A. M. Buswell, E. C. Dunlop, W. H. Rodebush, and J. B. Swartz, *ibid.*, **62**:325 (1940).
973. T. M. Sugden, A. D. Walsh, and W. C. Price, *Nature*, **148**:372 (1941).
974. L. K. Evans and A. E. Gillam, *J. Chem. Soc.*, **1943**:565.
975. A. D. Walsh, *Trans. Faraday Soc.*, **41**:498 (1945).
976. K. T. Holman, W. O. Lundberg, and G. O. Burr, *J. Am. Chem. Soc.*, **67**:1386 (1945).
977. F. E. Blacet, *J. Phys. Colloid Chem.*, **52**:534 (1948).
978. A. A. Dobrinskaya and M. B. Neiman, *Izv. Akad. Nauk SSSR, Ser. Fiz.*, **14**:520 (1950); *Chem. Abstr.*, **45**:3240 (1951).
979. F. Korte, *Angew. Chem.*, **63**:370 (1951).
980. R. C. Cookson and S. H. Dandegaonker, *J. Chem. Soc.*, **1955**:1651.
981. R. Hauschild and J. Petit, *Bull. Soc. Chim. France*, **1956**:878.
982. D. Buck and G. Scheibe, *Z. Elektrochem.*, **61**:901 (1957).
983. K. Watanabe, *J. Chem. Phys.*, **26**:542 (1957).
984. W. F. Forbes and R. Shilton, (a) *J. Am. Chem. Soc.*, **81**:786 (1959); (b) *J. Org. Chem.*, **24**:436 (1959).
985. S. Nagakura, *Mol. Phys.*, **3**:105 (1960).
986. R. Mecke and K. Noack, *Chem. Ber.*, **93**:210 (1960).
987. K. Inuzuka, *Bull. Chem. Soc. Japan*, (a) **33**:678 (1960); (b) **34**:729 (1961). *Chem. Abstr.*, **54**:23727 (1960); **56**:5547 (1962).
988. A. V. Krishna Rao, *J. Sci. Ind. Res.*, **21B**:446 (1962); *Chem. Abstr.*, **57**:15999 (1962).
989. J. F. Horwood and J. R. Williams, *Spectrochim. Acta*, **19**(8):1351 (1963).
990. J. M. Hollas, *ibid.*, **19**(9):1425 (1963).
991. J. C. D. Brand and D. G. Williamson, *Discussions Faraday Soc.*, **1963**(35):184.
992. E. Eastwood and C. P. Snow, *Proc. Roy. Soc. (London)*, **A149**:446 (1935).
993. P. Lambert and J. Lecomte, *Compt. Rend.*, **208**:740 (1939).
994. H. L. McMurry, *J. Chem. Phys.*, **9**:241 (1941).
995. G. M. Barrow, *ibid.*, **21**:2008 (1953).
996. W. H. T. Davison and G. R. Bates, *J. Chem. Soc.*, **1953**:2607.
997. R. H. Pierson, A. N. Fletcher, and E. St. C. Gantz, *Anal. Chem.*, **28**:1218 (1956).
998. R. K. Harris, *Spectrochim. Acta*, **20**(7):1129 (1964).
999. W. G. Fateley, R. K. Harris, F. A. Miller, and R. E. Witkowski, *ibid.*, **21**(2):231 (1965).

1000. A. Bertoluzza, G. Fabbri, and G. Farne, *Ann. Chim. (Rome)*, **55**(1-2):46 (1965); *Chem. Abstr.*, **62**:16020 (1965).

1001. M. Bourguel and L. Piaux, *Bull. Soc. Chim. France*, (5), **2**:1958 (1935).

1002. E. Scrocco and P. Chiorboli, *Atti Accad. Nag. Lincei, Rend., Cl. Sci. Fis. Mat. Nat.*, **8**:248 (1950); *Chem. Abstr.*, **44**:7147 (1950).

1003. M. Harrand and H. Martin, *Bull. Soc. Chim. France*, **1956**:1383.

1004. L. H. Meyer, A. Saika, and H. S. Gutowsky, *J. Am. Chem. Soc.*, **75**:4567 (1953).

1005. B. A. Arbuzov, A. I. Konovalov, and Y. Y. Samitov, *Dokl. Akad. Nauk SSSR*, **143**:109 (1962); through *Chem. Abstr.*, **57**:4521 (1962).

1006. J. B. Stothers and P. C. Lauterbur, *Can. J. Chem.*, **42**(7):1563 (1964).

1007. D. H. Marr and J. B. Stothers, *ibid.*, **43**(3):596 (1965).

1008. A. W. Douglas and J. H. Goldstein, *J. Mol. Spectry.*, **16**(1):1 (1965).

1009. R. E. Klinck and J. B. Stothers, *Can. J. Chem.*, **44**(1):45 (1966).

1010. J. D. Morrison and A. J. C. Nicholson, *J. Chem. Phys.*, **20**:1021 (1952).

1011. I. Omura, K. Higasi, and H. Baba, *Bull. Chem. Soc. Japan*, **29**:504 (1956); *Chem. Abstr.*, **51**:844 (1957).

1012. R. K. Harris and R. E. Witkowski, *Spectrochim. Acta*, **20**(11):1651 (1964).

1013. D. R. Davis, R. P. Lutz, and J. D. Roberts, *J. Am. Chem. Soc.*, **83**:246 (1961).

1014. R. A. Hoffman and S. Gronowitz, *Arkiv. Kemi*, **16**:471 (1960); *Chem. Abstr.*, **55**:21810 (1961).

1015. L. K. Evans and A. E. Gillam, *J. Chem. Soc.*, **1941**:815.

1016. C. J. Timmons, B. P. Straughan, W. F. Forbes, and R. Shilton, *Proc. Intern. Meeting Mol. Spectrosc., 4th, Bologna*, **2**:934 (1959) (Pub. 1962).

1017. C. Cherrier, *Compt. Rend.*, **225**:997 (1947).

1018. E. B. Lous, *Compt. Rend. 27th Congr. Intern. Chim. Ind., Brussels, 1954*, **3**; *Ind. Chim. Belge*, **20**, spec. no., 417 (1955).

1019. E. V. Sobolev and V. T. Aleksanyan, *Izv. Akad. Nauk SSSR, Ser. Khim.*, **1963**(7):1336; through *Chem. Abstr.*, **59**:13478 (1963).

1020. L. J. Bellamy and R. J. Pace, *Spectrochim. Acta*, **19**(11):1831 (1963).

1021. K. W. F. Kohlrausch, *Chem. Ber.*, **72B**:2054 (1939).

1022. G. Michel and G. Duyckaerts, *Spectrochim. Acta*, **10**:259 (1958).

1023. M. Martin and G. Martin, *Compt. Rend.*, **249**:884 (1959).

1024. S. Castellano and J. S. Waugh, *J. Chem. Phys.*, **37**:1951 (1962).

1025. J. Niwa and H. Kashiwagi, *Bull. Chem. Soc. Japan*, **36**(11):1414 (1963); *Chem. Abstr.*, **60**:6367 (1964).

1026. K. Takahashi, *ibid.*, **37**(7):963 (1964); *Chem. Abstr.*, **61**:11498 (1964).

1027. R. Goto and H. Inokawa, *Nippon Kagaku Zasshi*, **84**(8):650 (1963); *Chem. Abstr.*, **60**:444 (1964).

1028. M. S. De Groot and J. Lamb, (a) *Trans. Faraday Soc.*, **51**:1676 (1955); (b) *Nature*, **177**:1231 (1956); (c) *Proc. Roy. Soc. (London)*, **A242**:36 (1957).

1029. F. Gallais and J. F. Labarre, *J. Chim. Phys.*, **61**(5):717 (1964).

1030. G. G. Stoner and J. S. McNulty, *J. Am. Chem. Soc.*, **72**:1531 (1950).

1031. W. F. Gresham (to E. I. du Pont de Nemours & Co.), U.S. Patent 2,504,680, Apr. 18, 1950; *Chem. Abstr.*, **44**:6878 (1950).

1032. J. Howlett and H. R. Archer (to Distillers Co. Ltd.), (a) German Patent 924,629, Feb. 3, 1955; (b) U.S. Patent 2,800,434, Jul. 23, 1957. *Chem. Abstr.*, **52**:10149 (1958); **51**:18396 (1957).

1033. J. W. Mecorney and E. C. Shokal (to Shell Development Co.), U.S. Patent 2,886,493, May 12, 1959; *Chem. Abstr.*, **53**:18866 (1959).

1034. J. W. Mecorney (to Shell Oil Co.), U.S. Patent 2,939,882, Jun. 7, 1960; *Chem. Abstr.*, **54**:20878 (1960).

1035. British Patent 489,634 (to I. G. Farbenindustrie A.-G.), Jul. 29, 1938; French Patent 832,255, Sept. 23, 1938. *Chem. Abstr.*, **33**:647, 2541 (1939).

1036. E. M. McMahon (to Eastman Kodak Co.), U.S. Patent 2,324,101, Jul. 13, 1943; *Chem. Abstr.*, **38**:121 (1944).

1037. B. V. Ioffe and K. N. Zelenin, (a) *Izv. Vyssh. Ucheb. Zaved., Khim. Khim. Tekhnol.*, **6**(1):78 (1963); (b) *Dokl. Akad. Nauk SSSR*, **141**:1369 (1961); through *Chem. Abstr.*, **59**:6244 (1963); **56**:14038 (1962).

1038. R. L. Shriner and A. G. Sharp, *J. Am. Chem. Soc.*, **62**:2245 (1940).

1039. H. D. Finch, Reaction with Nitrogen Compounds, in ref. 549, p. 96.

1040. R. S. Corley, S. G. Cohen, M. S. Simon, and H. T. Wolosinski, *J. Am. Chem. Soc.*, **78**:2608 (1956).

1041. T. Shono and R. Oda, *J. Chem. Soc. Japan, Ind. Chem. Sect.*, **58**:276 (1955); *Bull. Inst. Chem. Res., Kyoto Univ.*, **33**:58 (1955); through *Chem. Abstr.*, **50**:4102, 14681 (1956).

1042. Y. K. Yur'ev, Z. V. Belyalcova, and V. P. Volkov, *Zh. Obshch. Khim.*, **28**:2372 (1958); through *Chem. Abstr.*, **53**:3135 (1959).

1043. F. L. Scott, J. C. Riordan, and A. F. Hegarty, *Tetrahedron Letters*, **1963**(9): 537.

1044. B. T. Gillis and K. F. Schimmel, *J. Org. Chem.*, **25**:2187 (1960).

1045. G. B. Payne and P. H. Williams, *ibid.*, **24**:284 (1959).

1046. I. N. Nazarov, L. A. Kazitsyna, and I. I. Zaretskaya, *Zh. Obshch. Khim.*, **27**:606 (1957); through *Chem. Abstr.*, **51**:16383 (1957).

1047. M. F. Hawthorne, *J. Org. Chem.*, **21**:1523 (1956).

1048. H. J. Shine, *ibid.*, **24**:1790 (1959).

1049. E. A. Braude, R. P. Linstead, and K. R. H. Wooldridge, *J. Chem. Soc.*, **1956**: 3070.

1050. K. Kratzl, H. Däubner, and U. Siegens, *Monatsh. Chem.*, **77**:146 (1947).

1051. C. W. Smith, D. G. Norton, and S. A. Ballard, *J. Am. Chem. Soc.*, **75**:3316 (1953).

1052. M. Simalty-Siemiatycki, J. Carretto, and F. Malbec, *Bull. Soc. Chim. France*, **1962**:125.

1053. R. C. Morris, Reaction with Alcohols, Mercaptans and Phenols, in ref. 549, p. 110.

1054. R. H. Hall and E. S. Stern, *J. Chem. Soc.*, (a) **1954**:3388 (b) **1955**:2657.

1055. S. Searles, Jr., R. Liepins, and H. M. Kash, *J. Org. Chem.*, **20**:36 (1960).

1056. R. F. Fischer and C. W. Smith (to Shell Oil Co.), U.S. Patent 3,043,851, Jul. 10, 1962; *Chem. Abstr.*, **57**:15116 (1962); *J. Org. Chem.*, **28**:594 (1963).

1057. C. K. Ikeda, R. A. Braun, and B. E. Sorenson, *J. Org. Chem.*, **29**:286 (1964).

1058. Belgian Patent 638,619 (to Farbwerke Hoechst A.-G.), Apr. 14, 1964; *Chem. Abstr.*, **62**:16261 (1965).

1059. F. Brown, D. E. Hudgin, and R. J. Kray, *J. Chem. Eng. Data*, **4**:182 (1959).

1060. R. J. Kray and F. Brown (to Celanese Corp. of America), U.S. Patent 2,915,530, Dec. 1, 1959; *Chem. Abstr.*, **54**:4393 (1960).

1061. H. R. Guest and B. W. Kiff (to Union Carbide Corp.), U.S. Patent 2,951,826, Sept. 6, 1960; *Chem. Abstr.*, **55**:1072 (1961).

1062. G. N. Koshel and M. I. Farberov, *Izv. Vyssh. Ucheb. Zaved., Khim. Khim. Tekhnol.*, **7**(4):639 (1964); through *Chem. Abstr.*, **62**:3997 (1965).

1063. A. N. Churbakov, *J. Gen. Chem. USSR*, **10**:977 (1940); through *Chem. Abstr.*, **35**:2469 (1941).

1064. W. Huber and A. Businger (to Hoffmann–La Roche, Inc.), U.S. Patent 2,540,116, Feb. 6, 1951; *Chem. Abstr.*, **45**:7584 (1951).

1065. G. F. Hennion and D. J. Lieb, *J. Am. Chem. Soc.*, **66**:1289 (1944).

1066. J. Cymerman, I. M. Heilbron, and E. R. H. Jones, *J. Chem. Soc.*, **1945**:90.

1067. W. Oroshnik and A. D. Mebane, *J. Am. Chem. Soc.*, **71**:2062 (1949).

1068. O. F. Beumel, Jr. and R. F. Harris, *J. Org. Chem.*, **29**(7):1872 (1964).

1069. Y. Kurihara, T. Higa, K. Sata, and S. Abe, *Bull. Chem. Soc. Japan*, **38**(1):29 (1965); through *Chem. Abstr.*, **62**:10326 (1965).

1070. W. Bauer (to Rohm and Haas Co.), U.S. Patent 2,153,406, Apr. 4, 1939; *Chem. Abstr.*, **33**:5006 (1939).

1071. W. Dietrich and B. Ritzenthaler (to Chemische Werke Hüls GmbH), German Patent 869,950, Mar. 9, 1953; *Chem. Abstr.*, **50**:8711 (1956).

1072. C. W. Smith and R. T. Holm, *J. Org. Chem.*, **22**:746 (1957).

1073. W. F. Brill and F. Lister, *ibid.*, **26**:565 (1961).

1074. J. R. Bethell and D. J. Hadley (to Distillers Co. Ltd.), British Patent 878,802, Oct. 4, 1961; *Chem. Abstr.*, **56**:11448 (1962).

1075. J. D. Idol, Jr., J. L. Callahan, and R. W. Foreman (to Sohio), U.S. Patents 2,881,212 and 2,881,213, Apr. 7, 1959; *Chem. Abstr.*, **57**:8441 (1962).

1076. J. R. Bethell, D. J. Hadley, E. J. Gasson, and R. F. Neale (to Distillers Co. Ltd.), British Patent 903,034, Aug. 9, 1962; *Chem. Abstr.*, **58**:8901 (1963).

1077. M. Kitahara, T. Mitsui, and T. Hirayama, *Rika Gaku Kenkyusho Hokoku*, **38**:81 (1962); through *Chem. Abstr.*, **58**:13788 (1963).

1078. J. L. Callahan, E. C. Milberger, and R. E. Utter (to Sohio), Belgian Patent 625,848, Mar. 29, 1963; *Chem. Abstr.*, **59**:9803 (1963).

1079. G. Calvin, B. Wood, and R. H. Jenkins (to Distillers Co. Ltd.), British Patent 924,532, Apr. 24, 1963; *Chem. Abstr.*, **59**:9803 (1963).

1080. B. J. Barone and W. F. Brill (to Petro-Tex Chemical Corp.), U.S. Patent 3,114,769, Dec. 17, 1963; *Chem. Abstr.*, **60**:6752 (1964).

1081. French Patent 1,346,534 (to Stamicarbon NV), Dec. 20, 1963; *Chem. Abstr.*, **61**:4223 (1964).

1082. G. N. Koshel, M. I. Farberov, and Y. A. Moskvichev, *Zh. Prikl. Khim.*, **37**(10):2287 (1964); through *Chem. Abstr.*, **62**:417 (1965).

1083. A. C. Shotts and J. R. Motes (to Cities Service Research & Development Co.), U.S. Patent 3,155,729, Nov. 3, 1964; *Chem. Abstr.*, **62**:2710 (1965).

1084. I. Schlossman (to Halcon International, Inc.), French Patent 1,395,951, Apr. 16, 1965; *Chem. Abstr.*, **63**:9817 (1965).

1085. F. Lanos and G. Clement (to Inst. Français du Petrole des Carburants et Lubrifiants), French Patent 1,401,176, May 21, 1965; *Chem. Abstr.*, **63**:13083 (1965).

1086. M. I. Farberov and G. N. Koshel, *Kinetika i Kataliz*, **6**(4):666 (1965); through *Chem. Abstr.*, **63**:16163 (1965).

1087. British Patent 1,007,405 (to Deutsche Gold- und Silber-Scheideanstalt vorm. Roessler), Oct. 13, 1965; *Chem. Abstr.*, **64**:1965 (1966).

1088. R. O. Kerr (to Petro-Tex Chemical Corp.), U.S. Patent 3,238,254, Mar. 1, 1966; *Chem. Abstr.*, **64**:17428 (1966).

1089. C. E. Castro and J. K. Kochi (to Shell Oil Co.), U.S. Patent 3,075,000, Jan. 22, 1963; *Chem. Abstr.*, **58**:11223 (1963).

1090. Netherlands Appl. 6,412,904 (to Halcon International, Inc.), May 13, 1965; U.S. Appl., Nov. 12, 1963; *Chem. Abstr.*, **63**:13085 (1965).

1091. K. Nakajima and M. Kitahara, *Kogyo Kagaku Zasshi*, **68**(10):1822 (1965); through *Chem. Abstr.*, **64**:3345 (1966).

1092. G. B. Payne, *J. Am. Chem. Soc.*, **81**:4901 (1959).

1093. G. W. Hearne, D. S. La France, and H. D. Finch (to Shell Development Co.), U.S. Patent 2,887,498, May 19, 1959; *Chem. Abstr.*, **53**:17148 (1959).

1094. A. S. Carter (to E. I. du Pont de Nemours & Co.), U.S. Patent 2,145,388, Jan. 31, 1939; *Chem. Abstr.*, **33**:3400 (1939).

1095. H. Berg and F. Leiss (to Alexander Wacker Ges. für Elektrochem. Ind. GmbH), German Patent 670,782, Jan. 27, 1939; *Chem. Abstr.*, **33**:6350 (1939).

1096. D. K. Sembaev, B. V. Suvorov, and S. R. Rafikov, *Dokl. Akad. Nauk SSSR*, **155**(4):868 (1964); through *Chem. Abstr.*, **60**:15728 (1964).

1097. N. C. Yang and R. A. Finnegan, *J. Am. Chem. Soc.*, **80**:5845 (1958).

1098. I. G. Tischenko, *Zhidkofazne Okislenie Nepredel'n Org. Soedin.*, **1961**(1): 73; through *Chem. Abstr.*, **58**:3306 (1963).

1099. E. Dyer, O. A. Pickett, Jr., S. F. Strause, and H. W. Worrell, Jr., *J. Am. Chem. Soc.*, **78**:3384 (1956).

1100. R. W. Foreman, L. S. Szabo, and F. Veatch (to Sohio), U.S. Patent 2,981,667, Apr. 25, 1961; *Chem. Abstr.*, **55**:17312 (1961).

1101. R. W. Goetz and M. Orchin, *J. Am. Chem. Soc.*, **85**(18):2782 (1963).

1102. W. B. Howsmon, Jr. (to Sohio), U.S. Patent 3,056,840, Oct. 2, 1962; *Chem. Abstr.*, **58**:11221 (1963).

1103. H. D. Finch, Hydrogenation and Reduction, in ref. 549, p. 88.

1104. R. W. Foreman (to Sohio), U.S. Patent 3,109,865, Nov. 5, 1963; *Chem. Abstr.*, **60**:2764 (1964).

1105. A. C. Shotts and B. M. Lloyd (to Cities Service Research & Development Co.), U.S. Patent 3,162,682, Dec. 22, 1964; *Chem. Abstr.*, **62**:11691 (1965).

1106. K. Yamagishi, S. Hamada, H. Arai, and H. Yokoo, *Yuki Gosei Kagaku Shi*, **24**(1):54 (1966); through *Chem. Abstr.*, **64**:8022 (1966).

1107. C. Chien, C. Hsieh, and L. Tai, *Hua Hsueh Hsueh Pao*, **31**(5):376 (1965); through *Chem. Abstr.*, **64**:8022 (1966).

1108. Netherlands Patent 65,211 (to NV de Bataafsche Petroleum Maatschappij), Feb. 15, 1950; *Chem. Abstr.*, **44**:6424 (1950).

1109. H. Brendlein (to Deutsche Gold- und Silber-Scheideanstalt vorm. Roessler), German Patent 858,247, Dec. 4, 1952; *Chem. Abstr.*, **50**:1891 (1956).

1110. H. G. Kuivila and O. F. Beumel, Jr., *J. Am. Chem. Soc.*, **80**:3798 (1958); *ibid.*, **83**:1246 (1961); (to Research Corp.), U.S. Patent 2,997,485, Aug. 22, 1961; *Chem. Abstr.*, **57**:866 (1962).

1111. J. G. Noltes and G. J. M. van der Kerk, *Chem. Ind.* (*London*), **1959**:294.

1112. A. Auerhahn and R. Stadler (to I. G. Farbenindustrie A.-G.), German Patent 724,668, Jul. 16, 1942; *Chem. Abstr.*, **37**:5737 (1943).

1113. J. Wiemann, M. R. Monot, and J. Gardan, *Compt. Rend.*, **245**:172 (1957).

1114. J. Wiemann, *Bull. Soc. Chim. France*, **1960**:1454.

1115. J. Wiemann and J. Gardan (to Centre National de la Recherche Scientifique), French Patent 1,177,602, Apr. 28, 1959; J. Wiemann and R. Jon, French Addn. 74,672 (to above pat.), May 5, 1961; *Chem. Abstr.*, **55**:27059 (1961); **57**:16406 (1962).

1116. L. Holleck and D. Marquarding, *Naturwissenschaften*, **49**:468 (1962).

1117. J. Kossanyi, *Bull. Soc. Chim. France*, **1965**(3):714.

1118. M. M. Baizer and J. D. Anderson, *J. Org. Chem.*, **30**(9):3138 (1965).

1119. T. Fueno, K. Asada, K. Morokuma, and J. Furukawa, *J. Polymer Sci.*, **40**:511 (1959).
1120. P. Zuman and J. Michl, *Nature*, **192**:655 (1961).
1121. R. W. Fourie and G. H. Riesser, Thermal Dimer, in ref. 549, p. 181.
1122. C. W. Smith, Diels-Alder Reactions, in ref. 549, p. 211.
1123. R. R. Whetstone (to Shell Development Co.), U.S. Patents 2,479,283 and 2,479,284, Aug. 16, 1949; *Chem. Abstr.*, **44**:667, 668 (1950).
1124. J. Habeshaw and R. W. Rae (to Anglo-Iranian Oil Co. Ltd.), British Patent 689,568, Apr. 1, 1953; *Chem. Abstr.*, **48**:7056 (1954).
1125. H. van B. Joy and J. B. Rust (to Lyndhurst Chemical Corp.), U.S. Patent 2,373,568, Apr. 10, 1945; *Chem. Abstr.*, **39**:3886 (1945).
1126. R. C. Morris, A. V. Snider, and P. H. Williams (to Shell Development Co.), U.S. Patent 2,450,765, Oct. 5, 1948; *Chem. Abstr.*, **44**:2021 (1950).
1127. M. Mousseron-Canet and M. Mousseron, *Bull. Soc. Chim. France*, **1956**:391.
1128. N. M. Bortnick (to Rohm and Haas Co.), U.S. Patents 2,473,497, Jun. 21, 1949, and 2,577,445, Dec. 4, 1951; *Chem. Abstr.*, **44**:1142 (1950); **46**:6142 (1952).
1129. R. R. Whetstone, W. J. Raah, and S. A. Ballard (to Shell Development Co.), U.S. Patent 2,562,849, Jul. 31, 1951; *Chem. Abstr.*, **46**:1584 (1952).
1130. R. H. Hall, *J. Chem. Soc.*, **1953**:1398.
1131. C. W. Smith, D. G. Norton, and S. A. Ballard, *J. Am. Chem. Soc.*, **73**: (a) 5267, (b) 5270, (c) 5273 (1951).
1132. British Patent 653,764 (to NV de Bataafsche Petroleum Maatschappij), May 23, 1951; *Chem. Abstr.*, **47**:5451 (1953).
1133. F. Tamura, K. Uehra, Y. Kubota, and N. Murata, *Kogyo Kagaku Zasshi*, **67**(10):1566 (1964); through *Chem. Abstr.*, **62**:10400 (1965).
1134. E. D. Bergmann, H. Davies, and R. Pappo, *J. Org. Chem.*, **17**:1331 (1952).
1135. A. A. Petrov, *J. Gen. Chem. USSR*, **11**:309 (1941); through *Chem. Abstr.*, **35**:5873 (1941).
1136. H. Hopff and C. W. Rautenstrauch (to I. G. Farbenindustrie A.-G.), U.S. Patent 2,262,002, Nov. 11, 1941; *Chem. Abstr.*, **36**:1046 (1942).
1137. R. C. Morris and T. W. Evans (to Shell Development Co.), U.S. Patent 2,435,403, Feb. 3, 1948; *Chem. Abstr.*, **42**:5049 (1948).
1138. K. Alder and W. Vogt, *Ann. Chem.*, **564**:109 (1949).
1139. A. A. Petrov and N. P. Sopov, *Zh. Obshch. Khim.*, **22**:591 (1952); through *Chem. Abstr.*, **47**:2735 (1953).
1140. M. Tanaka, *Kogyo Kagaku Zasshi*, **60**:1509 (1957); through *Chem. Abstr.*, **53**:18925 (1959).
1141. A. Etienne and E. Toromanoff, *Compt. Rend.*, **230**:306 (1950).
1142. A. Etienne, A. Spire, and E. Toromanoff, *Bull. Soc. Chim. France*, **1952**:750.
1143. (a) L. L. Placek and W. G. Bickford, *J. Am. Oil Chem. Soc.*, **36**:463 (1959).
(b) L. L. Placek, F. C. Magne, and W. G. Bickford, *ibid.*, **36**:651 (1959).
(c) L. L. Placek, H. P. Pastor, J. P. Hughes, and W. G. Bickford, *ibid.*, **37**:307 (1960).
1144. A. A. Petrov, *J. Gen. Chem. USSR*, **17**:538 (1947); through *Chem. Abstr.*, **42**:881 (1948).
1145. A. A. Petrov and A. V. Tumanova, *Zh. Obshch. Khim.*, **26**:2744 (1956); through *Chem. Abstr.*, **51**:7325 (1957).
1146. E. A. Prill, *J. Am. Chem. Soc.*, **69**:62 (1947).
1147. E. F. Lutz and G. M. Bailey, *ibid.*, **86**(18):3899 (1964).

1148. G. A. Ropp and E. C. Coyner, *ibid.*, **71**:1832 (1949).
1149. L. Reich and E. I. Becker, *ibid.*, **71**:1834 (1949).
1150. K. Alder, H. Offermanns, and E. Ruden, *Chem. Ber.*, **74B**: (*a*) 905, (*b*) 926 (1941).
1151. F. Tamura, Y. Shimodori, and N. Murata, *Kogyo Kagaku Zasshi*, **66**(9):1344, 1348 (1963); through *Chem. Abstr.*, **60**:10635 (1964).
1152. G. Opitz and H. Holtmann, *Ann. Chem.*, **684**:79 (1965).
1153. Y. A. Arbuzov and N. N. Bulatova, *Zh. Obshch. Khim.*, **33**(6):2045 (1963); through *Chem. Abstr.*, **59**:9862 (1963).
1154. C. F. H. Allen, A. C. Bell, A. Bell, and J. Van Allan, *J. Am. Chem. Soc.*, **62**:656 (1940).
1155. D. D. Coffman, P. L. Barrick, R. D. Cramer, and M. S. Raasch, *ibid.*, **71**:490 (1949).
1156. P. L. Barrick (to E. I. du Pont de Nemours & Co.), U.S. Patent 2,462,345, Feb. 22, 1949; *Chem. Abstr.*, **43**:4294 (1949).
1157. H. N. Cripps, J. K. Williams, and W. H. Sharkey, *J. Am. Chem. Soc.*, **80**:751 (1958).
1158. P. Pino and R. Ercoli, *Gazz. Chim. Ital.*, **81**:757 (1951); through *Chem. Abstr.*, **46**:7042 (1952).
1159. E. M. Kosower and G.-S. Wu, *J. Org. Chem.*, **28**:633 (1963).
1160. K. M. Taylor (to Monsanto Chemical Co.), French Patent 1,317,753, Feb. 8, 1963; *Chem. Abstr.*, **59**:7377 (1963).
1161. S. Malinowski, *Rocz. Chem.*, **29**:37 (1955); through *Chem. Abstr.*, **50**:3292 (1956).
1162. L. Horner and E. Lingnau, *Ann. Chem.*, **591**:21 (1955).
1163. K. A. Ogloblin and A. A. Potekin, *Zh. Obshch. Khim.*, **34**(8):2688 (1964); through *Chem. Abstr.*, **61**:14519 (1964).
1164. G. W. Hearne and H. D. Finch, Claisen and Michael Type Condensations, in ref. 549, p. 174.
1165. D. T. Warner and O. A. Moe (to General Mills, Inc.), U.S. Patents 2,546,958 and 2,546,960, Apr. 3, 1951; *Chem. Abstr.*, **45**:8036 (1951).
1166. O. A. Moe, D. T. Warner, and M. I. Buckley, *J. Am. Chem. Soc.*, **73**:1062 (1951).
1167. H. J. Gunst, M. Tobkes, and E. I. Becker, *ibid.*, **76**:3595 (1954).
1168. G. N. Walker, *ibid.*, **77**:3664 (1955).
1169. E. D. Bergmann and R. Corett, *J. Org. Chem.*, (a) **21**:107 (1956); (b) **23**:1507 (1958).
1170. H. Beyer, W. Lassig, and G. Schudy, *Chem. Ber.*, **90**:592 (1957).
1171. M. Tanaka and N. Murata, *Kogyo Kagaku Zasshi*, **59**:1181 (1956); through *Chem. Abstr.*, **52**:14607 (1958).
1172. L. I. Vereshchagin and I. L. Kotlyarevskii, *Izv. Akad. Nauk SSSR, Otd. Khim. Nauk*, **1960**:1632; through *Chem. Abstr.*, **55**:8404 (1961).
1173. H. Arai and N. Murata, *Kogyo Kagaku Zasshi*, **63**:107 (1960); through *Chem. Abstr.*, **56**:8585 (1962).
1174. J. Wiemann and Y. Dubois, *Bull. Soc. Chim. France*, **1962**:1813.
1175. T. A. Spencer, M. D. Newton, and S. W. Baldwin, *J. Org. Chem.*, **29**(4):787 (1964).
1176. N. C. Ross and R. Levine, *ibid.*, **29**(8): (a) 2341; (b) 2346 (1964).
1177. D. J. Goldsmith and J. A. Hartman, *ibid.*, **29**(12):3520 (1964).
1178. H. G. O. Becker, U. Fratz, G. Klose, and K. Heller, *J. Prakt. Chem.*, **29**(3–6):142 (1965).

1179. S. M. Abdullah, *J. Indian Chem. Soc.*, **12**:62 (1935).
1180. C. F. H. Allen and S. C. Overbaugh, *J. Am. Chem. Soc.*, **57**:1322 (1935).
1181. J. Wiemann and J. Dupayrat, *Bull. Soc. Chim. France*, **1961**(2):209.
1182. A. Treibs and R. Derra, *Ann. Chem.*, **589**:176 (1954).
1183. British Patent 666,623 (to NV de Bataafsche Petroleum Maatschappij), Feb. 13, 1952; *Chem. Abstr.*, **46**:11230 (1952).
1184. O. A. Moe and D. T. Warner (to General Mills, Inc.), U.S. Patent 2,599,653, Jun. 10, 1952; *Chem. Abstr.*, **47**:3339 (1953).
1185. British Patent 671,412 (to British Celanese Ltd.), May 7, 1952; *Chem. Abstr.*, **47**:2198 (1953).
1186. H. Shechter, D. L. Ley, and L. Zeldin, *J. Am. Chem. Soc.*, **74**:3664 (1952).
1187. H. Feuer and C. N. Aguilar, *J. Org. Chem.*, **23**:607 (1958).
1188. S. S. Novikov, I. S. Korsakova, and N. N. Bulatova, *Zh. Obshch. Khim.*, **29**:3659 (1959); through *Chem. Abstr.*, **54**:19465 (1960).
1189. C. W. Smith, U.S. Patent 2,600,275, Jun. 10, 1952; *Chem. Abstr.*, **47**:3337 (1953).
1190. R. H. Hall and E. S. Stern, *J. Chem. Soc.*, **1952**:4083; British Patent 695,789 (to Distillers Co. Ltd.), Aug. 19, 1953; *Chem. Abstr.*, **48**:8816 (1954).
1191. R. H. Hall and R. N. Lacey (to Distillers Co. Ltd.), British Patent 688,553, Mar. 11, 1953; *Chem. Abstr.*, **48**:2765 (1954).
1192. T. Bewley and B. K. Howe (to Distillers Co. Ltd.), British Patent 693,843, Jul. 8, 1953; *Chem. Abstr.*, **48**:10057 (1954).
1193. T. Bewley, R. H. Hall, B. K. Howe, and R. N. Lacey (to Distillers Co. Ltd.), British Patent 706,176, Mar. 24, 1954; *Chem. Abstr.*, **49**:9030 (1955).
1194. B. Thompson (to Eastman Kodak Co.), U.S. Patent 2,725,387, Nov. 29, 1955; *Chem. Abstr.*, **50**:13083 (1956).
1195. R. C. Elderfield, B. M. Pitt, and I. Wempen, *J. Am. Chem. Soc.*, **72**:1334 (1950).
1196. To Consortium für Elektrochemische Ind. GmbH: French Patents (a) 842,724, Jun. 19, 1939; (b) 834,192, Nov. 15, 1938. *Chem. Abstr.*, **34**:5859 (1940); **33**:3399 (1939).
1197. M. Kühn, *J. Prakt. Chem.*, **156**:103 (1940).
1198. N. Murata and H. Arai, *J. Chem. Soc. Japan, Ind. Chem. Sect.*, (a) **56**:628 (1953); (b) **57**:578 (1954); (c) **58**:387 (1955); (d) **59**:129 (1956). Through *Chem. Abstr.*, **49**:7517 (1955); **50**:205, 4150 (1956); **51**:1039 (1957).
1199. J. E. Dubois and R. Luft, *Compt. Rend.*, **242**:905 (1956).
1200. H. Arai, K. Saito, and N. Murata, *Kogyo Kagaku Zasshi*, **63**:319 (1960); through *Chem. Abstr.*, **59**:6362 (1963).
1201. B. P. Geyer, Reaction with Water, in ref. 549, p. 144.
1202. E. E. Gilbert and J. J. Donleavy, *J. Am. Chem. Soc.*, **60**:1737 (1938).
1203. E. R. White, Reaction with Organic Acids and Anhydrides, in ref. 549, p. 136.
1204. I. N. Nazarov, S. G. Matsoyan, and S. A. Vartanyan, *Zh. Obshch. Khim.*, **27**:1818 (1957); through *Chem. Abstr.*, **52**:4619 (1958).
1205. R. F. Fischer and C. W. Smith (to Shell Development Co.), U.S. Patent 2,857,422, Oct. 21, 1958; *Chem. Abstr.*, **53**:5204 (1959).
1206. H. D. Finch, E. A. Peterson, and S. A. Ballard, *J. Am. Chem. Soc.*, **74**:2016 (1952).
1207. H. Schesinger (to Kalle A.-G.), German Patent 1,147,482, Apr. 18, 1963; *Chem. Abstr.*, **59**:2281 (1963).
1208. H. Bestian, J. Heyna, A. Bauer, G. Ehlers, B. Hirsekorn, T. Jacobs, W. Noll, W. Weibezahn, and F. Römer, *Ann. Chem.*, **566**:210 (1950).

1209. H. Hopff and H. Spänig (to Badische Anilin- & Soda-Fabrik A.-G.), German Patent 840,546, Jun. 3, 1952; *Chem. Abstr.*, **47**:1739 (1953).
1210. S. Tamura and M. Yamasaki, *J. Pharm. Soc. Japan*, **76**:915 (1956); through *Chem. Abstr.*, **51**:2782 (1957).
1211. H. Arai, S. Shima, and N. Murata, *Kogyo Kagaku Zasshi*, **62**:825 (1959); through *Chem. Abstr.*, **57**:8555 (1962).
1212. H. Larramona, *Compt. Rend.*, **238**:488 (1954).
1213. H. Uchino, *Bull. Chem. Soc. Japan*, **32**:1009, 1012 (1959); through *Chem. Abstr.*, **54**:18502, 18503 (1960).
1214. D. J. Casey and C. S. Marvel, *J. Org. Chem.*, **24**:1022 (1959).
1215. O. A. Moe and D. T. Warner, *J. Am. Chem. Soc.*, **71**:1251 (1949).
1216. H. Feuer and R. Harmetz, *ibid.*, **80**:5877 (1958).
1217. H. Arai, S. Shima, and N. Murata, *Kogyo Kagaku Zasshi*, **62**:82 (1959); through *Chem. Abstr.*, **58**:5659 (1963).
1218. L. W. Kissinger and M. Schwartz, *J. Org. Chem.*, **23**:1342 (1958).
1219. British Patent 850,360 (to Société des Usines Chimique Rhône-Poulenc), Oct. 5, 1960; *Chem. Abstr.*, **55**:14385 (1961).
1220. S. A. Miller and R. Robinson, *J. Chem. Soc.*, **1934**:1535.
1221. K. Alder and C-H. Schmidt, *Chem. Ber.*, **76B**:183 (1943).
1222. I. D. Webb and G. T. Borcherdt, *J. Am. Chem. Soc.*, **73**:752 (1951).
1223. J. Szmuszkovicz, *ibid.*, **79**:2819 (1957).
1224. H. Arai and N. Murata, *Kogyo Kagaku Zasshi*, **61**:563 (1958); through *Chem. Abstr.*, **55**:10371 (1961).
1225. W. H. Vinton (to E. I. du Pont de Nemours & Co.), U.S. Patent 2,427,582, Sept. 16, 1947; *Chem. Abstr.*, **42**:212 (1948).
1226. H. Böhme and P. Heller, *Chem. Ber.*, **86**:443 (1953).
1227. M. Tanaka, M. Sashio, Y. Shimodoi, and N. Murata, *Kogyo Kagaku Zasshi*, **59**:577 (1956); through *Chem. Abstr.*, **52**:3680 (1958).
1228. K. Yamagishi, *Nippon Kagaku Zasshi*, **80**:764 (1959); through *Chem. Abstr.*, **55**:3493 (1961).
1229. G. W. Hearne, Reaction with Inorganic and Organometallic Compounds, in ref. 549, p. 154.
1230. E. L. Hoegberg (to American Cyanamid Co.), U.S. Patent 2,632,020, Mar. 17, 1953; *Chem. Abstr.*, **48**:2759 (1954).
1231. M. Tanaka and N. Murata, *Kogyo Kagaku Zasshi*, **60**:433 (1957); through *Chem. Abstr.*, **53**:9043 (1959).
1232. H. Hopff and W. Rapp (to I. G. Farbenindustrie A.-G.), U.S. Patent 2,265,165, Dec. 9, 1941; *Chem. Abstr.*, **36**:1615 (1942).
1233. British Patent 503,623 (to I. G. Farbenindustrie A.-G.), Apr. 12, 1939 (addition to ref. 657a); *Chem. Abstr.*, **33**:7435 (1939).
1234. German Patent 720,269 (to I. G. Farbenindustrie A.-G.), Apr. 2, 1942; *Chem. Abstr.*, **37**:2102 (1943).
1235. H. Hopff, A. Weickmann, and R. Kern (to General Aniline & Film Corp.), U.S. Patent 2,352,387, Jun. 27, 1944; *Chem. Abstr.*, **38**:5619 (1944).
1236. D. P. Tate, A. A. Buss, J. M. Augl, B. L. Ross, J. G. Grasselli, W. M. Ritchey, and F. J. Knoll, *Inorg. Chem.*, **4**(9):1323 (1965).
1237. S. Kawaguchi and T. Ogura, *ibid.*, **5**(5):844 (1966).
1238. F. Bottino and G. Purrello, *Gaz. Chim. Ital.*, **95**(6):693 (1965); through *Chem. Abstr.*, **64**:17583 (1966).
1239. R. F. Fischer, Polymers from Acrolein, in ref. 549, p. 225.

1240. R. Hank (to Deutsche Dunlop Gummi Co. A.-G.), German Patent 1,172,041, Jun. 11, 1964; *Chem. Abstr.*, **61**:7194 (1964).

1241. M. F. Shostakovskii, V. I. Belyaev, and L. T. Ivanova, *Izv. Sib. Otd. Akad. Nauk SSSR, Ser. Khim. Nauk*, **1964**(3):110; through *Chem. Abstr.*, **63**:4411 (1965).

1242. E. Ajisaka and H. Nakanishi (to Japan Catalytic Chemical Industry Co. Ltd.), Japanese Patent 1,235, Feb. 19, 1963; through *Chem. Abstr.*, **59**:7672 (1963).

1243. M. F. Shostakovskii, V. I. Belyaev, Z. A. Okladnikova, L. V. Vasil'eva, and E. V. Serebrennikova, *Izv. Sib. Otd. Akad. Nauk SSSR, Ser. Khim. Nauk*, **1965**(1):88; through *Chem. Abstr.*, **63**:13423 (1965).

1244. E. I. Finkel'shtein and A. D. Abkin, *Dokl. Akad. Nauk SSSR*, **161**(5):1098 (1965); through *Chem. Abstr.*, **63**:4397 (1965).

1245. M. M. Koton, I. V. Andreeva, and Y. P. Getmanchuk, *Vysokomol. Soedin.*, **4**:1537 (1962); through *Chem. Abstr.*, **59**:1764 (1963).

1246. R. C. Schulz and W. Passmann, *Makromol. Chem.*, **60**:139 (1963).

1247. Y. Toi and Y. Hachihama, *Bull. Chem. Soc. Japan*, **37**(3):302 (1964); through *Chem. Abstr.*, **60**:15986 (1964).

1248. To Shell Internationale Research Maatschappij NV: (*a*) British Patent 916,614, Jan. 23, 1963; (*b*) and (*c*) Netherlands Appls. 6,415,218 and 6,415,219, Jul. 1, 1965; *Chem. Abstr.*, **58**:11534 (1963); **63**:18294 ,18480 (1965).

1249. H. Brendlein, E. Baeder, H. Leyerzapf, and K. H. Rink (to Deutsche Gold- und Silber-Scheideanstalt vorm. Roessler), U.S. Patent 3,142,661, Jul. 28, 1964; *Chem. Abstr.*, **61**:12110 (1964).

1250. E. E. Ryder, Jr. and P. Pezzaglia, *J. Polymer Sci., Pt. A*, **3**(10):3459 (1965).

1251. K. H. Rink and O. Schweitzer (to Deutsche Gold- und Silber-Scheideanstalt vorm. Roessler), German Patent 1,138,546, Oct. 25, 1962; *Chem. Abstr.*, **58**:4661 (1963).

1252. V. A. Campanile (to Shell Oil Co.), U.S. Patent 3,081,244, Mar. 12, 1963; *Chem. Abstr.*, **58**:11492 (1963).

1253. E. E. Ryder, Jr. (to Shell Internationale Research Maatschappij NV), Belgian Patent 627,451, Jul. 23, 1963; *Chem. Abstr.*, **61**:1974 (1964).

1254. A. T. Stewart, Jr. and A. C. Nixon (to same assignee), French Patent 1,353,101, Feb. 21, 1964; *Chem. Abstr.*, **61**:8435 (1964).

1255. British Patent 990,263 (to Deutsche Gold- und Silber-Scheideanstalt vorm. Roessler), Apr. 28, 1965; *Chem. Abstr.*, **63**:4413 (1965).

1256. E. Baeder, K. H. Rink, and H. Trautwein, *Makromol. Chem.*, **92**:198 (1966).

1257. F. J. Welch (to Union Carbide Corp.), U.S. Patent 3,069,389, Dec. 18, 1962; *Chem. Abstr.*, **58**:4661 (1963).

1258. R. F. Fischer (to Shell Oil Co.), U.S. Patent 3,079,357, Feb. 26, 1963; *Chem. Abstr.*, **59**:11683 (1963).

1259. E. R. Bell, V. A. Campanile, and F. Bergman (to Shell Oil Co.), U.S. Patent 3,105,801, Oct. 1, 1963; *Chem. Abstr.*, **59**:14133 (1963).

1260. I. Sobolev (to Shell Oil Co.), U.S. Patent 3,215,674, Nov. 2, 1965; *Chem. Abstr.*, **64**:840 (1966).

1261. H. Schilling (to Deutsche Dunlop Gummi Co. A.-G.), German Patent 1,199,979, Sept. 2, 1965; *Chem. Abstr.*, **63**:16559 (1965).

1262. R. F. Fischer and A. T. Stewart, Jr., *J. Polymer Sci., Pt. A*, **3**(10):3495 (1965).

1263. H. G. Hammon (to C. L. Wilson), German Patent 1,117,304, Nov. 16, 1961; *Chem. Abstr.*, **56**:14484 (1962).

1264. Y. Hachihama and Y. Toi, *Technol. Rept. Osaka Univ.*, **13**(537–562):237 (1963); through *Chem. Abstr.*, **59**:10244 (1963).

1265. Y. Toi and Y. Hachihama, *Kogyo Kagaku Zasshi*, **63**(9):1654 (1960); through *Chem. Abstr.*, **61**:7108 (1964).

1266. French Patent 1,375,851 (to Deutsche Gold- und Silber-Scheideanstalt vorm. Roessler), Oct. 23, 1964; *Chem. Abstr.*, **62**:11935 (1965).

1267. E. Bergman (to Shell Oil Co.), U.S. Patent 3,121,700, Feb. 18, 1964; *Chem. Abstr.*, **60**:13446 (1964).

1268. E. E. Gruber and E. F. Kalafus (to General Tire and Rubber Co.), U.S. Patent 3,177,171, Apr. 6, 1965; *Chem. Abstr.*, **62**:16470 (1965).

1269. W. Kern, O. Schweitzer, and R. Schulz (to Deutsche Gold- und Silber-Scheideanstalt vorm. Roessler), U.S. Patent 3,234,164, Feb. 8, 1966; *Chem. Abstr.*, **64**:12843 (1966).

1270. E. E. Ryder, Jr. and P. Pazzaglia (to Shell Internationale Research Maatschappij NV), Belgian Patent 628,187, Aug. 7, 1963; *Chem. Abstr.*, **60**:14732 (1964).

1271. British Patent 1,010,883 (to Deutsche Gold- und Silber-Scheideanstalt vorm. Roessler), Nov. 24, 1965; *Chem. Abstr.*, **64**:3786 (1966).

1272. T. L. Dawson and F. E. Welch, *J. Am. Chem. Soc.*, **86**(22):4791 (1964).

1273. R. C. Schulz, J. Kovacs, and W. Kern, *Makromol. Chem.*, **67**:187 (1963).

1274. Netherlands Appl. 6,505,495 (to Dynamit-Nobel A.-G.), Nov. 3, 1965; *Chem. Abstr.*, **64**:14312 (1966).

1275. G. Bier, H. Hartel, and I. U. Nebel, *Makromol. Chem.*, **92**:240 (1966).

1276. "Acrolein Copolymers," Technical Bull. SC: 60–126, Shell Chemical Co., 1960.

1277. F. J. Welch (to Union Carbide Corp.), U.S. Patent 3,225,000, Dec. 21, 1965; *Chem. Abstr.*, **64**:8408 (1966).

1278. G. S. Whitby, M. D. Gross, J. R. Miller, and A. J. Costanza, *J. Polymer Sci.*, **16**:549 (1955).

1279. W. Kern and R. C. Schulz, *Angew. Chem.*, **69**:153 (1957).

1280. British Patent 829,601 (to E. I. du Pont de Nemours & Co.), Mar. 2, 1960; *Chem. Abstr.*, **54**:13743 (1960).

1281. M. M. Koton, I. V. Andreeva, and Y. P. Getmanchuk, *Dokl. Akad. Nauk SSSR*, **144**:1091 (1962); through *Chem. Abstr.*, **57**:11377 (1962).

1282. R. C. Schulz, S. Suzuki, H. Cherdron, and W. Kern, *Makromol. Chem.*, **53**:145 (1962).

1283. British Patent 916,089 (to Deutsche Gold- und Silber-Scheideanstalt vorm. Roessler), Jan. 16, 1963; *Chem. Abstr.*, **58**:9249 (1963).

1284. R. L. Eifert and B. M. Marks (to E. I. du Pont de Nemours & Co.), U.S. Patent 3,118,860, Jan. 21, 1964; *Chem. Abstr.*, **61**:744 (1964).

1285. I. V. Andreeva, M. M. Koton, Y. P. Getmanchuk, and M. G. Tarasova, *Zh. Prikl. Khim.*, **38**(12):2740 (1965); through *Chem. Abstr.*, **64**:11325 (1966).

1286. V. M. Zhulin, M. P. Pen'kova, A. A. Konkin, and M. G. Gonikberg, *Izv. Akad. Nauk SSSR, Ser. Khim.*, **1964**(8):1497; through *Chem. Abstr.*, **64**:15988 (1966).

1287. M. M. Koton, I. V. Andreeva, and Y. P. Getmanchuk, *Dokl. Akad. Nauk SSSR*, **155**(4):836 (1964); through *Chem. Abstr.*, **61**:1946 (1964).

1288. M. M. Koton, I. V. Andreeva, Y. P. Getmanchuk, L. Y. Madorskaya, E. I. Pokrovskii, A. I. Kol'tsov, and V. A. Filatova, *Vysokomol. Soedin.*, **7**(12):2039 (1965); through *Chem. Abstr.*, **64**:11321 (1966).

1289. I. V. Andreeva, M. M. Koton, and K. A. Kovaleva, *ibid.*, **4**:528 (1962); through *Chem. Abstr.*, **57**:12707 (1962).

1290. Y. Nagai and T. Nakajima, *Kogyo Kagaku Zasshi*, **66**(12):1905 (1963); through *Chem. Abstr.*, **61**:8181 (1964).

1291. W. Weiler (to Badische Anilin- & Soda-Fabrik A.-G.), German Patent 889,227, Sept. 7, 1953; *Chem. Abstr.*, **50**:13508 (1956).

1292. Netherlands Appl. 6,411,783 (to Deutsche Gold- und Silber-Scheideanstalt vorm. Roessler), Jun. 14, 1965; *Chem. Abstr.*, **63**:18299 (1965).

1293. R. L. Eifert and B. M. Marks (to E. I. du Pont de Nemours & Co.), German Patent 1,081,231, May 5, 1960; British Patent 872,331, Jul. 5, 1961; *Chem. Abstr.*, **55**:21673 (1961); **58**:3522 (1963).

1294. G. W. Hearne, D. S. La France, and E. C. Shokal (to Shell Development Co.), U.S. Patent 2,809,185, Oct. 8, 1957; *Chem. Abstr.*, **52**:10648 (1958).

1295. R. C. Schulz, E. Kaiser, and W. Kern, *Makromol. Chem.*, **76**:99 (1964).

1296. T. W. Evans (to Shell Development Co.), British Patent 596,620, Jan. 7, 1948; *Chem. Abstr.*, **42**:3997 (1948).

1297. G. E. van Gils (to General Tire and Rubber Co.), German Patent 1,109,898, appl. Mar. 29, 1960; *Chem. Abstr.*, **56**:7479 (1962).

1298. A. Y. Korotkova, Y. G. Kryazhev, and Z. A. Rogovin, *Vysokomol. Soedin.*, **6**(11):1980 (1964); through *Chem. Abstr.*, **62**:13317 (1965).

1299. E. L. Kropa (to American Cyanamid Co.), U.S. Patent 2,496,097, Jan. 31, 1950; *Chem. Abstr.*, **44**:5640 (1950).

1300. British Patent 908,511 (to Chemische Werke Albert), Oct. 17, 1962; *Chem. Abstr.*, **58**:3564 (1963).

1301. British Patent 803,053 (to General Tire and Rubber Co.), Oct. 15, 1958; *Chem. Abstr.*, **53**:7616 (1959).

1302. E. E. Gruber and E. F. Kalafus (to General Tire and Rubber Co.), U.S. Patent 2,999,830, appl. Apr. 21, 1958; *Chem. Abstr.*, **56**:2586 (1962).

1303. E. C. Chapin and R. I. Longley, Jr. (to Monsanto Chemical Co.), U.S. Patent 2,893,979, Jul. 7, 1959; *Chem. Abstr.*, **53**:17535 (1959).

1304. H. L. Cohen and L. M. Minsk, *J. Org. Chem.*, **24**:1404 (1959).

1305. R. L. Eifert and B. M. Marks (to E. I. du Pont de Nemours & Co.), U.S. Patent 3,000,862, appl. Jan. 13, 1958; *Chem. Abstr.*, **56**:1610 (1962).

1306. E. Tschunkur and W. Brock (to I. G. Farbenindustrie A.-G.), German Patent 555,859, Jun. 8, 1927; *Chem. Abstr.*, **26**:6165 (1932).

1307. K. Meisenburg (to I. G. Farbenindustrie A.-G.), U.S. Patent 2,005,295, Jun. 18, 1935; *Chem. Abstr.*, **29**:5203 (1935).

1308. C. S. Marvel and C. L. Levesque, *J. Am. Chem. Soc.*, (*a*) **60**:280 (1938); (*b*) **61**:3234 (1939).

1309. H. W. Melville, T. T. Jones, and R. F. Tuckett, *Chem. Ind.* (*London*), **1940**:267.

1310. A. Voss and K. Billig (to I. G. Farbenindustrie A.-G.), German Patent 743,319, Nov. 4, 1943; *Chem. Abstr.*, **39**:3180 (1945).

1311. T. T. Jones and H. W. Melville, *Proc. Roy. Soc.* (*London*), **A187**:19 (1946).

1312. H. Yoshinaga and M. Morikawa, *J. Chem. Soc. Japan, Ind. Chem. Sect.*, **51**:41 (1948); through *Chem. Abstr.*, **44**:9189 (1950).

1313. S. Kawada, *J. Chem. Soc. Japan*, **50**:128, 129 (1947); through *Chem. Abstr.*, **44**:9189 (1950).

1314. R. E. Davies (to Celanese Corp. of America), U.S. Patent 2,599,616, Jun. 10, 1952; *Chem. Abstr.*, **46**:11774 (1952).

1315. R. E. Davies and S. B. McFarlane (to Celanese Corp. of America), U.S. Patent 2,631,987, Mar. 17, 1953; *Chem. Abstr.*, **47**:6699 (1953).

1316. I. Skeist and S. B. McFarlane (to Celanese Corp. of America), U.S. Patent 2,626,943, Jan. 27, 1953; *Chem. Abstr.*, **47**:6700 (1953).

1317. W. C. Hollyday, Jr. (to Esso Research and Engineering Co.), U.S. Patent 2,769,783, Nov. 6, 1956; *Chem. Abstr.*, **51**:3988 (1957).

1318. R. C. Schulz, H. Vielhaber, and W. Kern, *Kunststoffe*, **50**:500 (1960).

1319. T. Matsuda, H. Yamakita, and S. Fujii, *Kobunshi Kagaku*, **21**(231):415 (1964); through *Chem. Abstr.*, **61**:12093 (1964).

1320. Y. Goto, Y. Tabata, and H. Sobue, *Kogyo Kagaku Zasshi*, **67**(8):1276 (1964); through *Chem. Abstr.*, **62**:7869 (1965).

1321. L. Horner, W. Jurgeleit, and K. Klüpfel, *Ann. Chem.*, **591**:108 (1955).

1322. R. Fujio, T. Tsuruta, and J. Furukawa, *Makromol. Chem.*, **52**:233 (1962).

1323. T. Tsuruta and J. Furukawa, *Bull. Inst. Chem. Res., Kyoto Univ.*, **40**:151 (1962); through *Chem. Abstr.*, **57**:16858 (1962).

1324. S. Iwatsuki, Y. Yamashita, and Y. Ishii, *J. Polymer Sci., Pt. B*, **1**(10):545 (1963).

1325. S. Iwatsuki, Y. Yamashita, and Y. Ishii, *Kogyo Kagaku Zasshi*, **66**(8):1162 (1963); through *Chem. Abstr.*, **60**:9360 (1964).

1326. G. Wasai, T. Tsuruta, J. Furukawa, and R. Fujio, *Kogyo Kagaku Zasshi*, **66**(19):1339 (1963); through *Chem. Abstr.*, **60**:13332 (1964).

1327. J. Furukawa, T. Tsuruta, and R. Fujio (to Research Institute for Synthetic Fibers), Japanese Patent 18,944, Sept. 4, 1964; through *Chem. Abstr.*, **62**:5359 (1965).

1328. T. Tsuruta, R. Fujio, and J. Furukawa, *Makromol. Chem.*, **80**:172 (1964).

1329. L. Trossarelli, M. Guaita, and A. Priola, *Ric. Sci., Rend., Sez. A*, **7**(2):451 (1964); through *Chem. Abstr.*, **63**:3045 (1965).

1330. R. Fujio, T. Tsuruta, and J. Furukawa, *Kogyo Kagaku Zasshi*, **66**(3):365 (1963); through *Chem. Abstr.*, **63**:5749 (1965).

1331. G. K. Hoeschele, J. B. Andelman, and H. P. Gregor, *J. Phys. Chem.*, **62**:1239 (1958).

1332. R. M. Joshi, *J. Polymer Sci.*, **56**:313 (1962).

1333. J. N. Hay, *Makromol. Chem.*, **67**:31 (1963).

1334. J. A. Blanchette and J. D. Cotman, Jr., *J. Org. Chem.*, **23**:1117 (1958).

1335. J. A. Blanchette (to Monsanto Chemical Co.), U.S. Patents (*a*) 2,962,477, Nov. 29, 1960, and (*b*) 2,862,911, Dec. 2, 1958; *Chem. Abstr.*, **55**:8945 (1961); **53**:7672 (1959).

1336. Y. Shimodoi and N. Murata, *Kogyo Kagaku Zasshi*, **65**:591 (1962); through *Chem. Abstr.*, **57**:11366 (1962).

1337. K. A. Kun and H. G. Cassidy, *J. Polymer Sci.*, **44**:383 (1960).

1338. C. S. Marvel and J. C. Wright, *ibid.*, **8**:495 (1952).

1339. British Patent 787,358 (to Rohm and Haas Co.), Dec. 4, 1957; *Chem. Abstr.*, **52**:6670 (1958).

1340. C. W. Mortenson (to E. I. du Pont de Nemours & Co.), U.S. Patent 2,396,963, Mar. 19, 1946; *Chem. Abstr.*, **40**:3937 (1946).

1341. J. F. Jones (to B. F. Goodrich Co.), U.S. Patent 2,978,436, Apr. 4, 1961; *Chem. Abstr.*, **55**:18194 (1961).

1342. S. Tsuruta, S. Nakamuta, and M. Tsukui (to Hitachi Ltd.), Japanese Patent 4,237, Apr. 25, 1960; through *Chem. Abstr.*, **55**:4044 (1961).

1343. British Patent 497,031 (to I. G. Farbenindustrie A.-G.), Dec. 12, 1938; *Chem. Abstr.*, **33**:3606 (1939).

1344. C. H. Greenewalt (to E. I. du Pont de Nemours & Co.), U.S. Patent 2,063,158, Dec. 8, 1936; *Chem. Abstr.*, **31**:779 (1937).
1345. W. Cooper and E. Catterall, (*a*) *Chem. Ind.* (*London*), **1954**:1514; (*b*) *Can. J. Chem.*, **34**:387 (1956).
1346. Netherlands Appl. 6,412,088 (to Badische Anilin- & Soda-Fabrik A.-G.), Apr. 20, 1965; *Chem. Abstr.*, **63**:8565 (1965).
1347. J. E. Guillet and R. G. W. Norrish, *Proc. Roy. Soc.* (*London*), **A23**:153 (1955).
1348. Y. Sawa, T. Miki, M. Ryang, and S. Tsutsumi, *Technol. Rept. Osaka Univ.*, **13**(537–562):229 (1963); through *Chem. Abstr.*, **59**:10097 (1963).
1349. W. Mertens (to Attorney General of the U.S.A.), U.S. Patent 2,418,978, Apr. 15, 1947; *Chem. Abstr.*, **41**:5754 (1947).
1350. M. F. Shostakovskii, G. G. Skvortsova, M. Y. Samoilova, and N. I. Shergina, *Izv. Akad. Nauk SSSR, Otd. Khim. Nauk*, **1962**:1447; through *Chem. Abstr.*, **58**:2509 (1963).
1351. V. A. Campanile, W. Tsatsos, R. F. Fischer, A. T. Stewart, Jr., and W. H. Houff (to Shell Internationale Research Maatschappij NV), Belgian Patent 611,688, Jun. 18, 1962; *Chem. Abstr.*, **58**:2521 (1963).
1352. R. C. Shulz, E. Kaiser, and W. Kern, *Makromol. Chem.*, **58**:160 (1962).
1353. W. T. Tsatos (to Shell Oil Co.), U.S. Patents 3,093,506, Jun. 11, 1963, and 3,231,538, Jan. 25, 1966; *Chem. Abstr.*, **59**:10264 (1963); **64**:12969 (1966).
1354. T. Harada and K. Sakaba (to Nitto Spinning Co.), Japanese Patent 10,871, Aug. 11, 1962; *Chem. Abstr.*, **59**:10294 (1963).
1355. R. C. Schulz and W. Passmann, *Makromol. Chem.*, **72**:198 (1964).
1356. British Patent 955,819 (to Farbwerke Hoechst A.-G.), Apr. 22, 1964; *Chem. Abstr.*, **61**:5813 (1964).
1357. R. K. June, A. De Benedictis, and P. H. Williams (to Shell Oil Co.), U.S. Patent 3,129,195, Apr. 14, 1964; *Chem. Abstr.*, **61**:5814 (1964).
1358. P. W. Sherwood, *Petroleum* (*London*), **27**(10):485 (1964).
1359. British Patent 992,037 (to Deutsche Gold- und Silber-Scheideanstalt vorm. Roessler), May 12, 1965; *Chem. Abstr.*, **63**:4420 (1965).
1360. M. F. Shostakovskii, G. G. Skvortsova, K. V. Zapunnaya, and V. G. Kozyrev, *Dokl. Akad. Nauk SSSR*, **162**(1):124 (1965); through *Chem. Abstr.*, **63**:7113 (1965).
1361. V. S. Ivanov and G. S. Buslaev, U.S.S.R. Patent 172,995, Jul. 7, 1965; through *Chem. Abstr.*, **64**:2259 (1966).
1362. W. Goeltner and P. Schlack (to Deutsche Gold- und Silber-Scheideanstalt vorm. Roessler), German Patent 1,201,556, Sept. 23, 1965; *Chem. Abstr.*, **64**:17738 (1966).
1363. H. S. Nutting and P. S. Petrie (to The Dow Chemical Co.), U.S. Patent 2,256,152, Sept. 16, 1941; *Chem. Abstr.*, **36**:192 (1942).
1364. H. T. Neher and C. F. Woodward (to Rohm and Haas Co.), U.S. Patent 2,416,536, Feb. 25, 1947; *Chem. Abstr.*, **41**:4006 (1947).
1365. H. C. Haas and M. S. Simon, *J. Polymer Sci.*, **9**:309 (1952).
1366. R. H. Reinhard (to Monsanto Chemical Co.), U.S. Patent 2,833,743, May 6, 1958; *Chem. Abstr.*, **52**:13323 (1958).
1367. E. C. Shokal and P. A. Devlin (to Shell Oil Co.), U.S. Patent 2,940,955, Jun. 14, 1960; *Chem. Abstr.*, **54**:21871 (1960).
1368. H. Fikentscher, H. Wilhelm, and H. Bolte (to Badische Anilin- & Soda-Fabrik A.-G.), German Patent 1,082,359, May 25, 1960; *Chem. Abstr.*, **55**:15958 (1961).

1369. A. R. Kol'k, A. A. Konkin, and Z. A. Rogovin, (a) *Khim. Volokna*, **1963**(4): 12; (b) *Eesti NSV Tead. Akad. Toim., Fuus.-Mat. Tehnikatead. Seer.*, **13**(3):241 (1964); through *Chem. Abstr.*, **59**:11670 (1963); **62**:7915 (1965).

1370. Z. M. Rumyantseva, A. A. Golitsyna, M. A. Farberov, V. G. Epshtein, E. G. Lazaryants, D. P. Emel'yanov, and L. V. Kosmodem'yanskii, *Kauch. Rezina*, **23**(7):7 (1964); through *Chem. Abstr.*, **61**:12169 (1964).

1371. W. C. Wiley (to General Tire and Rubber Co.), U.S. Patent 3,181,610, May 4, 1965; *Chem. Abstr.*, **63**:341 (1965).

1372. S. D. Zimmerman and W. S. Thompson (to Columbian Carbon Co.), U.S. Patent 3,208,976, Sept. 28, 1965; *Chem. Abstr.*, **63**:18388 (1965).

1373. C. N. Wolf (to Ethyl Corp.), French Patent 1,392,124, Mar. 12, 1965; *Chem. Abstr.*, **63**:11732 (1965).

1374. H. Hopff, W. Starck, and K. Billig (to General Aniline & Film Corp.), U.S. Patent 2,228,270, Jan. 14, 1941; *Chem. Abstr.*, **35**:2641 (1941).

1375. French Patent 847,203 (to I. G. Farbenindustrie A.-G.), Oct. 5, 1939; *Chem. Abstr.*, **35**:5220 (1941).

1376. W. Heuer (to I. G. Farbenindustrie A.-G.), German Patent 710,374, Jul. 31, 1941; *Chem. Abstr.*, **37**:3860 (1943).

1377. R. R. Dreisbach (to The Dow Chemical Co.), U.S. Patent 2,374,589, Apr. 24, 1945; *Chem. Abstr.*, **39**:4774 (1945).

1378. H. Wolthan and W. Becher (to I. G. Farbenindustrie A.-G.), German Patent 745,032, Nov. 25, 1943; *Chem. Abstr.*, **40**:490 (1946).

1379. D. E. Strain (to E. I. du Pont de Nemours & Co.), U.S. Patent 2,404,817, Jul. 30, 1946; *Chem. Abstr.*, **40**:5958 (1946).

1380. R. R. Dreisbach, E. C. Britton, and W. J. Le Fevre (to The Dow Chemical Co.), U.S. Patent 2,407,953, Sept. 17, 1946; *Chem. Abstr.*, **41**:617 (1947).

1381. H. W. Arnold (to E. I. du Pont de Nemours & Co.), U.S. Patent 2,408,402, Oct. 1, 1946; *Chem. Abstr.*, **41**:625 (1947).

1382. H. W. Starkweather, P. O. Bare, A. S. Carter, F. B. Hill, Jr., V. R. Hurka, C. J. Mighton, P. A. Sanders, H. W. Walker, and M. A. Youker, *Ind. Eng. Chem.*, **39**:210 (1947).

1383. F. M. Lewis, C. Walling, W. Cummings, E. R. Briggs, and W. J. Wenisch, *J. Am. Chem. Soc.*, **70**:1527 (1948).

1384. R. L. Meier and W. E. Elwell (to California Research Corp.), U.S. Patent 2,469,295, May 3, 1949; *Chem. Abstr.*, **43**:5231 (1949).

1385. H. Hopff and C. Rautenstrauch, *Makromol. Chem.*, **6**:39 (1951).

1386. J. C. H. Hwa (to Rohm and Haas Co.), U.S. Patent 2,597,491, May 20, 1952; *Chem. Abstr.*, **47**:9526 (1953).

1387. W. G. Lowe, L. M. Minsk, and W. O. Kenyon (to Eastman Kodak Co.), U.S. Patent 2,632,704, Mar. 24, 1953; *Chem. Abstr.*, **47**:6807 (1953).

1388. C. S. Marvel and J. C. Wright, *J. Polymer Sci.*, **8**:255 (1952).

1389. F. C. Foster and J. L. Binder, *J. Am. Chem. Soc.*, **75**:2910 (1953).

1390. H. Gilbert and F. F. Miller (to B. F. Goodrich Co.), U.S. Patent 2,654,724, Oct. 6, 1953; *Chem. Abstr.*, **48**:1070 (1954).

1391. R. W. Laundrie, M. Feldon, and A. L. Rodde, *Ind. Eng. Chem.*, **48**:9738 (1954).

1392. W. K. Taft and G. F. Tiger, Diene Polymers and Copolymers Other Than GR-S and the Specialty Rubbers, in G. S. Whitby (ed.), "Synthetic Rubber," p. 682, Wiley, New York, 1954.

1393. G. E. Ham (to Chemstrand Corp.), U.S. Patent 2,740,763, Apr. 3, 1956; *Chem. Abstr.*, **50**:12495 (1956).

1394. L. S. Cohen, A. P. Giraitis, and J. R. Zietz (to Ethyl Corp.), U.S. Patent 2,773,052, Dec. 4, 1956; *Chem. Abstr.*, **51**:4055 (1957).

1395. C. G. Overberger and F. W. Michelotti, *J. Am. Chem. Soc.*, **80**:988 (1958).

1396. S. Matsumoto, M. Yano, and T. Osugi (to Kurashiki Rayon Co.), Japanese Patent 9,493, Nov. 6, 1956; through *Chem. Abstr.*, **52**:15916 (1958).

1397. K. Jost and W. Schmidt (to Badische Anilin- & Soda-Fabrik A.-G.), German Patent 947,205, Aug. 9, 1956; *Chem. Abstr.*, **52**:17744 (1958).

1398. J. A. Blanchette (to Monsanto Chemical Co.), U.S. Patent 2,865,889, Dec. 23, 1958; *Chem. Abstr.*, **53**:5758 (1959).

1399. H. Suda, S. Inoue, and R. Oda, *Kogyo Kagaku Zasshi*, **60**:508 (1957); through *Chem. Abstr.*, **53**:7462 (1959).

1400. K. Lohs (to VEB Farbenfabriken Wolfen), German (East) Patent 14,282, Jan. 6, 1958; *Chem. Abstr.*, **53**:7465 (1959).

1401. T. Tiba and T. Yonezawa (to Fuji Photo Film Co. Ltd.), Japanese Patents 2,496, May 23, 1962, and 20,498, Oct. 4, 1963; through *Chem. Abstr.*, **58**:11543 (1963); **60**:4274 (1964).

1402. T. Tsuruta and R. Fujio, *Makromol. Chem.*, **64**:219 (1963).

1403. P. Colombo, M. Steinberg, and D. Macchia, *J. Polymer Sci., Pt. B*, **1**(9):483 (1963).

1404. A. Yamamoto and M. Ishihara (to Toyo Spinning Co. Ltd.), Japanese Patent 26,097, Dec. 9, 1963; through *Chem. Abstr.*, **60**:6981 (1964).

1405. M. Steinberg and P. Colombo, *Ind. Uses of Large Radiation Sources, Proc. Conf., Salzburg, Austria*, **1**:121 (1963); *Chem. Abstr.*, **60**:10791 (1964).

1406. P. Colombo and M. Steinberg, *U.S. At. Energy Comm. BNL-6849* (1963); *Chem. Abstr.*, **60**:14615 (1964).

1407. F. Ida, K. Uemura, and S. Abe, *Kagaku To Kogyo (Osaka)*, **38**(4):215 (1964); through *Chem. Abstr.*, **61**:7105 (1964).

1408. L. Strzelecki and J. Petit, *Compt. Rend.*, **259**(18):3019 (1964).

1409. C. E. Barnes (to E. I. du Pont de Nemours & Co.), U.S. Patent 2,278,635, Apr. 7, 1942; *Chem. Abstr.*, **36**:4934 (1942).

1410. T. R. E. Kressman (to Permutit Co. Ltd.), British Patents 804,782, Nov. 26, 1958, and 810,027, Mar. 11, 1959; *Chem. Abstr.*, **53**:5547, 11718 (1959).

1411. J. R. Millar (to Permutit Co. Ltd.), British Patent 849,122, Sept. 21, 1960; *Chem. Abstr.*, **55**:12705 (1961).

1412. R. C. Schulz, H. Cherdron, and W. Kern, *Makromol. Chem.*, **28**:197 (1958).

1413. W. Kern, private communication cited in ref. 315.

1414. Ref. 131, p. 35.

1415. F. Leavitt, V. Stannett, and M. Szwarc, *J. Polymer Sci.*, **31**:193 (1958).

1416. L. Herk, A. Stefani, and M. Szwarc, *J. Am. Chem. Soc.*, **83**:3008 (1961).

1417. W. Kern, R. C. Schulz, and D. Braun, *Chemiker-Ztg.*, **84**:385 (1960).

1418. H. Leyerzapf (to Deutsche Gold- und Silber-Scheideanstalt vorm. Roessler), U.S. Patent 3,089,861, May 14, 1963; *Chem. Abstr.*, **59**:4065 (1963).

1419. I. V. Andreeva, P. F. Andreev, and E. M. Rogozina, *Radiokhimiya*, **5**(1):103 (1963); through *Chem. Abstr.*, **60**:3704 (1964).

1420. G. H. Swart (to General Tire and Rubber Co.), U.S. Patent 2,651,624, Sept. 8, 1953; *Chem. Abstr.*, **48**:7332 (1954).

1421. H. Piotrowski and G. E. Van Gils (to General Tire and Rubber Co.), U.S. Patent 3,055,854, Sept. 25, 1962; *Chem. Abstr.*, **58**:1587 (1963).

1422. F. R. Millhiser (to E. I. du Pont de Nemours & Co.), U.S. Patent 2,837,501, Jun. 3, 1958; *Chem. Abstr.*, **52**:15964 (1958).

1423. Belgian Patent 553,517 (to Kodak SA), Jan. 15, 1957; *Chem. Abstr.*, **53**:17738 (1959).

1424. L. M. Minsk (to Eastman Kodak Co.), U.S. Patent 2,882,156, Apr. 14, 1959; *Chem. Abstr.*, **53**:22971 (1959).

1425. Y. Izumi, Y. Mizutani, S. Sasao, R. Yamane, and Y. Onoue, *J. Electrochem. Soc. Japan, Overseas Ed.*, **30**(1):39 (1962).

1426. C. E. Reid and H. G. Spencer, *J. Appl. Polymer Sci.*, **4**:354 (1960).

1427. G. M. Doyle (to Dunlop Rubber Co. Ltd.), German Patent 1,112,235, appl. Oct. 26, 1960; *Chem. Abstr.*, **56**:3618 (1962).

1428. F. W. Semmler, *Ann. Chem.*, **241**:90 (1887).

1429. W. L. Ruigh and A. E. Erickson, *J. Am. Chem. Soc.*, **61**:915 (1939).

1430. E. P. Kohler, *Am. Chem. J.*, **19**:728 (1897); *ibid.*, **20**:680 (1898); *ibid.*, **21**:349 (1899).

1431. D. Strömholm, *Chem. Ber.*, **33**:823 (1900).

1432. (*a*) W. Davies, *J. Chem. Soc.*, **117**:297 (1920); (*b*) O. B. Helfrich and E. E. Reid, *J. Am. Chem. Soc.*, **42**:1208 (1920); (*c*) S. H. Bales and S. A. Nickelson, *J. Chem. Soc.*, **121**:2137 (1922); (*d*) W. E. Lawson and E. E. Reid, *J. Am. Chem. Soc.*, **47**:2821 (1925).

1433. J. von Braun and G. Kirschbaum, *Chem. Ber.*, **53B**:1399 (1920).

1434. T. C. Whitner, Jr. and E. E. Reid, *J. Am. Chem. Soc.*, **43**:636 (1921).

1435. P. W. Clutterbuck and J. B. Cohen, *J. Chem. Soc.*, **121**:120 (1922).

1436. L. N. Levin, *J. Prakt. Chem.*, **127**:77 (1930).

1437. A. E. Kretov, *J. Russ. Phys.-Chem. Soc.*, **62**:1 (1930); through *Chem. Abstr.*, **24**:4257 (1930).

1438. French Patent 777,427 (to I. G. Farbenindustrie A.-G.), Feb. 20, 1935; W. Reppe and F. Nicolai (to I. G. Farbenindustrie A.-G.), German Patent 617,543, Aug. 26, 1935; *Chem. Abstr.*, **29**:4024 (1935); **30**:733 (1936).

1439. Patents assigned to I. G. Farbenindustrie A.-G.: French Patent 789,947, Nov. 8, 1935; British Patent 442,524, Feb. 10, 1936; German Patent 635,396, Sept. 22, 1936; H. Ufer, U.S. Patent 2,163,180, Jun. 20, 1939. *Chem. Abstr.*, **30**:2992 (1936); **30**:4512 (1936); **31**:112 (1937); **33**:8030 (1939).

1440. French Patent 783,884 (to I. G. Farbenindustrie A.-G.), Jul. 19, 1935; *Chem. Abstr.*, **29**:8356 (1935).

1441. Patents assigned to I. G. Farbenindustrie A.-G.: French Patent 794,022, Feb. 6, 1936; British Patent 445,434, Apr. 9, 1936; W. Reppe, H. Ufer, and E. Kühn, U.S. Patent 2,125,649, Aug. 2, 1938. *Chem. Abstr.*, **30**:4594 (1936); **30**:6853 (1936); **32**:7615 (1938).

1442. T. Abend, O. A. Stamm, and H. Zollinger, *Helv. Chim. Acta*, **49**(4):1391 (1966).

1443. M. F. Shostakovskii, *Chemie (Prague)*, **10**:273 (1958); M. F. Shostakovskii, E. P. Gracheva, and N. K. Kul'bovskaya, *Uspekhi Khim.*, **30**:493 (1961); through *Chem. Abstr.*, **54**:1250 (1960); **55**:20914 (1961).

1444. H. Distler, *Angew. Chem.*, **77**(7):291 (1965).

1445. W. Reppe et al., *Ann. Chem.*, **601**:111 (1956).

1446. A. Kutner and D. S. Breslow, Ethylenesulfonic Acid Polymers, in N. M. Bikales (ed.), "Encyclopedia of Polymer Science and Technology," vol. VI, p. 455, Interscience, New York, 1967.

1447. V. V. Alderman and W. E. Hanford (to E. I. du Pont de Nemours & Co.), U.S. Patent 2,348,705, May 16, 1944; *Chem. Abstr.*, **39**:711 (1945).

1448. S. Kunichika and T. Katagiri, *Kogyo Kagaku Zasshi*, **64**:929 (1961); through *Chem. Abstr.*, **57**:7480 (1962).

1449. S. Ohba, T. Yonezawa, and A. Kumai (to Fuji Photo Film Co. Ltd.), U.S. Patent 3,022,172, Feb. 20, 1962; *Chem. Abstr.*, **57**:325 (1962).

1450. W. Neugebauer and H. Mengel (to Kalle and Co., A.-G.), German Patent 677,843, Jul. 3, 1939; *Chem. Abstr.*, **33**:9325 (1939).

1451. D. S. Breslow, R. R. Hough, and J. T. Fairclough, *J. Am. Chem. Soc.*, **76**:5361 (1954).

1452. D. S. Breslow and G. E. Hulse, *ibid.*, **76**:6399 (1954).

1453. D. S. Breslow and R. R. Hough, *ibid.*, **79**:5000 (1957).

1454. H. Distler and K. Küspert (to Badische Anilin- & Soda-Fabrik A.-G.), German Patents 1,090,656, Oct. 13, 1960, and 110,628, Jul. 13, 1961; U.S. Patent 3,048,625, Aug. 7, 1962; *Chem. Abstr.*, **55**:23344 (1961); **56**:3358 (1962); **57**:16404 (1962).

1455. O. Nicodemus and W. Schmidt (to I. G. Farbenindustrie A.-G.), German Patent 678,730, Jul. 24, 1939; *Chem. Abstr.*, **33**:9326 (1939).

1456. W. Schmidt (to Farbwerke Hoechst A.-G.), German Patent 836,491, Apr. 15, 1952; *Chem. Abstr.*, **47**:2766 (1953).

1457. W. F. Whitmore and E. F. Landau, *J. Am. Chem. Soc.*, **68**:1797 (1946).

1458. J. A. Anthes and J. R. Dudley (to American Cyanamid Co.), U.S. Patent 2,597,696, May 20, 1952; *Chem. Abstr.*, **47**:2196 (1953).

1459. H. F. Park (to Monsanto Chemical Co.), U.S. Patents (*a*) 2,727,057, Dec. 13, 1955; (*b*) 2,772,307, Nov. 27, 1956; (*c*) 2,700,055, Jan. 18, 1955, 2,715,142, Aug. 9. 1955; (*d*) 2,709,707, May 31, 1955; (*e*) 2,754,287, Jul. 10, 1956. *Chem. Abstr.*, **50**:10758 (1956); **51**:5815 (1957); **50**:1074 (1956); **50**:12096 (1956); **50**:5723 (1956); **50**:15128 (1956).

1460. H. Distler and A. Wegerich (to Badische Anilin- & Soda-Fabrik A.-G.), German Patent 1,171,419, Jun. 4, 1964; *Chem. Abstr.*, **61**:14532 (1964).

1461. J. Preston and J. K. Lawson, Jr., *J. Polymer Sci.*, *Pt. A*, **2**(12):5364 (1964).

1462. E. F. Landau, *J. Am. Chem. Soc.*, **69**:1219 (1947).

1463. G. T. Esayan, A. G. Vardanyan, and E. E. Oganesyan, *Izv. Akad. Nauk Arm. SSR, Khim. Nauki*, **12**:221 (1959); through *Chem. Abstr.*, **54**:22614 (1960).

1464. Patents assigned to Badische Anilin- & Soda-Fabrik A.-G.: (*a*) H. Distler and K. Küspert, German Patent 1,091,104, Oct. 20, 1960; (*b*) H. Distler and W. Mueller, German Patent 1,163,797, Feb. 27, 1964; (*c*) British Patent 910,377, Nov. 14, 1962. *Chem. Abstr.*, **55**:19787 (1961); **60**:14440 (1964); **64**:2015 (1966).

1465. H. W. Coover, Jr. and J. B. Dickey (to Eastman Kodak Co.), U.S. Patent 2,728,749, Dec. 27, 1955; *Chem. Abstr.*, **50**:8251 (1956).

1466. Belgian Patent 634,902 (to Farbwerke Hoechst A.-G.), Jan. 13, 1964; French Patent 1,367,818, Jul. 24, 1964. *Chem. Abstr.*, **61**:2974 (1964); **62**:446 (1965).

1467. H. Distler (to Badische Anilin- & Soda-Fabrik A.-G.), German Patent 1,119,253, Dec. 14, 1961; *Chem. Abstr.*, **57**:7174 (1962).

1468. S. M. McElvain, A. Jelinek, and K. Rorig, *J. Am. Chem. Soc.*, **67**:1578 (1945).

1469. J. Preston and H. G. Clark III (to Chemstrand Corp.), U.S. Patent 2,928,859, Mar. 15, 1960; *Chem. Abstr.*, **55**:3522 (1961).

1470. A. Lambert and J. D. Rose, *J. Chem. Soc.*, **1949**:46.

1471. A. S. Matlack, *J. Org. Chem.*, **23**:729 (1959).

1472. D. Ludsteck and H. Distler (to Badische Anilin- & Soda-Fabrik A.-G.), Belgian Patent 615,081, Sept. 14, 1962; *Chem. Abstr.*, **58**:10082 (1963).

1473. H. F. Park and R. I. Longley, Jr. (to Monsanto Chemical Co.), U.S. Patent 2,710,882, Jun. 14, 1955; *Chem. Abstr.*, **49**:13695 (1955).

1474. A. A. Goldberg, *J. Chem. Soc.*, 1945:464.
1475. J. S. H. Davies and A. E. Oxford, *ibid.*, 1931:224.
1476. J. R. Alexander and H. McCombie, *ibid.*, 1931:1913.
1477. H. Mohler and J. Sorge, *Helv. Chim. Acta*, 23:1200 (1940).
1478. British Patent 562,269 (to Wingfoot Corp.), Jun. 26, 1944; *Chem. Abstr.*, 40:492 (1946).
1479. G. Schrader (to Farbenfabriken Bayer A.-G.), German Patent 1,089,749, Sept. 29, 1960; *Chem. Abstr.*, 55:17500 (1961).
1480. L. Brandsma and J. F. Arens, *Rec. Trav. Chim. Pays-Bas*, 81:33 (1962).
1481. K. K. Georgieff and D. Dupré, *Can. J. Chem.*, 37:1104 (1959).
1482. A. E. Kretov and S. M. Kliger, *J. Gen. Chem. USSR*, 2:322 (1932); through *Chem. Abstr.*, 27:952 (1933).
1483. C. C. Price and O. H. Bullitt, Jr., *J. Org. Chem.*, 12:238 (1947).
1484. A. H. Ford-Moore, R. A. Peters, and R. W. Wakelin, *J. Chem. Soc.*, 1949:1754.
1485. C. G. Overberger, D. L. Schoene, P. M. Kamath, and I. Tashlick, *J. Org. Chem.*, 19:1486 (1954).
1486. H. Distler, G. Faber, and N. von Kutepew (to Badische Anilin- & Soda-Fabrik A.-G.), German Patent 1,076,670, Mar. 3, 1960; *Chem. Abstr.*, 55:12432 (1961).
1487. A. K. Khomenko, *Izv. Akad. Nauk SSSR, Otd. Khim. Nauk*, 1951:280; through *Chem. Abstr.*, 46:884 (1952).
1488. (a) E. N. Prilezhaeva and M. F. Shostakovskii, *Akad. Nauk SSSR, Inst. Org. Khim., Sintezy Org. Soedin., Sbornik*, 2:54 (1952); through *Chem. Abstr.*, 48:552 (1954). (b) M. F. Shostakovskii, E. N. Prilezhaeva, and N. I. Uvarova, *Izv. Akad. Nauk SSSR, Otd. Khim. Nauk*, 1954:526; *Bull. Acad. Sci. USSR, Div. Chem. Sci.*, 1954:447. (c) M. F. Shostakovskii, E. N. Prilezhaeva, and N. I. Uvarova, *ibid.*, 1955:906; through *Chem. Abstr.*, 50:9278 (1956).
1489. E. D. Holly and S. H. Vasicek (to The Dow Chemical Co.), U.S. Patent 2,832,806, Apr. 29, 1958; *Chem. Abstr.*, 52:14682 (1958).
1490. J. J. Nedwick and J. R. Snyder (to Rohm and Haas Co.), U.S. Patent 2,930,815, Mar. 29, 1960; *Chem. Abstr.*, 54:15245 (1960).
1491. British Patent 444,689 (to I. G. Farbenindustrie A.-G.), Mar. 24, 1936; *Chem. Abstr.*, 30:5594 (1936).
1492. H. J. Schneider (to Rohm and Haas Co.), U.S. Patent 2,890,224, Jun. 9, 1959; *Chem. Abstr.*, 54:295 (1960). H. J. Schneider, J. J. Bagnell, and G. C. Murdock, *J. Org. Chem.*, 26:1980, 1982 (1961).
1493. H. J. Schneider (to Rohm and Haas Co.), U.S. Patent 3,062,892, Nov. 6, 1962; *Chem. Abstr.*, 58:10082 (1963).
1494. J. Loevenich, J. Losen, and A. Dierichs, *Chem. Ber.*, 60B:950 (1927).
1495. French Patent 814,852 (to I. G. Farbenindustrie A.-G.), Jul. 1, 1937; *Chem. Abstr.*, 32:956 (1938).
1496. J. Flynn, Jr., V. V. Badiger, and W. E. Truce, *J. Org. Chem.*, 28:2298 (1963).
1497. R. Brown and R. C. G. Moggridge, *J. Chem. Soc.*, 1946:816.
1498. (a) C. C. Price and J. Zomlefer, *J. Am. Chem. Soc.*, 72:14 (1950). (b) C. C. Price and H. Morita, *ibid.*, 75:4747 (1953).
1499. F. Montanari, *Boll. Sci. Fac. Chim. Ind. Bologna*, 14:55 (1956); through *Chem. Abstr.*, 51:5723 (1957). E. Angeletti, F. Montanari, and A. Negrini, *Gazz. Chim. Ital.*, 87:1115 (1957); through *Chem. Abstr.*, 52:9986 (1958).
1500. J. F. Harris (to E. I. du Pont de Nemours & Co.), U.S. Patent 3,048,569,

Aug. 7, 1962; *Chem. Abstr.*, **57**:16886 (1962). British Patent 926,573 (to E. I. du Pont de Nemours & Co.), May 22, 1963; *Chem. Abstr.*, **60**:1596 (1964).

1501. W. E. Parham, F. D. Blake, and D. R. Thiessen, *J. Org. Chem.*, **27**:2415 (1962).
1502. V. S. Tsivunin, G. K. Kamai, and V. V. Kormachev, *Probl. Organ. Sinteza, Akad. Nauk SSSR, Otd. Obshch. i Tekhn. Khim.*, **1965**:40; through *Chem. Abstr.*, **64**:8065 (1966).
1503. J. F. Arens and T. Doornbos, *Rec. Trav. Chim. Pays-Bas*, **75**:481 (1956). Netherlands Patent 87,522 (to NV Organon), Feb. 15, 1958; *Chem. Abstr.*, **53**:16963 (1959).
1504. H. Böhme and H. Bentler, *Chem. Ber.*, **89**:1464 (1956).
1505. M. Protiva, *Collect. Czech. Chem. Commun.*, **14**:354 (1949); through *Chem. Abstr.*, **45**:575 (1951).
1506. W. E. Parham and R. F. Motter, *J. Am. Chem. Soc.*, **81**:2146 (1959).
1507. T. F. Doumani (to Union Oil Co. of California), U.S. Patent 2,402,878, Jun. 25 1946; *Chem. Abstr.*, **40**:6496 (1946).
1508. C. C. Price and R. G. Gillis, *J. Am. Chem. Soc.*, **75**:4750 (1953).
1509. J. F. Arens, H. C. Volger, T. Doornbos, J. Bonnema, J. W. Greidanus, and J. H. van den Hende, *Rec. Trav. Chim. Pays-Bas*, **75**:1459 (1956).
1510. E. M. LaCombe and B. Stewart (to Union Carbide Corp.), U.S. Patent 3,121,110, Feb. 11, 1964; *Chem. Abstr.*, **60**:13146 (1964).
1511. G. Sosnovsky, *Tetrahedron*, **18**:15, 903 (1962).
1512. E. F. Landau and E. P. Irany (to Celanese Corp. of America), U.S. Patent 2,490,875, Dec. 13, 1949; *Chem. Abstr.*, **44**:4502 (1950). British Patent 660,054 (to British Celanese Ltd.), Oct. 31, 1951; *Chem. Abstr.*, **46**:9126 (1952).
1513. R. J. Charnock and R. C. G. Moggridge, *J. Chem. Soc.*, **1946**:815.
1514. H. J. Schneider (to Rohm and Haas Co.), U.S. Patent 3,061,648, Oct. 30, 1962; *Chem. Abstr.*, **58**:3318 (1963).
1515. D. J. Foster and E. Tobler, *J. Am. Chem. Soc.*, **83**:851 (1961); (to Union Carbide Corp.), U.S. Patent 3,087,952, Apr. 30, 1963; *Chem. Abstr.*, **59**:12843 (1963).
1516. D. Swern and E. F. Jordan, Jr., in N. Rabjohn (ed.), "Organic Syntheses," coll. vol. IV, p. 977, Wiley, New York, 1963.
1517. E. K. Wilip (to W. R. Grace & Co.), U.S. Patent 2,976,326, Mar. 21, 1961; *Chem. Abstr.*, **55**:15348 (1961).
1518. V. A. Dorokhov and B. M. Mikhailov, *Izv. Akad. Nauk SSSR, Ser. Khim.*, **1966**:364; through *Chem. Abstr.*, **64**:17624 (1966).
1519. E. N. Prilezhaeva, L. V. Tsymbal, O. N. Domnina, T. N. Shkurina, and M. F. Shostakovskii, *Izv. Akad. Nauk SSSR, Otd. Khim. Nauk*, **1960**:724; through *Chem. Abstr.*, **54**:22323 (1960).
1520. C. C. Price and R. D. Gilbert, *J. Am. Chem. Soc.*, **74**:2073 (1952).
1521. A. H. Ford-Moore, *J. Chem. Soc.*, **1949**:2126.
1522. G. D. Buckley, J. L. Charlish, and J. D. Rose, *ibid.*, **1947**:1514.
1523. E. A. Fehnel and M. Carmack, *J. Am. Chem. Soc.*, **71**:231 (1949).
1524. A. H. Ford-Moore, *J. Chem. Soc.*, **1949**:2433.
1525. A. C. Cope, D. E. Morrison, and L. Field, *J. Am. Chem. Soc.*, **72**:59 (1950).
1526. L. I. Smith and H. R. Davis, Jr., *J. Org. Chem.*, **15**:824 (1950).
1527. H. R. Snyder, H. V. Anderson, and D. P. Hallada, *J. Am. Chem. Soc.*, **73**:3258 (1951).
1528. R. B. Thompson (to Universal Oil Products Co.), U.S. Patent 2,677,617, May 4, 1954; *Chem. Abstr.*, **48**:10332 (1954).

1529. L. Ramberg and B. Bäcklund, *Ark. Kemi, Mineral. Geol.*, **13A**: no. 27 (1940); *Chem. Abstr.*, **34**:4725 (1940).

1530. E. F. Landau and E. P. Irany (to Celanese Corp. of America), U.S. Patent 2,554,576, May 29, 1951; *Chem. Abstr.*, **46**:3065 (1952). British Patent 661,783 (to British Celanese Ltd.), Nov. 28, 1951; *Chem. Abstr.*, **46**:6138 (1952).

1531. G. Kränzlein, W. Schuhmacher, and J. Heyna (to Farbwerke Hoechst A.-G.), German Patent 877,607, Jul. 2, 1953; *Chem. Abstr.*, **52**:9206 (1958).

1532. G. Kränzlein, J. Heyna, and W. Schuhmacher (to Farbwerke Hoechst A.-G.), German Patent 842,198, Jun. 23, 1952; *Chem. Abstr.*, **47**:11244 (1953).

1533. D. L. Schoene (to U.S. Rubber Co.), U.S. Patent 2,474,808, Jul. 5, 1949; *Chem. Abstr.*, **43**:7952 (1949).

1534. (*a*) M. F. Shostakovskii, E. N. Prilezhaeva, V. A. Azovskaya, and G. V. Dmitrieva, *Zh. Obshch. Khim.*, **30**:1123 (1960); through *Chem. Abstr.*, **55**:414 (1961). (*b*) M. F. Shostakovskii, E. N. Prilezhaeva, and A. V. Sviridova, *Dokl. Akad. Nauk SSSR*, **146**:837 (1962); through *Chem. Abstr.*, **58**:3512 (1963).

1535. E. Siegel (to Farbenfabriken Bayer A.-G.), German Patent 1,127,889, Apr. 19, 1962; *Chem. Abstr.*, **57**:9666 (1962).

1536. V. Z. Sharf, L. K. Freidlin, E. N. Prilezhaeva, A. V. Sviridova, and R. Y. Tolchinskaya (to N. D. Zelinskii Inst. of Org. Chem., Acad. of Sciences, USSR), British Patent 1,027,984, May 4, 1966; *Chem. Abstr.*, **65**:2126 (1966).

1537. British Patent 659,779 (to British Celanese Ltd.), Oct. 24, 1951; *Chem. Abstr.*, **46**:9120 (1952).

1538. J. Heyna and W. Riemenschneider (to Farbwerke Hoechst A.-G.), German Patent 932,488, Sept. 1, 1955; *Chem. Abstr.*, **50**:7859 (1956).

1539. E. Tobler and D. J. Foster, *Z. Naturforsch.*, **17b**(2):135 (1962).

1540. G. D. Jones and C. E. Barnes (to General Aniline & Film Corp.), U.S. Patents 2,515,714, Jul. 18, 1950, and 2,619,452, Nov. 25, 1952; *Chem. Abstr.*, **44**:11176 (1950); **47**:9342 (1953).

1541. W. V. Farrar, *J. Chem. Soc.*, **1960**:3058.

1542. P. C. Aichenegg and C. D. Emerson (to Chemagro Corp.), U.S. Patent 3,144,383, Aug. 11, 1964; *Chem. Abstr.*, **61**:9404 (1964).

1543. M. F. Shostakovskii, E. N. Prilezhaeva, and V. M. Karavaeva, *Vysokomol. Soedin.*, **1**:582, 594, 781 (1959); through *Chem. Abstr.*, **54**:17241, 17242, 14893 (1960).

1544. T. Kaneko and T. Inui, *Nippon Kagaku Zasshi*, **76**:306 (1955); through *Chem. Abstr.*, **51**:17748 (1957).

1545. W. E. Truce and E. Wellisch, *J. Am. Chem. Soc.*, **74**:2881 (1952).

1546. F. C. Foster, *ibid.*, **74**:2299 (1952).

1547. N. B. Hannay and C. P. Smyth, *ibid.*, **68**:1005 (1946).

1548. V. Baliah and S. Shanmuganathan, *Trans. Faraday Soc.*, **55**:232 (1959).

1549. M. A. Stahmann, C. Golumbic, W. H. Stein, and J. S. Fruton, *J. Org. Chem.* **11**:719 (1946).

1550. D. Cordier and G. Cordier, *C. R. Soc. Biol.*, **144**:868 (1950); through *Chem. Abstr.*, **45**:263 (1951).

1551. C. R. Stahl, *Anal. Chem.*, **34**:980 (1962).

1552. L. Fishbein and J. Fawkes, *J. Chromatog.*, **22**(2):323 (1966).

1553. C. E. Scott and C. C. Price, *J. Am. Chem. Soc.*, **81**:2672 (1959).

1554. J. Sicé, *J. Phys. Chem.*, **64**:1573 (1960).

1555. V. T. Aleksanyan, Y. M. Kumel'fel'd, S. M. Shostakovskii, and A. I. L'vov

Zh. Prikl. Spektrosk., Akad. Nauk Belorussk. SSR, **3**(4):355 (1965); through *Chem. Abstr.*, **64**:10613 (1966).

1556. W. R. Feairheller, Jr. and J. E. Katon, *Spectrochim. Acta*, **20**:1099 (1964).

1557. K. C. Schreiber, *Anal. Chem.*, **21**:1168 (1949).

1558. R. T. Hobgood, G. S. Reddy, and J. H. Goldstein, *J. Phys. Chem.*, **67**:110 (1963).

1559. F. Bohlmann, C. Arndt, and J. Starnick, *Tetrahedron Letters*, **1963**:1605.

1560. P. P. Shorygin, M. F. Shostakovskii, E. N. Prilezhaeva, T. N. Shkurina, L. G. Stolyarova, and A. P. Genich, *Izv. Akad. Nauk SSSR, Otd. Khim. Nauk*, **1961**:1571; *Bull. Acad. Sci. USSR, Div. Chem. Sci.*, **1961**:1468.

1561. G. Maccagnani, F. Taddei, and C. Zauli, *Boll. Sci. Fac. Chim. Ind. Bologna*, **21**:131 (1963); through *Chem. Abstr.*, **60**:3640 (1964).

1562. M. T. Rogers, G. M. Barrow, and F. G. Bordwell, *J. Am. Chem. Soc.*, **78**:1790 (1956).

1563. V. Baliah and S. Shanmuganathan, *J. Indian Chem. Soc.*, **35**:31 (1958).

1564. R. G. Gillis and J. L. Occolowitz, *Tetrahedron Letters*, 1997 (1966).

1565. T. Kitao, N. Kuroki, and K. Konishi, *Kogyo Kagaku Zasshi*, **62**:825 (1959); through *Chem. Abstr.*, **57**:8481 (1962).

1566. H. U. Werner and H. Distler (to Badische Anilin- & Soda-Fabrik A.-G.), German Patent 1,098,202, Jan. 26, 1961; *Chem. Abstr.*, **55**:24106 (1961).

1567. W. Mueller, H. Distler, and A. Palm (to Badische Anilin- & Soda-Fabrik A.-G.), German Patent 1,135,898, Sept. 6, 1962; *Chem. Abstr.*, **58**:1402 (1963).

1568. W. Riemenschneider (to Farbwerke Hoechst A.-G.), German Patent 1,010,063, Jun. 13, 1957; *Chem. Abstr.*, **54**:2177 (1960).

1569. D. L. Schoene (to U.S. Rubber Co.), U.S. Patents (*a*) 2,493,364, Jan. 3, 1950, and (*b*) 2,505,366, Apr. 25, 1950; *Chem. Abstr.*, **44**:5643, 6676 (1950).

1570. J. W. Schappel (to American Viscose Corp.), U.S. Patent 2,623,035, Dec. 23, 1952; *Chem. Abstr.*, **47**:3038 (1953).

1571. W. E. Parham, M. A. Kalvins, and D. R. Theissen, *J. Org. Chem.*, **27**:2698 (1962).

1572. C. K. Ingold and H. G. Smith, *J. Chem. Soc.*, **1931**:2742.

1573. E. P. Prilezhaeva, N. P. Petukhova, and M. F. Shostakovskii, *Dokl. Akad. Nauk SSSR*, **144**:1059 (1962); through *Chem. Abstr.*, **57**:13632 (1962).

1574. E. Siegel and S. Peterson, *Angew. Chem.*, **74**:873 (1962).

1575. C. Nakeshima, S. Tanimoto, and R. Oda, (*a*) *Kogyo Kagaku Zasshi*, **67**:1705 (1964); through *Chem. Abstr.*, **62**:10357 (1965). (*b*) *Nippon Kagaku Zasshi*, **86**:442 (1965); through *Chem. Abstr.*, **63**:8239 (1965).

1576. M. F. Shostakovskii, A. V. Bogdanova, and T. M. Ushakova, *Dokl. Akad. Nauk SSSR*, **118**:520 (1958); through *Chem. Abstr.*, **52**:11789 (1958).

1577. K. Alder, H. F. Rickert, and E. Windemuth, *Chem. Ber.*, **71B**:2451 (1938).

1578. C. W. Smith, D. G. Norton, and S. A. Ballard, *J. Am. Chem. Soc.*, **73**:5267 (1951).

1579. F. G. Bordwell and W. H. McKellin, *ibid.*, **73**:2251 (1951).

1580. J. G. Noltes and G. J. M. van der Kerk, *Chem. Ind.* (*London*), **1959**:294.

1581. C. W. Johnson, C. G. Overberger, and W. J. Seagers, *J. Am. Chem. Soc.*, **75**:1495 (1953).

1582. W. Kern and R. C. Schulz, *Angew. Chem.*, **69**:153 (1957).

1583. H. Distler, W. Mueller, and H. Werner (to Badische Anilin- & Soda-Fabrik A.-G.), German Patent 1,122,064, Jan. 18, 1962; *Chem. Abstr.*, **58**:3357 (1963).

1596 VINYL AND DIENE MONOMERS

1584. D. S. Breslow and A. Kutner, *J. Polymer Sci.*, (*a*) **27**:295 (1958); (*b*) **38**:274 (1959).
1585. W. Kern, V. V. Kale, and B. Schering, *Makromol. Chem.*, **32**:37 (1959).
1586. S. Yoshikawa, O. K. Kim, and T. Hori, *Bull. Chem. Soc. Japan*, **39**:1937 (1966).
1587. R. H. Wiley and D. E. Gensheimer, *J. Polymer Sci.*, **42**:119 (1960).
1588. T. Alfrey, Jr. and C. R. Pfeifer, *J. Polymer Sci.*, *Pt. A-1*, **4**:2447 (1966).
1589. H. Ito and S. Suzuki, *Kogyo Kagaku Zasshi*, **60**:1056 (1957); through *Chem. Abstr.*, **53**:10837 (1959); Japanese Patent 3,897 (to East Asia Synthetic Chemical Industries Co.), May 19, 1958; *Chem. Abstr.*, **53**:8715 (1959).
1590. J. Bourdais, *Compt. Rend.*, **246**:2374 (1958).
1591. A. Kutner (to Hercules Powder Co.), U.S. Patent 2,961,431, Nov. 22, 1960; *Chem. Abstr.*, **55**:9965 (1961).
1592. H. Distler, W. Mueller, and H. U. Werner (to Badische Anilin- & Soda-Fabrik A.-G.), British Patent 922,388, Mar. 27, 1963; *Chem. Abstr.*, **59**:12770 (1963).
1593. N. Fujisaki and K. Shibata (to Asahi Chemical Industry Co.), Japanese Patents (*a*) 2,143 and (*b*) 2,144, Mar. 28, 1958; *Chem. Abstr.*, **52**:21147 (1958).
1594. R. M. Hedrick (to Monsanto Chemical Co.), U.S. Patent 2,801,991, Aug. 6, 1957; *Chem. Abstr.*, **51**:16003 (1957).
1595. C. G. Overberger, D. E. Baldwin, and H. P. Gregor, *J. Am. Chem. Soc.*, **72**:4864 (1950).
1596. C. S. Marvel, V. C. Menikheim, H. K. Inskip, W. K. Taft, and B. G. Labbe, *J. Polymer Sci.*, **10**:39 (1953).
1597. K. Ashida (to Hodogaya Chemical Co.), Japanese Patent 4,195, May 28, 1958; *Chem. Abstr.*, **52**:21230 (1958).
1598. W. Bauer (to Röhm & Haas GmbH), German Patent 888,768, Sept. 3, 1953; *Chem. Abstr.*, **51**:2327 (1957).
1599. J. E. Pritchard (to Phillips Petroleum Co.), U.S. Patent 2,769,802, Nov. 6, 1956; *Chem. Abstr.*, **51**:6220 (1957).
1600. E. Y. C. Chang and C. C. Price, *J. Am. Chem. Soc.*, **83**:4650 (1961).
1601. G. B. Butler and R. B. Kasat, *J. Polymer Sci.*, *Pt. A*, **3**(12):4205 (1965).
1602. M. F. Shostakovskii, E. N. Prilezhaeva, and V. M. Karavaeva, *Izv. Akad. Nauk SSSR, Otd. Khim. Nauk*, **1957**:621, 650; through *Chem. Abstr.*, **51**:15459 (1957).
1603. M. F. Shostakovskii, E. N. Prilezhaeva, and N. I. Uvarova, *Soobshcheniya, Nauch. Rabot. Vsesoyuz. Khim. Obshch. im. Mendeleeva*, **1955**(3):21; through *Chem. Abstr.*, **53**:6678 (1959).
1604. D. W. J. Young and W. J. Sparks (to Standard Oil Development Co.), U.S. Patent 2,462,703, Feb. 22, 1949; *Chem. Abstr.*, **43**:3660 (1949).
1605. W. R. Conard and C. E. Best (to Firestone Tire & Rubber Co.), U.S. Patent 2,605,256, Jul. 29, 1952; *Chem. Abstr.*, **46**:10688 (1952).
1606. E. D. Holly, *J. Polymer Sci.*, **36**:329 (1959).
1607. M. H. Opheim and B. Franzus (to Phillips Petroleum Co.), U.S. Patent 2,877,214, Mar. 10, 1959; *Chem. Abstr.*, **53**:11891 (1959).
1608. (a) C. J. Mighton (to E. I. du Pont de Nemours & Co.), U.S. Patent 2,472,672, Jun. 7, 1949; *Chem. Abstr.*, **43**:7266 (1949). (b) C. G. Overberger and A. M. Schiller, *J. Org. Chem.*, **26**:4230 (1961).
1609. E. P. Irany and E. F. Landau (to Celanese Corp. of America), U.S. Patent 2,538,100, Jan. 16, 1951; *Chem. Abstr.*, **45**:3653 (1951).

MISCELLANEOUS VINYL MONOMERS

1597

1610. W. Starck, G. Bier, and G. Lorentz (to Farbwerke Hoechst A.-G.), German Patent 937,616, Jan. 12, 1956; *Chem. Abstr.*, **52**:21249 (1958).

1611. J. M. Judge and C. C. Price, *J. Polymer Sci.*, **41**:435 (1959).

1612. K. W. Doak, private communication to F. R. Mayo and C. Walling, *Chem. Rev.*, **46**:191 (1950).

1613. R. L. Hill, private communication cited in ref. 1588.

1614. D. S. Breslow and A. Kutner, *J. Polymer Sci.*, **31**:253 (1960).

1615. W. Heuer (to I. G. Farbenindustrie A.-G.), German Patent 724,889, Jul. 23, 1942; *Chem. Abstr.*, **37**:5808 (1943). W. Heuer (to General Aniline & Film Corp.), U.S. Patent 2,300,920, Nov. 3, 1942; *Chem. Abstr.*, **37**:2105 (1943).

1616. E. Penning, H. Wilhelm, and H. Wolf (to Badische Anilin- & Soda-Fabrik A.-G.), Belgian Patent 612,753, Jul. 17, 1962; *Chem. Abstr.*, **57**:16892 (1962).

1617. British Patent 910,136 (to Farbenfabriken Bayer A.-G.), Nov. 7, 1962; *Chem. Abstr.*, **58**:6944 (1963).

1618. R. F. Tietz (to E. I. du Pont de Nemours & Co.), U.S. Patent 3,020,265, Feb. 6, 1962; *Chem. Abstr.*, **60**:13379 (1964). L. D. Grandine, Jr. and W. K. Wilkinson (to E. I. du Pont de Nemours & Co.), U.S. Patent 3,075,934, Jan. 29, 1963; *Chem. Abstr.*, **60**:12165 (1964).

1619. W. Starck, K. Dietz, and R. Gauglitz (to Farbwerke Hoechst A.-G.), German Patent 1,162,253, Jan. 30, 1964; *Chem. Abstr.*, **60**:10369 (1964).

1620. H. Pohlemann and H. Spoor (to Badische Anilin- & Soda-Fabrik A.-G.), German Patent 1,164,095, Feb. 27, 1964; *Chem. Abstr.*, **60**:14694 (1964).

1621. R. Yamane, Y. Mizutani, and Y. Onone, *J. Electrochem. Soc. Japan, Overseas Ed.*, **30**:335 (1962); *Chem. Abstr.*, **61**:11367 (1964).

1622. (a) J. B. Conant, A. D. MacDonald, and A. M. Kinney, *J. Am. Chem. Soc.*, **43**:1928 (1921). (b) J. B. Conant and B. B. Coyne, *ibid.*, **44**:2530 (1922).

1623. L. A. Hamilton (to E. I. du Pont de Nemours & Co.), U.S. Patents 2,365,466, Dec. 19, 1944, and 2,382,309, Aug. 14, 1945; *Chem. Abstr.*, **39**:4619 (1945).

1624. Patents assigned to Monsanto Chemical Co.: (a) G. M. Kosolapoff, U.S. Patent 2,389,576, Nov. 20, 1945; *Chem. Abstr.*, **40**:1536 (1946). (b) G. M. Kosolapoff, U.S. Patent 2,486,657, Nov. 1, 1949; *Chem. Abstr.*, **44**:2008 (1950).

1625. A. H. Ford-Moore and J. H. Williams, *J. Chem. Soc.*, **1947**:1465.

1626. M. I. Kabachnik, *Izv. Akad. Nauk SSSR, Otd. Khim. Nauk*, **1947**:233; through *Chem. Abstr.*, **42**:4132 (1948).

1627. R. W. Upson (to E. I. du Pont de Nemours & Co.), U.S. Patent 2,557,805, Jun. 19, 1951; *Chem. Abstr.*, **45**:8298 (1951). R. W. Upson, *J. Am. Chem. Soc.*, **75**:1763 (1953).

1628. J. F. Allen and O. H. Johnson, *ibid.*, **77**:2871 (1955).

1629. G. M. Kosolapoff, "Organophosphorus Compounds," Wiley, New York, 1950.

1630. (a) M. I. Kabachnik and P. A. Rossiiskaya, *Izv. Akad. Nauk SSSR, Otd. Khim. Nauk*, **1946**:403; through *Chem. Abstr.*, **42**:7242 (1948). (b) M. I. Kabachnik and P. A. Rossiiskaya, *ibid.*, **1947**:389; through *Chem. Abstr.*, **42**:1558 (1948). (c) M. I. Kabachnik and P. A. Rossiiskaya, *ibid.*, **1946**:295; through *Chem. Abstr.*, **42**:7241 (1948). (d) M. I. Kabachnik and P. A. Rossiiskaya, *ibid.*, **1946**:515; through *Chem. Abstr.*, **42**:7242 (1948).

1631. M. I. Kabachnik, P. A. Rossiiskaya, and N. N. Novikova, *ibid.*, **1947**:97; through *Chem. Abstr.*, **42**:4132 (1948).

1632. G. M. Kosolapoff, *J. Am. Chem. Soc.*, **70**:1971 (1948).

1633. E. L. Gefter, *Plast. Massy*, **1961**(11):38; through *Chem. Abstr.*, **56**:10184 (1962).

1634. E. L. Gefter and M. I. Kabachnik, *ibid.*, **1961**(1):63; through *Chem. Abstr.*, **56**:501 (1962).

1635. E. L. Gefter, *Khim. Nauka i Prom.*, **3**:544 (1958); through *Chem. Abstr.*, **53**:4179 (1959).

1636. A. N. Pudovik and M. G. Imaev, *Izv. Akad. Nauk SSSR, Otd. Khim. Nauk*, **1952**:916; through *Chem. Abstr.*, **47**:10463 (1953).

1637. C. L. Arcus and R. J. S. Matthews, *J. Chem. Soc.*, **1956**:4607.

1638. H. Zenftman and D. Calder (to Imperial Chemical Industries Ltd.), British Patent 812,983, May 6, 1959; *Chem. Abstr.*, **53**:15647 (1959).

1639. G. S. Kolsnikov and E. F. Rodionova, *Vysokomol. Soedin.*, **1**(4):641 (1959); through *Chem. Abstr.*, **54**:17244 (1960).

1640. J. I. G. Cadogan, *J. Chem. Soc.*, **1957**:4154.

1641. R. G. Gillis, J. F. Horwood, and G. L. White, *J. Am. Chem. Soc.*, **80**:2999 (1958).

1642. G. S. Kolsnikov, E. F. Rodionova, and L. S. Fedorova, *Vysokomol. Soedin.*, **1**(3):367 (1959); through *Chem. Abstr.*, **54**:7214 (1960).

1643. A. Yuldashev, *Uzb. Khim. Zh.*, **7**(5):69 (1963); through *Chem. Abstr.*, **60**:5544 (1964).

1644. A. Y. Yakubovich, L. Z. Soborovskii, L. I. Muler, and V. S. Faermark, *Zh. Obshch. Khim.*, **28**:317 (1958); through *Chem. Abstr.*, **52**:13613 (1958).

1645. E. L. Gefter and P. A. Moshkin, *Plast. Massy*, **1960**(4):54; through *Chem. Abstr.*, **54**:24345 (1960).

1646. E. O. Leupold and H. Zorn (to Farbwerke Hoechst A.-G.), U.S. Patent 2,959,609, Nov. 8, 1960; *Chem. Abstr.*, **55**:7288 (1961).

1647. D. H. Chadwick, C. H. Campbell, and S. H. Metzger (to Monsanto Chemical Co.), U.S. Patent 3,064,030, Nov. 13, 1962; *Chem. Abstr.*, **58**:9139 (1963).

1648. M. I. Kabachnik and T. Y. Medved, *Izv. Akad. Nauk SSSR, Otd. Khim. Nauk*, **1959**:2142; through *Chem. Abstr.*, **54**:10834 (1960).

1649. G. S. Kolesnikov, I. G. Safaralieva, and E. F. Rodionova, *Vysokomol. Soedin.*, **6**(4):615 (1964); through *Chem. Abstr.*, **61**:1943 (1964).

1650. L. Z. Soborovskii, Y. M. Zinovév, and L. I. Muler, *Dokl. Akad. Nauk SSSR*, **109**:98 (1956); through *Chem. Abstr.*, **51**:1825 (1957).

1651. M. I. Kabachnik, J.-Y. Chang, and E. N. Tsvetkov, *Zh. Obshch. Khim.*, **32**:3351 (1962); through *Chem. Abstr.*, **58**:9126 (1963).

1652. M. I. Kabachnik, T. A. Mastryukova, and T. A. Melent'eva, *ibid.*, **33**:382 (1963); through *Chem. Abstr.*, **59**:1677 (1963).

1653. Patents assigned to Farbwerke Hoechst A.-G.: (a) K. Schimmelschmidt and W. Denk, German Patent 1,020,019, Nov. 28, 1957; *Chem. Abstr.*, **54**:1304 (1960). (b) K. Schimmelschmidt and W. Denk, German Patent 1,023,033, Jan. 23, 1958; *Chem. Abstr.*, **54**:5466 (1960). (c) K. Schimmelschmidt and W. Denk, German Patent 1,023,034, Jan. 23, 1958; *Chem. Abstr.*, **54**:5466 (1960).

1654. R. Rabinowitz, *J. Org. Chem.*, **28**:2975 (1963).

1655. C. M. Welch, E. J. Gonzales, and J. D. Guthrie, *ibid.*, **26**:3270 (1961).

1656. F. F. Blicke and S. Raines, *ibid.*, **29**:2036 (1964).

1657. K. N. Anisimov and N. E. Kolobova, *Izv. Akad. Nauk SSSR, Otd. Khim. Nauk*, **1956**:923; through *Chem. Abstr.*, **51**:4933 (1957).

1658. A. N. Pudovik and I. V. Konovalova, *Zh. Obshch. Khim.*, **31**:1693 (1961); through *Chem. Abstr.*, **55**:24540 (1961).

1659. R. C. De Selms and T.-W. Lin, *J. Org. Chem.*, **32**:2023 (1967).

1660. J. P. Schroeder, unpublished results.

1661. J. F. Allen, S. K. Reed, O. H. Johnson, and N. J. Brunsvold, *J. Am. Chem. Soc.*, **78**:3715 (1956).

1662. W. Perkow, K. Ullerich, and F. Meyer, *Naturwissenschaften*, **39**:353 (1952).

1663. F. W. Lichtenthaler, *Chem. Rev.*, **61**:607 (1961).

1664. R. R. Whetstone and D. Harman (to Shell Development Co.), U.S. Patent 2,765,331, Oct. 2, 1956; *Chem. Abstr.*, **51**:5816 (1957).

1665. (a) British Patent 784,985 (to Food Machinery and Chemical Corp.), Oct. 23, 1957; *Chem. Abstr.*, **52**:8195 (1958). (b) British Patent 784,986 (to Food Machinery and Chemical Corp.), Oct. 23, 1957; *Chem. Abstr.*, **52**:8195 (1958).

1666. P. S. Magee, *Tetrahedron Letters*, no. 45, 3995 (1965).

1667. M. Zief and C. H. Schramm, *Chem. Ind.* (*London*), **1964**:660.

1668. E. L. Gefter and M. I. Kabachnik, *Dokl. Akad. Nauk SSSR*, **114**:541 (1957); through *Chem. Abstr.*, **52**:295 (1958).

1669. Y. G. Gololobov, T. F. Dmitrieva, and L. Z. Soborovskii, *Probl. Organ. Sinteza, Akad. Nauk SSSR, Otd. Obshch. i Tekhn. Khim.*, **1965**:314; through *Chem. Abstr.*, **64**:6683 (1966).

1670. "Bis(*beta*-chloroethyl) Vinylphosphonate," Technical Data Sheet, Monsanto Chemical Co., St. Louis, Mo., 1956.

1671. "Bis(*beta*-chloroethyl) Vinylphosphonate," Product Data Sheet, Specialty Chemical Division, Stauffer Chemical Co., New York, 1966.

1672. M. I. Rizpolozhenski, L. V. Boiko, and M. A. Zvereva, *Dokl. Akad. Nauk SSSR*, **155**(5):1137 (1964); through *Chem. Abstr.*, **61**:1817 (1964).

1673. G. L. Kenyon and F. H. Westheimer, *J. Am. Chem. Soc.*, **88**:3557 (1966).

1674. Ref. 1629, p. 6.

1675. E. M. Popov, E. N. Tsvetkov, J.-Y. Chang, and T. Y. Medved, *Zh. Obshch. Khim.*, **32**:3255 (1962); through *Chem. Abstr.*, **58**:5165 (1963).

1676. F. A. Cotton and R. A. Schunn, *J. Am. Chem. Soc.*, **85**:2394 (1963).

1677. N. A. Slovokhotova, K. N. Anisimov, G. M. Kunitskaya, and N. E. Kolobova, *Izv. Akad. Nauk SSSR, Otd. Khim. Nauk*, **1961**:71; through *Chem. Abstr.*, **55**:23049 (1961).

1678. J. L. Occolowitz and J. M. Swan, *Aust. J. Chem.*, **19**(7):1187 (1966).

1679. M. M. Baizer and J. D. Anderson, *J. Org. Chem.*, **30**:3138 (1965).

1680. Patents assigned to U.S. Rubber Co.: (a) P. O. Tawney, U.S. Patents 2,535,172 and 2,535,174, Dec. 26, 1950; *Chem. Abstr.*, **45**:3408, 3409 (1951). (b) P. O. Tawney, U.S. Patent 2,570,503, Oct. 9, 1951; *Chem. Abstr.*, **46**:3556 (1952).

1681. A. N. Pudovik and R. G. Kuzovleva, *Zh. Obshch. Khim.*, **33**:2755 (1963); through *Chem. Abstr.*, **60**:543 (1964).

1682. A. N. Pudovik and G. M. Denisova, *ibid.*, **23**:263 (1953); through *Chem. Abstr.*, **48**:2572 (1954).

1683. A. N. Pudovik and O. N. Grishina, *ibid.*, **23**:267 (1953); through *Chem. Abstr.*, **48**:2573 (1954).

1684. (a) E. C. Ladd and M. P. Harvey (to U.S. Rubber Co.), U.S. Patent 2,651,656, Sept. 8, 1953; *Chem. Abstr.*, **48**:10052 (1954). (b) E. C. Ladd (to U.S. Rubber Co.), U.S. Patent 2,622,096, Dec. 16, 1952; *Chem. Abstr.*, **47**:9344 (1953).

1685. Y. Wada and R. Oda, *Kogyo Kagaku Zasshi*, **67**:2093 (1964); through *Chem. Abstr.*, **62**:13177 (1965).

1686. J. B. Dickey, H. W. Coover, Jr., and N. H. Shearer, Jr. (to Eastman Kodak Co.), U.S. Patent 2,550,651, Apr. 24, 1951; *Chem. Abstr.*, **45**:8029 (1951).

1687. W. M. Daniewski and C. E. Griffin, *J. Org. Chem.*, **31**:3236 (1966).
1688. Patents assigned to Monsanto Chemical Co.: (a) D. H. Chadwick, U.S. Patent 2,784,206, Mar. 5, 1957; *Chem. Abstr.*, **51**:10559 (1957). (b) D. H. Chadwick, U.S. Patent 2,951,086, Aug. 30, 1960; *Chem. Abstr.*, **55**:4362 (1961).
1689. H. W. Coover, Jr. and J. B. Dickey (to Eastman Kodak Co.), U.S. Patent 2,632,768, Mar. 24, 1953; *Chem. Abstr.*, **48**:2084 (1954).
1690. F. W. Lichtenthaler and F. Cramer, *Chem. Ber.*, **95**:1971 (1962).
1691. C. L. Arcus and R. J. S. Matthews, *Chem. Ind. (London)*, **1958**:890.
1692. V. E. Shashoua (to E. I. du Pont de Nemours & Co.), U.S. Patent 2,888,434, May 26, 1959; *Chem. Abstr.*, **53**:16554 (1959).
1693. R. M. Pike and R. A. Cohen, *J. Polymer Sci.*, **44**:531 (1960).
1694. V. A. Abramov and V. S. Tsivunin, *Trudy Kazan Khim.-Tekhnol. Inst. im. S. M. Kirova*, **1959**(26):96; through *Chem. Abstr.*, **54**:25952 (1960).
1695. M. H. Bride, W. A. W. Cummings, and W. Pickles, *J. Appl. Chem. (London)*, **11**:352 (1961).
1696. (a) V. A. Orlov and O. G. Tarakanov, *Plast. Massy*, **1962** (4):6; through *Chem. Abstr.*, **57**:4859 (1962). (b) V. A. Orlov and O. G. Tarakanov, *ibid.*, **1964**(6):6; through *Chem. Abstr.*, **61**:8412 (1964).
1697. T. Tsuda and Y. Yamashita, *Kogyo Kagaku Zasshi*, **65**:811 (1962); through *Chem. Abstr.*, **57**:15344 (1962).
1698. G. S. Kolesnikov, E. F. Rodionova, and I. G. Safaralieva, *Izv. Akad. Nauk SSSR, Ser. Khim.*, **1963**(11):2028; through *Chem. Abstr.*, **60**:6933 (1964).
1699. S. Konya and M. Yokoyama, *Kogyo Kagaku Zasshi*, **68**(6):1080 (1965); through *Chem. Abstr.*, **64**:11326 (1966).
1700. E. F. Rodionova, G. S. Kolesnikov, and L. A. Gavrikova, *Vysokomol. Soedin.*, **7**(3):377 (1965); through *Chem. Abstr.*, **63**:3046 (1965).
1701. E. L. Gefter and A. Yuldashev, *Plast. Massy*, **1962**(2):49; through *Chem. Abstr.*, **57**:1039 (1962).
1702. F. Marktscheffel, A. F. Turbak, and Z. W. Wilchinsky, *J. Polymer Sci., Pt. A-1*, **4**(10):2423 (1966).
1703. H. W. Coover, Jr. and M. A. McCall (to Eastman Kodak Co.), U.S. Patent 3,043,821, Jul. 10, 1962; *Chem. Abstr.*, **57**:12731 (1962).
1704. H. Krämer, G. Messwarb, and W. Denk (to Farbwerke Hoechst A.-G.), German Patent 1,032,537, Jun. 19, 1958; *Chem. Abstr.*, **54**:16010 (1960).
1705. F. Rochlitz, H. Vilcsek, and G. Koch (to Farbwerke Hoechst A.-G.), German Patent 1,135,176, Aug. 23, 1962; *Chem. Abstr.*, **57**:15371 (1962).
1706. F. Rochlitz and H. Vilcsek (to Farbwerke Hoechst A.-G.), German Patent 1,130,177, May 24, 1962; *Chem. Abstr.*, **57**:15358 (1962).
1707. C. S. Marvel and J. C. Wright, *J. Polymer Sci.*, **8**:255 (1952).
1708. R. V. Lindsey, Jr. (to E. I. du Pont de Nemours & Co.), U.S. Patent 2,439,214, Apr. 6, 1948; *Chem. Abstr.*, **42**:4795 (1948).
1709. B. B. Levin and I. N. Fetin, U.S.S.R. Patent 179,469, Feb. 8, 1966; *Chem. Abstr.*, **65**:2371 (1966).
1710. G. S. Kolesnikov, A. S. Tevlina, and A. B. Alovitdinov, *Vysokomol. Soedin.*, **7**(11): 1913 (1965); through *Chem. Abstr.*, **64**:6764 (1966).
1711. G. S. Kolesnikov, A. S. Tevlina, S. P. Novikova, and S. N. Sividova, *ibid.*, **7**(12):2160 (1965); through *Chem. Abstr.*, **64**:11323 (1966).
1712. G. S. Kolesnikov, E. F. Rodionova, B. B. Levin, and I. N. Fetin, U.S.S.R. Patent 179,922, Feb. 28, 1966; *Chem. Abstr.*, **65**:7310 (1966).

1713. A. B. Alovitdinov, A. S. Tevlina, and G. S. Kolesnikov, *Plast. Massy*, **1966**(8): 21; *Chem. Abstr.*, **65**:17131 (1966).

1714. K. A. Andrianov, B. B. Levin, E. F. Rodionova, and I. N. Fetin, U.S.S.R. Patent 178,985, Feb. 3, 1966; *Chem. Abstr.*, **65**:2373 (1966).

1715. H. W. Coover, Jr. and J. B. Dickey (to Eastman Kodak Co.), U.S. Patent 2,743,261, Apr. 24, 1956; *Chem. Abstr.*, **50**:12537 (1956).

1716. E. V. Kuznetsov, V. I. Gusev, L. S. Semenova, and L. A. Shurygina, U.S.S.R. Patent 181,290, Apr. 15, 1966; *Chem. Abstr.*, **65**:9055 (1966).

1717. V. V. Korshak, I. A. Gribova, M. A. Andreeva, and G. M. Popova, *Vysokomol. Soedin.*, **4**(1):58 (1962); through *Chem. Abstr.*, **57**:16852 (1962).

1718. A. F. Lewis and L. J. Forrestal, *ASTM Spec. Tech. Publ. No. 360*, **1963**:59; *Chem. Abstr.*, **62**:5389 (1965).

1719. E. L. Gefter, A. B. Pashkov, and E. I. Lyustgarten, *Khim. Nauka i Prom.*, **3**:825 (1958); through *Chem. Abstr.*, **53**:11281 (1959).

1720. C. Liebermann, *Chem. Ber.*, **27**:283 (1894).

1721. C. Moureu, C. Dufraisse, and J. R. Johnson, *Ann. Chim. (Paris)*, **7**:14 (1927).

1722. R. Kuhn and O. Dann, *Ann. Chem.*, **547**:293 (1941).

1723. R. L. Frank, C. E. Adams, J. R. Blegen, P. V. Smith, A. E. Juve, C. H. Schroeder, and M. M. Goff, *Ind. Eng. Chem.*, **40**:420 (1948).

1724. (a) W. S. Emerson, *Chem. Rev.*, **45**(2):183 (1949). (b) W. S. Emerson, *ibid.*, **45**(2):347 (1949).

1725. J. R. Johnson, The Perkin Reaction, in R. Adams (ed.), "Organic Reactions," vol. I, p. 210, Wiley, New York, 1942.

1726. M. M. Koton, A. P. Votinova, and F. S. Florinskii, *J. Appl. Chem. USSR*, **14**:181 (1941); through *Chem. Abstr.*, **36**:1604 (1942).

1727. L. Galimberti, *Boll. Sci. Fac. Chim. Ind. Bologna*, **1940**:351; through *Chem. Abstr.*, **37**:3410 (1943).

1728. A. Mora and J. Infiesta, *Combustibles (Zaragoza)*, **7**(37):27 (1947); through *Chem. Abstr.*, **42**:8519 (1948).

1729. I. Simek and M. Hanus, *Chem. Zvesti*, **13**:108 (1959); through *Chem. Abstr.*, **53**:17994 (1959).

1730. R. Paul and S. Tchelitcheff, *Bull. Soc. Chim. France*, **1947**:453.

1731. Y. K. Yur'ev, N. S. Zefirov, and V. M. Gurevich, *Zh. Obshch. Khim.*, **31**:3531 (1961); through *Chem. Abstr.*, **57**:7204 (1962).

1732. C. R. Wagner (to Phillips Petroleum Co.), U.S. Patent 2,431,216, Nov. 18, 1947; *Chem. Abstr.*, **42**:1609 (1948).

1733. L. Levi and R. V. V. Nicholls, *Ind. Eng. Chem.*, **50**:1005 (1958).

1734. Patents assigned to Eastman Kodak Co.: (a) H. J. Hagemeyer, U.S. Patent 2,466,420, Apr. 5, 1949; *Chem. Abstr.*, **43**:5037 (1949). (b) H. J. Hagemeyer, U.S. Patent 2,469,110, May 3, 1949; *Chem. Abstr.*, **43**:5415 (1949).

1735. R. Paul, *Compt. Rend.*, **200**:1118 (1935); *Bull. Soc. Chim.*, **2**:2220 (1935).

1736. C. R. Wagner (to Phillips Petroleum Co.), U.S. Patents 2,560,610, Jul. 17, 1951, and 2,689,855, Sept. 21, 1954; *Chem. Abstr.*, **46**:3572 (1952) and **49**:11720 (1955).

1737. A. A. Balandin, G. M. Marukyan, R. G. Seimovich, T. K. Lavrovskaya, and I. I. Levitskii, *Zh. Obshch. Khim.*, **30**:321 (1960); through *Chem. Abstr.*, **54**:22561 (1960).

1738. W. S. Emerson and T. M. Patrick, Jr., *J. Org. Chem.*, **13**:729 (1948); in N. Rabjohn (ed.), "Organic Syntheses," coll. vol. IV, p. 980, Wiley, New York,

1963. W. S. Emerson and T. M. Patrick, Jr. (to Monsanto Chemical Co.), U.S. Patent 2,547,905, Apr. 3, 1951; *Chem. Abstr.*, **45**:9084 (1951).

1739. J. W. Schick and H. D. Hartough, *J. Am. Chem. Soc.*, **70**:1645 (1948). J. W. Schick (to Socony-Vacuum Oil Co., Inc.), U.S. Patent 2,492,663, Dec. 27, 1949; *Chem. Abstr.*, **44**:3032 (1950).

1740. J. F. Scully and E. V. Brown, *J. Am. Chem. Soc.*, **75**:6329 (1953).

1741. W. Davies and Q. N. Porter, *J. Chem. Soc.*, **1957**:4958.

1742. (a) I. V. Andreeva and M. M. Koton, *Dokl. Akad. Nauk SSSR*, **110**:75 (1956); through *Chem. Abstr.*, **51**:5039 (1957). (b) I. V. Andreeva and M. M. Koton, *Zh. Obshch. Khim.*, **27**:997 (1957); through *Chem. Abstr.*, **52**:4598 (1958). (c) I. V. Andreeva and M. M. Koton, *Zh. Fiz. Khim.*, **32**:1847 (1958); through *Chem. Abstr.*, **53**:4805 (1959).

1743. M. M. Koton, *J. Polymer Sci.*, **30**:331 (1958).

1744. P. M. Arnold (to Phillips Petroleum Co.), U.S. Patent 2,512,596, Jun. 27, 1950; *Chem. Abstr.*, **44**:9986 (1950).

1745. R. T. Nazzaro and J. L. Bullock, *J. Am. Chem. Soc.*, **68**:2121 (1946).

1746. D. T. Mowry, M. Renoll, and W. F. Huber, *ibid.*, **68**:1105 (1946).

1747. G. Van Zyl, R. J. Langenberg, H. H. Tan, and R. N. Schut, *ibid.*, **78**:1955 (1956).

1748. E. M. Smolin, K. Matsuda, and D. S. Hoffenberg, *Ind. Eng. Chem., Prod. Res. Develop.*, **3**(1):16 (1964).

1749. R. W. Strassburg, R. A. Gregg, and C. Walling, *J. Am. Chem. Soc.*, **69**:2141 (1947).

1750. U. Hasserodt and F. Korte, *Angew. Chem.*, **75**(2):138 (1963).

1751. N. Dost and K. van Nes, *Rec. Trav. Chim. Pays-Bas*, **70**:403 (1951).

1752. J. D. A. Johnson and G. A. R. Kon, *J. Chem. Soc.*, **1926**:2748.

1753. S. Sabetay, *Bull. Soc. Chim.* (*4*), **47**:614 (1930).

1754. D. J. Cram, *J. Am. Chem. Soc.*, **74**:2137 (1952).

1755. C. G. Overberger and D. Tanner, *ibid.*, **77**:369 (1955).

1756. K. Y. Yuldashev and I. P. Tsukervanik, *Uzb. Khim. Zh.*, **1961**(6):40; through *Chem. Abstr.*, **57**:16443 (1962).

1757. A. Modestinu-Nicolescu, *Stud. Cercet. Chim.*, **7**:221 (1959); through *Chem. Abstr.*, **54**:14156 (1960).

1758. A. Z. Shikhmamedbekova, N. A. Sevost'yanova, and S. I. Sadykh-Zade, *Azerb. Khim. Zh.*, **1960**(5):37; through *Chem. Abstr.*, **55**:23386 (1961).

1759. S. Searles, Jr., D. G. Hummel, S. Nukina, and P. E. Throckmorton, *J. Am. Chem. Soc.*, **82**:2928 (1960).

1760. S. Searles, Jr., R. G. Nickerson, and W. K. Witsiepe, *J. Org. Chem.*, **24**:1839 (1959).

1761. P. Landrieu, F. Baylocq, and J. R. Johnson, *Bull. Soc. Chim.*, **45**:36 (1929).

1762. E. Breault and O. C. Derner, *Proc. Oklahoma Acad. Sci.*, **28**:82 (1948); *Chem. Abstr.*, **43**:2615 (1949).

1763. I. V. Andreeva and M. M. Koton, *Zh. Obshch. Khim.*, **27**:671 (1957); through *Chem. Abstr.*, **51**:16409 (1957).

1764. S. Murahasi, S. Nozakura, and K. Hatada, *Bull. Chem. Soc. Japan*, **34**:939 (1961); through *Chem. Abstr.*, **56**:1590 (1962).

1765. D. J. Cram and M. R. V. Sahyun, *J. Am. Chem. Soc.*, **85**:1257 (1963).

1766. K. Fries, H. Bestian, and W. Klauditz, *Chem. Ber.*, **69B**:715 (1936).

1767. R. R. Dreisbach and R. A. Martin, *Ind. Eng. Chem.*, **41**:2875 (1949).

1768. R. R. Dreisbach and S. A. Shrader, *ibid.*, **41**:2879 (1949).

1769. French Patent 682,569 (to I. G. Farbenindustrie A.-G.), Oct. 1, 1929; *Chem. Abstr.*, **24**:4523 (1930).

1770. D. Gauthier and P. Gauthier, *Bull. Soc. Chim.*, **53**:323 (1933).

1771. M. Sulzbacher and E. Bergmann, *J. Org. Chem.*, **13**:303 (1948).

1772. C. G. Overberger, C. Frazier, J. Mandelman, and H. F. Smith, *J. Am. Chem. Soc.*, **75**:3326 (1953).

1773. S. Tsutsumi, K. Sato, and W. Kawai, *J. Fuel Soc. Japan*, **35**:395 (1956); through *Chem. Abstr.*, **51**:3479 (1957).

1774. D. Sianesi, *Gazz. Chim. Ital.*, **89**:1749 (1959); through *Chem. Abstr.*, **55**:4394 (1961).

1775. C. G. Overberger, D. Tanner, and E. M. Pearce, *J. Am. Chem. Soc.*, **80**:4566 (1958).

1776. A. Klages and R. Keil, *Chem. Ber.*, **36**:1632 (1903).

1777. A. Klages, *ibid.*, **35**:2245 (1902).

1778. E. Matsui, *J. Soc. Chem. Ind. Japan*, suppl. binding, **44**:107 (1941); through *Chem. Abstr.*, **38**:3748 (1944).

1779. C. S. Marvel, G. E. Innskeep, R. Deanin, D. W. Hein, P. V. Smith, J. D. Young, A. E. Juve, C. H. Schroeder, and M. M. Goff, *Ind. Eng. Chem.*, **40**:2371 (1948).

1780. V. D. Ryabov and V. L. Vaiser, *Dokl. Akad. Nauk SSSR*, **118**:964 (1958); through *Chem. Abstr.*, **52**:12812 (1958).

1781. A. N. Pudovik and I. V. Konovalova, *ibid.*, **149**(5):1091 (1963); through *Chem. Abstr.*, **59**:6434 (1963).

1782. H. Suida, Austrian Patent 132,042, Sept. 15, 1932; *Chem. Abstr.*, **27**:2691 (1933).

1783. R. R. Dreisbach (to The Dow Chemical Co.), U.S. Patent 2,385,696, Sept. 25, 1945; *Chem. Abstr.*, **40**:602 (1946).

1784. J. L. Amos and F. J. Soderquist (to The Dow Chemical Co.), U.S. Patent 2,443,217, Jun. 15, 1948; *Chem. Abstr.*, **42**:7791 (1948).

1785. A. A. Balandin, N. I. Shuikin, G. M. Marukyan, I. I. Brusov, R. G. Seimovich, T. K. Lavrovskaya, and V. K. Mikhailovskii, *Zh. Prikl. Khim.*, **32**:2566 (1959); through *Chem. Abstr.*, **54**:8676 (1960).

1786. A. Pop, R. Popescu, and P. Dumitrescu, *Rev. Chim. (Bucharest)*, **16**(10):510 (1965); through *Chem. Abstr.*, **64**:11836 (1966).

1787. W. E. Parham, E. L. Wheeler, R. M. Dodson, and S. W. Fenton, *J. Am. Chem. Soc.*, **76**:5380 (1954).

1788. N. F. Usmanova, A. V. Golubeva, and A. A. Vansheidt, *Khim. Nauka i Prom.*, **3**:833 (1958); through *Chem. Abstr.*, **53**:10077 (1959).

1789. J. D. Matlack, S. N. Chinai, R. A. Guzzi, and D. W. Levi, *J. Polymer Sci.*, **49**:533 (1961).

1790. D. Braun and H. G. Keppler, *Monatsh. Chem.*, **94**(6):1250 (1963).

1791. A. M. Kuliev, V. M. Farazaliev, and A. M. Levshina, *Azerb. Khim. Zh.*, **1966**(2):85; through *Chem. Abstr.*, **65**:13579 (1966).

1792. N. Tokura, M. Matsuda, and K. Arakawa, *J. Polymer Sci.*, *Pt. A*, **2**(7):3355 (1964).

1793. E. Matsui, *J. Soc. Chem. Ind. Japan*, suppl. binding, **44**:284 (1941); through *Chem. Abstr.*, **44**:7580 (1950).

1794. C. S. Marvel, R. E. Allen, and C. G. Overberger, *J. Am. Chem. Soc.*, **68**:1088 (1946).

1795. L. L. Ferstandig, J. C. Butler, and A. E. Straus, *ibid.*, **76**:5779 (1954).

1796. E. Levine and W. E. Elwell (to California Research Corp.), U.S. Patent 2,723,261, Nov. 8, 1955; *Chem. Abstr.*, **50**:3009 (1956).

1797. J. G. Noltes, H. A. Budding, and G. J. M. van der Kerk, *Rec. Trav. Chim. Pays-Bas*, **79**:1076 (1960).

1798. E. M. Filachione, J. H. Lengel, and W. P. Ratchford, *J. Am. Chem. Soc.*, **72**:839 (1950).

1799. C. L. Arcus and N. S. Salomons, *J. Chem. Soc.*, **1962**:1515.

1800. J. T. Clarke and A. H. Hamerschlag (to Ionics, Inc.), U.S. Patent 2,780,604, Feb. 5, 1957; *Chem. Abstr.*, **51**:12144 (1957).

1801. British Patents 792,859–60 (to The Dow Chemical Co.), Apr. 2, 1958; *Chem. Abstr.*, **53**:2156 (1959).

1802. D. S. Hoffenberg, *Ind. Eng. Chem., Prod. Res. Develop.*, **3**(2):113 (1964). D. S. Hoffenberg (to American Cyanamid Co.), U.S. Patent 2,981,758, Apr. 25, 1961; *Chem. Abstr.*, **55**:25859 (1961).

1803. O. Wichterle and J. Cerny, Czechoslovakian Patent 83,721, Jan. 3, 1955; *Chem. Abstr.*, **50**:5738 (1956).

1804. L. A. Brooks, *J. Am. Chem. Soc.*, **66**:1295 (1944).

1805. (a) M. M. Koton, E. P. Moskvina, and F. S. Florinskii, *Zh. Obshch. Khim.*, **21**:1843 (1951); through *Chem. Abstr.*, **46**:8027 (1952). (b) M. M. Koton, E. P. Moskvina, and F. S. Florinskii, *ibid.*, **21**:1847 (1951); through *Chem. Abstr.*, **46**:8028 (1952).

1806. C. S. Marvel and D. W. Hein, *J. Am. Chem. Soc.*, **70**:1895 (1948).

1807. C. G. Overberger and J. H. Saunders, in E. C. Horning (ed.), "Organic Syntheses," coll. vol. III, p. 204, Wiley, New York, 1955.

1808. C. S. Marvel, G. E. Inskeep, R. Deanin, A. E. Juve, C. H. Schroeder, and M. M. Goff, *Ind. Eng. Chem.*, **39**:1486 (1947).

1809. M. W. Renoll, *J. Am. Chem. Soc.*, **68**:1159 (1946).

1810. G. B. Bachman and L. L. Lewis, *ibid.*, **69**:2022 (1947).

1811. G. Olah, A. Pavlath, and I. Kuhn, *Acta Chim. Acad. Sci. Hung.*, **7**:65 (1955); through *Chem. Abstr.*, **50**:11262 (1956).

1812. S. G. Malkevich and L. V. Chereshkevich, *Plast. Massy*, **1960**(4):1.

1813. M. Kinoshita, *Kogyo Kagaku Zasshi*, **66**(7):982 (1963); through *Chem. Abstr.*, **61**:721 (1964).

1814. L. S. Fosdick, O. Fancher, and K. F. Urbach, *J. Am. Chem. Soc.*, **68**:840 (1946).

1815. P. P. Shorygin and N. V. Shorygina, *J. Gen. Chem. USSR*, **9**:845 (1939); through *Chem. Abstr.*, **34**:389 (1940).

1816. C. S. Marvel and N. S. Moon, *J. Am. Chem. Soc.*, **62**:45 (1940).

1817. J. H. Brown and C. S. Marvel, *ibid.*, **59**:1176 (1937).

1818. C. Walling and K. B. Wolfstirn, *ibid.*, **69**:852 (1947).

1819. K. Ziegler and P. Tiemann, *Chem. Ber.*, **55B**:3406 (1922).

1820. R. Quelet, *Bull. Soc. Chim.*, **45**:75 (1929).

1821. E. Matsui, *J. Soc. Chem. Ind. Japan*, suppl. binding, **45**:412 (1942); through *Chem. Abstr.*, **44**:9186 (1950).

1822. G. B. Bachman, C. L. Carlson, and M. Robinson, *J. Am. Chem. Soc.*, **73**:1964 (1951).

1823. J. Cazes, *Compt. Rend.*, **247**:1874 (1958).

1824. B. Houel, *J. Rech. Centre Natl. Rech. Sci., Lab. Bellevue (Paris)*, **56**:227 (1961).

1825. K. C. Frisch, *J. Polymer Sci.*, **41**:359 (1959).

1826. J. von Braun and J. Nelles, *Chem. Ber.*, **66B**:1464 (1933).

1827. D. Sontag, *Ann. Chim.*, (*11*), 1:359 (1934).
1828. G. B. Bachman, H. Hellman, K. R. Robinson, R. W. Finholt, E. J. Kahler, L. J. Filar, L. V. Heisey, L. L. Lewis, and D. D. Micucci, *J. Org. Chem.*, 12:108 (1947).
1829. A. Einhorn, *Chem. Ber.*, 16:2208 (1883).
1830. R. H. Wiley and N. R. Smith, *J. Am. Chem. Soc.*, 72:5198 (1950).
1831. G. Lo Vecchio and P. Monforte, *Atti. Soc. Peloritana Sci. Fis. Mat. Nat.*, 2:111 (1955–56); through *Chem. Abstr.*, 51:7328 (1957).
1832. B. Lüning, *Acta Chem. Scand.*, 11:959 (1957); through *Chem. Abstr.*, 52:10918 (1958).
1833. C. S. Marvel, C. G. Overberger, R. E. Allen, and J. H. Saunders, *J. Am. Chem. Soc.*, 68:736 (1946).
1834. E. Matsui, *J. Soc. Chem. Ind. Japan*, suppl. binding, 45:437 (1942); through *Chem. Abstr.*, 44:9187 (1950).
1835. C. L. Arcus, *J. Chem. Soc.*, 1958:2428.
1836. G. Prausnitz, *Chem. Ber.*, 17:595 (1884).
1837. R. H. Wiley and N. R. Smith, *J. Am. Chem. Soc.*, 70:2295 (1948).
1838. R. H. Wiley and N. R. Smith, in N. Rabjohn (ed.), "Organic Syntheses," coll. vol. IV, p. 731, Wiley, New York, 1963.
1839. H. B. Hass and M. L. Bender, *J. Am. Chem. Soc.*, 71:3482 (1949).
1840. K. B. Everard, L. Kumar, and L. E. Sutton, *J. Chem. Soc.*, 1951:2807.
1841. M. M. Koton, Y. V. Mitin, and F. S. Florinskii, *Zh. Obshch. Khim.*, 25:1469 (1955); through *Chem. Abstr.*, 50:4823 (1956).
1842. W. H. Saunders, Jr. and R. A. Williams, *J. Am. Chem. Soc.*, 79:3712 (1957).
1843. A. M. Shur and N. A. Barba, *Zh. Obshch. Khim.*, 33(5):1504 (1963); through *Chem. Abstr.*, 59:12676 (1963).
1844. A. Basler, *Chem. Ber.*, 16:3001 (1883).
1845. A. M. Shur and N. A. Barba, *Zh. Org. Khim.*, 1(2):260 (1965); through *Chem. Abstr.*, 62:16091 (1965).
1846. A. M. Shur and N. A. Barba, *Uch. Zap., Kishinevsk. Gos. Univ.*, 68:79 (1964); through *Chem. Abstr.*, 64:3384 (1966).
1847. C. S. Marvel, J. H. Saunders, and C. G. Overberger, *J. Am. Chem. Soc.*, 68:1085 (1946).
1848. L. H. Schwartzman and B. B. Corson, *ibid.*, 78:322 (1956).
1849. J. V. Harispe, *Ann. Chim.*, 6:249 (1936).
1850. A. V. Bondarenko, V. P. Dolinkina, A. M. Kut'in, M. I. Farberov, and S. V. Eigin, *Uch. Zap. Yaroslavsk. Tekhnol. Inst.*, 6:11 (1961); through *Chem. Abstr.*, 58:2383 (1963).
1851. M. M. Koton and T. G. Smolynk, *Dokl. Akad. Nauk SSSR*, 102:305 (1955); through *Chem. Abstr.*, 50:6340 (1956).
1852. M. G. Sturrock and T. Lawe (to Dominion Tar & Chemical Co. Ltd.), U.S. Patents 2,420,688–89, May 20, 1947, 2,422,318, Jun. 17, 1947, and 2,519,719, Aug. 22, 1950; *Chem. Abstr.*, 41:5146, 5147, 6282 (1947) and 45:2503 (1951).
1853. A. V. Bondarenko, V. P. Dolinkina, A. M. Kut'in, and M. I. Farberov, *Neftekhimiya*, 2:585 (1962).
1854. W. Kleeberg and G. Suchardt, *Naturwissenschaften*, 44:584 (1957).
1855. L. H. Schwartzman and B. B. Corson (to Koppers Co., Inc.), U.S. Patent 2,913,440, Nov. 17, 1959; *Chem. Abstr.*, 54:6198 (1960).
1856. E. Matsui, *J. Soc. Chem. Ind. Japan*, suppl. binding, 45:300 (1942); through *Chem. Abstr.*, 44:9186 (1950).

1857. V. A. Zasova, A. V. Bondarenko, R. I. Ledneva, and M. I. Farberov, *Zh. Prikl. Khim.*, **39**(1):194 (1966); through *Chem. Abstr.*, **64**:11106 (1966).

1858. F. E. Salt, W. Webster, and G. Galitzenstein (to Distillers Co. Ltd.), British Patent 634,587, Mar. 22, 1950; *Chem. Abstr.*, **44**:6880 (1950).

1859. J. K. Dixon and K. W. Saunders, *Ind. Eng. Chem.*, **46**:652 (1954).

1860. H. Kröper and R. Platz (to Badische Anilin- & Soda-Fabrik A.-G.), German Patent 1,034,620, Jul. 24, 1958; *Chem. Abstr.*, **54**:10954 (1960).

1861. J. K. Dixon, E. M. Smolin, and K. W. Saunders (to American Cyanamid Co.), U.S. Patents 2,976,333–4, Mar. 21, 1961; *Chem. Abstr.*, **55**:16478 (1961).

1862. W. E. Elwell (to California Research Corp.), U.S. Patent 2,531,327, Nov. 21, 1950; *Chem. Abstr.*, **45**:3417 (1951).

1863a. Y. G. Mamedaliev, G. M. Mamedaliev, S. M. Aliev, and N. I. Guseinov, *Dokl. Akad. Nauk Azerb. SSR*, **19**(1):13 (1963); through *Chem. Abstr.*, **59**:6279 (1963). (b) Y. G. Mamedaliev, G. M. Mamedaliev, S. M. Aliev, and N. I. Guseinov, *Azerb. Khim. Zh.*, **1962**(3):11; through *Chem. Abstr.*, **58**: 11241 (1963).

1864. British Patent 616,751 (to Dominion Tar & Chemical Co. Ltd.), Jan. 26, 1949; *Chem. Abstr.*, **43**:5636 (1949).

1865. British Patent 807,603 (to The Dow Chemical Co.), Jan. 21, 1959; *Chem. Abstr.*, **53**:14054 (1959).

1866. F. R. Buck, K. F. Coles, G. T. Kennedy, and F. Morton, *J. Chem. Soc.*, **1949**: 2377.

1867. A. Klages and P. Allendorff, *Chem. Ber.*, **31**:998 (1898).

1868. C. O. Guss, *J. Am. Chem. Soc.*, **75**:3177 (1953).

1869. R. Y. Levina, V. N. Kostin, P. A. Gembitskii, S. M. Shostakovskii, and E. G. Treshchova, *Zh. Obshch. Khim.*, **32**:1377 (1962); through *Chem. Abstr.*, **58**:4442 (1963).

1870. J. C. Michalek and C. C. Clark, *Chem. Eng. News*, **22**:1559 (1944); *Colloid Chem.*, **6**:1010 (1946).

1871. C. S. Marvel, C. G. Overberger, R. E. Allen, H. W. Johnston, J. H. Saunders, and J. D. Young, *J. Am. Chem. Soc.*, **68**:861 (1946).

1872. M. M. Koton, I. N. Samsonova, and F. S. Florinskii, *Zh. Obshch. Khim.*, **22**:489 (1952); through *Chem. Abstr.*, **47**:2717 (1953).

1873. V. V. Korshak and G. S. Kolesnikov, *Akad. Nauk SSSR, Inst. Org. Khim., Sintezy Org. Soedin., Sbornik*, **2**:92 (1952); through *Chem. Abstr.*, **48**:617 (1954).

1874. G. S. Kolesnikov, *Izv. Akad. Nauk SSSR, Otd. Khim. Nauk*, **1959**:1333; through *Chem. Abstr.*, **54**:1367 (1960).

1875. British Patent 609,482 (to Monsanto Chemical Co.), Oct. 1, 1948; *Chem. Abstr.*, **43**:2636 (1949).

1876. H. Pollack and H. W. Davis, *Ind. Eng. Chem.*, **45**:2552 (1953).

1877. J. C. Michalek, U.S. Patent 2,916,523, Dec. 8, 1959; *Chem. Abstr.*, **54**:8726 (1960).

1878. W. M. Quattlebaum, Jr. (to Carbide & Carbon Chemicals Corp.), U.S. Patents 2,482,207–8, Sept. 20, 1949; *Chem. Abstr.*, **44**:1137 (1950). British Patent 616,844 (to Carbide & Carbon Chemicals Corp.), Jan. 27, 1949; *Chem. Abstr.*, **43**:6230 (1949).

1879. S. Barkovic, *J. Org. Chem.*, **20**:1322 (1955).

1880. C. H. Basdekis (to Monsanto Chemical Co.), U.S. Patent 2,485,524, Oct. 18, 1949; *Chem. Abstr.*, **44**:3019 (1950).

1881. H. A. Dutcher (to Phillips Petroleum Co.), U.S. Patent 2,501,382, Mar. 21, 1950; *Chem. Abstr.*, **44**:5386 (1950).

1882. L. K. J. Tong and W. O. Kenyon, *J. Am. Chem. Soc.*, **69**:1402 (1947).

1883. E. R. Erickson (to Mathieson Chemical Corp.), U.S. Patent 2,519,125, Aug. 15, 1950; *Chem. Abstr.*, **45**:2503 (1951).

1884. H. E. Ramsden (to Metal and Thermit Corp.), British Patent 820,084, Sept. 16, 1959; *Chem. Abstr.*, **54**:6510 (1960).

1885. H. Ratz (to Dynamit A.-G. vorm. Alfred Nobel and Co.), German Patent 1,022,215, Jan. 9, 1958; *Chem. Abstr.*, **54**:4493 (1960).

1886. E. R. Erickson and J. C. Michalek (to Mathieson Chemical Corp.), U.S. Patent 2,432,737, Dec. 16, 1947; *Chem. Abstr.*, **42**:2279 (1948).

1887. F. Reicheneder and F. Stolp (to Badische Anilin- & Soda-Fabrik A.-G.), German Patent 1,076,119, Feb. 25, 1960; *Chem. Abstr.*, **55**:16479 (1961).

1888. F. J. Soderquist and J. L. Amos (to The Dow Chemical Co.), U.S. Patent 2,979,536, Apr. 11, 1961; *Chem. Abstr.*, **56**:4674 (1962).

1889. S. R. Ross, M. Markarian, and M. Nazzewski, (a) *J. Am. Chem. Soc.*, **69**:1914 (1947); (b) *ibid.*, **69**:2468 (1947).

1890. M. Markarian (to Sprague Electric Co.), U.S. Patent 2,569,131, Sept. 25, 1951; *Chem. Abstr.*, **46**:1810 (1952).

1891. N. A. Glukhov, M. M. Koton, and Z. A. Koroleva, *Zh. Obshch. Khim.*, **28**:3277 (1958); through *Chem. Abstr.*, **53**:14028 (1959).

1892. A. A. Levine and O. W. Cass (to E. I. du Pont de Nemours & Co.), U.S. Patent 2,193,823, Mar. 19, 1940; *Chem. Abstr.*, **34**:4746 (1940). E. I. du Pont de Nemours & Co. (to Imperial Chemical Industries Ltd.), British Patent 539,667, Sept. 19, 1941; *Chem. Abstr.*, **36**:4133 (1942).

1893. J. W. Churchill and J. J. Hayes (to Mathieson Chemical Corp.), British Patent 645,065, Oct. 25, 1950; *Chem. Abstr.*, **45**:4092 (1951).

1894. T. Alfrey, Jr. and W. H. Ebelke, *J. Am. Chem. Soc.*, **71**:3235 (1949).

1895. F. W. Kay and W. H. Perkin, *J. Chem. Soc.*, **87**:1066 (1905).

1896. M. Tiffeneau, *Ann. Chim. Phys.*, **10**:145 (1907).

1897. A. P. Sabatier and M. Murat, *Compt. Rend.*, **156**:184 (1913); *Ann. Chim.*, **4**:253 (1915).

1898. H. Pines, D. R. Strehlau, and V. N. Ipatieff, *J. Am. Chem. Soc.*, **72**:5521 (1950).

1899. V. V. Korshak, A. M. Polyakova, and I. M. Stoletova, *Izv. Akad. Nauk SSSR, Otd. Khim. Nauk*, **1959**:1471,1477; through *Chem. Abstr.*, **54**:1368 (1960).

1900. I. I. Lapkin and M. N. Rybakova, *Zh. Obshch. Khim.*, **30**:1227 (1960); through *Chem. Abstr.*, **55**:442 (1961).

1901. M. J. Murray and W. S. Gallaway, *J. Am. Chem. Soc.*, **70**:3867 (1948).

1902. W. H. Perkin and G. Tattersall, *J. Chem. Soc.*, **87**:1083 (1905).

1903. K. Gollnick, G. Schade, and S. Schroeter, *Tetrahedron*, **22**(1):139 (1966).

1904. A. I. Zakharova and V. A. Bezel-Sycheva, *J. Gen. Chem. USSR*, **11**:67 (1941); through *Chem. Abstr.*, **35**:5457 (1941).

1905. W. H. Perkin and S. S. Pickles, *J. Chem. Soc.*, **87**:639 (1905).

1906. K. von Auwers, *Chem. Ber.*, **45**:2764 (1912).

1907. R. C. Palmer, C. H. Bibb, and W. T. McDuffee, Jr. (to Newport Industries, Inc.), U.S. Patent 2,291,915, Aug. 4, 1942; *Chem. Abstr.*, **37**:656 (1943).

1908. R. C. Palmer and C. H. Bibb (to Newport Industries, Inc.), U.S. Patent 2,345,625, Apr. 4, 1944; *Chem. Abstr.*, **38**:4620 (1944).

1909. P. A. J. Janssen, Belgian Patent 577,977, May 15, 1959; *Chem. Abstr.*, **54**:4629

(1960). P. A. J. Janssen, British Patent 881,893, Nov. 8, 1961; *Chem. Abstr.*, **57**:2198 (1962).

1910. J. I. G. Cadogan, D. H. Hey, and W. A. Sanderson, *J. Chem. Soc.*, **1960**: 4897

1911. H. Sodomann, B. Hauschulz, and M. Hanke (to Phenolchemie GmbH), German Patent 1,186,047, Jan. 28, 1965; *Chem. Abstr.*, **62**:9056 (1965).

1912. A. A. Balandin, G. M. Marukyan, and R. G. Seimovich, *Dokl. Akad. Nauk SSSR*, **41**:71 (1943); through *Chem. Abstr.*, **38**:5207 (1944). U.S.S.R. Patent 64,222, Jan. 31, 1945; *Chem. Abstr.*, **40**:4745 (1946).

1913. K. A. Kobe and R. T. Romans, *Ind. Eng. Chem.*, **43**:1755 (1951).

1914. K. Takubo and T. Murayama, *Bull. Govt. Forest Expt. Sta.*, **78**:84 (1955); through *Chem. Abstr.*, **50**:2187 (1956). *Nippon Mokuzai Gakkaishi*, **9**:52 (1963); through *Chem. Abstr.*, **59**:7394 (1963).

1915. S. Tsutsumi, H. Akatsuka, Y. Morimura, and K. Taga, *J. Fuel Soc. Japan*, **35**:96 (1956); through *Chem. Abstr.*, **50**:17400 (1956).

1916. S. Sakuyama and Z. Iwao (to Nippon Catalyst Chemical Industries Co.), Japanese Patent 2,373, Apr. 18, 1957; *Chem. Abstr.*, **52**:4069 (1958).

1917. J. F. Garcia de la Banda and J. B. Alvarez (to Instituto de Quimica Fisica "Rocasolano"), Spanish Patent 318,934, May 1, 1966; through *Chem. Abstr.*, **65**:12135 (1966).

1918. C. Walling, D. Seymour, and K. B. Wolfstirn, *J. Am. Chem. Soc.*, **70**:1544 (1948).

1919. G. Errera, *Gazz. Chim. Ital.*, **21**:76 (1891).

1920. G. B. Bachman and H. M. Hellman, *J. Am. Chem. Soc.*, **70**:1772 (1948).

1921. J. H. Elliott and E. V. Cook, *Ind. Eng. Chem., Anal. Ed.*, **16**:20 (1944).

1922. J. K. Dixon (to American Cyanamid Co.), U.S. Patents 2,376,309–10, May 15, 1945, and 2,387,836, Oct. 30, 1945; *Chem. Abstr.*, **39**:4629 (1945) and **40**:1174 (1946).

1923. B. V. Suvorov, S. R. Rafikov, and L. K. Tuturova, *Izv. Akad. Nauk Kazakh. SSR, Ser. Khim.*, **1955**(8):133; through *Chem. Abstr.*, **50**:1676 (1956).

1924. L. Tomaszewska and W. Zacharewicz, *Rocz. Chem.*, **35**:1511, 1597 (1961); through *Chem. Abstr.*, **57**:7313, 9884 (1962).

1925. A. C. Johnston (to Hercules Powder Co.), U.S. Patent 2,366,409, Jan. 2, 1945; *Chem. Abstr.*, **39**:2002 (1945).

1926. G. S. Serif, C. F. Hunt, and A. N. Bourns, *Can. J. Chem.*, **31**:1229 (1953).

1927. E. Grishkevich-Trokhimovskii, *J. Russ. Phys. Chem. Soc.*, **42**:1543 (1910); through *Chem. Zentr.*, **82**(II):1511 (1911).

1928. I. Matzurevich, *ibid.*, **41**:56 (1909); through *Chem. Abstr.*, **3**:1861 (1909).

1929. J. Hukki, *Acta Chem. Scand.*, **3**:279 (1949).

1930. D. T. Mowry, W. F. Huber, and E. L. Ringwald, *J. Am. Chem. Soc.*, **69**:851 (1947).

1931. G. B. Bachman and R. W. Finholt, *ibid.*, **70**:622 (1948).

1932. A. L. J. Beckwith and J. E. Goodrich, *Aust. J. Chem.*, **18**(7):1023 (1965).

1933. R. A. Benkeser, R. A. Hickner, and D. I. Hoke, *J. Am. Chem. Soc.*, **80**:2279 (1958).

1934. E. Bergmann and A. Weizmann, *Trans. Faraday Soc.*, **32**:1327 (1936).

1935. D. Seymour and K. B. Wolfstirn, *J. Am. Chem. Soc.*, **70**:1177 (1948).

1936. C. W. Roberts and N. F. Neunke, *J. Org. Chem.*, **24**:1907 (1959).

1937. Y. G. Mamedaliev and S. V. Veliev, *Dokl. Akad. Nauk SSSR*, **96**:531 (1954); through *Chem. Abstr.*, **49**:8835 (1955).

1938. J. Slosar and V. Sterba, *Chem. Prum.*, **15**(4):206 (1965); through *Chem. Abstr.*, **63**:4195 (1965).

1939. H. K. F. Hermans, A. G. Knaeps, and C. A. M. van der Eycken, *Ind. Chim. Belge*, **24**:1467 (1959).

1940. G. H. Stempel, Jr. (to General Tire & Rubber Co.), U.S. Patent 2,816,934, Dec. 17, 1957; *Chem. Abstr.*, **52**:11917 (1958).

1941. Y. G. Mamedaliev and R. A. Babakhanov, *Dokl. Akad. Nauk Azerb. SSR*, **17**:467 (1961); through *Chem. Abstr.*, **56**:4600 (1962).

1942. G. H. Stempel, Jr. (to General Tire & Rubber Co.), British Patent 652,618, Apr. 25, 1951; *Chem. Abstr.*, **46**:1037 (1952).

1943. M. M. Otto and H. H. Wenzke, *J. Am. Chem. Soc.*, **57**:294 (1935).

1944. V. Y. Rusin and V. A. Filov, see *Chem. Abstr.*, **54**:17704 (1960); V. Y. Rusin, see *Chem. Abstr.*, **54**:25286 (1960); **55**:7675, 8653 (1961).

1945. H. F. Smyth, Jr. and C. P. Carpenter, *J. Ind. Hyg. Toxicol.*, **30**:63 (1948).

1946. F. Boehm, *Z. Physiol. Chem.*, **260**:1 (1939).

1947. N. E. Stepovaya and L. G. Sarkisova, see *Chem. Abstr.*, **60**:12576 (1964).

1948. J. L. Simousen and A. R. Todd, *J. Chem. Soc.*, **1942**:188.

1949. E. Blasius and H. Lohde, *Talanta*, **13**(5):701 (1966).

1950. G. Schwachula and F. Wolf, *Chem. Tech. (Berlin)*, **18**(7):421 (1966).

1951. V. D. Bezuglyi and Y. P. Ponomarev, *Zh. Anal. Khim.*, **20**(7):842 (1965); through *Chem. Abstr.*, **63**:15561 (1965).

1952. K. Heyns, R. Stute, and H. Paulsen, *Carbohyd. Res.*, **2**(2):132 (1966); K. Heyns, R. Stute, and H. Scharmann, *Tetrahedron*, **22**(7):2223 (1966).

1953. E. C. Hughes and J. R. Johnson, *J. Am. Chem. Soc.*, **53**:737 (1931).

1954. H. A. Laitenen, F. A. Miller, and T. D. Parks, *ibid.*, **69**:2707 (1947).

1955. R. Andrisano and A. Tundo, *Atti Accad. Naz. Lincei, Rend., Cl. Sci. Fis. Mat. Nat.*, **13**:158 (1952); through *Chem. Abstr.*, **48**:13443 (1954).

1956. K. Takano, *Nippon Kagaku Zasshi*, **82**:373 (1961); through *Chem. Abstr.*, **56**:10071 (1962).

1957. K. Han, *Bull. Chem. Soc. Japan*, **11**:701 (1936); through *Chem. Abstr.*, **31**:4901 (1937).

1958. E. V. Sobolev, V. T. Aleksanyan, R. A. Karakhanov, I. F. Bel'skii, and V. A. Ovodova, *Zh. Strukt. Khim.*, **4**(3):358 (1963); through *Chem. Abstr.*, **59**:5958 (1963).

1959. A. Hidalgo, *Compt. Rend.*, **239**:253 (1954).

1960. M. J. Kamlet and D. J. Glover, *J. Am. Chem. Soc.*, **77**:5696 (1955).

1961. K. C. Bryant, G. T. Kennedy, and E. M. Tanner, *J. Chem. Soc.*, **1949**:2389.

1962. P. Grammaticakis and J. Chauvelier, *Compt. Rend.*, **238**:1232 (1954).

1963. A. de Pauw and G. Smets, *Bull. Soc. Chim. Belges*, **59**:629 (1950).

1964. J. Weinstock and V. Boekelheide, *J. Am. Chem. Soc.*, **75**:2546 (1953).

1965. C. Moureu, C. Dufraisse, and J. R. Johnson, *Bull. Soc. Chim. France*, **43**:586 (1928).

1966. J. G. M. Bremmer, F. Starkey, and D. A. Dowden (to Imperial Chemical Industries Ltd.), British Patent 596,880, Jan. 13, 1948, and U. S. Patent 2,519,631, Aug. 22, 1950; *Chem. Abstr.*, **42**:4203 (1948) and **44**:10737 (1950).

1967. D. G. Jones (to Imperial Chemical Industries Ltd.), British Patent 621,744, Apr. 19, 1949; *Chem. Abstr.*, **44**:3030 (1950).

1968. J. G. M. Bremmer (to Imperial Chemical Industries Ltd.), British Patent 627,492, Aug. 10, 1949; *Chem. Abstr.*, **44**:2565 (1950).

1969. A. A. Balandin and A. A. Ponomarev, *Dokl. Akad. Nauk SSSR*, **100**:917 (1955); through *Chem. Abstr.*, **50**:1747 (1956).

1970. N. I. Shuikin and I. F. Bel'skii, *ibid.*, **125**:345 (1959); through *Chem. Abstr.*, **53**:20015 (1959).

1971. T. Kariyone, *J. Pharm. Soc. Japan*, **515**:1 (1925); through *Chem. Abstr.*, **20**:412 (1926).

1972. R. Paul, *Compt. Rend.*, **208**:1028 (1939); *Bull. Soc. Chim.*, **10**:163 (1943).

1973. K. Kuwata, H. Kawazura, and K. Hirota, *Nippon Kagaku Zasshi*, **81**:1770 (1960); through *Chem. Abstr.*, **56**:4928 (1962).

1974. K. Hirota, K. Kuwata, and H. Kawaomo (to Japan Synthetic Chemical Industry Co. Ltd.), Japanese Patent 14,494, Sept. 19, 1962; through *Chem. Abstr.*, **59**:1781 (1963).

1975. D. O. Coffman, P. L. Barrick, R. D. Cramer, and M. S. Raasch, *J. Am. Chem. Soc.*, **71**:490 (1949). P. L. Barrick (to E. I. du Pont de Nemours & Co.), U.S. Patent 2,462,345, Feb. 22, 1949; *Chem. Abstr.*, **43**:4294 (1949).

1976. K. Y. Novitskii, L. V. Brattseva, and Y. K. Yur'ev, *Zh. Org. Khim.*, **1**(6):1097 (1965); through *Chem. Abstr.*, **63**:11471 (1965).

1977. E. Lukevics and M. G. Voronkov, *Khim. Geterotsikl. Soedin.*, *Akad. Nauk Latv. SSR*, **1965**(4):490; through *Chem. Abstr.*, **64**:3584 (1966).

1978. C. Aso, T. Kunitake, and Y. Tanaka, *Bull. Chem. Soc. Japan*, **38**(4):675 (1965).

1979. E. V. Brown, *Iowa State College J. Sci.*, **11**:227 (1937); through *Chem. Abstr.*, **31**:8529 (1937).

1980. British Patent 349,442 (to Imperial Chemical Industries Ltd.), Jan. 23, 1929; *Chem. Abstr.*, **26**:2071 (1932).

1981. B. E. Sorenson (to E. I. du Pont de Nemours & Co.), U.S. Patent 1,911,722, May 30, 1933; *Chem. Abstr.*, **27**:3952 (1933).

1982. C. Aso and Y. Tanaka, *Kobunshi Kagaku*, **21**(6):373 (1964); through *Chem. Abstr.*, **62**:9239 (1965).

1983. Patents assigned to I. G. Farbenindustrie A.-G.: (a) British Patent 487,604, Jun. 22, 1938; *Chem. Abstr.*, **33**:283 (1939). (b) French Patent 830,122, Jul. 21, 1938; *Chem. Abstr.*, **33**:1417 (1939). (c) W. Bock, German Patent 677,868, Jul. 4, 1939; *Chem. Abstr.*, **33**:9496 (1939).

1984. Patents assigned to E. I. du Pont de Nemours & Co.: (a) C. J. Mighton, U.S. Patent 2,390,446, Dec. 4, 1945; *Chem. Abstr.*, **40**:1688 (1946). (b) C. J. Mighton, U.S. Patent 2,401,769, Jun. 11, 1946; *Chem. Abstr.*, **40**:4898 (1946).

1985. I. V. Andreeva and M. M. Koton, *Zh. Fiz. Khim.*, **32**:991 (1958); through *Chem. Abstr.*, **53**:1295 (1959).

1986. Y. Nakamura and M. Saito, *Kogyo Kagaku Zasshi*, **62**:1168 (1959); through *Chem. Abstr.*, **57**:13985 (1962).

1987. G. Goutiere and J. Gole, *Bull. Soc. Chim. France*, **1965**(1):153.

1988. A. Trifonov and I. Panaiotov, *C. R. Acad. Bulg. Sci.*, **10**:301, 363 (1957); *Izv. Khim. Inst. Bulg. Akad. Nauk*, **5**:433 (1957); through *Chem. Abstr.*, **52**:16857 (1958); **55**:15079 (1961).

1989. A. Mishina, *Nippon Nogei Kagaku Kaishi*, **34**:649 (1960); through *Chem. Abstr.*, **58**:14109 (1963).

1990. G. Goutiere, J. B. Leonetti, and J. Gole, *Compt. Rend.*, **257**(17):2485 (1963).

1991. V. O. Reikhsfel'd, *T'ien Tsin Ta Hsueh Hsueh Pao*, **4**/**5**:200 (1957); *Voprosy Ispol'zovan. Pentozansodevzhashego Syr'ya Trudy Vsesoyuz. Soveshchaniya Riga*, **1955**:355 (Pub. 1958); through *Chem. Abstr.*, **56**:7477 (1962); **53**:11871 (1959).

1992. M. Imoto, *J. Soc. Chem. Ind. Japan*, **45**:1065 (1942); through *Chem. Abstr.*, **42**:6161 (1948).
1993. Patents assigned to Wingfoot Corp.: (a) A. M. Clifford, U.S. Patent 2,478,860, Aug. 9, 1949; *Chem. Abstr.*, **43**:9532 (1949). (b) A. M. Clifford, U. S. Patent 2,483,182, Sept. 27, 1949; *Chem. Abstr.*, **43**:9532 (1949). (c) A. M. Clifford, U.S. Patent 2,444,807, Jul. 6, 1948; *Chem. Abstr.*, **42**:7106 (1948). (d) A. M. Clifford, U.S. Patent 2,419,057, Apr. 15, 1947; *Chem. Abstr.*, **41**:4964 (1947).
1994. S. Kamenar, I. Simek, and E. Regensbogenova, *Chem. Zvesti*, **14**:581 (1960); through *Chem. Abstr.*, **55**:15450 (1961).
1995. W. Bock (to General Aniline & Film Corp.), U.S. Patent 2,442,283, May 25, 1948; *Chem. Abstr.*, **42**:6165 (1948).
1996. E. C. Pitzer (to Standard Oil Co. of Indiana), U.S. Patent 2,505,204, Apr. 25, 1950; *Chem. Abstr.*, **44**:6670 (1950).
1997. M. M. Koton, *Khim. Prom.*, **1961**:371; through *Chem. Abstr.*, **56**:1588 (1962).
1998. T. E. Davies, *Brit. Plastics*, **32**(6):283 (1959).
1999. F. R. Buck, G. T. Kennedy, F. Morton, and E. M. Tanner, *Nature*, **162**:103 (1948).
2000. V. I. Isazuljanc and A. Desukki, *Kem. Ind.*, **15**(4):209 (1966); through *Chem. Abstr.*, **65**:16844 (1966).
2001. O. V. Kallistov, *Vysokomol. Soedin.*, **2**:797 (1960); through *Chem. Abstr.*, **55**:6999 (1961).
2002. M. M. Koton, E. P. Moskvina, and F. S. Florinskii, *Zh. Obshch. Khim.*, **22**:789 (1952); through *Chem. Abstr.*, **47**:3253 (1953).
2003. G. T. Kennedy and F. Morton, *J. Chem. Soc.*, **1949**:2383.
2004. F. S. Dainton, K. J. Ivin, and D. A. G. Walmsley, *Trans. Faraday Soc.*, **56**:1784 (1960).
2005. G. Natta, F. Danusso, and D. Sianesi, (*a*) *Makromol. Chem.*, **28**:253 (1958); (*b*) *ibid.*, **30**:238 (1959).
2006. D. Sianesi, R. Serra, and F. Danusso, *Chim. Ind.* (*Milan*), **41**:515 (1959); through *Chem. Abstr.*, **53**:20897 (1959).
2007. C. G. Overberger and S. Nozakura, *J. Polymer Sci.*, *Pt. A*, **1**, 1439 (1963).
2008. G. B. Bachman, L. J. Filar, R. W. Finholt, L. V. Heisey, H. M. Hellman, L. L. Lewis, and D. D. Micucci, *Ind. Eng. Chem.*, **43**:997 (1951).
2009. D. Sianesi, G. Pajaro, and F. Danusso, *Chim. Ind.* (*Milan*), **41**:1176 (1959); *Chem. Abstr.*, **55**:19320 (1961).
2010. D. Braun and H. G. Keppler, *Makromol. Chem.*, **78**:100 (1964).
2011. A. R. Gantmakher, Y. L. Spirin, and S. S. Medvedev, *Vysokomol. Soedin.*, **1**:1526 (1959); through *Chem. Zentr.*, **132**:10581 (1961).
2012. C. Walling, E. R. Briggs, K. B. Wolfstirn, and F. R. Mayo, *J. Am. Chem. Soc.*, **70**:1537 (1948).
2013. C. G. Overberger, L. H. Arond, D. Tanner, J. J. Taylor, and T. Alfrey, Jr., *ibid.*, **74**:4848 (1952).
2014. M. Imoto, M. Kinoshita, and M. Nishigaki, *Makromol. Chem.*, **94**:238 (1966).
2015. W. Kern and D. Braun, *ibid.*, **27**:23 (1958).
2016. G. Smets and A. Rackers, *Rec. Trav. Chim. Pays-Bas*, **68**:983 (1949).
2017. A. De Pauw and G. Smets, *Bull. Soc. Chim. Belges*, **59**:629 (1950).
2018. M. Shima, D. N. Bhattacharyya, J. Smid, and M. Szwarc, *J. Am. Chem. Soc.*, **85**:1306 (1963).
2019. G. Goldfinger and M. Steidlitz, *J. Polymer Sci.*, **3**:786 (1948).

2020. T. Alfrey, Jr., A. I. Goldberg, and W. P. Hohenstein, *J. Am. Chem. Soc.*, **68**:2464 (1946).

2021. F. T. Wall, R. W. Powers, G. D. Sands, and G. S. Stent, *ibid.*, **70**:1031 (1948).

2022. F. Leonard, W. P. Hohenstein, and E. Merz, *ibid.*, **70**:1283 (1948).

2023. P. Agron, T. Alfrey, Jr., J. Bohrer, H. Haas, and H. Wechsler, *J. Polymer Sci.*, **3**:157 (1948).

2024. R. Hess, M.S. thesis, Brooklyn Polytechnic Institute, 1947, cited in ref. 131, p. 36.

2025. R. H. Wiley and B. Davis, *J. Polymer Sci.*, **62**(174):S 140 (1962).

2026. T. Alfrey, Jr., E. Merz, and H. Mark, *ibid.*, **1**:37 (1946).

2027. (a) R. E. Florin, *J. Am. Chem. Soc.*, **71**:1867 (1949). (b) R. E. Florin, *ibid.*, **73**:4468 (1951).

2028. Ref. 2024; cited in ref. 131, p. 38.

2029. H. Gilbert, F. F. Miller, S. J. Averill, E. J. Carlson, V. L. Folt, H. J. Heller, F. D. Stewart, R. F. Schmidt, and H. L. Trumbull, *J. Am. Chem. Soc.*, **78**:1669 (1956).

2030. G. V. Tkachenko, V. S. Etlis, L. V. Stupen, and L. P. Kofman, *Zh. Fiz. Khim.*, **33**:25 (1959); through *Chem. Abstr.*, **54**:11557 (1960).

CUMULATIVE INDEX
PARTS I–III